# Improving dairy herd health

It is widely recognised that agriculture is a significant contributor to global warming and climate change. Agriculture needs to reduce its environmental impact and adapt to current climate change whilst still feeding a growing population, i.e. become more 'climate-smart'. Burleigh Dodds Science Publishing is playing its part in achieving this by bringing together key research on making the production of the world's most important crops and livestock products more sustainable.

Based on extensive research, our publications specifically target the challenge of climate-smart agriculture. In this way we are using 'smart publishing' to help achieve climate-smart agriculture.

Burleigh Dodds Science Publishing is an independent and innovative publisher delivering high quality customer-focused agricultural science content in both print and online formats for the academic and research communities. Our aim is to build a foundation of knowledge on which researchers can build to meet the challenge of climate-smart agriculture.

For more information about Burleigh Dodds Science Publishing simply call us on +44 (0) 1223 839365, email info@bdspublishing.com or alternatively please visit our website at www.bdspublishing.com.

**Related titles:**

Achieving sustainable production of milk Volume 3: Dairy herd management and welfare
Print (ISBN 978-1-78676-052-4); Online (ISBN 978-1-78676-054-8, 978-1-78676-055-5)

Advances in breeding of dairy cattle
Print (ISBN 978-1-78676-296-2); Online (ISBN 978-1-78676-298-6, 978-1-78676-299-3)

Improving rumen function
Print (ISBN 978-1-78676-332-7); Online (ISBN 978-1-78676-334-1, 978-1-78676-335-8)

Understanding the behaviour and improving the welfare of dairy cattle
Print (ISBN 978-1-78676-459-1); Online (ISBN 978-1-78676-461-4, 978-1-78676-462-1)

Chapters are available individually from our online bookshop:
https://shop.bdspublishing.com

climate•SMART•publishing
IN AGRICULTURAL SCIENCE

BURLEIGH DODDS SERIES IN AGRICULTURAL SCIENCE
NUMBER 102

# Improving dairy herd health

Edited by Professor Emeritus Émile Bouchard, University of Montreal, Canada

burleigh dodds
SCIENCE PUBLISHING

Published by Burleigh Dodds Science Publishing Limited
82 High Street, Sawston, Cambridge CB22 3HJ, UK
www.bdspublishing.com

Burleigh Dodds Science Publishing, 1518 Walnut Street, Suite 900, Philadelphia, PA 19102-3406, USA

First published 2021 by Burleigh Dodds Science Publishing Limited


Library of Congress Control Number: 2021931544

British Library Cataloguing in Publication Data
A catalogue record for this book is available from the British Library

ISBN 978-1-78676-467-6 (Print)
ISBN 978-1-78676-470-6 (PDF)
ISBN 978-1-78676-469-0 (ePub)
ISSN 2059-6936 (print)
ISSN 2059-6944 (online)

DOI 10.19103/AS.2021.0086

Typeset by Deanta Global Publishing Services, Dublin, Ireland

# Contents

## Part 4 Particular health issues

# Series list

| Title | Series number |
|---|---|
| **Achieving sustainable cultivation of maize - Vol 1**<br>From improved varieties to local applications<br>Edited by: Dr Dave Watson, CGIAR Maize Research Program Manager, CIMMYT, Mexico | 001 |
| **Achieving sustainable cultivation of maize - Vol 2**<br>Cultivation techniques, pest and disease control<br>Edited by: Dr Dave Watson, CGIAR Maize Research Program Manager, CIMMYT, Mexico | 002 |
| **Achieving sustainable cultivation of rice - Vol 1**<br>Breeding for higher yield and quality<br>Edited by: Prof. Takuji Sasaki, Tokyo University of Agriculture, Japan | 003 |
| **Achieving sustainable cultivation of rice - Vol 2**<br>Cultivation, pest and disease management<br>Edited by: Prof. Takuji Sasaki, Tokyo University of Agriculture, Japan | 004 |
| **Achieving sustainable cultivation of wheat - Vol 1**<br>Breeding, quality traits, pests and diseases<br>Edited by: Prof. Peter Langridge, The University of Adelaide, Australia | 005 |
| **Achieving sustainable cultivation of wheat - Vol 2**<br>Cultivation techniques<br>Edited by: Prof. Peter Langridge, The University of Adelaide, Australia | 006 |
| **Achieving sustainable cultivation of tomatoes**<br>Edited by: Dr Autar Mattoo, USDA-ARS, USA; and Prof. Avtar Handa, Purdue University, USA | 007 |
| **Achieving sustainable production of milk - Vol 1**<br>Milk composition, genetics and breeding<br>Edited by: Dr Nico van Belzen, International Dairy Federation (IDF), Belgium | 008 |
| **Achieving sustainable production of milk - Vol 2**<br>Safety, quality and sustainability<br>Edited by: Dr Nico van Belzen, International Dairy Federation (IDF), Belgium | 009 |
| **Achieving sustainable production of milk - Vol 3**<br>Dairy herd management and welfare<br>Edited by: Prof. John Webster, University of Bristol, UK | 010 |
| **Ensuring safety and quality in the production of beef - Vol 1**<br>Safety<br>Edited by: Prof. Gary Acuff, Texas A&M University, USA; and Prof. James Dickson, Iowa State University, USA | 011 |
| **Ensuring safety and quality in the production of beef - Vol 2**<br>Quality<br>Edited by: Prof. Michael Dikeman, Kansas State University, USA | 012 |
| **Achieving sustainable production of poultry meat - Vol 1**<br>Safety, quality and sustainability<br>Edited by: Prof. Steven C. Ricke, University of Arkansas, USA | 013 |
| **Achieving sustainable production of poultry meat - Vol 2**<br>Breeding and nutrition<br>Edited by: Prof. Todd Applegate, University of Georgia, USA | 014 |

# Acknowledgements

We wish to acknowledge the following for their help in reviewing particular chapters:

- Chapter 1: Dr Caroline Ritter, University of Prince Edward Island, Canada
- Chapter 8: Dr David Kelton, University of Guelph, Canada; and Dr Sam Rowe, The University of Sydney, Australia

# Introduction

Increasing concerns about animal welfare, the need for better use of antibiotics to reduce antimicrobial resistance, as well as consumer and economic pressures, have put greater emphasis on preventative measures in maintaining the health of farm animals. Herd health management programmes take a population approach based on quantitative epidemiology which makes it possible to assess disease risk and, as a result, prevent and manage diseases more effectively. This volume reviews key challenges in dairy herd health management.

Part 1 covers the principles of dairy herd health management, such as the key issues in herd health management and challenges in disease surveillance of dairy cattle as well as the advances in techniques for health monitoring and disease detection in dairy herds. Part 2 focuses on the prerequisites of dairy herd health management, specifically the advances in understanding immune response and the relationship between dairy cattle welfare and health. Chapters in Part 3 cover herd health at different stages of the life cycle. Discussions on optimising reproduction and transition cow management to maximise dairy herd health are included. Chapters also examine managing calves and managing replacement and culling to optimise dairy herd health. The final part of the book examines various ways to optimise dairy herd health, covering areas such as optimising udder health, hoof health, preventing bacterial diseases and the ways data-driven decision support tools can be used in dairy herd health.

## Part 1 Principles

Chapter 1 reviews key issues in dairy herd health management. Dairy herd health management is assessing, monitoring and improving the health of dairy cows at a population level. Good herd health management takes a holistic approach and is ongoing and cyclical. All members of the dairy farm team and their advisors are involved, decisions are informed by data generated by the herd. These data may come from numerous sources. The data are processed and analysed to monitor cow health, target investigations and evaluate progress. To make lasting change on farms, advisors must communicate appropriately with farm managers to understand behaviour and motivate change. This chapter reviews these aspects of dairy herd health management, giving practical suggestions on how to get started, how to incorporate herd health management into business models and how to maintain momentum.

The next chapter provides an overview on the different aspects concerning disease surveillance programs. Chapter 2 describes a specific and conceptual framework related to disease surveillance of production diseases within the individual herd, including both animals and farmers. Regarding farmers, this chapter focus on the justification and purposes for doing disease surveillance as well as the possible decisions and actions they can take to enhance the efficiency of the disease surveillance programs. It also discusses some of the most novel biomarkers that can be potentially used to identify pre-clinical disease states, which will have the potential to minimize the negative effects of production diseases. Finally, the chapter looks into the future perspectives and possible challenges that future automated disease surveillance systems will need to face in order to keep an optimal animal health, performance and welfare within the individual herd.

The subject of Chapter 3 is advances in techniques for health monitoring and disease detection in dairy cattle. It starts by reviewing how the focus in the veterinary profession has shifted from the treatment of acutely diseased animals to more proactive management, which includes the use of epidemiological tools to identify risk factors for animal health, welfare and production. The chapter then reviews information management systems and the different on-farm diagnostic tests that can be performed to provide the necessary data on dairy herd health. A section discussing the use of electronic devices and precision livestock farming techniques is also provided. The chapter also provides a case study which describes how subclinical ketosis was detected in dairy cows.

The final chapter of Part 1 examines the use of data-driven support tools in dairy herd health. Chapter 4 begins by describing the development process of data-driven decision support tools for dairy herd management with an emphasis on real-time continuous data integration and its applications on dairy herd health. It includes concepts on big data analysis, expert systems, and artificial intelligence towards more sustainable dairy farm production systems.

## Part 2 Prerequisites

Part 2 opens with a chapter that reviews advances in understanding immune response in dairy cattle. Chapter 5 begins by analysing the genetics for dairy health, specifically focusing on the importance of identifying the most appropriate measure of disease resistance to ensure the desired dairy health outcome. The chapter also discusses epigenetics and how epigenetic mechanisms are integral to improving dairy immune responses. A section on environmental stresses that dairy cattle encounter is also provided, specifically heat stress and climate. Crossbreeding and immunity in dairy cattle is also discussed. The chapter also examines the importance of colostrum in calf health and emphasises the importance of ensuring early colostrum ingestion for calf survival. The chapter concludes by stressing how cattle have played a key role in immunology and why it is critical for dairy producers to identify cattle with a high immune response.

Chapter 6 discusses the intimate partnership between dairy cattle welfare and health. It begins by examining the welfare implications of common dairy cow diseases such as lameness, mastitis, metritis, acidosis, ketosis and other production-related diseases. It also addresses the impact of subclinical diseases as well. Stress and immune function is also discussed, followed by a review of how mental health can impact the welfare of dairy cattle. A case study on the health and welfare of cows in Indian shelters is also included.

## Part 3 Health at different stages in the life cycle

Chapter 7 reviews optimising reproductive management to maximise dairy herd health and production. Reproduction is central to the operation of a dairy herd through initiating lactation and providing replacement animals and offspring for sale. This chapter outlines the steps required to detect reproductive problems and limitations in cows and herds. It then elaborates some strategies to overcome limits to reproductive health and production concentrating on detecting cows requiring attention through record monitoring, analysing herd records to monitor reproduction and identify areas that limit performance and providing strategies to deal with these limits. The main limits addressed are anoestrus and inadequate oestrus detection that are addressed through education of farm workers, aids to oestrus detection and the use of planned breeding programs to induce and control oestrus and breeding. Nutritional causes of limited reproductive performance are also considered as well as strategies to limit abortion.

The next chapter assesses managing dry cow udder health. The dry period lays a foundation for a successful next lactation, especially from the udder health perspective. It is a high-risk period for acquiring new intramammary infections (IMI), but it also provides an excellent opportunity for eliminating existing subclinical infections. The way cows are dried off and milking is halted at the end of lactation impacts the involution process, mammary health and cow comfort. Chapter 8 reviews the current knowledge about the impact of milk cessation methods (abrupt vs. gradual dry-off) on mammary involution, udder health and cow comfort. The importance of dry cow therapy is discussed, especially in the light of current global concerns related to antibiotic resistance.

Chapter 9 focuses on managing calves to optimise dairy herd health. The chapter demonstrates how calve management can play a critical role in optimising herd health. It starts by discussing the costs of heifer rearing and how good early life management can reduce the costs of heifer rearing. The chapter also discusses the importance of setting targets for heifer rearing, focusing specifically on data management, data recording and benchmarking. It then goes on to discuss managing dairy cattle at the start of the herd lifecycle and how this can have significant effects on calf health. Sections on the impact of calfhood diseases and nutritional management on dairy herd health are also provided. The chapter also addresses the role of vets in communicating best practice in calve management, then concludes by highlighting the impact of better calf management on overall herd health.

The subject of Chapter 10 is managing replacement and culling in dairy herds. Approximately one third of dairy cows are replaced every year. Replacement of dairy cattle is an important part of the cost of dairy production and an environmental sustainability concern. Primary culling reasons are reduced health and fertility. Reduced welfare often proceeds culling. The chapter focuses on factors that affect replacement and culling in dairy herds with a focus on cows. The act of culling is simple, but the risk factors and economic considerations are complex. The chapter first presents some data on culling risks and reasons, explores more in depth the effects of poor health on culling, and presents aspects of economic decision-making regarding culling and replacement decisions.

## Part 4 Particular health issues

The first chapter of Part 4 covers optimising udder health in dairy cattle. Chapter 11 begins by reviewing mastitis, inflammation of the mammary gland, which is generally caused by bacterial infections, is one of the most important and most studied diseases in dairy cattle. Diagnostic approaches are discussed with specific attention for the bacteriological causes of the disease. Subsequently immunological aspects of intramammary infections will be reviewed. Because treatment of mastitis in unavoidable at some point in time in most dairy herds, attention is given to treatment of mastitis with an emphasis on different types of antibiotics and antibiotic resistance. The most important part of udder health management, however, is the preventive management. From that perspective, breeding, housing and nutrition are shortly discussed, as are the milking machine and milking procedures. Finally, attention is given to problem solving once mastitis has led to a herd level problem and some future trends are discussed.

Chapter 12 examines optimising foot health in dairy cattle. The chapter begins by reviewing the importance of lameness then goes on to discuss claw horn disruption. It also reviews aetiopathogenesis of white line bruising and lesions, which is then followed by a section on aetiopathogenesis and control of digital dermatitis. A case study on an 800 cow Holstein herd with a sudden rise in sole ulcers and white line lesions is also included. The chapter concludes with an overview of the emerging diseases in dairy cattle.

The final chapter of the book discusses preventing bacterial diseases in dairy cattle. Chapter 13 begins by examining state of the art disease prevention in dairy cattle, focusing specifically on bovine respiratory disease (BRD). The chapter uses the disease triangle as a basis for discussion, emphasising how disrupting certain parts of the triangle can prevent diseases. It first focuses on bacterial and viral pathogens associated with BRD and the role of the host and the role of the environment in bacterial infection. The chapter then goes on to discuss the importance of disease detection and how various tools can be used to help prevent diseases such as BRD. A discussion on risk assessment tools is also provided. The chapter concludes by highlighting the importance of considering all factors of the disease triangle when looking at ways to prevent diseases.

# Part 1

## Principles

# Chapter 1

## Key issues in dairy herd health management

*John Remnant, James Breen, Peter Down, Chris Hudson and Martin Green, University of Nottingham, UK*



## 1 Introduction

Dairy herd health management involves assessing, monitoring and improving the health of dairy cows at a population level. This is an approach advocated internationally in areas with an industrialized dairy production (Alawneh et al., 2018; Ansari-Lari et al., 2010; Barkema et al., 2015; Cannas da Silva et al., 2006; Galon et al., 2010; Noordhuizen and Wentink, 2001). Since the benefits of maintaining a healthy, efficient productive dairy herd are so wide-ranging, it is difficult to understand why the concept of 'herd health' hasn't been more firmly embedded in dairy industries throughout the world. This chapter outlines some of the key beneficiaries and outcomes of adopting a successful herd health program.

### *1.1 The farmer*

The economic benefits of maintaining a healthy herd are clear. Substantial financial losses in dairy production are often associated with key endemic diseases such as mastitis, lameness and infectious conditions as well as suboptimal nutrition and reproduction (Geary et al., 2012; Kossaibati and

http://dx.doi.org/10.19103/AS.2020.0086.01

Esslemont, 1997; Liang et al., 2017; Mahnani et al., 2015). For dairies to be a sustainable business, farms must be profitable and good cow health is one major element of the financial equation (Edwards-Jones, 2006). However, the benefits of herd health to the farmer extend beyond just monetary considerations. Other reasons farmers may wish to have healthy livestock include: pride in a well-run business, the core belief that animal well-being is important, a dislike of wastage, increased ease of management through not having to deal with sick or under-performing stock and altruism – the knowledge that disease from their animals will not be passed to other animals on the farm or to other farms.

## 1.2 The environment

With concerns about climate change and its potential devastating impacts on much of the world, looking after our environment has become increasingly prominent. The global dairy sector is considered to be responsible for around 4% of total anthropogenic greenhouse gas (GHG) emissions although large variations are recognized between farms and regions (FAO, 2019). In many cases, improvements to herd health are acknowledged as one of several mitigation strategies to reduce emissions alongside fertilizer/slurry and soil management, optimal use of feeds and nutrition, improving the efficiency of energy and water consumption and genetics (Green et al., 2011). Health and productivity are known to vary considerably between herds, which suggests that there is often scope for improvement. Increasingly, the environmental impact attributed to individual farms is being measured and benchmarked and, although tools for such analyses have scope for substantial improvement, this trend is likely to continue. Indeed, it is likely in the relatively near future that benchmarking of environmental impact will play a major role in the auditing of dairy farms and the sales of dairy products, for example, the 'proAction' initiative by the Dairy Farmers of Canada (https://www.dairyfarmers.ca/proaction).

## 1.3 The cow

Beyond the importance of herd health in terms of farm economics and the environment, the most compelling reason to improve herd health is probably cow welfare. Dairy cows are sentient beings, and it is right that we should care for them throughout their lives by looking after their health and welfare. It is clear that good health plays a pivotal role in good welfare whilst poor health is often a reason for compromised welfare of dairy cattle. Most diseases and conditions have an important impact on cow welfare, but we draw particular attention to lameness, mastitis, periparturient disease, dystocia and delayed treatment as being potentially substantial welfare issues. Good welfare, however, extends beyond good health. There is increasing awareness that welfare should not only include minimizing negative experiences but also incorporate enhancement of positive aspects of the lives of cows. As an example, a 'quality of life framework' has been proposed by the UK Farm Animal Welfare Council (FAWC, 2009) identifying five opportunities for positive welfare in farmed animals: comfort, pleasure, confidence, interest and a healthy life. Although research into positive aspects of the welfare of dairy cattle is still limited, it is likely to become of increasing importance in the future. A modern dairy herd health program should include every aspect that influences the welfare of cows.

### 1.4 The citizen

The consumer and wider society have a legitimate interest in how food is produced, both from the perspective of whether the products they consume are produced in a manner they find acceptable as well as the extent to which farming itself affects the natural environment (Boogaard et al., 2008; Cardoso et al., 2017; Jackson et al., 2020). The fact that it involves care and management of live animals adds complexity to ethical assessments of livestock farming. Important issues that must be considered include food security, affordability and choice, animal welfare, impacts on local and wider landscapes and environments as well as the problem of antimicrobial resistance. The relative importance of these different factors depends on regional, economic and cultural differences. However, animal welfare should be an essential element underpinning any herd health program. Even though citizens in many parts of the world are separated from farming in terms of their own experience and understanding, it is still essential to take full account of wider social attitudes about how animals are farmed. An active, successful herd health program should help address these attitudes and provide a clear route in demonstrating high levels of health, welfare and husbandry on dairy farms. Improving health, welfare and husbandry are likely to improve other areas important to the citizen. As well as the environmental benefits from reduced emissions and antimicrobial use should be reduced by herd health management. Preventing, and therefore reducing, diseases is likely to reduce antimicrobial use on farms, thereby reducing the selection pressure for antimicrobial resistance (Hyde et al., 2017, 2019).

## 2 Key features of herd health management

In this section, we outline some key aspects of herd health management that are consistent between farms, countries and advisors. These features provide the foundation of herd health management. We also assess the particular role of veterinary practitioner within the herd health management team.

## 2.1 Population level focus

Herd health management is essentially applied epidemiology. The health and welfare of a herd are monitored at a population level, with interventions made to prevent disease and improve the health and welfare of the whole herd. This contrasts with considering health at an individual cow level by making a diagnosis and treatment plan for a sick animal. Both approaches are important. However, it is the population-level approach that distinguishes herd health management. This may not be immediately obvious to many veterinary practitioners whose training is traditionally weighted heavily toward the former rather than the latter. The population-level nature of herd health management tends to result in more preventive approaches compared to individual cow medicine although the two are linked. Individual animal diagnoses should commonly lead to considerations of disease issues at the herd level.

## 2.2 A holistic approach

In addition to being at a population level, herd health management considers the goals and motivations of farmers and the dairy business. A deep understanding of the owner's or manager's aspirations for the herd and farm as well as an understanding of the system in use are essential to ensure that recommendations are relevant to the dairy business being advised.

## 2.3 Data driven

Herd health management is informed by the regular and systematic collection and analysis of data from the farm. These data may consist of farmer records of management or treatments, external records such as those collated by milk quality laboratories from Dairy Herd Improvement (DHI) testing or data from cow and environmental sensors. These data are used to inform on-farm investigations and observations to identify and then manage risk factors. Effective herd health management relies on these data to facilitate on farm investigations and to drive high-quality decision-making.

## 2.4 Ongoing and cyclical

Effective herd health management is not a one-off intervention. Herd health management is a cyclical and iterative process. This distinguishes herd health management from both one-off problem-solving visits and occasional visits to audit compliance with standards and/or protocols for quality assurance. These alternative approaches have their merits and may be integrated into a wider herd health management program, but its continuous and iterative nature is what distinguishes herd health management. Herd health management relies on continuous monitoring and improvement of cattle health, welfare and production. This may begin with the identification and reduction of immediate disease-based challenges, but, over time, the focus should shift to continuous, incremental improvements in health, welfare and production. Ultimately, herd health management should be about the maintenance of good health and welfare through serial improvements in management as well as reacting quickly to any data providing an 'early warning' of deteriorating control.

## 2.5 Multifactorial

Herd health management addresses complex and multifactorial health and production issues on farms. Problems and their solutions are often complex. Monitoring, assessment and interventions span a range of husbandry practices on the farm. This will often result in some uncertainty, making an ongoing and cyclical process of analysis essential. Data are reviewed following management changes to monitor improvement and identify further opportunities for progress.

## 2.6 Team based

Due to the multifactorial nature of herd health management, it is unlikely that any one veterinarian or advisor will have the necessary expertise in all relevant areas. This necessitates a team approach where relevant advisors and experts work with the farm team in a coordinated way. Herd health management works best when these partners work together, sharing insight and expertise for the benefit of the farm.

## 2.7 Veterinary practitioners and herd health management

As noted above, many advisors are potentially involved in herd health management on dairy farms. A veterinarian practitioner will often be the key advisor for a farm on matters relating to dairy cow health. The demand for herd health management services, in parallel with increases in herd size and production in many dairy-producing countries, has resulted in a change in the role of cattle veterinary practitioners (Barkema et al., 2015). Historically, farm animal veterinary services were focused on the treatment of individual sick animals but, since the 1960s, herd health management has become an integral part of the farm veterinary practitioner's role (Noordhuizen and Wentink, 2001). It is expected to be increasingly important for cattle veterinary practitioners in the future (Woodward et al., 2019). Herd health management services are often delivered alongside more traditional veterinary work. This may be facilitated by regular farm visits by practitioners, with herd health management frequently being structured around routine fertility visits (Cannas da Silva et al., 2006).

These regular visits often result in a strong working relationship between the veterinary practitioner and farmer, with farmers considering the veterinary practitioner a trusted source of information (Hall and Wapenaar, 2012).

This evolving role of the veterinary practitioner can present some challenges to the implementation of herd health management. The changing role might not be appealing to particular practitioners and, as a result, motivation to provide herd health management services will vary between veterinary practitioners. Similarly, farmers may not perceive herd health management as important or may not think to ask their veterinary practitioner for help in its implementation. Attitudes to and motivations for engaging in herd health management programs have been reported to vary between farmers (Derks et al., 2013). Similarly, there are varying perceptions between veterinary practitioners and farmers about priorities in delivering herd health management services (Hall and Wapenaar, 2012). Veterinary business models may also need to change to adapt to the evolution in veterinary services. It can be challenging to incorporate herd health management into traditional veterinary business models, with pressure to deliver emergency care and fee structures that can present a barrier to further engagement by clients. There are also risks posed by changes in technology. In many countries, herd health management services are aligned to pregnancy diagnosis visits. It is important that when uptake of non-veterinary solutions to services such as pregnancy diagnosis increases, there is still an effective model for delivering herd health management to farms.

For high-quality herd health management services to be widely available, it is essential that training for veterinary practitioners, at both the undergraduate and postgraduate levels, keeps pace with changes in herd health concepts and technology. Equipping veterinary practitioners with the skills required to deliver herd health management, such as data analysis and communication skills, is essential.

## 3 Concepts in measuring disease and performance

As herd health management is applied epidemiology, it depends on some basic concepts and terminology in population medicine. These relate to describing disease or performance in a population and are fundamental to measuring the impact of improvements or changes put forward to reduce disease or improve performance.

### *3.1 The eligible population*

The concept of 'eligibility' is crucial in measuring disease and performance in populations. It describes the group of animals which *may* become affected.

When monitoring disease or performance, we use the term 'eligible population' or the *denominator*. As an example, those cows that are eligible to become pregnant in a 21-day period of risk could be considered as those cows with a reported calving event, those that are out with a voluntary wait period after calving, those that are not already reported as pregnant and those that are not already reported as due to be culled from the herd. Similarly, those cows that are eligible for a new intramammary infection of likely dry period origin, as measured by somatic cell count (SCC), may be those cows reported with an SCC <200 000 cells/mL for three consecutive milk recordings prior to a drying-off event and, in addition, not reported with a clinical mastitis event for the 3 months prior to the drying-off event.

### 3.2 The affected population

After describing or setting criteria for which cows are eligible, definitions are required for those cows which become affected or diseased, known as the *numerator*. This will depend upon clear guidance from the herd's veterinary surgeon if using event data. As an example, a clinical case of hypocalcemia ('milk fever') should not occur in any cow that receives parenteral calcium salts around the time of calving, while a clinical case of mastitis describes any abnormal changes to the milk (color, clots) irrespective of what treatment strategy may be chosen. A lack of guidance around which cows are affected may result in *over*-reporting the numerator (e.g. elective treatments administered to cows with sub clinical mastitis or routine preventive hoof trimming of cows) as well as *under*-reporting of the numerator (e.g. not reporting clinical mastitis events that are not treated with intramammary antibiotics).

### 3.3 Incidence rate

The incidence rate describes the rate at which *new* events (sometimes referred to as the 'index' case) occur in an eligible population over a period of time, for example, the rate at which clinical mastitis events are reported in a dairy herd. Once a cow becomes affected, she should be removed from the denominator population. As an example, a cow cannot be reported with a first case of clinical mastitis twice. An incidence rate is expressed per population-time, as cows are 'at risk' for a period of time. We, therefore, report an incidence rate of 30 cases of clinical mastitis per 100 cows per year or 0.3 cases per cow-year, for example. Generally speaking, the incidence rate of *new* cases will inform the herd health advisor about prevention of disease and the effectiveness of current control strategies, whereas the rate of recurrent cases will allow the effectiveness of treatment protocols to be evaluated.

### 3.4 Prevalence

The prevalence of disease in a population of dairy cows measures the *amount* of disease present at a particular point in time and describes the proportion of cows currently affected (e.g. lameness using mobility score data) or currently infected (e.g. subclinical mastitis using SCC data). Both prevalence data and incidence data are required when reviewing performance in a dairy herd. As an example, the prevalence of cows measured with an SCC of >200 000 cells/mL may be very low and yet the incidence rate of clinical mastitis could be above a herd target. Similarly, the rate of lameness treatments in a dairy herd may be very low and yet the prevalence of cows that are likely to benefit from treatment may be much higher.

## 4 Using data in herd health management

### 4.1 Sources of data

Using data to inform decision-making is a fundamental concept in dairy herd health. The data required varies with the specific aspect of herd health being evaluated. The key requirement for most areas of herd health is access to animal identity and event records. Such data will include, for example, calving dates, lactation numbers and the dates of reproductive and disease events. Although this is a relatively simple set of data, it allows analysis of health and performance across a wide range of areas. The way this data is recorded, stored and accessed varies but can generally be managed in two ways (Hudson et al., 2019).

The first way is for data to be recorded using on-farm management software. Farm staff enter data into their own computer software systems (some packages also allow entry via smartphone app), and this forms the main data source for the farm. In addition to event-type records recorded by staff, data from other sources is often added. Examples include data imported from a specified file format (e.g. of individual cow SCC data from a milk recording organization) or automatically using an application programming interface to acquire data from a different software system on the same machine or via the internet (e.g. transfer of daily milk yield records from milking plant software).

The common alternative is for farm records to be held remotely in a centralized database, which may be provided as a service by a commercial organization (e.g. a dairy herd improvement/milk recording organization) or by a government or industry-led service. Records are transcribed to the central database from paper farm records by technicians making farm visits. More recently, it has become common for such databases to allow farm staff to enter data directly via a web portal. Data from on-farm software can often also be imported to such centralized systems.

Data derived from sensor systems has become much more common in dairy herds over the last decade (Hudson et al., 2018). This can be broadly categorized into information from:

- on-animal sensors; and
- off-animal sensors.

The primary purpose of on-animal sensor technology is generally to inform real-time individual animal-level decision-making, such as the use of activity monitors to detect cows potentially in estrus. Amalgamated sensor data can also be used for herd-level monitoring, for example, by evaluating the distribution of rumination activity across cows before and after a diet change. Analysis of data from on-animal sensors is usually only practical using the computer software packaged with the sensors themselves. The potential to use this information to inform herd- or group-level decisions thus varies between products.

Environmental sensors are also becoming more common – either as stand-alone sensors or as part of an automatically controlled ventilation system. Most such systems come with very limited analytical capability built in, so analysis to inform herd-level decision is often reliant on the data manipulation skills of the user. Other important sources of data for dairy herd health include laboratory results (including those from post-mortem examinations) and veterinary practice records.

Amalgamation of data from these different sources is a common feature of herd health analysis. This can create problems, especially where animal identities are not standardized between data sources. Where there is some overlap between data sources (e.g. where some features are recorded in both datasets), the analyst also has to decide which is to be the 'master' dataset. This requires judgment on which is most likely to be more complete and accurate for the specific unit under consideration.

Irrespective of data source, data quality is always a key consideration in dairy herd health (Hermans et al., 2017). Missing data is a common feature of dairy herd data, for example, where event data is under-recorded. This can be random, for example, where a proportion of disease cases is not recorded, resulting in an underestimation of the herd's incidence rate. Missing data can also be systematic, for example, where only the last insemination of each lactation is recorded since the focus is on predicting calving dates rather than on measuring performance, resulting in an overestimation of herd conception risk and underestimation of submission rate.

Measurement error is an often overlooked contributor to poor data quality, although generally of lesser importance. This is typically because herd health analysis is more dependent on event-type data rather than on quantitative measurements. Measurement error can be at random, for example, where visual estimation of milk yields are used as the basis for recording and a lower degree of accuracy of measurement is possible. Error can also be systematic, for example, where sampling technique in a particular herd means that milk butterfat concentrations are consistently overestimated. Some systems for data analysis have features for assessment of data quality. However, in many cases, the most important safeguard is for the analyst to be mindful of the potential impacts of data quality and to apply critical thinking to the results.

## 4.2 Decision making with data

Data is only useful for dairy herd health if it has a genuine impact on decision-making. Using data to inform herd-level management changes has been a paradigm shift in dairy practice over the last 10–20 years. It has played a significant role in recent improvements in many aspects of health and performance (Hudson et al., 2018).

The key advantage of using data is the ability to target investigations and interventions more accurately, thereby improving the likelihood that changes will be successful and that there will be a return on investment. The way in which data analysis achieves these goals will vary, depending on the specific problem or area being investigated. As an example, when investigating a high incidence rate of clinical mastitis, an understanding of the likely origin of the infections leading to clinical cases is critical. All dairy herds will have some cow-to-cow 'contagious' spread of intramammary infections as well as the acquisition of infections from the environment (both during lactation and during the dry period). However, in most herds, one of these sources can be identified as being more important than others. The first step in a rational investigation of such a problem should, therefore, be an analysis of clinical mastitis data to determine whether further analysis should focus on the dry period environment, the lactating cow environment (and pre-milking udder preparation) or on prevention of cow-to-cow spread. This represents a key shift since, as well as being more likely to be successful, it is also a more efficient use of time to focus the investigation on a specific area (such as the dry period environment) rather than assess all factors across the farm that could influence mastitis risk. This approach has been fundamental to the success of the national mastitis control plan in the United Kingdom, and parallel examples exist in other areas including reproduction and foot health.

As well as insights into the likely epidemiology of a disease on a specific unit show that there are often features common to all aspects of health and performance which can be explored using herd data. These include seasonality and patterns relating to lactation number (or age of young stock) or stage of lactation. These features can be assessed for most aspects of dairy health and performance. They can provide information on likely underlying causes or risk factors, for example, a reduction in the rate of submission for artificial insemination during the summer months may relate to heat stress or to different observation practices whilst cows are at grass. They can also allow interventions to be focused on specific groups, for example, targeting infection control measures such as pre-milking teat disinfection or disinfectant foot bathing to specific groups of cows.

Unlocking these insights from data is often not straightforward and, in most cases, will rely on the use of specialized computer software. A key decision is whether to use the same system on which the data is recorded to perform the analysis or to export the data and analyze it using separate software. The former approach is clearly simpler but is reliant on the analytical capacity of the main recording system. This can be a problem where systems exist which are highly focused on data input, record keeping and day-to-day management tasks (with limited analysis functionality). It can also be a problem where there are a number of different software products in common use (requiring the analyst to become conversant with many different products and their individual strengths and limitations). These problems are largely avoided by using a software product specially designed for data analysis and able to import data easily from multiple different formats. However, such products are not always available and, in some cases, some data can be 'lost in translation' during the export/import process.

Tools to support decision-making in the dairy sector are also becoming more popular. A wide range of tools is available to inform herd-level management decisions, for example, by estimating the likely return on an investment from using estrus detection technology or a fixed-time insemination program to improve reproductive performance. These commonly allow input of some simple information relating to current herd performance and characteristics and return estimated outcomes (physical and/or economic) of an intervention. These tools can either be deterministic (returning a single 'central estimate' figure representing the outcome considered most likely) or stochastic (returning a range or distribution of outcomes and allowing a probabilistic approach to risk and reward). Decision support at the individual animal level is also relatively widely available. This is often integrated into herd management software, for example, by providing lists of potential candidates for culling.

Increasingly, predictive modeling techniques (such as machine learning) are used as the basis for individual animal decision support, especially where a large volume of data is being used as the basis for a prediction or recommendation. Perhaps the clearest example of this is in the use of on-animal sensor data, for example, in automated estrus detection systems (Roelofs and Van Erp-Van Der Kooij, 2015). Such products may use multiple sources of sensor and/or other data (such as activity, rumination, milk yield and previous insemination history), much of which is recorded at a very granular level. Machine learning algorithms (usually 'trained' on a set of data with known outcomes) are used to combine features from the multiple data sources to provide a prediction, for example, of the probability that a cow is in estrus at a particular time. Such methods can be highly useful and may predict outcomes very accurately, but it is important to understand the performance characteristics of a system in order to give context to decisions made using the output. There are some examples of the application of machine learning to herd-level decision support, and it is likely that this will become increasingly common in the future.

## 5 Herd health management in practice: initiating change

### 5.1 Making change on farms

The successful delivery of herd health relies on the ability of herd health advisors to work with and encourage farmers to implement specific intervention measures. This process of behavior change is probably the most daunting and difficult task of all facing herd health advisors. It is now widely acknowledged that technical knowledge alone is far from sufficient when it comes to influencing behavior. An understanding of what drives human behavior is essential in order for veterinary practitioners and other advisors to help farmers initiate and sustain the required changes in the management of the farm.

### 5.2 Barriers to herd health

A report by the Veterinary Development Council, UK, stated that 'The overriding delivery tool for delivering animal health and welfare benefits and improved food safety must be effective farm health planning based on the achievement of outputs' (VDC, 2012). The substantial business opportunity that the delivery of this sort of service offers to veterinary practitioners was also highlighted in the report. Farmers are known to hold their veterinary practitioners in high regard when it comes to the provision of herd health advice (Hall and Wapenaar, 2012). Despite this, it appears that provision and uptake of this type of service remains at a modest to a low level (Down et al., 2012).

Some of the reported barriers to veterinary practitioners providing a herd health service include a perception that there is no demand, that the cost cannot be justified, a lack of confidence and full understanding of herd health management, a fear of being challenged, a feeling that data analysis is not 'proper' veterinary work or simply a lack of financial incentive due to a ready supply of routine work and drug sales (Mee, 2007). Whilst some of these reasons may be valid, the failure of the profession to embrace the challenge risks it being marginalized and missing out on the opportunity to lead advances in dairy herd health (Down et al., 2012).

Possibly the biggest barrier to herd health is the ability of the veterinary practitioner or other advisors to effect change. There exists a vast array of information on improving most health problems that currently impact dairy cows but the challenge lies in how to translate that knowledge into a successful on-farm application (LeBlanc et al., 2006). An example of how this can be achieved is the Agriculture and Horticulture Development Board Dairy Mastitis Control Plan (www.mastitiscontrolplan.co.uk) which has been shown to reduce mastitis on-farm by an average of 20% in 1 year (Green et al., 2007). This demonstrates how a large body of research evidence can be combined into a single, structured approach to disease control, which has now been implemented in well over 2000 UK dairy herds. Whilst gaps in knowledge can still be an issue, it is the failure to engage and communicate effectively that represents the most significant barrier to the successful implementation of a herd health program (Down et al., 2012). An appreciation of the farmer's current level of knowledge, what motivates them, how they learn and apply their knowledge is vital if the advice given is to be effective on farm (Lam et al., 2011).

## 5.3 Understanding behavior on farms

Farmers are not a homogenous group. Each farmer has his/her own unique set of demographic factors, experiences, values and goals that will contribute to views about animal health, prevention and control strategies (Ritter et al., 2017). These socio-psychological factors have often been found to explain more variation in farm performance than the actual on-farm management practices implemented (Bigras-Poulin et al., 1985; van den Borne et al., 2014). Many theories have been put forward to explain behavior (Conner and Norman, 2005). The theory of planned behavior (Ajzen, 1985) is one of the most widely applied social cognition models (Green et al., 2012; Ritter et al., 2017). It is based on the premise that a person's intention to perform a given behavior is influenced by:

- attitude ('do I believe this is a good or bad thing to do?');
- social norms (typically derived from the perceived expectations of close social contacts); and
- belief in self-efficacy ('do I believe I can do it?).

Unlike some of the other intrinsic factors (such as gender, age and personality) and extrinsic factors (such as economics, legislation and personal circumstances), these so-called 'intrinsic' cognitive factors are amenable to change and can, therefore, be influenced by veterinary practitioners and other advisors. In a herd health context, establishing which of these factors is the most significant barrier can help to target interventions in a way that is more likely to result in behavior change.

It is worth noting that most of the research in this area has stemmed from the field of human health behavior, in particular, studies exploring how to modify a person's behavior to improve his/her own health (e.g. by stopping smoking or drinking less). This so-called 'selfish' health behavior is not directly applicable to the herd health context, where the health benefit is for another nonhuman species (the cow), and many of these theoretical frameworks have not been validated in the veterinary field. There are also some specific concerns about the limited predictive ability of the theory of planned behavior (Sniehotta et al., 2014). Despite these challenges, the theory of planned behavior is still widely perceived as helpful in the field of behavior change, although additional considerations around farmer habits and perceived personal moral responsibilities should also be considered when applying it to the herd health context.

Once farmers have expressed an intention to change, the theory of planned behavior suggests that 'extrinsic' circumstances are more likely to influence successful implementation of a control program (Ellis-Iversen et al., 2010). These extrinsic factors include (Panter-Brick et al., 2006):

- community and industry (i.e. norms, stigmas and traditions within the livestock industry/community);
- culture and society (laws/regulations and ethical beliefs, moral values and demands from consumers and society in general); and
- knowledge, skills and resources.

Whilst it is difficult as a veterinary practitioner to influence most of these extrinsic factors, there are some practical measures that could help the client in making a change. These could include helping to motivate the herd health team, reducing the 'hassle factor' through advice and support and helping the client to develop the relevant skills (Green et al., 2012).

After a specific behavior change or intervention has been achieved, sustaining a herd health program becomes important. There are several different 'motivators' that can help with this. Of importance here is measuring outcomes and providing rapid feedback, thereby helping farmers to quantify the fruit of their labor. As an example, measuring the incidence rate of *new* cases of a disease in the previous 3 months and noting a downward trend with the client is a powerful indicator of progress, whereas measuring a rolling 12-month average incidence rate of all disease events may act to demotivate the client. This may not be easy with longer-term disease management programs such as control of Johne's disease, and managing farmer expectation is critical in this context. Other motivators that can help sustain change include trial periods to create good habits, financial rewards, benchmarking, acknowledgment for effort rather than effect, effective information flow and audits (Ellis-Iversen et al., 2010). This reflects the cyclical and holistic nature of herd health management described earlier.

### 5.4 Motivating change

When it comes to effecting behavior change, it is essential that veterinary practitioners or other advisors first take the time to appreciate whether or not the client has reached the point where they intend to make a change. It is important to avoid making the assumptions that after they have taken the time to explain the importance of a disease, the producer will have necessarily reached the same conclusions as the advisor. A more proactive approach may be more successful in which farmers are encouraged to describe a disease using a series of open questions (Atkinson, 2010). After listening to their answers, the veterinary practitioner can help them to consider the pros and cons of potential solutions. This concept described, as 'ownership of change' (Whay and Main, 2009), is important as the co-creation of plans between invested parties has been shown to stimulate better engagement and commitment in the tackling of complex problems (Bard et al., 2019).

Even when farmers have a true understanding of a problem, it is entirely possible that they may still have no intention of doing anything about it. If this is the case, then reflecting on the three main constructs of the theory of planned behavior (as described previously) can be helpful in establishing the main reason for this position. When it comes to changing attitudes, this is normally achieved by providing further scientific information in a format that is most helpful to the client. A veterinary practitioner or other advisor must be prepared to utilize a wide variety of information sources to account for different information-seeking behaviors (Ritter et al., 2017). Social norms can be influenced through facilitated farmer meetings or local farmer discussion groups. Other approaches include practice newsletters, farmer 'testimonials' and benchmarking (Green et al., 2012). When it comes to influencing confidence/'self-efficacy', veterinary advisors should endeavor to convey a very positive attitude regarding the client's abilities, irrespective of any previous failures. Other tools that can help include sharing stories of successfully implemented management changes on other farms, giving farmers customized recommendations for on-farm improvements that are feasible and practical for their operation and the use of structured risk assessments and management plans to facilitate discussion about perceived constraints (Ritter et al., 2017).

Of all these underlying strategies influencing behavior change, the most important is effective communication. Research has suggested that veterinary practitioners would benefit from improving their communication skills, including active listening and eliciting farmers' opinions and values (Jansen, 2010). One recent study using role-play sessions discussing lameness and mastitis found that 'vets dominated the agenda, typically placed minimal value on eliciting the client's own motivations and ideas within a consultation, kept strictly to the topic of disease management at the expense of rapport building and prioritized instrumental support strategies' (Bard et al., 2017). There is now a considerable body of evidence demonstrating the wide variation in the values and motivations of farmers (Edwards-Jones, 2006; Kristensen and Jakobsen, 2011). Veterinary practitioners and other advisors should be careful not to assume that just because a clear economic case for a particular intervention can't be made, it won't be implemented. In some cases, the farmer may be motivated more by professional pride and willing to invest in a change that may not necessarily result in a direct economic return.

A more collaborative consultation works with farmer priorities, motivations and goals which can then be used to inform advisory messages as advocated by Bard et al. (2019). These could establish a meaningful culture of change within the herd health context through a shared understanding of a farmer's world view and the co-creation of plans between interested parties. Client-oriented communication methodologies such as 'motivational interviewing' (Miller and Rollnick, 2012) have been shown to be effective in stimulating client behavior change in other health sectors (Hettema et al., 2005; Lundahl et al., 2010) and could also be beneficial to advisors delivering a herd health service (Svensson et al., 2020).

Despite all this, there will probably always be a group of farmers that seem immune to argument and who choose to have very little veterinary input. These so-called 'hard to reach' farmers (Jansen et al., 2010) may require a different communication strategy to successfully engage with them. The use of material tailored specifically to their motivations and perceptions is likely to prove more fruitful (Noar et al., 2007). Such strategies may include the use of the internet, newsletters, one-to-one contact, cost-benefit information, demonstration farms and farm magazines depending on the client and the types of information sources they use (Jansen et al., 2010).

Clearly, there are many steps that need to be taken and many potential barriers to overcome, in order to achieve sustained change on the dairy farms that veterinary practitioners and other advisors work with. Despite the many challenges, there is a great deal of satisfaction to be gained from the provision of this sort of service. By being equipped with a better understanding of clients' aspirations, values and goals and a better understanding of human behavior, we can successfully implement and sustain herd health programs on more farms.

## 6 Herd health management in practice: implementing change

### *6.1 Getting started*

When commencing a herd health program on-farm, several key elements need to be taken into account. It is critical at the outset to obtain a clear picture of a client's values, aspirations, objectives and ways of working. Transparency of expectations by both parties is vital to ensure the program runs smoothly and delivers what is expected. The relationship between advisor and client is a crucial aspect of practical herd health delivery and trust from both parties is essential. Understanding mutual motivations and drivers is therefore important; a successful herd health program is largely built on the trust developed in this relationship. It is also important to note that, whilst the relationship between the owner or the manager of the farm is crucial, much of the implementation will fall to other farm staff. Wider farm staff should therefore be involved in discussions when relevant. The owner or the manager must be supported by the veterinary practitioner or other advisors in providing training for staff and practical standard operating procedures for new processes.

A second key element of commencing a herd program is to establish the current farm situation and performance. Understanding the existing position makes it possible to set a relevant and realistic set of farm-specific targets which need to be transparent to all parties. It is useful to follow the 'SMART' principle of making targets: specific, measurable, attainable, realistic and timed. Aspects of herd health management to consider include those related to cow health, reproduction and welfare, production and finances, nutrition, herd genetics and environmental impact metrics. Obtaining this information will identify where information is lacking and stimulate discussions about methods of data capture and storage. Increasingly, dairy farm data are being stored electronically which greatly improves opportunities for data interrogation to improve decision making – this widens the scope of a herd health program enormously. Retrieving data from third-party organizations (e.g. laboratory results, genetic indices) is important and the advisor may need to play a role in synthesizing information from a variety of sources.

An important consideration for an advisor at the outset is the development of an appropriate personal skill set needed to provide a high-quality herd health service. Holistic herd health management covers a wide range of topics, and it is often difficult for a single advisor to cover all areas. Identifying personal strengths and weaknesses, understanding where continued professional development (CPD) is needed and recognition of when to seek additional specialist advice are critical to ensure that a herd health management can be successfully delivered.

## 6.2 Business models

A variety of different business models have been used successfully to deliver herd health work in veterinary and other businesses. A key choice is whether such work is delivered on a straightforward charge-per-time basis, or by using a fixed fee system whereby a specified service is delivered for an agreed cost per month, or whether a hybrid of these models is applied. The former (charge by time) has the advantage of being completely transparent and leaves the onus on the individual delivering the work to report the time taken appropriately. However, time spent doing this type of work away from the farm (e.g. in analyzing data before a visit and/or putting findings in writing) is notoriously challenging to charge appropriately for. Charging by time is, therefore, perhaps more useful in situations where very limited time will be spent away from the client working on this service (e.g. where skills and infrastructure allow for data analysis in 'real time' with the client present). A simple hourly cost also has the advantage of fitting easily into most veterinary practice's existing business models - this is often the most straightforward way to begin delivering a herd health service.

A fixed-fee system (either a flat fee per herd or, e.g. based on herd size or milk production) has the advantage that it is predictable for client and advisor, and perhaps provides a better option where substantial time away from the farm is spent on delivery of herd health. This type of fee structure can also cover other services, such as additional work delivered by veterinary practitioners (such as routine reproductive examination work) or technicians (such as mobility scoring) on farm.

A similar choice exists when considering how such work fits into working life within a veterinary practice - herd health work can be added to existing visits (usually routine fertility visits), with a small amount of time spent at some point during the visit on examining data, evaluating management and facilitating change. Conversely, some veterinary practitioners prefer to deliver herd health as a completely separate visit, generally on a less frequent basis, but with time clearly set aside from clinical work to focus on preventive health and performance. A hybrid of these two approaches can be very effective. Different options will suit particular clients, advisors and businesses better. An appropriate structure for a 200-cow herd might, for example, feature a routine fertility and health visit every 2 weeks, coupled with a herd health management meeting every month as a good starting point.

## 6.3 Maintaining momentum

Whilst facilitating behavior change is clearly challenging, perhaps even more difficult is sustaining any given change over the longer term. It is important, therefore, to have a strategy for supporting and maintaining positive behavior change and to not simply consider the work done once a farmer appears to have acted on advice. This strategy should consider a number of different elements reflecting the different socio-psychological factors affecting behavior. Of critical importance is feedback, which should ideally be based on rapidly observable outcomes that are directly attributable to the intervention (Green et al., 2012). Advisors should also celebrate any success with the whole farm team which, in turn, will help to motivate them to keep up the good work. Farmer meetings and benchmarking groups are a well-established method for facilitating the exchange of ideas and peer-peer learning, and they can also provide a trusted forum for the sharing of any recent successes which can also help with maintaining momentum.

Veterinary practitioners and advisors not only need a strategy for maintaining behavior change in clients, but they also need to consider how to provide herd health services over the longer term within a sustainable business model. This is likely to require significant investment in specialist CPD, according to the needs and interests of the staff. Fostering a team environment where all members feel comfortable in asking for help is important. The provision of ancillary services such as foot trimming can provide a useful additional income stream as well as a way into farms that are not yet willing to pay for a more comprehensive herd health package. Finally, as relationships are at the heart of any successful herd health service, it is vital that farmers are allocated their own veterinary practitioners that they see frequently and consistently which will result in familiarity with each other's situation, personal characteristics, preferences, beliefs, aspirations and competencies as well as the level of trust that can only develop over a period of time (Jansen and Lam, 2012).

## 7 Summary

Embracing dairy herd health management is important for the sustainability of both individual dairy farms and the dairy sector in society. To deliver herd health management, advisors need appropriate technical skills and confidence, for example, in handling and analyzing on-farm data and identifying changes to be made to the management to improve health and welfare. Whilst these technical skills are essential, in many situations the limiting factor in dairy herd health management is not what we know but whether changes are made on farms. An ability to communicate and motivate the need for change is also key for successful implementation of herd health management, to the advantage of dairy cows, dairy farmers, the environment and society.

## 8 Where to look for further information

- Brand, A., Noordhuizen, J. P. T. M. and Schukken, Y. H. 1996. *Herd Health and Production Management in Dairy Practice*. Wageningen Academic Press.
- Green, M., Bradley, A., Breen, J., Higgins, H., Hudson, C., Huxley, J., Statham, J., Green, L. and Hayton, A. 2012. *Dairy Herd Health*. M. J. Green, ed. CABI.

## 9 References

Ajzen, I. (1985). From intentions to actions: a theory of planned behavior. In: Kuhl, J. and Beckmann, J. (Eds) *Action Control*. SSSP Springer Series in Social Psychology. Berlin Heidelberg: Springer, pp. 11–39. DOI: 10.1007/978-3-642-69746-3_2.

Alawneh, J. I., Henning, J. and Olchowy, T. W. J. (2018). Functionality and interfaces of a herd health decision support system for practising dairy cattle veterinarians in New Zealand. *Frontiers in Veterinary Science* 5: 21.

Ansari-Lari, M., Kafi, M., Sokhtanlo, M. and Ahmadi, H. N. (2010). Reproductive performance of Holstein dairy cows in Iran. *Tropical Animal Health and Production* 42(6): 1277–1283. DOI: 10.1007/s11250-010-9561-y.

Atkinson, O. (2010). Communication in farm animal practice 1. Farmer-vet relationships. *In Practice*. British Medical Journal Publishing Group 32(3): 114–117. DOI: 10.1136/inp.c836.

Bard, A. M., Main, D. C. J., Haase, A. M., Whay, H. R., Roe, E. J. and Reyher, K. K. (2017). The future of veterinary communication: partnership or persuasion? A qualitative investigation of veterinary communication in the pursuit of client behaviour change. *PLoS ONE*. Weary, D. (Ed) 12(3): e0171380.

Bard, A. M., Main, D., Roe, E., Haase, A., Whay, H. R. and Reyher, K. K. (2019). To change or not to change? Veterinarian and farmer perceptions of relational factors influencing the enactment of veterinary advice on dairy farms in the United Kingdom. *Journal of Dairy Science*. Elsevier Inc. 102(11): 10379–10394. DOI: 10.3168/jds.2019-16364.

Barkema, H. W., von Keyserlingk, M. A. G., Kastelic, J. P., Lam, T. J., Luby, C., Roy, J. P., LeBlanc, S. J., Keefe, G. P. and Kelton, D. F. (2015). Invited review: changes in the dairy industry affecting dairy cattle health and welfare. *Journal of Dairy Science* 98(11): 7426–7445. DOI: 10.3168/jds.2015-9377.

Bigras-Poulin, M., Meek, A. H., Martin, S. W. and McMillan, I. (1985). Attitudes, management practices, and herd performance – a study of Ontario dairy farm managers. II. Associations. *Preventive Veterinary Medicine*. Elsevier 3(3): 241–250. DOI: 10.1016/0167-5877(85)90019-4.

Boogaard, B. K., Oosting, S. J. and Bock, B. B. (2008). Defining sustainability as a socio-cultural concept: citizen panels visiting dairy farms in the Netherlands. *Livestock Science* 117(1): 24–33. DOI: 10.1016/j.livsci.2007.11.004.

Cannas da Silva, J., Noordhuizen, J. P. T. M., Vagneur, M., Bexiga, R., Gelfert, C. C. and Baumgartner, W. (2006). Veterinary dairy herd health management in Europe: constraints and perspectives. *The Veterinary Quarterly* 28(1): 23-32. DOI: 10.1080/01652176.2006.9695203.

Cardoso, C. S., Von Keyserlingk, M. A. G. and Hötzel, M. J. (2017). Brazilian citizens: expectations regarding dairy cattle welfare and awareness of contentious practices. *Animals*7(12). DOI: 10.3390/ani7120089.

Conner, M. and Norman, P. (2005). *Predicting Health Behaviour: Research and Practice with Social Cognition Models*. Milton Keynes: Open University Press. DOI: 10.1016/S0925-7535(97)81483-X.

Derks, M., van Werven, T., Hogeveen, H. and Kremer, W. D. (2013). Veterinary herd health management programs on dairy farms in the Netherlands: use, execution, and relations to farmer characteristics. *Journal of Dairy Science* 96(3): 1623–1637. DOI: 10.3168/jds.2012-6106.

Down, P. M., Kerby, M., Hall, J., Statham, J. E., Green, M. J. and Hudson, C. D. (2012). Providing herd health management in practice – how does it work on farm? *Cattle Practice* 20(2), 112–119.

Edwards-Jones, G. (2006). Modelling farmer decision-making: concepts, progress and challenges. *Animal Science* 82(6): 783–790. DOI: 10.1017/ASC2006112.

Ellis-Iversen, J., Cook, A. J. C., Watson, E., Nielen, M., Larkin, L., Wooldridge, M. and Hogeveen, H. (2010). Perceptions, circumstances and motivators that influence implementation of zoonotic control programs on cattle farms. *Preventive Veterinary Medicine*. Elsevier B.V. 93(4): 276–285. DOI: 10.1016/j.prevetmed.2009.11.005.

FAO. (2019). Results | global livestock environmental assessment model (GLEAM). Food and Agriculture Organization of the United Nations. Available at: http://www.fao.org/gleam/results/en/.

FAWC. (2009). Farm animal welfare in Great Britain: past, present and future. *Farm Animal Welfare Council* London, UK. Available at: https://www.gov.uk/government/publications/fawc-report-on-farm-animal-welfare-in-great-britain-past-present-and-future.

Galon, N., Zeron, Y. and Ezra, E. (2010). Factors affecting fertility of dairy cows in Israel. *The Journal of Reproduction and Development* 56 (Suppl.): S8–14. DOI: 10.1262/jrd.1056s08.

Geary, U., Lopez-Villalobos, N., Begley, N., McCoy, F., O'Brien, B., O'Grady, L. and Shalloo, L. (2012). Estimating the effect of mastitis on the profitability of Irish dairy farms. *Journal of Dairy Science* 95(7): 3662–3673. DOI: 10.3168/jds.2011-4863.

Green, M., Bradley, A., Breen, J., et al. (2012). *Dairy Herd Health* (M. J. Green, ed.). Wallingford UK: CABI Publishing. DOI: 10.3168/jds.S0022-0302(55)94982-1. Available at: https://www.cabi.org/bookshop/book/9781845939977/.

Green, M., Husband, J., Huxley, J. and Statham, J. (2011). Role of the veterinary surgeon in managing the impact of dairy farming on the environment. *In Practice* 33(8): 366-373. DOI: 10.1136/inp.d5348.

Green, M. J., Leach, K. A., Breen, J. E., Green, L. E. and Bradley, A. J. (2007). National intervention study of mastitis control in dairy herds in England and Wales. *Veterinary Record* 160(9): 287–293. DOI: 10.1136/VR.160.9.287.

Hall, J. and Wapenaar, W. (2012). Opinions and practices of veterinarians and dairy farmers towards herd health management in the UK. *The Veterinary Record*. BMJ Publishing Group Limited 170(17): 441. DOI: 10.1136/vr.100318.

Hermans, K., Waegeman, W., Opsomer, G., Van Ranst, B., De Koster, J., Van Eetvelde, M. and Hostens, M. (2017). Novel approaches to assess the quality of fertility data stored in dairy herd management software. *Journal of Dairy Science* 100(5): 4078–4089. DOI: 10.3168/jds.2016-11896.

Hettema, J., Steele, J. and Miller, W. R. (2005). Motivational interviewing. *Annual Review of Clinical Psychology* 1: 91–111. DOI: 10.1146/annurev.clinpsy.1.102803.143833.

Hudson, C., Kaler, J. and Down, P. (2018). Using big data in cattle practice. *In Practice*. British Medical Journal Publishing Group 40(9): 396–410. DOI: 10.1136/inp.k4328.

Hudson, C., Cook, J. G. and Laven, R. (2019). Veterinary control of herd fertility in intensively managed dairy herds. In: Noakes, D. E., Parkinson, T. J. and England, G. C. W. (Eds) *Veterinary Reproduction and Obstetrics* (10th edn). W.B. Saunders, pp. 467–484. DOI: 10.1016/B978-0-7020-7233-8.00025-2.

Hyde, R. M., Remnant, J. G., Bradley, A. J., Breen, J. E., Hudson, C. D., Davies, P. L., Clarke, T., Critchell, Y., Hylands, M., Linton, E., Wood, E. and Green, M. J. (2017). Quantitative analysis of antimicrobial use on British dairy farms. *Veterinary Record* 181(25): 683. DOI: 10.1136/vr.104614.

Hyde, R., Tisdall, D., Gordon, P. and Remnant, J. (2019). Reducing antimicrobial use on dairy farms using a herd health approach. *In Practice* 41(8): 368–382. DOI: 10.1136/inp.l5518.

Jackson, A., Green, M., Millar, K. and Kaler, J. (2020). Is it just about grazing? UK citizens have diverse preferences for how dairy cows should be managed. *Journal of Dairy Science* 103(4): 3250–3263. DOI: 10.3168/jds.2019-17111.

Jansen, J. (2010). *Mastitis and Farmer Mindset: Towards Effective Communication Strategies to Improve Udder Health Management on Dutch Dairy Farms. Mastitis and Farmer Mindset: Toward Effective Communication Strategies to Improve Udder Health Management on Dutch Dairy Farms*. Wageningen, The Netherlands: Wageningen University.

Jansen, J. and Lam, T. J. G. M. (2012). The role of communication in improving udder health. *Veterinary Clinics of North America. Food Animal Practice*. Elsevier 28(2): 363–379. DOI: 10.1016/j.cvfa.2012.03.003.

Jansen, J., Steuten, C. D. M., Renes, R. J., Aarts, N. and Lam, T. J. (2010). Debunking the myth of the hard-to-reach farmer: effective communication on udder health. *Journal of Dairy Science*. Elsevier 93(3): 1296–1306. DOI: 10.3168/jds.2009-2794.

Kossaibati, M. A. and Esslemont, R. J. (1997). The costs of production diseases in dairy herds in England. *Veterinary Journal* 154(1): 41–51. DOI: 10.1016/S1090-0233(05)80007-3.

Kristensen, E. and Jakobsen, E. B. (2011). Challenging the myth of the irrational dairy farmer; understanding decision-making related to herd health. *New Zealand Veterinary Journal* 59(1): 1–7. DOI: 10.1080/00480169.2011.547162.

Lam, T. J. G. M., Jansen, J., van den Borne, B. H. P., Renes, R. J. and Hogeveen, H. (2011). What veterinarians need to know about communication to optimise their role as advisors on udder health in dairy herds. *New Zealand Veterinary Journal*. Taylor & Francis 59(1): 8–15.

LeBlanc, S. J., Lissemore, K. D., Kelton, D. F., Duffield, T. F. and Leslie, K. E. (2006). Major advances in disease prevention in dairy cattle. *Journal of Dairy Science*. Elsevier 89(4): 1267–1279. DOI: 10.3168/jds.S0022-0302(06)72195-6.

Liang, D., Arnold, L. M., Stowe, C. J., Harmon, R. J. and Bewley, J. M. (2017). Estimating US dairy clinical disease costs with a stochastic simulation model. *Journal of Dairy Science* 100(2): 1472–1486. DOI: 10.3168/jds.2016-11565.

Lundahl, B. W., Kunz, C., Brownell, C., Tollefson, D. and Burke, B. L. (2010). A meta-analysis of motivational interviewing: twenty-five years of empirical studies. *Research on Social Work Practice* 20(2): 137–160. DOI: 10.1177/1049731509347850.

Mahnani, A., Sadeghi-Sefidmazgi, A. and Cabrera, V. E. (2015). Consequences and economics of metritis in Iranian Holstein dairy farms. *Journal of Dairy Science* 98(9): 6048–6057. DOI: 10.3168/jds.2014-8862.

Mee, J. F. (2007). The role of the veterinarian in bovine fertility management on modern dairy farms. *Theriogenology* 68 (Suppl. 1): S257–S265. DOI: 10.1016/j.theriogenology.2007.04.030.

Miller, W. R. and Rollnick, S. (2012). *Motivational Interviewing (3rd edn.): Helping People Change*. The Effects of Brief Mindfulness Intervention on Acute Pain Experience: An Examination of Individual Difference 1, pp. 1–73. Available at: https://www.guilford.com/books/Motivational-Interviewing/Miller-Rollnick/9781609182274/contents.

Noar, S. M., Benac, C. N. and Harris, M. S. (2007). Does tailoring matter? Meta-analytic review of tailored print health behavior change interventions. *Psychological Bulletin* 133(4): 673–693.

Noordhuizen, J. P. T. M. and Wentink, G. H. (2001). Developments in veterinary herd health programmes on dairy farms: a review. *Veterinary Quarterly*. Taylor & Francis 23(4): 162–169. DOI: 10.1080/01652176.2001.9695106.

Panter-Brick, C., Clarke, S. E., Lomas, H., Pinder, M. and Lindsay, S. W. (2006). Culturally compelling strategies for behaviour change: A social ecology model and case study in malaria prevention. *Social Science and Medicine*. Pergamon 62(11): 2810–2825. DOI: 10.1016/j.socscimed.2005.10.009.

Ritter, C., Jansen, J., Roche, S., Kelton, D. F., Adams, C. L., Orsel, K., Erskine, R. J., Benedictus, G., Lam, T. J. G. M. and Barkema, H. W. (2017). Invited review: determinants of farmers' adoption of management-based strategies for infectious disease prevention and control. *Journal of Dairy Science* 100(5): 3329–3347. DOI: 10.3168/jds.2016-11977.

Roelofs, J. B. and Van Erp-Van Der Kooij, E. (2015). Estrus detection tools and their applicability in cattle: recent and perspectival situation. *Animal Reproduction* 12(3): 498–504.

Sniehotta, F. F., Presseau, J. and Araújo-Soares, V. (2014). Time to retire the theory of planned behaviour. *Health Psychology Review* 8(1): 1–7. DOI: 10.1080/17437199.2013.869710.

Svensson, C., Wickström, H., Emanuelson, U., Bard, A. M., Reyher, K. K. and Forsberg, L. (2020). Training in motivational interviewing improves cattle veterinarians' communication skills for herd health management. *Veterinary Record*. British Veterinary Association 187(5): 191. DOI: 10.1136/vr.105646.

van den Borne, B. H. P., Jansen, J., Lam, T. J. G. M. and van Schaik, G. (2014). Associations between the decrease in bovine clinical mastitis and changes in dairy farmers' attitude, knowledge, and behavior in the Netherlands. *Research in Veterinary Science*. Elsevier B.V. 97(2): 226–229. DOI: 10.1016/j.rvsc.2014.06.017.

VDC. (2012). *The Veterinary Development Council Report*. Available at: https://www.vetfutures.org.uk/resource/veterinary-development-council-report-2012/.

Whay, H. R. and Main, D. C. J. (2009). Improving animal welfare: practical approaches for achieving change. In: Temple Grandin (Ed) *Improving Animal Welfare: Practical Approaches for Achieving Change*. Wallingford, UK: CABI Publishing.

Woodward, H., Cobb, K. and Remnant, J. (2019). The future of cattle veterinary practice: insights from a qualitative study. *Veterinary Record* 185(7): 205. DOI: 10.1136/vr.105321.

# Chapter 2

# Key issues and challenges in disease surveillance in dairy cattle

*Lorenzo E. Hernández-Castellano, Klaus L. Ingvartsen and Mogens A. Krogh, Aarhus University, Denmark*



## 1 Introduction

The United Nations estimates that the world population will increase to 9.7 billion in 2050. Global changes in food consumption and development of some countries and regions around the world will increase the demand for livestock-derived protein (Zarrin et al., 2020). In order to meet these demands, it is likely that industrial farming will need to be intensified in the near future. Because of such intensification, it is also expected that the incidence of animal diseases will increase in the future. Compromised animal health and welfare lead to suboptimal animal performance, which in turn reduces productivity and increases the use of veterinary treatments such as antibiotics (de Almeida et al., 2019). In this regard, an efficient disease surveillance will play an essential role in animal production, increasing animal health and welfare, and reducing antibiotic treatments, which are currently major concerns for society. In addition, diseases in dairy cattle can also have negative consequences on food safety, national economies, biodiversity and rural environment.

This chapter provides an overview on the different aspects concerning disease-surveillance programs. This chapter also describes a specific and conceptual framework related to disease surveillance of production diseases

http://dx.doi.org/10.19103/AS.2020.0086.03

within the individual herd, including both animals and farmers. Regarding farmers, this chapter focuses on the justification and purposes for doing disease surveillance as well as the possible decisions and actions they can take to enhance the efficiency of the disease-surveillance programs. Thus, this chapter details some of the most novel biomarkers that can be potentially used to identify pre-clinical disease states, which will have the potential to minimize the negative effects of production diseases. Finally, this chapter looks into the future perspectives and possible challenges that future automated disease-surveillance systems will need to face in order to keep an optimal animal health, performance and welfare within the individual herd.

## 2 Theory of disease surveillance

### *2.1 Purpose of disease surveillance*

Disease surveillance can at least have three different aims or purposes. The first purpose of disease surveillance is to identify infected animals to limit disease dissemination and ultimately achieve eradication. The second purpose is to use disease surveillance in disease/health management within herd. The last purpose is to use disease observations for breeding purposes to achieve resilient animals. These three purposes will be addressed separately to show the continuous need for disease surveillance.

#### *2.1.1 Disease eradication*

Disease surveillance has been essential to achieve eradication of infectious diseases in dairy cows. The World Organisation for Animal Health (2019) defines animal health surveillance as a tool to monitor disease trends, to facilitate the control of infection or infestation, to provide data to be used in risk analysis for animal or public health purposes, to support the rationale for sanitary measures and for providing assurances to trading partners. This definition of surveillance has a strict focus on the control of infections in animal populations. To some extent, this definition can be applied on all structural levels (animals, groups of animals, herd, herds within region, national and international level), but the main focus is at higher levels and also where disease control could lead to eradication from populations. Some of the first efforts made to eradicate infectious diseases have been performed in highly contagious diseases with fatal outcomes such as rinderpest (Roeder et al., 2013) and also in other diseases such as foot and mouth disease. This last disease reduces animal performance and therefore contributes to starvation and malnutrition in humans. Another group of diseases to be eradicated are those infectious diseases that have mostly a serious zoonotic impact on human populations such as echinococcosis

(Craig et al., 2007) and salmonellosis, and diseases that affect production outcomes such as bovine virus diarrhea and infectious bovine rhinotracheitis. Houe et al. (2014) provided an overview on the different steps that are needed to control and eradicate endemic infectious diseases in cattle based primarily on Danish experiences, but also from a higher-level perspective with the ultimate, long-term intention of achieving regional/national eradication. Today, the increasing public concerns about animal welfare and the development of antimicrobial resistance are becoming important future drivers for increased disease surveillance, at least, in the most developed parts of the world.

### 2.1.2 Management of infectious and metabolic diseases

The potential for eradication has been a key driver for surveillance of infectious diseases. However, for some infectious diseases eradication is still not possible. Examples of such diseases are *Staphylococcus aureus* mastitis and bacterial pneumonia in calves, where pathogens are highly opportunistic and can be found frequently in both healthy and diseased animals. For these types of infectious diseases as well as for metabolic diseases, surveillance is still relevant from farmer and herd perspectives due to reduced animal performance and welfare as well as increased production costs related to diseases. These types of diseases are often called production diseases, which are characterized by having a multifactorial origin highly related to the different stages of the production cycle (e.g. around calving and dry-off) and by reduced productivity. The purpose of monitoring production diseases is to quickly treat the disease and thereby reduce associated production losses. Most losses are reflected as reduced milk yield, but losses due to secondary diseases or fertility problems and premature culling are also associated with production diseases. Also, the extra workload associated to the disease should be considered. From a disease management perspective, the incidence of these diseases will always vary over time. Therefore, it is important to monitor them in a standardized way to manage them efficiently. Some of the central questions that need to be addressed by the herd manager are: (1) is the current level of this production disease acceptable? This involves the quantification of the incidence, but also a decision on which level of a production disease is acceptable, (2) is there a development in the incidences of production disease? Since production diseases are closely linked to high-risk periods in lactation and different risk for different parity groups, fluctuations in underlying herd demographics (e.g. more first parity in one period than in the next) can be important and should be included in the evaluation.

For most production diseases, it is assumed that for each clinical case there is a substantial number of subclinical cases as well. At the herd level, these subclinical cases increase the accumulated production losses substantially. As it will be described later in this chapter, quantification of the incidence of either subclinical diseases or the relation between clinical and subclinical cases is very difficult and highly related to local herd factors.

### 2.1.3 Breeding for resilience

Resilience or robustness can be defined as the animal's ability to cope with diverse challenges and genetics play an important role in these. However, there is a wide range of different challenges such as environmental conditions, social stressors, suboptimal nutrition, and infections, among others, that are most likely interconnected. Production diseases are directly related to robustness due to the close link to longevity. Breeding on robustness implies that animals can be identified that have a high production level for a wide range of climatic conditions and production systems, accompanied with a high level of animal welfare (Rauw and Gomez-Raja, 2015). For some infectious diseases (i.e. mastitis), differences among breeds has shown the potential role of genetics in the improvement of robustness (Begley et al., 2009). During recent years, direct health traits are included into selection indices, which improve traits being linked to resilience against diseases. So far, genetic evaluation for health traits is primarily based on farmer observation of disease (treatment records), with disease-trait heritabilities in a low-to-moderate range (König and May, 2019). Using the information from disease-surveillance systems or disease monitoring can be used in breeding for health traits, most likely with much higher heritabilities than seen for the current health traits due to a higher degree of standardization.

## 2.2 Framework for disease surveillance

It is essential to have a clear justification and purpose for initiating disease surveillance. In Fig. 1, the justification - purpose - context - target - condition - performance - utility complex is shown as a diagram. Figure 1 was initially designed to describe the required components needed to quantify the utility of a diagnostic test. This figure illustrates that the utility of observing an entity systematically (i.e. a disease) is dependent on the initial justification.

The initial justification could be related to the production losses or welfare problems associated with a disease. The combination of justification and the herd-specific context leads to a purpose for doing disease surveillance. The herd-specific context could be the incidence of a specific disease or the importance of each disease incidence. This target condition could be, for instance, the incidence of either a clinical or subclinical disease or a biomarker. Whatever the target condition is, there will always be some uncertainty associated to its observation. In case of clinical diseases, there will be some

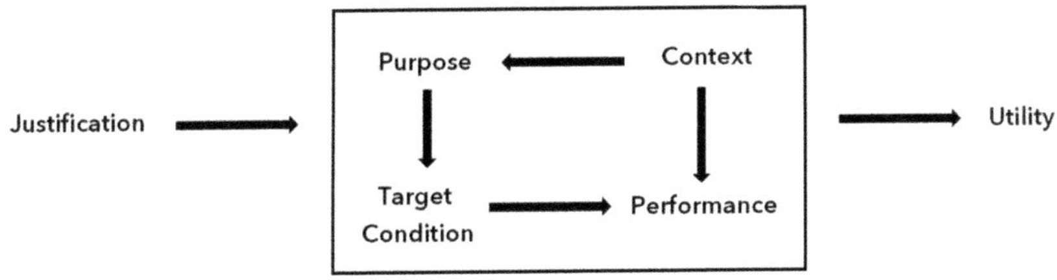


**Figure 1** The justification-purpose-context-target-condition-performance-utility complex (modified from Kostoulas et al., 2017).

level of misclassification. In case of biomarkers, there will be situations where the observed value of the biomarker will not reflect the expected condition. The performance of the disease monitoring needs to be evaluated within this context, considering justification, the herd-specific context, the purpose and the target condition as well as the uncertainty in observing the target conditions. These are the core elements of the monitoring system. If the purpose and context are known, they can also be evaluated in terms of performance and misclassification. However, the utility of the monitoring system relies on the decisions that can be made from the information drawn from the system.

## 3 From disease surveillance toward disease prevention

Over the past decades, there has been a huge structural development with fewer but much larger cattle farms The fewer herds with still increasing number of cows have to cope with new challenges. The most obvious challenge is that the ratio between skilled personnel (professional farmers/owners) and cows has decreased, which results in less time to observe individual cows. This occurs not only due to the increased number of cows, but also because some of the tasks are now automated (i.e. automatic milking systems) or taken over by unskilled staff. In the past, farmers could perform a lot of manual disease monitoring and intervention based on a broad foundation of experience-based reasoning. Nowadays, staff do not often have a broad foundation of experience-based skills and it is increasingly difficult for the people working directly with cows to acquire the requested experience in disease monitoring and intervention. These challenges highlight the need for on-farm automated surveillance systems where new technology (i.e. sensors and models) can be integrated into disease surveillance. However, surveillance or monitoring is not sufficient because, as stated in the previous section, the utility of the surveillance system depends on the actions taken, which focus on disease prevention. There are two different areas of interest regarding disease prevention, one in which animals observed benefit directly from the actions taken and another in which animals are observed and future animals benefit from the actions taken:

- **Disease prevention in future animals.** In this situation, the surveillance system detects an increased risk for disease in the herd. This provides an indication that future animals are at increased risk of disease development. Therefore, this area of the disease surveillance focuses on finding and eliminating the factors that cause the disease, with the acceptance that limited actions can be taken with the animals that have been exposed to such factors and have already developed the disease. Process monitoring and statistical process control (SPC) charts may be useful tools. De Vries and Reneau (2010) as well as Mertens et al. (2011) reviewed the application control charts in animal production and found that quantification of performance is scarce and mostly related to case studies. This is how most disease prevention is performed today, where, for instance, increased somatic cell counts in bulk milk is mitigated with increased hygiene (i.e. pre-milking teat disinfection). A recent exception of an application of SPC is the disease detection in preweaned calves based on feeding behavior (Knauer et al., 2018).
- **Prevention of animal diseases and production losses in the animals observed.** Figure 2 illustrates the relation between health and production and/or reproduction. The green triangle illustrates health status, where cows can be anywhere in the continuum between healthy and clinically diseased. The red triangle illustrates production losses related to health status. It is assumed that the development of clinical diseases are preceded by either a short- or a long-time period where cows are in a physiological imbalance or subclinical diseased. Physiological imbalance, as defined by Ingvartsen (2006), is a condition where cows are challenged and hence, have an increased risk of progression to the state of subclinical disease. In addition to the production losses associated with the subclinical state, some cows will be at a high risk of progression to clinical disease. The blue arrows illustrate that cows can change the health state over time and they do not necessarily need to progress from imbalanced to subclinical, but they can also move into the healthy state again. The idea behind disease prevention at the individual cow level is to detect one of the pre-clinical disease states (i.e. physiological imbalanced or subclinical state) and do some intervention that will reduce the risk of progression into a more severe disease state with larger production losses and welfare issues.

Disease prevention on the individual level has great potential for improving health, welfare and performance of dairy cows. However, if the disease progression from healthy to clinical is fast then it is unlikely that the pre-clinical states can be detected early enough to manage the situation effectively. Furthermore, an intervention that eliminates or reduces the risk of disease progression is also needed. If such intervention is not performed, then the approach actually returns

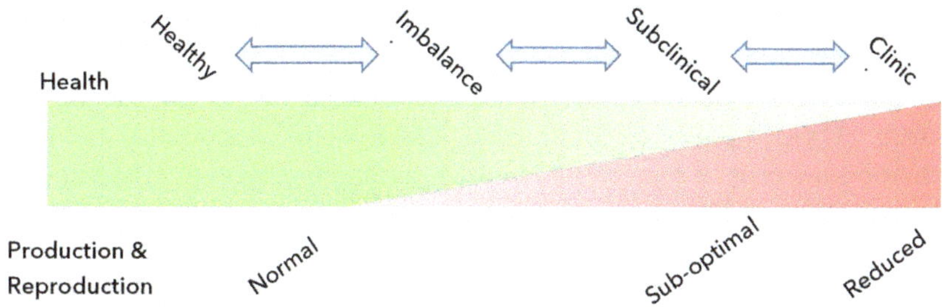


**Figure 2** Conceptual relation between health states and production and/or reproduction.

to disease prevention on future animals and the possible benefits associated with the identification of pre-clinical disease states are lost.

There are several examples of possible strategies for disease prevention. Three different examples are given below about the different aspects of disease prevention described above. In 1970, the Compton Metabolic Profile test was introduced by Payne et al. (1970). This test aimed to detect metabolic and nutritional disorders based on variables measured in blood. However, this test was based on seven blood samples from each of three cow groups tested (dry cows, middle-yielding cows and high-yielding cows). In this study, the cut-offs were based on a statistical theory about deviations from normality. Later studies by Lee et al. (1978) and Adams et al. (1978) described the difficulties of using the results obtained from these blood metabolites on both individual and group levels. During the next decade, more blood metabolites were included and Ingraham and Kappel (1988) concluded that metabolic profiles could be useful in some situations, but most likely unnecessary on a regular basis. Since 2000, research has focused on associated changes in concentrations of blood and milk metabolites with increased risks of metabolic and infectious diseases, proving knowledge of cow-level cut-off values for different concentrations of metabolites (Ospina et al., 2013; Suthar et al., 2013). These metabolic profiles can be used as a tool to monitor metabolic diseases. However, the number of samples that needs to be analyzed to establish a herd-level prevalence within groups of cows can increase the costs associated to these measurements.

A second example of a different disease surveillance was developed in Israel using systematic clinical examinations of all cows in high-risk periods and cow-side tests (Nir, 2003). Systematic clinical examinations were able to detect more subclinical signs of disease, reducing the appearance of more severe clinical signs. In addition, the systematic clinical examinations provided a consistent basis for further analyses of health data from the herd. As described above, a disease-surveillance scheme like this has elements from both types of disease prevention. Nevertheless, it is a limitation that the cows should be at least subclinical diseased. In addition, some analytical experience is required to be able to use the results from these examinations. In this way, Krogh and Enevoldsen (2014) evaluated the monitoring and intervention scheme for endometritis and found that after initiation of disease monitoring, the negative effects of the disease on production losses were considerably reduced. However, the cost of such labor-intensive monitoring systems is substantial, which limits its application in dairy farms.

A third example of disease surveillance is the herd navigator-system (www.herdnavigator.com), which is commercially available to farmers around the world. This system uses in-line measurements of biomarkers in milk to monitor cow health status (Chagunda et al., 2006; Nielsen et al., 2005), specifically mastitis and ketosis (further information is detailed in Section 5.2). The models developed focus on both individual concentrations of biomarkers per cow and changes in those concentrations over the time, providing information at early stages of the disease (i.e. mastitis and ketosis). In addition to the information obtained from biomarkers, these models also incorporate other sources of information (e.g. breed, previous diseases, days in milk, among others) in the calculation of the disease risk (i.e. mastitis and ketosis). Early disease detection is seen as one of the major goal areas within precision livestock farming (Berckmans, 2017). Despite the advantages previously described, the running costs associated with herd navigator are still substantial. Recent developments in Fourier-transform mid-infrared spectrometry on milk have demonstrated the ability to predict milk metabolites (Grelet et al., 2016) that could potentially reduce the costs of running systems for early disease detection using biomarkers.

## 4 High-risk periods for dairy cows

In dairy cows, major physiological, nutritional, metabolic, and immunological changes occur around parturition and during early lactation. During this period, cows shift from a gestational and non-lactating state to the onset of relatively intense milk synthesis and secretion. Due to the sudden demand of nutrients for milk production, the period from late pregnancy to early lactation represents an important metabolic challenge and risk for the modern dairy cow (Hernández-Castellano et al., 2017b; Ingvartsen, 2006). During this period, animals are in negative energy balance (NEB) in support of lactation. Consequently, high amounts of fat from the adipose tissue are mobilized to support lactation. During lipolysis, circulating free fatty acids (FFA) released from adipose tissue enter the liver and either (1) are completely oxidized for energy via the Kreb's cycle, (2) converted to β-hydroxybutyrate (BHB), or (3) re-synthesized to triglycerides (TG) where they can either be exported via very low-density lipoproteins (VLDL), or they can be stored in the liver (Ingvartsen and Moyes, 2013). However, the liver capacity for export of TG via VLDL is limited, and therefore, especially during early lactation, the rapid and extensive influx of circulating FFA into the liver results in TG accumulation in the liver as well as an increased export of BHB into the bloodstream.

Previous studies have shown that physiological imbalance and decreased immune competence are key risk factors for disease (Ingvartsen et al., 2003; Kehrli et al., 2006). Ingvartsen et al. (2003) argued that high milk yield as such does not increase disease incidence in early lactation, but more likely some animals are not able to adapt to the rapid acceleration in milk yield during this period. The high disease incidence in early lactation is partly caused by the fact that some cows have difficulties in adapting to lactation and consequently suffer from physiological imbalance (Ingvartsen 2006). He described physiological imbalance as a situation where the homeorhetic- and homeostatic-regulating mechanisms are insufficient for the animal to function optimally. Cows in physiological imbalance are therefore defined as cows whose variables reflecting the function of the digestive tract, metabolic state and immune state deviate from the normal, and who consequently have an increased risk of developing subclinical or clinical production diseases and reduced production and/or reproduction. In line with this, studies performed in the last years showed that metabolic disorders are more related to the ability of the individual cow to adapt to the new metabolic conditions (i.e. onset of lactation) rather than the high amount of milk produced (Sundrum, 2015; Abuelo et al., 2019). Most cows adapt to the challenge of sudden increase in milk yield but some cows do not. These animals can be either high-yielding dairy cows or low-yielding cows with either reduced or slow capacity to adapt to new physiological conditions, which in turn promotes the development of metabolic diseases. Therefore, there is a need for surveillance systems in modern dairy farms that are able to identify cows under physiological imbalance in order to implement appropriate management strategies and prevent the development of subclinical or clinical states. In such systems, biomarkers of physiological imbalance or disease are essential.

Metabolic diseases most related to energy and nutrient imbalances are ketosis, fatty liver and hypocalcemia. In addition, these diseases increase the susceptibility of cows to retained placenta, metritis, displaced abomasum, lameness and mastitis and cause reduced conception rate, increased embryonic mortality, and increased silent estrus (Zachut et al., 2020). Due to the physiological imbalance of cows with an increased rate of tissue mobilization, the immune system may also be affected (Ingvartsen and Moyes, 2013), increasing the risk of diseases during the transition period (Contreras and Sordillo, 2011; Cai et al., 1994; LeBlanc, 2010). The associations among various metabolic stressors and their relationships to other diseases, particularly infectious and inflammatory diseases in early lactation, have become a central focus of interest in metabolic diseases of dairy cattle (Herdt, 2013).

## 5 Biomarkers of disease risk

Biomarkers are biological molecules used either to detect a physiological imbalance or to diagnose an abnormal process or a disease. Good or suitable biomarkers for a physiological imbalance or a specific disease should meet the following criteria:

- **To be able to detect physiological imbalance or early stages of a disease.** If a biomarker can be detected before a disease is established and the animal show clinical symptoms, this may be used to prevent all negative consequences related to a specific disease as well as spreading to other animals (in case of infectious diseases).
- **Variability among animals.** A good biomarker needs to have a large between cow variation that is highly correlated to an important risk or disease. Further, a low-measuring error is advantageous.
- **Accessible, cost-effective, and measureable in practice.** It is preferable that a specific biomarker can be measured in accessible samples (e.g. milk or urine) than invasive samples (e.g. blood or biopsies) and that it can be measured automatically (e.g. milking robots). It is also preferable that the cost of measuring those biomarkers is low and do not exceed the benefits from intervening.

Based on these criteria, the following sections describe novel behavioral and milk biomarkers for the diagnosis of metabolic and infectious diseases in dairy cows.

### *5.1 Behavioral biomarkers*

As described above, if the cow is not able to adapt quickly to the challenge, infectious agents and/or increased demand of nutrients may cause major disturbances in the organism. Behavioral reactions are a direct reflection of the organism trying to compensate for a specific imbalance. Some of the most common behavioral biomarkers used to detect cows under physiological imbalance include reduced physical activity, feed intake, and social interaction (Dantzer and Kelley, 2007). In the last decade, multiple sensors have been implemented to record animal behavior in dairy farms. Some of these sensors are based on systems that record acceleration in the three axes (i.e. X, Y and Z). These sensors have been used to recognize various behavioral patterns such as those related to physical activity or the time spent laying, standing, eating, drinking, or rumination time (Ledgerwood et al., 2010). Therefore, continuous data recording from these acceleration sensors (also called accelerometers) can be used to identify cows having health problems (e.g. metabolic and infectious diseases).

Several studies have described behavioral changes in cows suffering ketosis after calving (Rodriguez-Jimenez et al., 2018; Goldhawk et al., 2009; Kaufman et al., 2016; Sahar et al., 2020). Behavioral patterns in these animals before developing ketosis are characterized by increased laying time and reduced standing and eating time either prepartum (Rodriguez-Jimenez et al., 2018) or postpartum (Goldhawk et al., 2009). Contrary to this behavior, cows suffering from rumen acidosis are characterized by increased standing and eating time as well as reduced laying and ruminating time (DeVries et al., 2009). Behavioral changes have also been described in cows suffering hypocalcemia postpartum (Jawor et al., 2012; Sepulveda-Varas et al., 2014). Behavioral changes in those animals are characterized by reduced physical activity, increased laying time and decreased time spent at the feeders and water ponds postpartum (Jawor et al., 2012; Sepulveda-Varas et al., 2014). Due to the fast development of hypocalcemia (within the first 48 h postpartum), the usefulness of behavioral biomarkers postpartum is limited. Instead, future research should focus on the identification of behavioral changes prepartum in those cows that will develop either subclinical or clinical hypocalcemia postpartum.

Similar studies have also been performed regarding the behavioral changes of dairy cows during mastitis. Using challenges with lipopolysaccharides (LPS) to mimic a mastitis caused by *Escherichia coli*, Siivonen et al. (2011) observed that the behavioral changes in cows were characterized by decreased time spent lying, lying on the affected quarter, ruminating and drinking water. Further, these animals spent more time eating silage. In a similar study performed by Fogsgaard et al. (2012), LPS-challenged cows decreased time spent on feeding, ruminating, and self-grooming, and increased time standing for the first 48 h after challenge. The results from the experimentally induced mastitis studies were confirmed by Sepulveda-Varas et al. (2016) in cows with natural occurring mastitis, although in this last study authors did not identify the agent causing mastitis. These animals had decreased feed intake and reduced competitive behavior at the feeder during the days before mastitis was diagnosed. This behavior was completely restored (i.e. increased feed intake and competitive behavior at the feeder) when cows were treated with intramammary antibiotics.

According to these results, it seems that sensors recording behavioral changes in dairy cows are useful tools to detect those cows facing any type of metabolic or infectious disease. However, changes such as reduced feed intake, increased standing time or reduced rumination activity are common for many of these diseases and are not able to diagnose a specific disease. Despite this fact, the combination of these sensors providing behavioral changes with the determination of specific biomarkers present in milk could provide enough information to support the diagnosis of individual metabolic and infectious diseases.

## 5.2 Milk biomarkers

The potential of milk biomarkers for diagnosis of metabolic and infectious diseases is enormous as sensors for these specific biomarkers could be implemented in milking parlors and milking robots to provide constant information about the energy and health status of each individual dairy cow in the herd.

Free glucose content in milk seems to be a suitable candidate as biomarker for NEB diagnosis. As for many other tissues in the organism, glucose is essential for the mammary epithelial cells. Due to the lack of the enzyme glucose-6-phosphatase, mammary epithelial cells are not able to synthesize glucose through the gluconeogenesis pathway (Scott et al., 1976). Therefore, glucose concentrations in mammary epithelial cells, and consequently in milk, are strictly dependent on glucose transferred from the bloodstream (Faulkner et al., 1981; Zhao, 2014). Glucose-6-phosphate (G6P) has also been proposed as a biomarker for NEB diagnosis as it is a central metabolite during synthesis of lactose (Zhao, 2014). Based on these facts, Larsen and Moyes (2015) and Zachut et al. (2016) analyzed free glucose and G6P concentrations in milk from dairy cows. Both studies found that free glucose concentrations in milk increased progressively through lactation, whereas G6P concentrations decreased progressively. Therefore, milk G6P/glucose ratio has been suggested as a biomarker of the oxidative stress in the mammary epithelial cells (Zachut et al., 2016). Both glucose and G6P concentrations seem to be useful biomarkers for NEB diagnosis. As described by Zachut et al. (2016), G6P/glucose ratio in milk is highly correlated to plasma FFA concentrations ($r^2$=0.81).

Large amounts of ketone bodies (i.e. acetone, acetoacetate and BHB) are released into the bloodstream during fat mobilization. As described by Geishauser et al. (1998) and Koeck et al. (2014), BHB measured in milk correlates with BHB measured in blood. Based on this principle, modern milking robots incorporate BHB measurements in milk, which provides constant information about the energy status of the animal (Nielsen et al., 2005). Besides BHB, fatty acids present in milk could also be considered as potential biomarkers to assess NEB in dairy cows. In milk, fatty acids are originated from four major sources: (1) directly from the feed, (2) formation in the rumen by biohydrogenation or bacterial degradation, (3) fat depots, and (4) de novo synthesis in the mammary gland (Stoop et al., 2009). Therefore, changes in the energy status during lactation will also affect milk fatty acids composition (Gross et al., 2011). During NEB, increased body fat mobilization reduces the de novo synthesis of fatty acids by the mammary gland (i.e. C6:0 to C14:0) (van Knegsel et al., 2005). Oleic acid (C18:1-cis9) is the predominant fatty acid in adipocytes, and it is released primarily through lipolysis during NEB (Rukkwamsuk et al., 2000). Therefore, under severe NEB, the content of long-chain fatty acids in milk is increased while short-chain and medium-chain fatty acids contents are reduced (Nogalski et al., 2012). Actually, Gross et al. (2011) found a correlation between NEB and the proportion of C18:1-cis9 in milk ($r^2 = 0.77$). Therefore, the proportion of C18:1-cis9 in milk seems to be a potential biomarker to assess NEB in dairy cows during lactation.

When these high dietary requirements are reached by feeding diets with high amounts of rapid fermentable carbohydrates (i.e. starch) and low fiber content, rumen bacterial populations are altered. Consequently, glucose and acids such as lactate accumulate in the rumen cause decreased ruminal pH, leading to ruminal acidosis (pH<4.8) or subacute ruminal acidosis (SARA). This metabolic disease damages the ruminal and intestinal wall and decreases blood pH, leading to metabolic acidosis and causing decreased DMI, milk yield, milk fat content. If this condition persists, it will cause poor body condition and will affect animal health (Nocek, 1997). During rumen acidosis, growth of specific microbial populations (i.e. *Streptococcus bovis*, *Lactobacillus* spp.) is stimulated and consequently large amounts of D-lactate are produced in the rumen (Larsen, 2017) and then transferred into the bloodstream through the rumen wall. The organism does not efficiently metabolize D-lactate and therefore excess of D-lactate in blood is excreted through urine or milk (Larsen, 2017). Based on this fact, increased D-lactate concentrations in milk have been proposed as a biomarker for ruminal acidosis (Hernández-Castellano et al., 2018). Milk fat content and milk fatty acid composition are closely related to ruminal pH. Indeed, Enjalbert et al. (2008) showed how cows with low ruminal pH produced milk with reduced fat content. In addition, these animals also produced milk with higher content of odd-chain fatty acids and increased ratio of trans-10 C18:1/trans-11 C18:1. Such changes in both milk fat content as well as milk fatty acid profile suggested the potential use for non-invasive diagnosis of acute and subacute ruminal acidosis.

Biomarkers in milk can also be used for monitoring calcium homeostasis in early lactation. In dairy cows, the endocrine pathways controlling calcium homeostasis during non-lactating and non-pregnant physiological states are unable to fully support the sudden high demand of calcium required for milk synthesis at the onset of lactation, especially in those animals with high milk yields during the first days postpartum. In the first milkings postpartum, when the incidence of hypocalcemia is the highest, milk calcium concentrations does not reflect the systemic calcemia in dairy cows (Salgado-Hernandez et al., 2014). Therefore, calcium in milk does not seem to be a good candidate to monitor hypocalcemia postpartum.

One of the main hormones regulating calcium homeostasis in the organism is the parathyroid hormone (Kronfeld, 1973). Parathyroid hormone-related peptide (PTHrP) is a 60–173 amino acids peptide that binds the PTH receptor and triggers the same biological activity as PTH (Hernandez-Castellano et al., 2020). During lactation, the mammary gland secretes PTHrP into milk at concentrations exceeding 10,000 times the amount of PTHrP present in the blood (Riond et al., 1995; Kovacs, 2011). Although the function of PTHrP in milk remains unknown, the concentration of this peptide in colostrum and milk is positively correlated to circulating calcium concentrations in cows (Kocabagli et al., 1995). Therefore, cows having low concentrations of PTHrP in milk postpartum could be expected to be at high risk of suffering hypocalcemia.

Despite the importance of metabolic diseases associated to NEB and hypocalcemia, mastitis remains as one of the most relevant diseases in dairy cows, causing important economic losses to the dairy industry (Hogeveen et al., 2019). In most cases, mastitis is caused by bacterial pathogens invading the mammary gland. Currently, somatic cell counts (SCC) and electrical conductivity (EC) measurements in milk are the most used techniques for clinical and subclinical mastitis detection at the farm level (Hernández-Castellano et al., 2017a; Hegde et al., 2013). However, as SCC and EC levels are not pathogen-specific, identification of the pathogen causing mastitis is not possible. Early detection of mastitis and characterization of the pathogen are essential to separate the milk from these infected animals, and to optimize treatments to improve the recovery rate and to minimize economic losses (Milner et al., 1997; Lehmann et al., 2015). Several studies have described the potential of using changes in milk pH for early detection of clinical and subclinical mastitis (El Badawy et al., 2015; Batavani et al., 2007; Qayyum et al., 2016). However, a recent study by Kandeel et al. (2019) showed that milk pH does not provide a clinically useful cow-side screening method for predicting subclinical mastitis or intramammary infections.

Common pathogens, namely *E. coli*, a gram-negative bacterium typically associated with acute clinical mastitis, and *Staph. aureus*, a gram-positive bacterium typically associated with subclinical chronic mastitis, can cause differential activation of the immune system, and consequently differential changes in milk composition (Kirkeby et al., 2020; Kusebauch et al., 2018; Wall et al., 2016a,b). Those changes in milk composition could be potentially used for early mastitis detection as well as for identification of the bacteria causing mastitis, allowing a more targeted mastitis treatment. As described above, those biomarker candidates should be easy, repeatable and cheap to measure in milk, but most important, their concentrations should be markedly changed rapidly after bacteria invade the mammary gland. In this field, Lehtolainen et al. (2004) proposed milk serum amyloid A (SAA) and tumor necrosis factor alpha (TNF-alpha) as candidates for mastitis caused by gram-negative bacteria. Authors described how TNF-alpha increased up to 100-fold 6 h after the intramammary administration LPS, a gram-negative cell wall component that mimics mastitis caused by this type of bacteria. Similarly, authors also found that milk SAA increased up to 350-fold 32 h after LPS administration. In agreement with these findings, Kusebauch et al. (2018) challenged cows with an intramammary infusion of either LPS or peptidoglycans (PGN; commonly used to mimic mastitis caused by gram-positive bacteria). Authors found how milk SAA was increased in cows receiving the LPS (1 mg/mL) compared to those receiving the PGN (0.1 mg/mL), although these differences were only observed 24 h after challenge. More interesting, other proteins such as CD14 increased rapidly 8 h after the challenge in the LPS group and remained higher for 24 h after the challenge. Similarly, interleukin 8 increased rapidly in milk from cows challenged with LPS 4 h after challenge and remained higher for 24 h after challenge. In another study, Hernández-Castellano et al. (2017a) described different correlations between SCC, lactate dehydrogenase (LDH) activity, and IgG concentration in milk from cows with mastitis caused by either gram-negative (i.e. *E. coli*) or gram-positive bacteria (i.e. *Corynebacterium bovis*, coagulase-negative *Staphylococcus*, *Staph. aureus* and *Streptococcus uberis*). Authors showed that despite SCC and LDH were similarly increased in mastitis caused by gram-positive and gram-negative bacteria, IgG concentrations were increased exclusively in mastitis caused by gram-negative bacteria.

## 6 Interventions and economic value of surveillance systems

The potential of using biomarkers for prediction of pre-clinical diseases is highly relevant. It is obvious that the efficiency of the biomarker surveillance system is highly dependent on the sensitivity and specificity of single or several biomarkers to predict the intended condition. The sensitivity describes the ability to classify the condition correctly, whereas the specificity describes the ability to classify cows as not having the condition (Greiner and Gardner, 2000). The prevalence of the condition is also very important for the performance of the surveillance system. Thus, the performance of a biomarker prediction, measured as sensitivity and specificity, cannot be evaluated without the prevalence of the condition. In fact, Dominiak and Kristensen (2017) described how sensors used in agriculture with high sensitivity and specificity still have poor performances characterized by large numbers of false positives, especially in diseases with low prevalence.

As illustrated in Fig. 2, the incidence of cows becoming imbalanced is much higher that the incidence of cows becoming subclinical diseased, as not all cows progress from imbalanced to subclinical. When the biomarker is intended to detect imbalanced cows and do some kind of treatment/intervention, this also implies that a proportion of cows will be treated unnecessarily. This is not an issue related to the ability of the biomarker to predict imbalanced cows, but instead related to the conceptual framework related to the changes between the different disease states. Assessment of the entire performance of the biomarker surveillance system should include the possible actions that can be taken based on the monitoring system. This is in line with the term Value of Information (VoI) taken from decision theory (Howard, 1966). This term is defined as how much a decision is improved by additional information (marginal) compared to the absence of that information. Recent studies of Rojo-Gimeno et al. (2019) and Ettema et al. (2020) have used forecasts of performance and prevalence to calculate the ability of systems for ketosis prevention. Both studies indicated that in case of high risk for ketosis, all cows should be treated, while in case of low risk for ketosis the subsequent appearance of the disease in some cows should be accepted. Based on the results from field trials performed by Tatone et al. (2016) and Østergaard et al. (2020), it seems that in certain circumstances, treating some of the cows under physiological imbalance does not result in improved health status and production.

The economic assessment of the negative consequences of a disease is commonly based on the failure costs alone. The failure costs are the treatment cost and the production losses, including labor, caused by the disease. For most production diseases the costs associated with the disease are the production loss and the premature culling (e.g. Hogeveen et al., 2019). These costs are often not obvious to the farmer and in addition they cannot be assumed to be equal in different herds. The consequence is that we do not only require good prevalence estimates of the pre-clinical condition, but also herd-specific estimates of the actual failure cost. However, the preventive cost should also be included in the economic assessment, especially in case of production diseases. Preventive cost is those related to the ongoing preventive measures implemented in the herd, such as the direct cost associated to post-milking teat dipping to prevent mastitis. This type of assessment was studied by van Soest et al. (2016) in cows with mastitis. These authors showed that approximately half of the total cost associated with subclinical and clinical mastitis was related to failure costs. This highlights the need for including the preventive costs associated to a disease when the evaluation of biomarker-based intervention systems is performed and that the total cost associated with a disease should be made herd-specific. In addition, most production diseases are interrelated, as for the postpartum disorders (Correa et al., 1993), which poses high risk of double-counting the negative production effects of the diseases; a challenge that can be overcome by stochastic modeling accounting for these interrelations (Raboisson et al., 2015; Liang et al., 2017).

For some more general literature on bridging the gap between disease effects and economic effects, and why this may be a challenge, we refer to Rushton (2017) and Hogeveen and Van Der Voort (2017).

## 7 Future perspectives

As described above, intensification and modernization of dairy herds demand a continuous need for disease surveillance, especially in those defined as production diseases. The strong correlation among production diseases suggests that a common pre-clinical condition exists where cows are in a physiological imbalance or at a high risk of progression to either subclinical or clinical metabolic and infectious diseases. As physiological imbalance in dairy cows is closely related to energy metabolism, monitoring of biomarkers involved in the metabolism of lipids and carbohydrates will be essential in future disease-surveillance programs. It seems that the one-dimensional approach defining physiological imbalance based on either the increased or decreased concentration of a single biomarker is not efficient. Instead, indexes similar to the physiological imbalance index (Moyes et al., 2013) or the development of new biomarkers and more sensitive devices will offer new possibilities. Combination of biomarkers (i.e. ratios between two or more biomarkers) or different sources of information and the variability and reactivity of individuals during an infection and/or metabolic challenge may have future potential. In agreement with this, Foldager et al. (2020) combined FFA, BHB, glucose and IGF-1 concentrations in blood from dairy cows in a cluster analysis. These authors concluded that different health status could be identified. Further research is needed in biomarkers to understand and improve physiological imbalance as a tool to prevent the progression to subclinical or clinical diseases or production losses. These biomarkers should not be measured in blood, but preferably be measured in, for example, milk as it can be collected automatically. In addition, the information obtained from these measurements could potentially be combined with behavioral biomarkers to provide a more complete picture of the entire status of the animal.

If physiological imbalance is precisely defined and the physiological mechanisms behind subclinical and clinical diseases are understood, the next step in the implementation of the future disease-surveillance programs would be to use cloud-based data integration from multiple farm sensors using advanced artificial intelligence methods to predict physiological imbalances. These implementations will improve animal monitoring accuracy by the identification of individual cows at an early imbalanced stage with an enhanced probability to develop specific diseases, improving disease surveillance at both individual and herd levels. Thus, insights from these sensing technologies generate data, which can be accessed remotely resulting in lower production cost and enhanced animal performance (Neethirajan, 2020; Morota et al., 2018). As concluded by Lokhorst et al. (2019), the real benefit of Big Data and artificial intelligence methods like deep learning is not yet utilized. Finally, future disease-surveillance systems should also provide herd-specific economic evaluations with both failure and preventive cost in order to help farmers to make the most appropriate decisions.

## 8 Where to look for further information

A wider introduction to animal disease surveillance in multiple species can be found in the textbook 'Animal Disease Surveillance and Survey Systems: Methods and Applications. Editor(s): M. D. Salman from 2003'. This book covers especially the surveillances and control of infectious disease between herds/populations from an epidemiological point of view. In addition, the book by Houe et al. (2014) introduces several central aspects of control and eradication of infectious diseases in cattle.

Recent developments in animal disease surveillance is the implementation of syndromic surveillance which is the systematic analysis of health data, clinical data, laboratory data and records of death, production records and other data sources that can be used to generate signals that may be indicative of changes in the occurrence of infection. Syndromic surveillance then directly link to Big Data and digital transition of animal production.

- A center of expertise includes the group at the National Veterinary Institute, Sweden. https://www.sva.se/en/research/strategic-research-areas/disease-preparedness/syndromic-surveillance/.

Precision livestock farming or smart farming also has elements of disease surveillance, mostly aiming at early disease detection using a variety of sensors.

- International resource is the European Association for Precision Livestock Farming. http://www.eaplf.eu/.

## 9 References

Abuelo, A., Hernandez, J., Benedito, J. L. and Castillo, C. 2019. Redox biology in transition Periods of dairy cattle: role in the health of periparturient and neonatal animals. *Antioxidants* 8(1), 20.

Adams, R. S., Stout, W. L., Kradel, D. C., Guss, S. B., Moser, B. L. and Jung, G. A. 1978. Use and limitations of profiles in assessing health or nutritional status of dairy herds. *Journal of Dairy Science* 61(11), 1671–1679.

Batavani, R. A., Asri, S. and Naebzadeh, H. 2007. The effect of subclinical mastitis on milk composition in dairy cows. *Iranian Journal of Veterinary Research* 8, 205–211.

Begley, N., Buckley, F., Pierce, K. M., Fahey, A. G. and Mallard, B. A. 2009. Differences in udder health and immune response traits of Holstein-Friesians, Norwegian Reds, and their crosses in second lactation. *Journal of Dairy Science* 92(2), 749–757.

Berckmans, D. J. A. F. 2017. General introduction to precision livestock farming. *Animal Frontiers* 7(1), 6–11.
Cai, T. Q., Weston, P. G., Lund, L. A., Brodie, B., Mckenna, D. J. and Wagner, W. C. 1994. Association between neutrophil functions and periparturient disorders in cows. *American Journal of Veterinary Research* 55(7), 934–943.
Chagunda, M. G., Friggens, N. C., Rasmussen, M. D. and Larsen, T. 2006. A model for detection of individual cow mastitis based on an indicator measured in milk. *Journal of Dairy Science* 89(8), 2980–2998.
Contreras, G. A. and Sordillo, L. M. 2011. Lipid mobilization and inflammatory responses during the transition period of dairy cows. *Comparative Immunology, Microbiology and Infectious Diseases* 34(3), 281–289.
Correa, M. T., Erb, H. and Scarlett, J. 1993. Path analysis for seven postpartum disorders of Holstein cows. *Journal of Dairy Science* 76(5), 1305–1312.
Craig, P. S., Mcmanus, D. P., Lightowlers, M. W., Chabalgoity, J. A., Garcia, H. H., Gavidia, C. M., Gilman, R. H., Gonzalez, A. E., Lorca, M., Naquira, C., Nieto, A. and Schantz, P. M. 2007. Prevention and control of cystic echinococcosis. *The Lancet. Infectious Diseases* 7(6), 385–394.
Dantzer, R. and Kelley, K. W. 2007. Twenty years of research on cytokine-induced sickness behavior. *Brain, Behavior, and Immunity* 21(2), 153–160.
De Almeida, A. M., Zachut, M., Hernandez-Castellano, L. E., Speranda, M., Gabai, G. and Mobasheri, A. 2019. Biomarkers of fitness and welfare in dairy animals: healthy living. *Journal of Dairy Research* 86(4), 379–387.
De Vries, A. and Reneau, J. K. 2010. Application of statistical process control charts to monitor changes in animal production systems. *Journal of Animal Science* 88(13) (Suppl.), E11–E24.
DeVries, T. J., Beauchemin, K. A., Dohme, F. and Schwartzkopf-Genswein, K. S. 2009. Repeated ruminal acidosis challenges in lactating dairy cows at high and low risk for developing acidosis: feeding, ruminating, and lying behavior. *Journal of Dairy Science* 92(10), 5067–5078.
Dominiak, K. N. and Kristensen, A. R. 2017. Prioritizing alarms from sensor-based detection models in livestock production–a review on model performance and alarm reducing methods. *Computers and Electronics in Agriculture* 133, 46–67.
El Badawy, S. A., Amer, A. M., Kamel, G. M., Eldeib, K. M. and Constable, P. D. 2015. Comparative pharmacokinetics using a microbiological assay and high performance liquid chromatography following intravenous administration of cefquinome in lactating goats with and without experimentally induced Staphylococcus aureus mastitis. *Small Ruminant Research* 133, 67–76.
Enjalbert, F., Videau, Y., Nicot, M. C. and Troegeler-Meynadier, A. 2008. Effects of induced subacute ruminal acidosis on milk fat content and milk fatty acid profile. *Journal of Animal Physiology and Animal Nutrition* 92(3), 284–291.
Ettema, J. F., Krogh, M. A., Østergaard, S. and GplusE Consortium 2020. Economic value of information from an alert system on physiological imbalance in fresh cows. *Preventive Veterinary Medicine* 181, 105039.
Faulkner, A., Chaiyabutr, N., Peaker, M., Carrick, D. T. and Kuhn, N. J. 1981. Metabolic significance of milk glucose. *Journal of Dairy Research* 48(1), 51–56.
Fogsgaard, K. K., Rontved, C. M., Sorensen, P. and Herskin, M. S. 2012. Sickness behavior in dairy cows during Escherichia coli mastitis. *Journal of Dairy Science* 95(2), 630–638.

Foldager, L., Gaillard, C., Sorensen, M. T., Larsen, T., Matthews, E., O'Flaherty, R., Carter, F., Crowe, M. A., Grelet, C., Salavati, M., Hostens, M., Ingvartsen, K. L., Krogh, M. A. and GplusE Consortium 2020. Predicting physiological imbalance in Holstein dairy cows by three different sets of milk biomarkers. *Preventive Veterinary Medicine* 179, 105006.

Geishauser, T., Leslie, K., Kelton, D. and Duffield, T. 1998. Evaluation of five cow side tests for use with milk to detect subclinical ketosis in dairy cows. *Journal of Dairy Science* 81(2), 438–443.

Goldhawk, C., Chapinal, N., Veira, D. M., Weary, D. M. and Von Keyserlingk, M. A. G. 2009. Prepartum feeding behavior is an early indicator of subclinical ketosis. *Journal of Dairy Science* 92(10), 4971-4977.

Greiner, M. and Gardner, I. A. 2000. Application of diagnostic tests in veterinary epidemiologic studies. *Preventive Veterinary Medicine* 45(1-2), 43–59.

Grelet, C., Bastin, C., Gelé, M., Davière, J. B., Johan, M., Werner, A., Reding, R., Fernandez Pierna, J. A., Colinet, F. G., Dardenne, P., Gengler, N., Soyeurt, H. and Dehareng, F. 2016. Development of Fourier transform mid-infrared calibrations to predict acetone, β-hydroxybutyrate, and citrate contents in bovine milk through a European dairy network. *Journal of Dairy Science* 99(6), 4816–4825.

Gross, J., Van Dorland, H. A., Bruckmaier, R. M. and Schwarz, F. J. 2011. Milk fatty acid profile related to energy balance in dairy cows. *Journal of Dairy Research* 78(4), 479–488.

Hegde, R., Isloor, S., Prabhu, K. N., Shome, B. R. Rathnamma, D., Suryanarayana, V. V., Yatiraj, S., Prasad, C. R., Krishnaveni, N., Sundareshan, S., Akhila, D. S., Gomes, A. R. and Hegde, N. R. 2013. Incidence of subclinical mastitis and prevalence of major mastitis pathogens in organized farms and unorganized sectors. *Indian Journal of Microbiology*, 53, 315–320.

Herdt, T. H. 2013. Metabolic diseases of dairy cattle. *Veterinary Clinics of North America. Food Animal Practice* 29(2), xi–xii.

Hernández-Castellano, L. E., Hernandez, L. L. and Bruckmaier, R. M. 2020. Review: endocrine pathways to regulate calcium homeostasis around parturition and the prevention of hypocalcemia in periparturient dairy cows. *Animal: An International Journal of Animal Bioscience* 14(2), 330–338.

Hernández-Castellano, L. E., Hernandez, L. L., Sauerwein, H. and Bruckmaier, R. M. 2017b. Endocrine and metabolic changes in transition dairy cows are affected by prepartum infusions of a serotonin precursor. *Journal of Dairy Science* 100(6), 5050–5057.

Hernández-Castellano, L. E., Storm, A. C., Larsen, T. and Larsen, M. 2018. D-lactate metabolism in dairy cows under risk for subacute ruminal acidosis. 2018 Annual Meeting. American Dairy Science Association®, Knoxville, United States, 321-321.

Hernández-Castellano, L. E., Wall, S. K., Stephan, R., Corti, S. and Bruckmaier, R. 2017a. Milk somatic cell count, lactate dehydrogenase activity, and immunoglobulin G concentration associated with mastitis caused by different pathogens: a field study. *Schweizer Archiv für Tierheilkunde* 159(5), 283–290.

Hogeveen, H., Steeneveld, W. and Wolf, C. A. 2019. Production diseases reduce the efficiency of dairy production: a review of the results, methods, and approaches regarding the economics of mastitis. *Annual Review of Resource Economics* 11(1), 289-312.

Hogeveen, H. and Van Der Voort, M. 2017. Assessing the economic impact of an endemic disease: the case of mastitis. *Revue Scientifique et Technique* 36(1), 217-226.

Houe, H., Nielsen, L. R. and Nielsen, S. S. 2014. *Control and Eradication of Endemic Infectious Diseases in Cattle*. London: College Publications.

Howard, R. A. 1966. Information value theory. *IEEE Transactions on Systems Science and Cybernetics* 2(1), 22-26.

Ingraham, R. H. and Kappel, L. C. 1988. Metabolic profile testing. *Veterinary Clinics of North America. Food Animal Practice* 4(2), 391-411.

Ingvartsen, K. L. 2006. Feeding- and management-related diseases in the transition cow. *Animal Feed Science and Technology* 126(3-4), 175-213.

Ingvartsen, K. L., Dewhurst, R. J. and Friggens, N. C. 2003. On the relationship between lactational performance and health: is it yield or metabolic imbalance that cause production diseases in dairy cattle? a position paper. *Livestock Production Science* 83(2-3), 277-308.

Ingvartsen, K. L. and Moyes, K. 2013. Nutrition, immune function and health of dairy cattle. *Animal: An International Journal of Animal Bioscience* 7 (Suppl. 1), 112-122.

Jawor, P. E., Huzzey, J. M., Leblanc, S. J. and Von Keyserlingk, M. A. 2012. Associations of subclinical hypocalcemia at calving with milk yield, and feeding, drinking, and standing behaviors around parturition in Holstein cows. *Journal of Dairy Science* 95(3), 1240-1248.

Kandeel, S. A., Megahed, A. A., Ebeid, M. H. and Constable, P. D. 2019. Ability of milk pH to predict subclinical mastitis and intramammary infection in quarters from lactating dairy cattle. *Journal of Dairy Science* 102(2), 1417-1427.

Kaufman, E. I., LeBlanc, S. J., McBride, B. W., Duffield, T. F. and DeVries, T. J. 2016. Short communication: association of lying behavior and subclinical ketosis in transition dairy cows. *Journal of Dairy Science* 99(9), 7473-7480.

Kehrli, M., Neil, J., Burvenich, C., Goff, J., Lippolis, J., Reinhardt, T. and Nonnecke, B. 2006. Energy and protein effects on the immune system. In: Sejrsen, K., Hvelplund, T. and Nielsen, M. O. (Eds) *Ruminant Physiology. Digestion, Metabolism and Impact of Nutrition on Gene Expression, Immunology and Stress*, pp. 455-471. The Netherlands: Wageningen Academic Publishers.

Kirkeby, C., Toft, N., Schwarz, D., Farre, M., Nielsen, S. S., Zervens, L., Hechinger, S. and Halasa, T. 2020. Differential somatic cell count as an additional indicator for intramammary infections in dairy cows. *Journal of Dairy Science* 103(2), 1759-1775.

Knauer, W. A., Godden, S. M., Dietrich, A., Hawkins, D. M. and James, R. E. 2018. Evaluation of applying statistical process control techniques to daily average feeding behaviors to detect disease in automatically fed group-housed preweaned dairy calves. *Journal of Dairy Science* 101(9), 8135-8145.

Kocabagli, N., Riond, J. L., Spichiger, U. E. and Wanner, M. 1995. Parathyroid hormone-related protein and calcium homeostasis during the periparturient period of dairy-cows. *American Journal of Veterinary Research* 56(3), 380-385.

Koeck, A., Jamrozik, J., Schenkel, F. S., Moore, R. K., Lefebvre, D. M., Kelton, D. F. and Miglior, F. 2014. Genetic analysis of milk beta-hydroxybutyrate and its association with fat-to-protein ratio, body condition score, clinical ketosis, and displaced abomasum in early first lactation of Canadian Holsteins. *Journal of Dairy Science* 97(11), 7286-7292.

König, S. and May, K. 2019. Invited review: phenotyping strategies and quantitative-genetic background of resistance, tolerance and resilience associated traits in dairy cattle. *Animal: An International Journal of Animal Bioscience* 13(5), 897-908.

Kostoulas, P., Nielsen, S. S., Branscum, A. J., Johnson, W. O., Dendukuri, N., Dhand, N. K., Toft, N. and Gardner, I. A. 2017. STARD-BLCM: standards for the reporting of diagnostic accuracy studies that use bayesian latent class models. *Preventive Veterinary Medicine* 138, 37-47.

Kovacs, C. S. 2011. Calcium and bone metabolism disorders during pregnancy and lactation. *Endocrinology and Metabolism Clinics of North America* 40(4), 795-826.

Krogh, M. A. and Enevoldsen, C. 2014. Evaluation of effects of metritis management in a complex dairy herd health management program. *Journal of Dairy Science* 97(1), 552-561.

Kronfeld, D. S. 1973. Calcium kinetics and parathyroid-hormone (PTH). *New Zealand Medical Journal* 78, 366-367.

Kusebauch, U., Hernandez-Castellano, L. E., Bislev, S. L., Moritz, R. L., Rontved, C. M. and Bendixen, E. 2018. Selected reaction monitoring mass spectrometry of mastitis milk reveals pathogen-specific regulation of bovine host response proteins. *Journal of Dairy Science* 101(7), 6532-6541.

Larsen, T. 2017. Fluorometric determination of d-lactate in biological fluids. *Analytical Biochemistry* 539, 152-157.

Larsen, T. and Moyes, K. M. 2015. Are free glucose and glucose-6-phosphate in milk indicators of specific physiological states in the cow? *Animal: An International Journal of Animal Bioscience* 9(1), 86-93.

LeBlanc, S. 2010. Monitoring metabolic health of dairy cattle in the transition period. *The Journal of Reproduction and Development* 56 (Suppl.), S29-S35.

Ledgerwood, D. N., Winckler, C. and Tucker, C. B. 2010. Evaluation of data loggers, sampling intervals, and editing techniques for measuring the lying behavior of dairy cattle. *Journal of Dairy Science* 93(11), 5129-5139.

Lee, A. J., Twardock, A. R., Bubar, R. H., Hall, J. E. and Davis, C. L. 1978. Blood metabolic profiles: their use and relation to nutritional status of dairy cows1. *Journal of Dairy Science* 61(11), 1652-1670.

Lehmann, M., Wall, S. K., Wellnitz, O. and Bruckmaier, R. M. 2015. Changes in milk L-lactate, lactate dehydrogenase, serum albumin, and IgG during milk ejection and their association with somatic cell count. *Journal of Dairy Research* 82(2), 129-134.

Lehtolainen, T., Rontved, C. and Pyorala, S. 2004. Serum amyloid A and TNF alpha in serum and milk during experimental endotoxin mastitis. *Veterinary Research* 35(6), 651-659.

Liang, D., Arnold, L. M., Stowe, C. J., Harmon, R. J. and Bewley, J. M. 2017. Estimating US dairy clinical disease costs with a stochastic simulation model. *Journal of Dairy Science* 100(2), 1472-1486.

Lokhorst, C., De Mol, R. M. and Kamphuis, C. 2019. Invited review: big data in precision dairy farming. *Animal: An International Journal of Animal Bioscience* 13(7), 1519-1528.

Mertens, K., Decuypere, E., De Baerdemaeker, J. and De Ketelaere, B. J. T. J. O. A. S. 2011. Statistical control charts as a support tool for the management of livestock production. *The Journal of Agricultural Science* 149(3), 369-384.

Milner, P., Page, K. L. and Hillerton, J. E. 1997. The effects of early antibiotic treatment following diagnosis of mastitis detected by a change in the electrical conductivity of milk. *Journal of Dairy Science* 80(5), 859-863.

Morota, G., Ventura, R. V., Silva, F. F., Koyama, M. and Fernando, S. C. 2018. Big data analytics and precision animal agriculture symposium: machine learning and data mining advance predictive big data analysis in precision animal agriculture. *Journal of Animal Science* 96(4), 1540-1550.

Moyes, K. M., Larsen, T. and Ingvartsen, K. L. 2013. Generation of an index for physiological imbalance and its use as a predictor of primary disease in dairy cows during early lactation. *Journal of Dairy Science* 96(4), 2161-2170.

Neethirajan, S. 2020. The role of sensors, big data and machine learning in modern animal farming. *Sensing and Bio-Sensing Research* 29, 100367.

Nielsen, N. I., Friggens, N. C., Chagunda, M. G. G. and Ingvartsen, K. L. 2005. Predicting risk of ketosis in dairy cows using in-line measurements of beta-hydroxybutyrate: a biological model. *Journal of Dairy Science* 88(7), 2441-2453.

Nir, O. 2003. What are production diseases, and how do we manage them? *Acta Veterinaria Scandinavica* 98, 21-32.

Nocek, J. E. 1997. Bovine acidosis: implications on laminitis. *Journal of Dairy Science* 80(5), 1005-1028.

Nogalski, Z., Wronski, M., Sobczuk-Szul, M., Mochol, M. and Pogorzelska, P. 2012. The effect of body energy reserve mobilization on the fatty acid profile of milk in high-yielding cows. *Asian-Australasian Journal of Animal Sciences* 25(12), 1712-1720.

Ospina, P. A., Mcart, J. A. T. R., Overton, T. R., Stokol, T. and Nydam, D. V. 2013. Using nonesterified fatty acids and β-hydroxybutyrate concentrations during the transition period for herd-level monitoring of increased risk of disease and decreased reproductive and milking performance. *Veterinary Clinics of North America. Food Animal Practice* 29(2), 387-412.

Østergaard, S., Krogh, M. A., Oliveira, V. H. S., Larsen, T., GplusE Consortium and Otten, N. D. 2020. Only few benefits from propylene glycol drench in early lactation for cows identified as physiologically imbalanced based on milk spectra analyses. *Journal of Dairy Science* 103(2), 1831-1842.

Payne, J. M., Dew, S. M., Manston, R. and Faulks, M. 1970. The use of a metabolic profile test in dairy herds. *Veterinary Record* 87(6), 150-158.

Qayyum, A., Khan, J. A., Hussain, R., Avais, M., Ahmad, N. and Khan, M. S. 2016. Investigation of milk and blood serum biochemical profile as an indicator of sub-clinical mastitis in Cholistani cattle. *Pakistan Veterinary Journal* 36, 275-279.

Raboisson, D., Mounie, M., Khenifar, E. and Maigne, E. 2015. The economic impact of subclinical ketosis at the farm level: tackling the challenge of over-estimation due to multiple interactions. *Preventive Veterinary Medicine* 122(4), 417-425.

Rauw, W. M. and Gomez-Raya, L. 2015. Genotype by environment interaction and breeding for robustness in livestock. *Frontiers in Genetics* 6, 310.

Riond, J. L., Kocabagli, N., Toromanoff, A. and Wanner, M. 1995. Parathyroid hormone-related protein and calcium homeostasis. *Schweizer Archiv Fur Tierheilkunde* 137(4), 117-123.

Rodriguez-Jimenez, S., Haerr, K. J., Trevisi, E., Loor, J. J., Cardoso, F. C. and Osorio, J. S. 2018. Prepartal standing behavior as a parameter for early detection of postpartal subclinical ketosis associated with inflammation and liver function biomarkers in peripartal dairy cows. *Journal of Dairy Science* 101(9), 8224–8235.

Roeder, P., Mariner, J. and Kock, R. 2013. Rinderpest: the veterinary perspective on eradication. *Philosophical Transactions of the Royal Society of London. Series B, Biological Sciences* 368(1623), 20120139.

Rojo-Gimeno, C., Van Der Voort, M., Niemi, J. K., Lauwers, L., Kristensen, A. R. and Wauters, E. 2019. Assessment of the value of information of precision livestock farming: a conceptual framework. *NJAS–Wageningen Journal of Life Sciences* 90-91, 100311.

Rukkwamsuk, T., Geelen, M. J. H., Kruip, T. A. M. and Wensing, T. 2000. Interrelation of fatty acid composition in adipose tissue, serum, and liver of dairy cows during the development of fatty liver postpartum. *Journal of Dairy Science* 83(1), 52-59.

Rushton, J. 2017. Improving the use of economics in animal health–challenges in research, policy and education. *Preventive Veterinary Medicine* 137(B), 130–139.

Sahar, M. W., Beaver, A., Weary, D. M. and Von Keyserlingk, M. A. G. 2020. Feeding behavior and agonistic interactions at the feed bunk are associated with hyperketonemia and metritis diagnosis in dairy cattle. *Journal of Dairy Science* 103(1), 783–790.

Salgado-Hernandez, E. G., Bouda, J., Villa-Godoy, A., Romano-Munoz, J. L., Gutierrez-Chavez, A. J. and Velasquez-Forero, F. 2014. Metabolites of vitamin D and minerals in blood and colostrum of primiparous and multiparous dairy cows postpartum. *Czech Journal of Animal Science* 59(1), 11–18.

Scott, R. A., Bauman, D. E. and Clark, J. H. 1976. Cellular gluconeogenesis by lactating bovine mammary tissue. *Journal of Dairy Science* 59(1), 50–56.

Sepulveda-Varas, P., Proudfoot, K. L., Weary, D. M. and Von Keyserlingk, M. A. G. 2016. Changes in behaviour of dairy cows with clinical mastitis. *Applied Animal Behaviour Science* 175, 8–13.

Sepulveda-Varas, P., Weary, D. M. and Von Keyserlingk, M. A. G. 2014. Lying behavior and postpartum health status in grazing dairy cows. *Journal of Dairy Science* 97(10), 6334–6343.

Siivonen, J., Taponen, S., Hovinen, M., Pastell, M., Lensink, B. J., Pyorala, S. and Hanninen, L. 2011. Impact of acute clinical mastitis on cow behaviour. *Applied Animal Behaviour Science* 132(3-4), 101–106.

Stoop, W. M., Bovenhuis, H., Heck, J. M. L. and Van Arendonk, J. A. M. 2009. Effect of lactation stage and energy status on milk fat composition of Holstein-Friesian cows. *Journal of Dairy Science* 92(4), 1469–1478.

Sundrum, A. 2015. Metabolic disorders in the transition period indicate that the dairy cows' ability to adapt is overstressed. *Animals (Basel)* 5(4), 978–1020.

Suthar, V. S., Canelas-Raposo, J., Deniz, A. and Heuweiser, W. 2013. Prevalence of subclinical ketosis and relationships with postpartum diseases in European dairy cows. *Journal of Dairy Science* 96(5), 2925–2938.

Tatone, E. H., Gordon, J. L., Hubbs, J., LeBlanc, S. J., DeVries, T. J. and Duffield, T. F. 2016. A systematic review and meta-analysis of the diagnostic accuracy of point-of-care tests for the detection of hyperketonemia in dairy cows. *Preventive Veterinary Medicine* 130, 18–32.

Van Knegsel, A. T. M., Van Den Branda, H., Dijkstra, J., Tamminga, S. and Kemp, B. 2005. Effect of dietary energy source on energy balance, production, metabolic disorders and reproduction in lactating dairy cattle. *Reproduction, Nutrition, Development* 45(6), 665–688.

Van Soest, F. J. S., Santman-Berends, I. M. G. A., Lam, T. J. G. M., Hogeveen, H. J. J. O. D. 2016. Failure and preventive costs of mastitis on Dutch dairy farms. *Journal of Dairy Science* 99(10), 8365–8374.

Wall, S. K., Hernandez-Castellano, L. E., Ahmadpour, A., Bruckmaier, R. M. and Wellnitz, O. 2016a. Differential glucocorticoid-induced closure of the blood-milk barrier during lipopolysaccharide- and lipoteichoic acid-induced mastitis in dairy cows. *Journal of Dairy Science* 99(9), 7544–7553.

Wall, S. K., Wellnitz, O., Hernandez-Castellano, L. E., Ahmadpour, A. and Bruckmaier, R. M. 2016b. Supraphysiological oxytocin increases the transfer of immunoglobulins and other blood components to milk during lipopolysaccharide- and lipoteichoic acid-induced mastitis in dairy cows. *Journal of Dairy Science* 99(11), 9165–9173.

World Organisation for Animal Health (OIE) 2019. *Terrestrial Animal Health Code*, 28th edn. OIE, Paris. Available at: www.oie.int/en/international-standard-setting/terrestrial-code/access-online/ (accessed on 23 March 2020).

Zachut, M., Kra, G., Portnik, Y., Shapiro, F. and Silanikove, N. 2016. Milk glucose-6-phosphate dehydrogenase activity and glucose-6-phosphate are associated with oxidative stress and serve as indicators of energy balance in dairy cows. *RSC Advances* 6(70), 65412–65417.

Zachut, M., Šperanda, M., De Almeida, A. M., Gabai, G., Mobasheri, A. and Hernandez-Castellano, L. E. 2020. Biomarkers of fitness and welfare in dairy cattle: healthy productivity. *Journal of Dairy Research* 87(1), 4–13.

Zarrin, M., Riveros, J. L., Ahmadpour, A., De Almeida, A. M., Konuspayeva, G., Vargas-Bello-Perez, E., Faye, B. and Hernandez-Castellano, L. E. 2020. Camelids: new players in the international animal production context. *Tropical Animal Health and Production* 52(3), 903–913.

Zhao, F. Q. 2014. Biology of glucose transport in the mammary gland. *Journal of Mammary Gland Biology and Neoplasia* 19(1), 3–17.

# Chapter 3

# Advances in techniques for health monitoring/disease detection in dairy cattle

*Michael Iwersen and Marc Drillich, University of Veterinary Medicine Vienna, Austria*



## 1 Introduction

In recent decades, livestock production has been characterized by the intensification and specialization of production processes leading to larger farms and, hence, to fewer farmers per animal. Thus, the animal-to-stockman ratio is increasing within the dairy sector and the available time for monitoring of an individual animal is reducing. Compared with the past, diversified family-owned farms have frequently switched to highly specialized companies that apply industrial production processes.

In order to earn a living, farmers have to house and care for a larger number of animals than in the past. Today's society, however, expects that animals are kept under high animal welfare and health standards that ensure 'a life worth living' (Webster, 2016), including that animals receive individual attention. However, in production-intensive sectors like cattle production, farmers have less time per animal as a consequence of the above-mentioned economy of scale. Within the production cycle, the farmer or stockperson has a unique role in ensuring high standards of animal welfare. Caring for livestock is, at least for the majority of farmers, a very positive experience, and positive human

http://dx.doi.org/10.19103/AS.2020.0086.04

contacts between farmer and animal have been shown to have a beneficial effect upon the welfare of both (Boivin et al., 2003). The quality of stockmanship, particularly the care and handling of animals, is responsible for some of the differences between farms and farmers in the level of animal welfare (Rushen et al., 2008). Additionally, there is evidence that good stockmanship improves productivity and is essential for an economically successful farm. Hence, reducing supervision time per animal might have negative socio-economic consequences for the stakeholders involved in the business, foremost not only for the animals and the farmer but also for their veterinarians, feed consultants, and consumers.

The observed economy of scale, as well as the growing demand for animal-derived food, will exert even stronger pressure on farm consolidation and process efficiency, inducing the creation of corporations with even larger farm units. On the other hand, these changes in farm structure offer the opportunity for the cost-effective use of sophisticated technologies, for example, automated feeding and milking systems, as well as animal health and welfare monitoring systems.

Traditional and increasingly sophisticated technologies support farmers and veterinarians in monitoring animals for the early detection of, for example, diseases or welfare issues and enable the collection of individual animal data as well as data from the entire herd and, thus, can bring the animals closer to the farmers again and assist them in their daily tasks. For instance, this includes the early and accurate detection of diseases, followed by timely veterinary intervention, which usually results in earlier recovery. Using these technologies has an impact not only on animal health and welfare but also on working routines and optimized resource management, increasing the competitiveness of the farm, in general.

## 2 Shift in the veterinary profession

With regard to animal health, the focus in the veterinary profession has shifted from the treatment of acutely diseased animals to more proactive health management of the entire herd (LeBlanc et al., 2006). This also includes ensuring high animal welfare standards. This shift in veterinary medicine has occurred in recent decades and is in accordance with changes in farmers' demands and expectations.

Currently, farmers' demand for veterinary advice goes beyond clinical activities and therapies (Pothmann et al., 2014). For example, farmers ask their veterinarians for profound support in areas of environmental protection, welfare, nutrition, grassland management, economics, and miscellaneous business management (Da Silva et al., 2006). Farmers' and veterinarians' decisions should be evidence based on the most reliable data available, preferably in real time. In this context, veterinary diagnoses contribute to the pool of farm data that can be used for herd (health) management decisions.

The above-mentioned paradigm shift in veterinary medicine and farmers' demands on ensuring animal health and productivity has brought proactive disease prevention strategies in the focus of herd health medicine. Realizing that most diseases have a multifactorial genesis led to the implementation of (farm) specific prevention programs, in particular, in early lactation. Epidemiological tools have increasingly been used to identify risk factors at the animal and herd level. As a result of this, research identified that an already subclinical form of a disease, for example, subclinical ketosis, hypocalcemia, mastitis, or endometritis, has a negative impact on individual animal health, productivity, and welfare (Walsh et al., 2007; Suthar et al., 2013; Rodriguez et al., 2017; Venjakob et al., 2017; Wagener et al., 2017; Gussmann et al., 2019) and on total herd performance. These findings led to the redefinitions of notorious and well-known diseases in a more complex and broader view (LeBlanc et al., 2006; Wagener et al., 2017).

## 2.1 Herd health management

Herd health management, also referred to as population medicine, is a multidisciplinary approach to optimize animal health, welfare, and productivity through systematic analyses involving planning, implementing, monitoring, and controlling production processes. Based on these systematic analyses, decisions are made as best and timely as possible to improve the general farm management over time (LeBlanc et al., 2006). For this, the strategies intend to gather information about the individual animal and the entire herd, striving for the final objective of assisting in decision-making processes. Typical tasks or events of a farmer and/or veterinarian, where decision-making processes are needed immediately are, for example, if and how to treat an animal with signs of diseases, or to intervene or not at parturition, but also in the case of physiological events, for example, in estrus detection.

After parturition, most larger dairies keep their fresh cows in specific pens to facilitate health monitoring and examinations (Guterbock, 2004), which is usually performed by farm personnel to identify sick cows. Not only do the training and skills but also the willingness of the personnel for fresh cow monitoring vary between farms (Espadamala et al., 2016). In a written survey of 429 farms, Heuwieser et al. (2010) reported that most dairy managers use subjective criteria to identify diseased animals, for example, general appearance (97%) and appetite (70%). Only the minority of responding participants monitor their cows by the use of objective (fever 34%) or semi-quantitative measures, for example, measuring ketones (3%) or body condition scoring (36%).

## 2.2 Epidemiology

Epidemiology deals with identifying potential causal associations between exposures and outcomes. Ultimately, research aims to 'make causal inferences about relationships between exposure and disease in the source population as a preliminary step toward developing policy and programs to maintain health and prevent disease' (Dohoo et al., 2009). From a veterinary standpoint, the exposures are often risk factors, leading to disease, or impaired animal performance as an outcome. In this context, farm-specific policies and programs consist of, for example, routine animal examinations and other prophylactic measures for disease prevention.

In recent decades, numerous large-scale observational studies have been conducted on commercial farms to identify risk factors for the development of diseases. As an example of early studies focusing on the interrelationship between animal nutrition, health, and reproductive performance of periparturient cows, Dohoo and Martin (1984), Curtis et al. (1985), and Cameron et al. (1998) should be mentioned here. These and other research groups have contributed to a profound knowledge base on the interrelationship between diseases and disorders, respectively, as well as their influence on animal performance. Additionally, Erb and Gröhn (1988), who postulated associations among metabolic disorders and risk factors [i.e. diseases and milk yield in previous lactation, dry cow nutrition, neonate factors (e.g. sex, size, survival, twins), dystocia, retained fetal membranes, and metritis], as presented in Figure 1, should also be mentioned here.

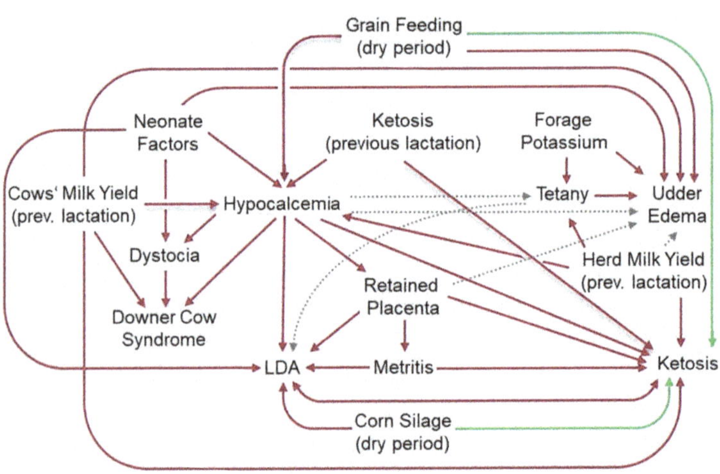


**Figure 1** Risk factors for clinical metabolic disorders in periparturient cows postulated by Erb and Gröhn (1988, modified figure). Projected effects are represented by dotted lines. Red lines indicate disease-promoting, green lines disease-inhibiting effects. LDA = left displaced abomasum.

The authors concluded that 'none of the six periparturient metabolic disorders occurs in isolation from the others' and that 'milk fever plays a central role linking to all of the other disorders' (Erb and Gröhn, 1988). Furthermore, the occurrence of ketosis was identified as a risk factor for the development of various diseases in the subsequent lactation. As ketosis can be a risk factor for and also a sequalae of the left displaced abomasum (LDA), the association between both was represented with a double-headed arrow.

Much recent research in dairy herd management focuses on the time period from late pregnancy to early lactation, which is also referred to as a 'transition period' and defined as the period from 3 weeks pre- until 3 weeks post-calving (Grummer, 1995). Dairy cows have to face many physiological changes during their transition from the dry to their lactating period. These include a temporary state of physiological immunosuppression around parturition, which, among others, is caused by a lowered bactericidal capacity of polymorphnuclear leukocytes and decreased responsiveness of lymphocytes to mitogenic agents (Detilleux et al., 1995; Nonnecke et al., 2003; Lacasse et al., 2018). Hence, animals in early lactation are at a greater risk for the occurrence of infectious diseases, such as mammary (Bradley and Green, 2004; Pyörälä, 2008) and uterine infections (Drillich and Wagener, 2018; Sheldon et al., 2020). Because of the before mentioned reasons, most health disorders occur during the transition period (Drackley, 1999; LeBlanc, 2010) and approximately 75% of all diseases take place in the first month of lactation (LeBlanc, 2006)

## 3 Information management systems

The above-mentioned findings of epidemiological research are now increasingly being applied in the daily practice of dairy farms. A fast and timely analysis is made possible by the use of information technologies, which consist of hard- and software solutions for capturing data as well as transforming them into information, which can support farmers and other stakeholders in decision making (Van Asseldonk, 1999). The active use of these systems is another cornerstone in the overall concept of early detection of problems and diseases at the herd and animal level.

The use of analog and electronic data collection systems has become a valuable tool in epidemiology as well as in herd health management. In particular, on-farm management software has been developed for data collection, integration, analyses, and easy presentation of results. While the integration of many cow-specific data, for example, on milk yield, milk solids, and veterinary diagnoses has already been possible for some decades using sophisticated software solutions, for example, DairyComp 305 (Valley Agricultural Software, Tulare, USA) and DSA Dairy-Veterinary (DSAHR,

St-Hyacinthe, Canada), the integration of external data resources, for example, on climate, feeding ration, and data exchange with other technical devices is often challenging. Hence, interfaces are needed to enable the sharing and uploading of these often heterogeneous data onto central data storage. This is one prerequisite of a more in-depth analysis of farm data and its environment in order to gather additional knowledge and enable further improvement of herd health management (Stewart et al., 1994; Wischenbart et al., 2017; Gusterer et al., 2019). Furthermore, access to a high-performance internet connection and the general availability of WLAN are further basic requirements for data exchange in real time, for linking production processes and increase our knowledge.

### 3.1 Data collection and definition of disease

In general, farmers and veterinarians do not need a huge amount of data to do their daily tasks. In this context, data collection should be as simple and timely as possible. Data collection that is too detailed may be interesting for profound in-depth analyses or from a scientific point of view, but for most dairy farms it does not seem to lead to an improvement in organizing daily tasks or in the identification and solving of problems. On the contrary, a data collection system that is too complex and that, for example, requires a detailed description or a selection from a several hundred diseases, tends to lead to frustration during data entry. This could be one reason that, quite often, only few or no data are collected on farms. Data collection should therefore be limited to a few and most common and relevant diseases occurring on farms. However, these well-defined, systematically recorded, and reliable data have to be entered consistently, in a timely manner and as easily as possible. For this further standardization of disease definition and an intuitive codification (i.e. avoiding entry of numerical or meaningless codes) is required.

Standard operating procedures (SOP), in which tasks are clearly defined and work steps are systematically described in detail, that is, who does what, when, and how, have been increasingly used on farms in recent years and contribute to improved data quantity and quality. Even if on-farm level SOPs are used for increasingly standardized data collection, there is often a lack of a common definition of diseases in veterinary practices with several employees.

In general, there is a lack of a uniform definition of diseases and production parameters that are applied across company borders, at national and international levels. Together with the variation in examination methods, this makes the comparison of, for example, disease incidences and key performance indicators between farms, regions, and countries, and, in science, of multicenter studies difficult or simply impossible. Even if increasing standardization in defining and reporting health data as well as production-related indices, for example, by published guidelines of the International Committee for Animal Recording (ICAR), is highly appreciated, there is still a need for further standardization as already requested for decades (Fetrow et al., 1990, 2006; Cook et al., 2006a; Pannwitz, 2015). One of the few examples of a cross-border standardization of health data recordings of dairy cows in a non-governmental database was developed by DSAHR Inc. (Association des médecins vétérinaires praticiens du Québec, St-Hyacinthe, Canada), which is available in Canada, France, Belgium, and Switzerland.

Considering animal health at the farm level, clinical disease is only the 'tip of the iceberg' underestimating the prevalence of disorders, which limit animal performance (LeBlanc, 2010). Furthermore, even subclinical forms of diseases, for example, subclinical ketosis, endometritis, hypocalcemia, and mastitis, can impair the health, well-being, and performance. Simply put, more accurate and cheaper measurement methods and diagnostic tools have led to this new understanding of diseases, where factors limiting performance at animal or herd level are considered as components of disease (LeBlanc et al., 2006). Currently, the routine and standardized examination of animals for the presence of subclinical diseases is recommended as an integral part of modern herd management (Cook et al., 2006a,b; LeBlanc et al., 2006). For this, several affordable diagnostic tools fostering herd management decisions have been developed.

### 3.2 National database

As a result of the BSE crisis at the end of the last century, Regulation No 1760/2000 was passed in Europe, which regulates the identification of cattle. For all bovines born on or after January 1, 1998, the regulation requires the lifelong application of two identical ear tags with a unique registration number and a passport enabling the animal to be traced back to the premises of birth. Furthermore, all animal movements must be documented in national databases. The unique identification and traceability of cattle enables the national and international monitoring of animal diseases and zoonosis so that, if necessary, measures can be taken as quickly as possible to prevent the outbreak or to implement control measures. Recent scenarios where databases were intensively used in Europe for tracking the spread of disease were the outbreak of Bluetongue starting in 2007 and Schmallenberg disease starting in 2011 (Wilson and Mellor, 2009; Koenraadt et al., 2014), which have led to the introduction of appropriate countermeasures.

Besides international monitoring of emerging and re-emerging infectious diseases, the unique identification of animals is a prerequisite for collecting reliable animal health data, at farm and national levels. In the Scandinavian countries, veterinary diagnoses have been collected and analyzed for a number of years (Nielsen et al., 2000; Forshell and Østerås, 2001; Philipsson and Lindhé, 2003; Negussie et al., 2010).

Under the leadership of the Federation of Austrian Cattle Breeders (ZAR, Vienna, Austria), a system for capturing veterinary diagnostic data in cattle was introduced throughout Austria in 2006. Various Austrian organizations involved in animal health, that is, the Ministry for Agriculture, Forestry, Environment and Water Management, the Ministry for Health, the University of Veterinary Medicine Vienna, the University of Natural Resources and Life Sciences Vienna, local animal health organizations, the Chamber of Agriculture, and the Chamber of Veterinarians, have contributed to further standardization of disease definition, data entry, and the development of the common database (Egger-Danner et al., 2012). Based on validated data sets of veterinary treatments, incidences in Simmental cows of 5.2% for metritis, 11.0% for mastitis, and 0.9% for ketosis were reported for 2010 (Egger-Danner et al., 2012). Additional information is presented in Figure 2.

An often-occurring problem is the underestimation of a disease at farm level. For instance, the above-mentioned incidence for subclinical and clinical ketosis in dairy cows based on veterinary treatments was 0.9%. This number, however, is not based on systematic monitoring of ketosis but on observations made by farmers or on occasional veterinary diagnoses. Particularly for subclinical diseases, this does not reflect the true incidence. This is illustrated by the fact that systematic ketosis monitoring on farms revealed incidences

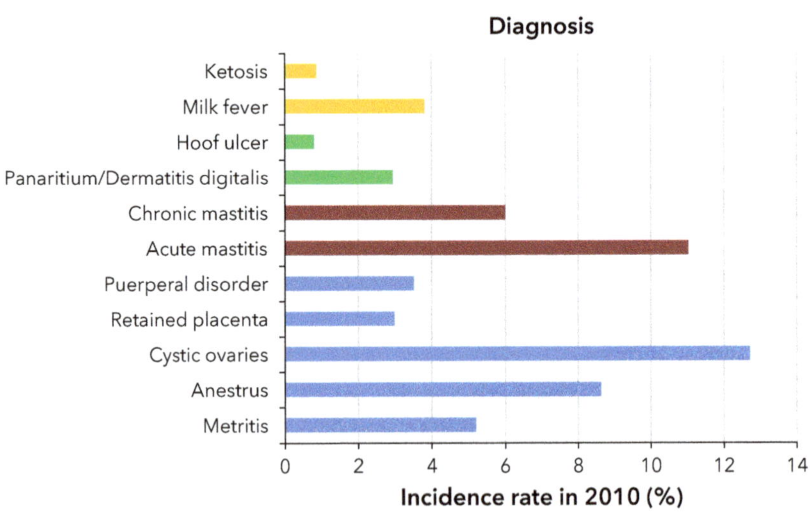


**Figure 2** Incidences for various cattle diseases in Austria 2010, based on recorded veterinary diagnoses and treatments (Egger-Danner et al., 2012).

between 11.2% and 36.6% in Europe (Suthar et al., 2013). With regard to subclinical ketosis, overseen cases do not necessarily result in obvious health problems for the animal but lead to a higher susceptibility to other diseases and to an impaired reproductive performance (Duffield et al., 2009; Chapinal et al., 2012).

## 4 On-farm diagnostic tests

In veterinary medicine and farm management, various tools, including diagnostic tests, are available for animal monitoring. In a broader sense, all (electronic) tools and devices that reduce the uncertainty about an animal's health and welfare status, as well as about its physiological state (e.g. pregnant vs. non-pregnant), can be considered as diagnostic tests. The variety of these tests is broad, ranging from a 'simple' combination of clinical examinations, which are indicative of a particular disease, to on-farm cow-side tests and the use of sophisticated laboratory methods. Examples of veterinary disciplines in which laboratory tests are used include clinical chemistry, cytology, hematology, histology, microbiology, serology, parasitology, toxicology, and fluid analyses. Describing the methodologies of these various tests in detail is beyond the scope of this chapter. About 30 years ago, the value of metabolic analyses, also with regard to their use in herd management, was recognized and triggered further activities in this field of research. In addition to technological developments, for example, cheaper devices for use on farms, this has led to an increasing interest of farmers, veterinarians, and consultants in using these tools, and it is expected that the 'rapid advancement of technologies that will enable real-time, automated measurement, will continue to revolutionize how we approach the monitoring of metabolic function and health in dairy cattle' (Overton et al., 2017). More recently and in addition to the previously described diagnostic approaches, farmers have increasingly been using sensor technologies for animal health monitoring, the potential of which should not be ignored by the veterinary profession. Several applications of sensor-based monitoring systems are merchandized by a rapidly increasing number of manufacturers and providers. In order to ensure that the technical tools and devices can be used successfully in veterinary practice as well as on dairy farms, however, they should be tested independently before commercialization with regard to their diagnostic reliability and economic benefits.

### *4.1 Validation and test performance of tools and devices*

Several diagnostic tests used in veterinary medicine have been derived from human medicine. Even if the same parameter is measured, various species-specific and environmental factors can alter the results. It is therefore important that only tests that have been previously validated (Abuelo and Alves-Nores, 2016) for the species to be examined, preferably in multicenter studies, are used. Strictly speaking, validations are reliable only for a single population of animals, considering their housing conditions and production purpose (i.e. beef and/or milk production), for instance. This also applies to reference limits which are only valid for a specific laboratory with given equipment, for example, personnel, devices, and reagents (Lumsden et al., 1980; Iwersen et al., 2013).

An ideal test would be able to perfectly distinguish between the presence and absence of disease, without giving any false-positive or false-negative results. A good clinical test should 'provide correct diagnoses, provide data to aid in prognosis, provide an indication of subclinical disease, provide for monitoring the effects of treatment and provide data that may indicate possible reoccurrence of chronic disease' (Galen and Gambino, 1975, cited in Reneau 1986).

In this context, the terminology 'true positive' (TP), 'false positive' (FP), 'true negative' (TN), and 'false negative' (FN) refers to a comparison of the predicted condition, for example, the test result, with the true condition, for example, the animal status, as presented in Figure 3.

With regard to animal health, correct outcomes include the cases where an animal is either diseased and with a positive test result (TP), or non-diseased with a negative test result (TN). A Type I error occurs in the case of a positive test result (i.e. diseased) when the animal is actually healthy. A Type II error exists when the test indicates a healthy animal when it is actually diseased. A Type I error can cause financial losses due to unnecessary treatment, or in a worst-case culling, a Type II error can constitute an animal welfare issue, because

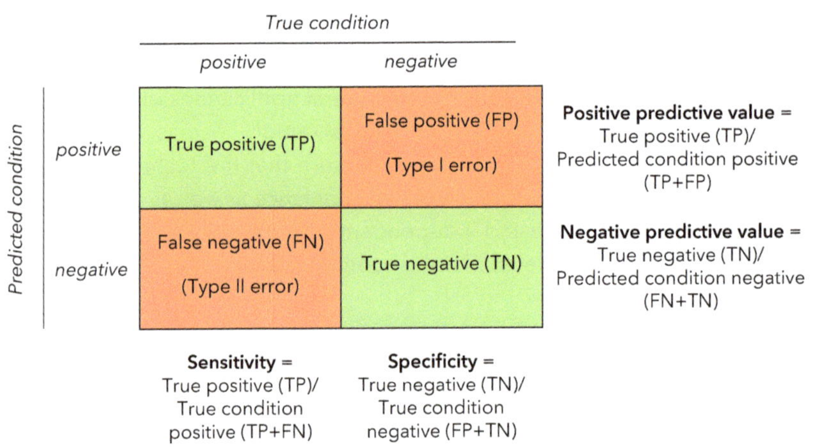


**Figure 3** Contingency table for the evaluation of a (diagnostic) test.

the diseased animals could have had the benefit of treatment or, in case of contagious diseases, the spread of disease. Hence, a possible treatment success leading to improved animal health and performance remains unrealized (Burfeind et al., 2010). Furthermore, an untreated animal bears a greater risk of becoming chronically diseased and/or spreading the disease in the herd.

For the quantification of the discriminate potential of a test system, sensitivities, specificities, as well as positive and negative predictive values are often reported in scientific validations. Detailed calculations of the above-mentioned test characteristics are presented in Figure 3. The diagnostic sensitivity (SE) is defined as the ability of a test to correctly classify an animal as diseased; diagnostic specificity (SP) is the ability of a test to correctly identify an animal as non-diseased. To determine SE and SP, the true condition of an animal must be known; that is, there must be a 'gold standard' (or a test combination) that always provides the correct information on the animal's condition. In practice, a gold standard does not always exist. Hence, sometimes methods that are believed as close enough to the true condition are considered as 'gold' (McKenna and Dohoo, 2006).

If the intention to test is to confirm a disease, one should use a test with a high SP, as there are few false positives. Otherwise, if the intention is to rule out disease, one should use a test with a high SE as there are few false negatives (Dohoo et al., 2009). The SE and SP of a test do not directly provide information on its performance when applied to animals with unknown disease status. The number of true and false test results is also dependent on the prevalence of the disease. The probability that an animal is or is not diseased, depending on whether the test is positive or not, is defined as the predictive value, which changes with different populations of tested animals (Dohoo et al., 2009). The positive predictive value (PPV) is the percentage of positive tested animals, which are actually suffering from the disease. The negative predictive value (NPV) is the percentage of animals with a negative test result, which do not have the disease.

Once again it should be mentioned that predictive values are dependent on the prevalence of the disease (see Tables 2 and 3). Hence, the direct comparison of the PPV and NPV of tests, captured in different populations and settings, is not valid. To get out of this dilemma, for comparison purposes, predictive values could be calculated for hypothetical levels of prevalence (Carrier et al., 2004; Iwersen et al., 2009).

### 4.1.1 Selecting a diagnostic test

In human medicine all clinical laboratory testing is regulated by federal legislation (Arnold et al., 2019). No equivalent regulation exists for veterinary medicine (Bertani et al., 2019). Therefore, independent analyses or approvals, as requested in human medicine from the manufacturers, are not required before introducing diagnostic tests into the veterinary market. The American Society for Veterinary Clinical Pathology (ASVCP) developed quality assurance guidelines for veterinary point-of-care tests (Flatland et al., 2013). Due to the lack of legal regulation, quality assurance (QA) and quality control of veterinary diagnostic tests should be in the own interest of the veterinary profession (Arnold et al., 2019).

The increase in diagnostic tests has led to various guidelines and recommendations toward testing procedures, and various publications on test performance were published. Evaluations of tests at farm conditions can reveal important features that were not revealed in optimal condition, for example, in a laboratory setting. For instance, test results may be influenced by the environment, the tested specimen, or the user's level of training (Iwersen et al., 2013; Pineda and Cardoso, 2015; Leal Yepes et al., 2018). Generic criteria which have to be considered besides SE and SP when selecting a diagnostic test are, the costs, user-friendliness, robustness, need of additional testing equipment, and time to perform the test, for example. However, these criteria need to be adapted to the specific needs of the farm and/or situation. Further information on monitoring and testing dairy herds can be found elsewhere (Oetzel, 2004; Cook et al., 2006a, b; Flatland et al., 2013).

## 4.2 Overview of available tests

In practice, it is often an advantage to have reliable test results available as quickly as possible, which are the basis for well-informed decisions, for example, for introducing preventive measures at herd level or for the treatment of individual animals. Next to feces, urine, blood, saliva, and rumen fluid, milk is an easily available specimen that has been used for decades in individual animal diagnostics.

An example of relatively inexpensive tests that have been used for decades in veterinary practice is the 'California Mastitis Test' (CMT) for (subclinical) mastitis detection (Schalm and Noorlander, 1957), sodium nitroprusside-based powders and tablets for diagnosing ketosis (Rothera, 1908), and a variety of liquid reagents, indicator papers, and test strips used for semi-quantitative measurement of single parameters, for example, ketones (Nielen et al., 1994; Geishauser et al., 2000; Osborne et al., 2002; Carrier et al., 2004), glucose (Rollin, 2006), calcium (Matsas et al., 1999; Fiore et al., 2020), pH, or a portfolio of metabolic parameters provided on one test strip (e.g. pH, specific gravity, leucocytes, nitrite, protein, glucose, ketones, bilirubin, blood). Many of the above-mentioned tests are for semi-quantitative analyses, which are based on increasing color intensity with increasing concentration of the substrate to be detected. Besides this, other relatively inexpensive instruments enabling quantitative measurement have become available for use in veterinary practices or on farms, for example, refractometer [reviewed by George (2001)] to analyze protein and urine solute concentration and to assess colostrum quality as well as the passive transfer of immunoglobulins in calves (McGuirk and Collins, 2004; Godden, 2008; Bartens et al., 2016).

Furthermore, on-farm tests for the rapid identification of infectious agents are on the market for many years, for example, for causative organisms of calf diarrhea (Luginbühl et al., 2005; Klein et al., 2009; Lichtmannsperger et al., 2019) and mastitis (Lago and Godden, 2018; Jones et al., 2019), or became recently of interest as a quick bacteriological test for a rough estimation of involved pathogens (Madoz et al., 2017).

# 5 Electronic devices and precision livestock farming technologies

## 5.1 Electronic devices

In recent years, various devices referred to as biosensors have been introduced in veterinary diagnostics and there is a growing number of hand-held devices (also referred to as point-of-care tests, on-farm tests, cow-side-tests) fostering animal health and herd management decisions. The first system that fits the definition of a biosensor, as provided later, was the introduction of a glucose oxidase sensor by Clark and Lyons (1962), a sensor which is, with modifications, still the most widely used (Grieshaber et al., 2008).

It is expected that in the veterinary diagnostics market point of care tests will increase at a compound annual growth rate of 18%, reaching US$ 6.71 billion by 2021 (Gattani et al., 2019).

Many of the hand-held devices used in veterinary medicine belong to the class of electrochemical biosensors, a subclass of biosensors, which convert biological information into a quantifiable electronic signal. In general, a biosensor consists of (a) a bio-receptor, which specifically binds to the analyte, (b) an interface architecture in which specific biological events take place and result in a signal that is picked up by, (c) the transducer element, converting the signal into an electronic signal which is amplified by a specific detector circuit enabling further processing, for example, by (d) computer software into a meaningful physical parameter, and (e) an interface with the operator where the quantified result is presented (Grieshaber et al., 2008). Elements used for bio-recognition of analytes typically consist of enzymes, nucleic acids, receptors, antibodies, and whole cells of which enzymes are most commonly used. Methods for the detection of changes in these biologic events often rely on amperometric (generating a measurable current), potentiometric (generating a measurable potential or charge accumulation), or conductometric (measurable alteration of conductive properties of a medium) procedures (Grieshaber et al., 2008).

Suitable on-farm devices can be used even under harsh conditions on farms, provide reliable results in a very short time and are relatively cheap. They usually require only small sample volumes and are easy to operate. The test media used ideally have a long shelf time, as, for example, for many dry chemistry tests. Furthermore, the provided (numerical) measures are easy to interpret, in particular, compared with color change-based semi-quantitative tests. In recent years, more and more devices evaluated in independent scientific studies often have significantly better test characteristics than the traditionally used test systems. These advantages lead to a steadily growing interest among farmers, veterinarians, and consultants in using these technologies.

## 5.2 Precision livestock farming technologies

Since the 1980s, sensors have been developed to capture parameters from individual farm animals. For dairy cows, the development started with individual cow identification and was followed by sensors that measured the electrical conductivity of milk for mastitis monitoring and pedometers to record the animal's activity for estrus detection. The introduction of automated milking systems in western Europe was accompanied by an increased focus on sensors to assist the farmer in herd and animal health management, leading to today's precision livestock technologies (Rutten et al., 2013). The core idea of precision livestock farming (PLF) is 'the use of information and communication technologies for improved control of fine-scale animal and physical resource variability to optimize economic, social, and environmental farm performance' (Eastwood et al., 2012). The continuous and automated monitoring of livestock aims to facilitate an early warning system and to optimize management procedures, leading to improved health and welfare of (individual) animals. While general herd management procedures in the past focused on larger groups or pens of animals, modern PLF technologies have the potential to pay more attention to individual animals within the herd. The techniques allow the creation of an individualized production management plan, a unique approach in livestock farming. The potential of PLF technologies to detect diseased animals earlier compared with classical diagnostic tools is important from an animal health and welfare perspective, as well as for the success of veterinary interventions. For managers of bigger farms, PLF technologies offer the opportunity to remove subjectivity from decision-making processes and reduce the need for skilled co-workers experienced in animal management (Borchers, 2015). Additionally, PLF technologies are able to monitor the feed efficacy, reducing waste of resources and minimizing the impact on the environment. For instance, a more targeted use of feedstuff and nutrients can lower phosphorus excretions. Besides maximizing production efficiency, PLF technologies reduce airborne pollution, improving the sustainability of livestock farming, and improving the occupational health of farm workers (Banhazi et al., 2010).

PLF technologies can continuously measure key parameters and provide quantitative information on the state of an animal, welfare, health, environmental load, or productivity. For animal monitoring various technologies are available, for instance, based on image, video, and audio analysis, or using innovative sensor systems attached to the animals (Berckmans, 2008; Berckmans and Guarino, 2008). In particular, video and audio technologies can monitor the animals without disturbing their natural behavior (Müller and Schrader, 2003). Sensors used in dairy cows can be divided into attached and nonattached systems. Attached sensors can either be placed inside the cow's body (in-cow), for example, as bolus or implant, or fixed on the cow's body (on-cow), for example, with collars or wristbands. Nonattached (off-cow) sensors were developed for measurements while cows pass by, through or over the device. Furthermore, nonattached sensors can be classified as in-line and on-line sensors. While in-line sensors were designed for continuous measurement of a fluid (e.g. milk) flowing through a pipe (e.g. in the milk line), on-line sensors take a sample for analyses (Rutten et al., 2013). From a practical and scientific point of view, the opportunity to collect big data using PLF technologies has the potential to provide knowledge on biology, physiology, and animal behavior at a far greater intensity than in the past. The automated milking and herd (health) monitoring systems increasingly used on farms are sources of big data that can be used by various stakeholders (e.g. farmers, veterinarians, consultants, companies). Inter alia, methods of artificial intelligence are used here to identify specific patterns in the data, which are associated with certain animal behaviors or diseases. At this point it should be pointed out that the knowledge newly generated from the multitude of farm data should be made available to the farmers for further business development.

Various PLF systems are on the market. The sheer size of the livestock market makes it possible to produce customized technologies at low unit cost (Berckmans, 2014). Most of the systems, however, have not yet been shown to give a clear benefit under different operating conditions. Research has frequently been conducted at the laboratory level or on research and/or single farms. There has been an emphasis on developing sophisticated sensor technologies, that is, the hardware, but the setup of a resilient background for the systems is often neglected, and there is a lack of customer service and training, easy to understand manuals, a reliable data storage and backup system, and user-friendly software (Rutten et al., 2013). Due to the lack of PLF systems installed on the farm level, it is hardly possible to evaluate the value or benefit of PLF technology. To date, the development of PLF technologies has largely been driven by European companies and the PLF systems were predominately used by European farmers (Banhazi et al., 2012), but currently there is a growing interest worldwide in the use and improvement of these technologies. The desire to increase milk production per cow while decreasing production costs is one of many reasons that farmers adopt technologies (De Koning, 2010; Steeneveld et al., 2012). PLF is one of the most powerful developments in modern farming, with the potential to revolutionize the livestock farming industry and improve the economic stability of rural areas.

## 6 Case study: detecting subclinical ketosis in dairy cows

Subclinical ketosis in early lactation is a widespread metabolic disease in dairy cows, which is often underestimated by farmers and/or veterinarians. Elevated concentrations of non-esterified fatty acids (NEFAs) and ketone bodies in blood in the absence of clinical symptoms are characteristic of the disease. The concentrations of NEFA and the ketone ß-hydroxybutyrate (BHB) are directly proportional to body fat mobilization, which is needed to support milk production in early lactation. A concentration of these metabolites beyond physiological limits is associated with an increased risk for sequalae (e.g., clinical ketosis, displaced abomasum, mastitis, metritis), decreased milk yield, and impaired reproductive performance, affecting the economics of a dairy farm (Sturm et al., 2020). Because of the subclinical nature of the disease, in particular, the absence of clinical symptoms, a routine testing of dairy cows in early lactation is recommended as part of a modern herd health management procedure.

Due to the widespread nature of this often underestimated disease and its influence of the economic success of the farm, the following chapter will further discuss the disease and its on-farm diagnosis, including PLF technologies.

### *6.1 Case definition, prevalence, and impact on animal health and performance*

As presented in Figure 4, in early lactation, the high energy requirements of milk production accompanied by a delayed increase in dry matter intake often lead to a negative energy balance (NEB) in dairy cows (Ingvartsen and Andersen, 2000; Grummer et al., 2004). According to Villa-Goday et al. (1988), approximately 80% of cows go through an at least temporary phase of NEB. In their review of 49 studies on energy balance published in 20 peer-reviewed scientific journals, Grummer and Rastani (2003) reported an average duration of NEB of 45 days (ranging from 7 to 105 days, standard deviation 21 days), with 90% of animals being in a positive energy balance by 63 days post-partum.

Successful adaptation to NEB is of central importance for ensuring animal health, high levels of reproductive performance, and milk yield (Drackley,

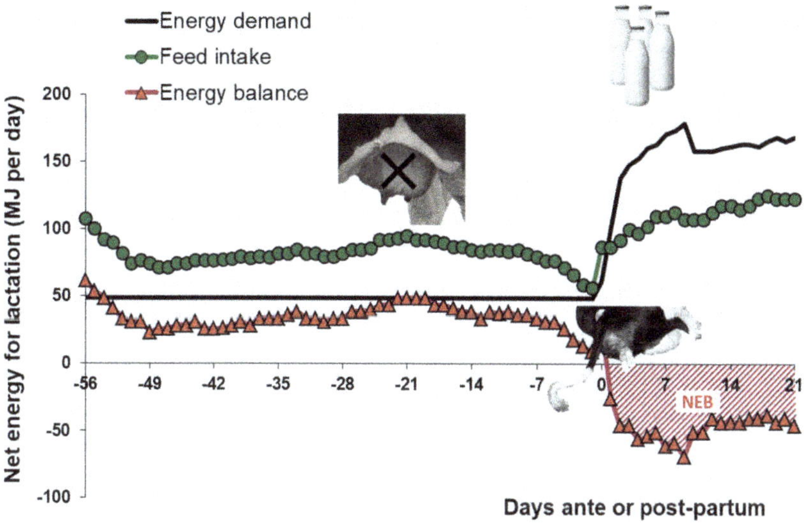


**Figure 4** Example of energy (MJ $NE_L$ per day) requirements, consumption, and the resulting energy balance during an experiment by Grummer (2008). Modified graph.

1999; Herdt, 2000; Ingvartsen, 2006). A maladaptation to this phase leads to excessive mobilization of body fat with increasing concentrations of NEFAs. If the metabolic capacity of the liver's hepatocytes is exceeded, ketogenesis occurs, leading to increased concentrations of ketones, that is, acetate (Ac), acetoacetate (AcAc), and ß-hydroxybutyrate in blood, urine, milk, and cerebrospinal fluid and other body fluids.

These increased concentrations of ketones lead to the disease being known as 'ketosis,' also referred to as acetonemia or hyperketonemia. This metabolic disease can occur not only as clinical ketosis but also as subclinical ketosis (SCK) (Duffield, 2000), that is, in the absence of any clinical symptoms, such as central nervous disorders, including salivation, dullness, and aggression (Andersson, 1988).

Acetoacetate decomposes rapidly to acetone and carbon dioxide and is therefore considered an unstable compound of low diagnostic value (Bergman, 1971; Bruss, 1989). In contrast to AcAc, BHB is a relatively stable compound, both in vivo and in vitro (Tyopponen and Kauppinen, 1980; Custer et al., 1983). Hence, the analysis of BHB in serum or plasma is recommended as the gold standard test for diagnosing ketosis not only at the animal level but also at the herd level (Duffield, 2000). Recommended thresholds to define SCK in cows are BHB concentrations in blood of 1.2 mmol/L and 1.4 mmol/L (Geishauser et al., 1998; Duffield et al., 2009), which have been used as a reference in recent research. Suthar et al. (2013) suggest the use of a uniform threshold to define SCK of 1.2 mmol/L BHB in blood. From the herd level perspective, an alarm level (i.e. the proportion of sampled animals above a certain threshold) for animals suffering from SCK of 15% in multiparous cows and 20% in heifers has been recommended, using a cut-off of the BHB concentration in serum of 1.4 mmol/L (Cook et al., 2006a; Ospina et al., 2010).

The occurrence of SCK in dairy cows in the periparturient period represents an important challenge for farmers. Based on BHB concentrations in blood of ≥1.2 mmol/L, Suthar et al. (2013) reported an overall prevalence of SCK for ten European countries of 21.8%, ranging from 11.2% to 36.6% within 2 weeks after calving. The reported prevalence for North American dairy herds ranged from 8.9% to 43.2% within the first two months of lactation (Dohoo and Martin, 1984; Geishauser et al., 1998; McArt et al., 2012).

Several studies have shown that SCK is associated with an increased risk of the occurrence of secondary diseases (LeBlanc et al., 2005; Duffield et al., 2009; Suthar et al., 2013). Additionally, a decrease in milk yield (Dohoo and Martin, 1984; Duffield, 1997) and impaired reproductive performance (Walsh et al., 2007; Chapinal et al., 2012) are associated with the occurrence of SCK. Raboisson et al. (2014) conducted a meta-analysis on 23 publications on the occurrence of sequalae related to SCK. In the presence of SCK (BHB >1.4 mmol/L in the blood), the authors determined relative risks for displaced abomasum, lameness, metritis, clinical mastitis, and retained fetal membranes of 3.33, 2.01, 1.75, 1.61, and 1.52, respectively.

Furthermore, an increased BHB concentration in pre-calving cows has been associated with a detrimental impact on milk yield in the next lactation and on animal health (Chapinal et al., 2011, 2012). Animals showing a BHB concentration in the serum of ≥0.7 mmol/L within the last week of gestation were at a greater risk of early culling (Roberts et al., 2012).

## *6.2 Advances in ketosis monitoring*

Considering the impact of hyperketonemia on animal health, welfare, and performance, monitoring of dairy herds for SCK is an appropriate measure for disease prevention and improvement of herd management efficiency in dairy farming (Cook et al., 2006b). Helpful guidelines, defining, for example, thresholds, alarm levels, and the animals of interest for metabolic testing, as presented in Table 1, have been published (Stokol and Nydam, 2005; Cook et al., 2006a,b).

However, the quantitative determination of BHB as the gold standard depends on special laboratory analyzers and requires blood sampling, centrifugation, freezing of plasma or serum samples, and shipping of frozen specimens to the laboratory. In order to minimize these inconveniences, in particular, to provide results immediately after sampling on farm and to reduce

**Table 1** Diagnostic tests, cut-off points, and alarm levels used in animals at risk to identify subclinical diseases

| Test | Cut-off point | Alarm level proportion | At-risk group | Associated disease |
|---|---|---|---|---|
| ß-Hydroxybutyrate (BHB) | ≥1.4 mmol/L | >10% | Lactating cows 50 to 50 DIM | Ketosis, displaced abomasum |
| Non-esterified fatty acid (NEFA) | ≥0.4 mmol/L | >10% | Dry cows 2 to 14 days before expected calving | Ketosis, displaced abomasum, fatty liver disease |
| Ruminal pH | ≤5.5 | >25% | Lactating cows 50 to 50 DIM in herds fed concentrate separately, 50 to 150 DIM in herds fed TMR | Subacute ruminal acidosis |
| Blood calcium | ≤2.0 mmol/L | >30% | Lactating multiparous cows 12 to 24 hours after calving | Clinical milk fever |

DIM: Days in milk, TMR: total mixed ration. Source: Cook et al. (2006a).

laboratory costs, various cow-side diagnostic tests, based on the reaction of AcAc and Ac with sodium nitroprusside leading to a ketone-dependent color change (Rothera, 1908; Adler et al., 1957), have been developed. Higher levels of ketones lead to a more intensive coloration; hence, these tests can be regarded as semi-quantitative.

### 6.2.1 Traditional semi-quantitative tests

To this day, these semi-quantitative tests are still used in the daily practice of farmers and veterinarians as tablets or dipsticks to detect AcAc, and to a lesser degree Ac in urine (e.g. Ketostix®, Bayer Corp.; AimTab™ Ketone Tablets, Germaine) or BHB in milk (PortaBHB®, PortaCheck Inc.; Keto-Test™, Elanco). An overview of scientifically evaluated semi-quantitative tests for the determination of ketones in milk and urine, respectively, are presented in Tables 2 and 3. An example of their application is shown in Figure 5.

Differences in SE, SP, and predictive values may be caused by variation in sampling or in the error-prone subjective interpretation of the color changes of the test strips (Carrier et al., 2004). Moreover, it has been demonstrated that high levels of somatic cell count (SCC) in milk (Jeppesen et al., 2006) and feeding of mal-fermented silages (Cook et al., 2006b) can distort the results of

**Table 2** Performance of different commercially available cow-side tests for use in milk for detecting subclinical ketosis at different cut-off levels

| Reference | Test and cut-offs (mmol/L) | Serum (mmol/L) | n[1] | SE[2] (%) | SP[3] (%) | $PV_{15}$[4] (%) pos | $PV_{15}$[4] (%) neg | $PV_{30}$[5] (%) pos | $PV_{30}$[5] (%) neg |
|---|---|---|---|---|---|---|---|---|---|
| Geishauser (1998) | Ketolac[6] | ≥1.2 | 529 | | | | | | |
| | ≥0.05 | | | 92 | 55 | 27 | 97 | 47 | 94 |
| | ≥0.1 | | | 72 | 89 | 54 | 95 | 74 | 88 |
| | ≥0.2 | | | 45 | 97 | 73 | 91 | 87 | 80 |
| | ≥0.5 | | | 17 | 100 | 100 | 87 | 100 | 74 |
| | ≥1.0 | | | 3 | 100 | 100 | 85 | 100 | 71 |
| Geishauser et al. (2000) | Ketolac | ≥1.4 | 469 | | | | | | |
| | ≥0.05 | | | 91 | 56 | 27 | 97 | 47 | 94 |
| | ≥0.1 | | | 80 | 76 | 37 | 96 | 59 | 90 |
| | ≥0.2 | | | 59 | 90 | 51 | 93 | 72 | 84 |
| | Pink[7] | | | | | | | | |
| | ≥0.1 | | | 76 | 93 | 66 | 96 | 82 | 90 |
| | ≥0.2 | | | 38 | 98 | 77 | 90 | 89 | 79 |
| | Uriscan[8] | | | | | | | | |
| | ≥500 | | | 13 | 100 | 100 | 87 | 100 | 73 |
| | Rapignost[9] | | | | | | | | |
| | ≥500 | | | 3 | 100 | 100 | 85 | 100 | 71 |
| Osborne et al. (2002) | KetoTest[10] | ≥1.4 | | | | | | | |
| | ≥0.1, 1 wk pp | | 121 | 95 | 71 | 37 | 99 | 58 | 97 |

| | | | | | | | | | |
|---|---|---|---|---|---|---|---|---|---|
| | ≥0.1, 2 wk pp | | 127 | 95 | 67 | 34 | 99 | 55 | 97 |
| Carrier et al. (2004) | KetoTest | ≥1.4 | 850 | | | | | | |
| | ≥0.05 | | | 88 | 90 | 61 | 98 | 79 | 95 |
| | ≥0.1 | | | 73 | 96 | 76 | 95 | 89 | 89 |
| | ≥0.2 | | | 27 | 99 | 83 | 88 | 92 | 76 |
| | ≥0.5 | | | 3 | 100 | 100 | 85 | 100 | 71 |
| | ≥1.0 | | | 2 | 100 | 100 | 85 | 100 | 70 |
| | KetoCheck[11] | ≥1.4 | 845 | | | | | | |
| | ≥Trace | | | 41 | 99 | 88 | 90 | 95 | 80 |
| | ≥Moderate | | | 10 | 100 | 100 | 86 | 100 | 72 |
| | ≥Large | | | 2 | 100 | 100 | 85 | 100 | 70 |
| Iwersen et al. (2009) | Ketolac[12] | ≥1.4 | 194 | | | | | | |
| | ≥0.1 | | | 90 | 94 | 73 | 98 | 87 | 96 |
| | ≥0.2 | | | 30 | 98 | 73 | 89 | 87 | 77 |

[1] Number of observations paired with a serum BHB measurement for each cow-side test.
[2] SE = Sensitivity: proportion of diseased cows that test positive.
[3] SP = Specificity: proportion of non-diseased cows that test negative.
[4] Positive and negative predicted values based on a theoretical prevalence of subclinical ketosis of 15%.
[5] Positive and negative predicted values based on a theoretical prevalence of subclinical ketosis of 30%.
[6] Hoechst, Unterschleißheim, Germany.
[7] Profs-Products, Wittibreut, Germany.
[8] Heiland, Hamburg, Germany.
[9] Behring, Marburg, Germany.
[10] Sanwa Kagaku, Nagoya, Japan.
[11] Great States Animal Health, St. Joseph, United States of America.
[12] Biolab, München, Germany.

**Table 3** Performance of different commercially available cow-side tests for use in urine for detecting subclinical ketosis at different cut-off levels

| Reference | Test and cut-offs (mmol/L) | Serum (mmol/L) | n[1] | SE[2] (%) | SP[3] (%) | $PV_{15}$[4] (%) | | $PV_{30}$[5] (%) | |
|---|---|---|---|---|---|---|---|---|---|
| | | | | | | pos | neg | pos | neg |
| Nielen et al. (1994) | Acetest[6] | ≥1.2 | 124 | 91 | 61 | 29 | 97 | 50 | 94 |
| | | ≥1.4 | 124 | 100 | 59 | 30 | 100 | 51 | 100 |
| Osborne et al. (2002) | KetoTest[7] | ≥1.4 | | | | | | | |
| | ≥0.1, 1 wk pp | | 71 | 93 | 54 | 26 | 98 | 46 | 95 |
| | ≥0.1, 2 wk pp | | 88 | 100 | 65 | 34 | 100 | 55 | 100 |
| Carrier et al. (2004) | Ketostix[8] | ≥1.4 | 710 | | | | | | |
| | ≥0.5 | | | 90 | 86 | 53 | 98 | 73 | 95 |
| | ≥1.5 | | | 78 | 96 | 77 | 96 | 89 | 91 |
| | ≥4.0 | | | 49 | 99 | 90 | 92 | 95 | 87 |
| | ≥8.0 | | | 12 | 100 | 100 | 87 | 100 | 73 |
| | ≥16.0 | | | 4 | 100 | 100 | 86 | 100 | 71 |
| Iwersen et al. (2009) | Ketostix[9] | ≥1.4 | 186 | | | | | | |
| | ≥0.5 | | | 78 | 92 | 63 | 96 | 81 | 91 |
| | ≥1.5 | | | 67 | 97 | 80 | 94 | 91 | 87 |
| | ≥4.0 | | | 67 | 100 | 100 | 94 | 100 | 88 |

| | | | | | | | |
|---|---|---|---|---|---|---|---|
| | ≥8.0 | 44 | 100 | 100 | 91 | 100 | 87 |
| | ≥16.0 | 22 | 100 | 100 | 88 | 100 | 75 |

[1] Number of observations paired with a serum BHB measurement for each cow-side test.
[2] SE = Sensitivity: proportion of diseased cows that test positive.
[3] SP = Specificity: proportion of non-diseased cows that test negative.
[4] Positive and negative predicted values based on a theoretical prevalence of subclinical ketosis of 15%.
[5] Positive and negative predicted values based on a theoretical prevalence of subclinical ketosis of 30%.
[6] Ames Devision, Glamorgan, United Kingdom.
[7] Sanwa Kagaku, Nagoya, Japan.
[8] Bayer Corporation, Elkhart, United States.
[9] Bayer, Leverkusen, Germany.

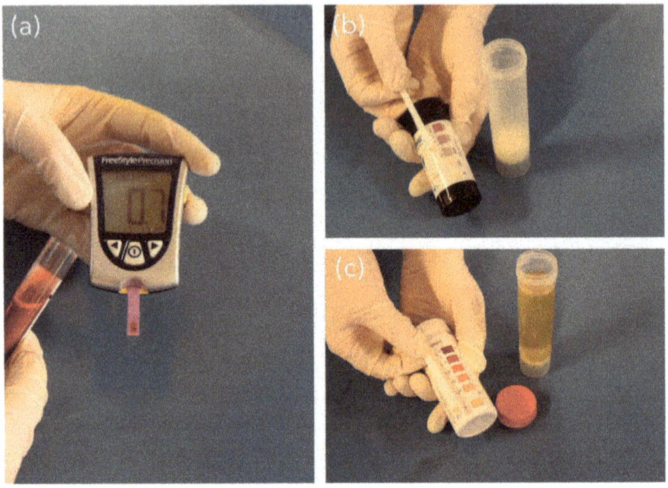


**Figure 5** Example for ketosis diagnostics by the use of (a) an electronic hand-held device in blood and semi-quantitative dipsticks for use in (b) milk and (c) urine.

milk ketone tests. Prolonged reaction times of the reagent with urine might also lead to FP results (Oetzel, 2004).

Even if the test characteristics vary between tests and studies, there is a general agreement that semi-quantitative tests are useful for ketosis monitoring, in particular, on herd level. Urine sampling, however, is often challenging, and the suboptimal specificities for urine testing and the inadequate sensitivities of milk testing were reported as drawbacks of the test systems (Oetzel, 2004).

### 6.2.2 Electronic hand-held devices and factors influencing the results

Electronic hand-held devices for measuring glucose (glucometer) and ketones (ketometer) are widely used in human medicine for diabetes monitoring (Guerci et al., 2005; Grieshaber et al., 2008). Ketometer systems consist of a hand-held meter and electrochemical test strips for BHB measurement (Figure 5). After inserting the test strip into the strip port of the device, the front edge of the application zone is brought into contact with, for example, a blood drop or serum sample. The specimen is drawn into the test strip, and for many devices, the chemical reaction starts as follows: the BHB in the specimen is oxidized to AcAc in the presence of the enzyme ß-hydroxybutyrate dehydrogenase, with the concomitant reduction of $NAD^+$ to NADH. The NADH is reoxidized to $NAD^+$ by a redox mediator. Electrons are released by this reaction leading to a small current, which is directly proportional to the BHB concentration in the specimen (Iwersen et al., 2009). For most devices, the BHB concentration is displayed as mmol/L within 10 seconds after specimen application on the test strip. For many devices, the minimum operating temperature is +4°C. This can lead to challenges when using the device during winter months. In this case, the device should be stored in a warm place (e.g. in a pocket near the body) until use (Iwersen et al., 2013).

Introducing ketometer for BHB measurements on farms has led to significant improvement in ketosis monitoring in dairy cows (Overton et al., 2017). In particular, presenting the results as a numerical value on a display and exact adherence to reaction times have improved the test interpretation and, hence, reduced the risk of misinterpretation, as reported for color-based semi-quantitative tests. Furthermore, only small amounts (0.8–1.5 μL) of whole blood, serum, or plasma are needed for measurements, and, in the case of using whole blood, further processing of samples (e.g. centrifugation, pipetting, storage) is not needed. The individually foil-packaged test strips meet the often unfavorable hygienic conditions on farms and reduce the potential negative impact on the quality of the test strips. If newer devices are used, calibration procedures prior to use are no longer necessary (Voyvoda and Erdogan, 2010; Iwersen et al., 2013).

Jeppesen et al. (2006) were among the first to study the use of an electronic hand-held device for BHB measurement (MediSense Precision, Abbott, Abingdon, UK) in dairy cows. The authors reported a high correlation of $r^2 = 0.99$, with spectrophotometrically determined BHB concentrations, which served as the gold standard in their study, and considered the test eligible for the monitoring of SCK in dairy cows. In recent decades, several hand-held devices have been tested under different farm conditions. For instance, in a survey of an electronic hand-held device tested by 35 veterinary practitioners in 77 farms, an SE of 96% and SP of 97% (based on a cut-off of 1.4 mmol/L BHB in serum) were reported (Iwersen et al., 2009). In the final evaluation by veterinarians (n = 30), 93% rated the diagnostic value of the device as 'good' or 'very good.' Of the respondents, 63% reported that they 'definitely' wanted to continue using the device and 23% would 'probably' continue using it. Today, for routine ketosis monitoring at animal and herd level, the use of hand-held electronic devices has become standard (Overton et al., 2017). An overview of evaluated hand-held devices for determining BHB in whole blood is presented in Tables 4 and 5. Reports on using other specimens for BHB testing using hand-held electronic devices are available elsewhere (Iwersen et al., 2009, 2013; Pineda and Cardoso, 2015).

In contrast to almost constant laboratory conditions, on-farm tests are used under varying environments, such as different temperatures in summer and winter. Therefore, various studies were carried out on the practical applications of hand-held electronic devices, for example, on the type of specimen, timing, and sampling procedures. Iwersen et al. (2013) determined

**Table 4** Performance of different commercially available hand-held devices for use with whole blood for detecting subclinical ketosis at a serum or plasma ß-hydroxybutyrate level of ≥1.2 mmol/L

| Reference | Device | Device cut-off[1] (mmol/L) | n | SE[2] (%) | SP[3] (%) |
|---|---|---|---|---|---|
| Iwersen et al. (2009b) | Precision Xtra[4] | 1.2 | 926 | 88 | 96 |
| Voyvoda and Erdogan (2010) | Optimum Xceed[4] | 1.2 | 78 | 85 | 94 |
| Panousis et al. (2011) | Precision Xceed[4] | 1.2 | 163 | 91 | 96 |
| Iwersen et al. (2013) | FreeStyle Precision[4] | 1.2 | 425 | 98 | 90 |
| | GlucoMen LX Plus[5] | 1.1 | 415 | 80 | 87 |
| Mahrt et al. (2014a) | Precision Xtra | 1.2 | 92 | 90 | 88 |
| Mahrt et al. (2014b) | NovaVet[6] | 1.2 | 155 | 97 | 82 |
| Kanz et al. (2015) | FreeStyle Precision | 1.2 | 240 | 100 | 93 |
| | GlucoMen LX Plus | 1.0 | 240 | 94 | 85 |
| | NovaVet | 1.0 | 240 | 100 | 83 |
| Bach et al. (2016) | Precision Xtra | 1.2 | 89 | 100 | 74 |
| | TaiDoc[7] | 1.2 | 89 | 100 | 74 |
| | NovaMax[6] | 1.2 | 89 | 75 | 100 |
| | NovaVet | 1.2 | 89 | 95 | 92 |
| Süss et al. (2016) | FreeStyle Precision Neo[4] | 1.1 | 240 | 100 | 95 |
| Macmillan et al. (2017) | FreeStyle Precision Neo | 1.2 | 441 | 98 | 95 |
| Zakian et al. (2017) | Precision Xtra[5] | 1.2 | 181 | 94 | 94 |
| Leal Yepes et al. (2018) | Precision Xtra | 1.2 | 100 | 100 | 87 |
| | TaiDoc | 1.2 | 100 | 100 | 76 |
| Khol et al. (2019) | WellionVet BELUA[8] | 1.2 | 243 | 89 | 99 |

[1] (Optimized) device threshold used for diagnosing subclinical ketosis.
[2] SE = Sensitivity: proportion of diseased cows that test positive.
[3] SP = Specificity: proportion of non-diseased cows that test negative.
[4] Abbott Diabetes Care Ltd., Witney, United Kingdom.
[5] A. Menarini, Vienna, Austria.
[6] Nova Biomedical, Waltham, United States.
[7] Pharmadoc, Lüdersdorf, Germany.
[8] MED TRUST Handels GmbH, Marz, Austria.

**Table 5** Performance of different commercially available hand-held devices for use with whole blood for detecting subclinical ketosis at a serum or plasma ß-hydroxybutyrate level of ≥1.4 mmol/L

| Reference | Device | Device cut-off[1] (mmol/L) | n | SE[2] (%) | SP[3] (%) |
|---|---|---|---|---|---|
| Iwersen et al. (2009) | Precision Xtra[4] (study 1) | 1.4 | 196 | 100 | 100 |
| | Precision Xtra (study 2) | 1.4 | 926 | 96 | 97 |
| Voyvoda and Erdogan (2010) | Optimum Xceed[4] | 1.4 | 78 | 90 | 98 |
| Panousis et al. (2011) | Precision Xceed[4] | 1.4 | 163 | 100 | 100 |
| Iwersen et al. (2013) | FreeStyle Precision[4] | 1.4 | 425 | 100 | 97 |
| | GlucoMen LX Plus[5] | 1.3 | 415 | 86 | 96 |
| Mahrt et al. (2014a) | Precision Xtra | 1.4 | 92 | 89 | 90 |
| Mahrt et al. (2014b) | NovaVet[6] | 1.3 | 155 | 96 | 85 |
| | NovaVet | 1.4 | 155 | 91 | 89 |
| Kanz et al. (2015) | FreeStyle Precision | 1.4 | 240 | 100 | 96 |
| | GlucoMen LX Plus | 1.3 | 240 | 85 | 97 |
| | NovaVet | 1.3 | 240 | 90 | 98 |
| Zakian et al. (2017) | Precision Xtra | 1.4 | 181 | 93 | 95 |

[1] (Optimized) device threshold used for diagnosing subclinical ketosis.
[2] SE = Sensitivity: proportion of diseased cows that test positive.
[3] SP = Specificity: proportion of non-diseased cows that test negative.
[4] Abbott Diabetes Care Ltd., Witney, United Kingdom.
[5] A. Menarini, Vienna, Austria.
[6] Nova Biomedical, Waltham, United States

similar BHB concentrations when either whole blood or EDTA-added whole blood was tested. Furthermore, similar results have also been reported in BHB measurements using serum and plasma. However, because BHB concentrations in serum and plasma are lower compared with whole blood, specimen and device-specific thresholds should be used for ketosis monitoring (Iwersen et al., 2013; Pineda and Cardoso, 2015). The influence of different temperatures during device and test strip storage or sample measurement, respectively, in a temperature range between +5°C and +32°C have been investigated (Iwersen et al., 2013). The storage conditions of the devices and test strips did not affect the test results, but analyzing a blood sample at different temperatures caused significant differences. Hence, to obtain reliable results, the sample temperature should be close to body temperature (i.e. immediately after sampling) while testing; otherwise, the temperature of the sample has to be considered. The effect of the sampling time and sampling location (tail vessel, jugular vein, and mammary vein) on BHB concentrations was tested by Mahrt et al. (2014). The authors reported that the sampling time in continuously fed dairy cows did not affect BHB concentrations. Mean BHB concentrations in the mammary vein were 0.3 mmol/L and 0.4 mmol/L lower compared with concentrations in the jugular vein and tail vessel, respectively. Hence, the authors concluded that blood samples could be taken at any time of the day in continuously fed cows, preferably from the tail vessels or jugular vein; the mammary vein should not be used. Other studies investigated whether capillary blood, obtained minimally invasively by the use of a lancet on the skin of the exterior vulva (Kanz et al., 2015) or the edge of an ear (Süss et al., 2016), was suitable for ketosis testing using hand-held electronic devices. The authors concluded that capillary blood is eligible for classifying cows suffering from SCK, but thresholds should be adjusted. Applying these minimally invasive procedures, farmers have the possibility of testing their cows using electronic devices in countries where national legislation prohibits conventional blood sampling by laypersons.

### 6.2.3 Milk composition

Both milk fat and protein composition are significantly influenced by hyperketonemia (Miettinen and Setala, 1993; Duffield et al., 1997), and, thus, the fat-to-protein ratio (F/P), as well as the protein-to-fat ratio, has been used as indicators for SCK (Grieve et al., 1986; Duffield et al., 1997; Heuer et al., 1999; Jenkins et al., 2015). Monthly reports from dairy herd improvement associations (DHIA) provide the farmer with individual animal information on milk yield, components (e.g. fat, protein, lactose, urea), and SCC, which provide a rough overview of the metabolic status at the herd level (Nelson and Redlus, 1989; Hamann and Krömker, 1997; Cook et al., 2006a). Because of inadequate SE and SP, fat and protein percentages alone or a combination thereof and their ratios were considered as not being useful in detecting SCK in individual animals (Duffield et al., 1997; Jenkins et al., 2015). Furthermore, because of the long sampling intervals of approximately 4 weeks and the limited factors analyzed in composite milk samples, DHIA records are considered as ineffective as a basis for modern health management (Hamann and Krömker, 1997).

Several validations of in-line methods for analyses of ketones in routine milk analyses have been performed (Nielsen et al., 2005). One of the modern methods within the quantitative analysis is Fourier-transform infrared (FTIR) spectrometry, which is currently used as a standard method for multi-component milk testing (Hansen, 1999; Pralle and White, 2020). De Roos et al. (2007) reported a correlation of approx. 80% between FTIR predicted Ac and BHB concentrations compared with their respective chemical tests. The authors considered the prediction of Ac and BHB by FTIR as 'valuable for screening cows for ketosis' with a reported SE for BHB of 69% and Ac of 70%, and SP of 95% for both. Similar results, presented in Table 6, were reported by Van Knegsel et al. (2010), Van der Drift et al. (2012), and Denis-Robichaud et al. (2014).

In a recent study, King et al. (2019) tested the accuracy of in-line F/P to detect SCK in the first 3 weeks of lactation on 484 cows kept in nine herds with automated milking systems. Blood samples (n = 1427) were taken at weekly intervals, and BHB concentrations of ≥1.2 mmol/L and ≥1.4 mmol/L, respectively, were used to define SCK. Using various F/P ratios to detect SCK, the authors reported poor SE (58% to 92%) and SP (65% to 69%), with high rates of false positives and negatives ranging from 31% to 39%. Similar proportions of FP- and FN-classified animals are reported in the studies presented in Table 6. Because of low PPV, Van Knegsel et al. (2010) had 'concerns about the practical applicability of FTIR predictions of acetone, BHB, and fat-to-protein ratio in milk to detect hyperketonemic cows.' Other authors considered the method an 'accurate diagnostic tool' (Denis-Robichaud et al., 2014) or as suitable for the pre-selection of animals to be subjected to further testing for SCK (Hansen, 1999; Van der Drift et al., 2012; Petersson et al., 2017).

In-line systems for use on farms, for example, in automated milking systems or parlors, are available for supporting herd health management. Two systems that have been commercially available for some time, the AfiLab (Afimilk, Kibbutz Afikim, Israel) and Herd Navigator (DeLaval, Tumba, Sweden), are presented here as examples.

The AfiLab is an optical in-line milk analyzer, which measures milk fat, protein, lactose, blood admixture in milk in real time and the coagulation potential in every milking of an individual cow for detecting, for example, ketosis, mastitis, and acidosis. Analyses are based on multivariate analysis of near-infrared spectra during milking, using algorithms to predict the total amount of milk components. In this system, an F/P ratio of 1.4 is used for pre-selecting animals that need further testing for ketosis, for example, by measuring blood or urine (Afikim, 2019).

The Herd Navigator automatically takes representative milk samples from specific cows during milking intended to be used, for example, for reproduction, ketosis, and mastitis management. Prior to milking, an algorithm selects individual cows and specific parameters to be tested. The fully automated analytic device, which is installed in a separate place, relies on dry stick analyses, using colorimetric reactions for the determination of lactate dehydrogenase, urea, and BHB and an immuno-assay for measuring progesterone (Mazeris, 2010). For both systems, analytical outputs can be presented as tables and graphs in herd management software. Thresholds, for example, used for generating alarms, can be adjusted by the user.

Although in-line systems have been used on farms for some years now, independent publications demonstrating the benefit of their use in animal

**Table 6** Performance of Fourier-transform infrared (FTIR) spectroscopy for predicting hyperketonemia from testing milk samples at different cut-off levels

| | | | | | Test characteristics (%) | | | |
|---|---|---|---|---|---|---|---|---|
| Reference | Parameter[1] | Threshold[2] | n[3] | Prev[4] (%) | SE[5] | SP[6] | PPV[7] | NPV[8] |
| De Roos et al.[9] (2007) | BHB (μmol/L) | ≥100 | 1080 | 17.2 | 69 | 95 | 75 | 93 |
| | Ac (μmol/L) | ≥150 | | | 70 | 95 | 73 | 94 |
| Van Knegsel et al.[10] (2010) | BHB (μmol/L) | ≥23 | 69 | 7.1 | 80 | 71 | 18 | 98 |
| | Ac (μmol/L) | ≥70 | | | 80 | 70 | 17 | 98 |
| | F/P | ≥1.5 | | | 66 | 71 | 15 | 96 |
| Van der Drift et al.[10] (2012) | BHB (μmol/L) | ≥76.3 | 1678 | 11.2 | 83 | 76 | 30 | 97 |
| | Ac (μmol/L) | ≥131.5 | | | 71 | 89 | 45 | 96 |
| Denis-Robichaud et al.[11] (2014) | BHB (μmol/L) | ≥200 | 163 | 21.0 | 84 | 96 | 84 | 96 |
| | Ac (μmol/L) | ≥80 | | | 87 | 95 | 83 | 96 |
| | F/P | ≥1.3 | | | 69 | 66 | 33 | 90 |

[1] BHB = ß-hydroybutyrate, Ac = Acetone, F/P = milk fat-to-protein ratio.
[2] (Optimized) device threshold used for diagnosing subclinical ketosis.
[3] Number of samples.
[4] Prev = Prevalence.
[5] SE = Sensitivity: proportion of diseased cows that test positive.
[6] SP = Specificity: proportion of non-diseased cows that test negative.
[7] PPV = Positive predictive value based on reported study prevalence.
[8] NPV = Negative predictive value based on reported study prevalence.
[9] Definition of subclinical ketosis based on BHB and Ac in milk.
[10] Definition of subclinical ketosis: BHB ≥1.2 mmol/L in blood serum or plasma.
[11] Definition of subclinical ketosis: BHB ≥1.4 mmol/L in blood serum or plasma.

health management as well as on economics are scarce, and future research is needed.

### 6.2.4 Sensor-derived animal behavior and data integration

Combining in-line data on milk yield and its components with sensor-derived animal behaviors might enhance ketosis monitoring by identifying cows at risk of SCK that could benefit from a cow-side test (Petersson et al., 2017).

As described in Section 5.1.1, PLF technologies are used inter alia to measure animal behaviors. As hyperketonemia has an impact on neurophysiological processes, it can be assumed that it also affects movement and behavior patterns of animals as well as the time spent in certain areas of the barn, that is, their time budget.

Recently, several sensor systems have been validated to accurately characterize distinct animal behaviors, for example, rumination (Schirmann et al., 2009; Burfeind et al., 2011; Borchers et al., 2016; Reiter et al., 2018), standing and lying times (Borchers et al., 2016; Van Erp-Van der Kooj et al., 2016), and animal position (Wolfger et al., 2016). Furthermore, changes in distinct behaviors were reported to be associated with animal disease (Titler et al., 2013; Itle et al., 2015; Liboreiro et al., 2015; Stangaferro et al., 2016a,b,c).

For cows suffering from ketosis, Edwards and Tozer (2004) reported an increase in walking activity and decrease in milk yield, which were already detectable in 8 and 6 days, respectively, before clinical diagnosis by use of sensor technology (Afikim system, Kibbutz Afikim, Israel). Furthermore, the time spent feeding, dry matter intake, and social behavior were associated with the occurrence of SCK. For every 10-min decrease in average daily time spent at the feeder and for every 1 kg decrease in dry matter intake during the week before calving, the risk of SCK increased by 1.9 and 2.2 times, respectively (Goldhawk et al., 2009). The previously mentioned parameters are predisposed to capture by the use of sensor technology and, hence, can contribute to an improvement of automated monitoring of ketosis. Stangaferro et al. (2016a,b,c) published a series of papers evaluating the performance of an automated health monitoring system (HR Tags, SCR Dairy, Israel) to identify cows suffering from metabolic and digestive disorders, mastitis, and metritis. In the final dataset, 1080 cows were included, of those 52% (n = 629) had at least one health disorder. Based on rumination time and physical activity, a daily 'health index score' (HIS) between 0 and 100 was calculated for individual cows by an algorithm. Defining a positive HIS outcome as a HIS of <86 for at least 1 day in the period of interest from 5 days before to 2 days after clinical diagnosis, an SE of the system for detecting ketosis (n = 54, i.e. an incidence of 5%) of 91% was reported. In the case of ketosis, the limit of HIS <86 was reached 1.6 days (mean) earlier before clinical diagnosis by farm personnel. For cows with ketosis, rumination and activity patterns were comparable to cows in the non-diseased group for most days in the study period and both parameters did not show a dramatic decline before clinical diagnosis. This study result once again demonstrates the potential of continuous sensor-based health monitoring, in which an animal acts as its own control. For the latter, the current rumination and activity are compared with past individual animal data and is not only compared with herd members.

In a recent approach, Sturm et al. (2020) developed a flexible classification algorithm for predicting ketosis based on various input variables, consisting of time series and features from different sources. In this study, individual cow rumination data, activity measures, standing/lying behavior, and the animal's position within the barn were captured by an accelerometer system (SMARTBOW, Smartbow GmbH, Austria) from 1 week before up to 1 week after parturition. Furthermore, barn climate (temperature and humidity), visually determined body condition score, back fat thickness measured by ultrasound, and NEFAs were included in the algorithm. Animals showing a BHB concentration in coccygeal blood of >1.2 mmol/L were defined as suffering from SCK. Depending on the model, SE varied between 63% and 67%, and SP between 59% and 74%. For all models, negative predictive values were high (>91%) but should be interpreted with caution as it was highly affected by an unbalanced data structure in this study. Although the quality of the algorithm is not yet sufficient for use in commercial farms for predicting ketosis at the animal level, it is a promising approach in terms of data integration. The flexible and easily adoptable algorithm will be further developed to include additional (heterogeneous) data from other sources. In this context, merging data from different sources is the next step toward a more 'holistic' analysis of data, which aims to provide new insights into the development and prevention of diseases.

## 7 Conclusion and future trends in research

The performance achieved in today's livestock farming is the result of decades of progress, in the fields of, for example, genetics, feeding, and animal husbandry, as well as in general farm and herd health management. In addition to profound technical knowledge, farmers increasingly need to possess managerial qualities, which are necessary to ensure the long-term economic success of their farms. Nowadays, for managing the complex farms, methods of (large-scale) production of other industrial sectors, such as SOPs and risk analyses, are used, and advice is sought from external consultants. In this context, veterinarians are the most important consultants for farmers in terms of securing and improving animal health and for ensuring high levels of animal welfare (Pothmann et al., 2014). Veterinary diagnoses contribute to the pool of farm data that can be used for, for example, herd (health) management. In the best case scenario, most reliable data, which can be collected in real time, are the basis for evidence-based decisions made by farmers and veterinarians.

A cornerstone of modern herd management is the routine examination of animals for the presence of disease. The findings of epidemiological research, that already subclinical forms of the disease (i.e. without externally detectable symptoms) have a detrimental impact on animal health, welfare, and performance, has led to an increased interest in the monitoring of subclinical disease. Hence, a close monitoring of dairy cows in the transition period, for example, for detecting subclinical hypocalcemia, ketosis, mastitis, and endometritis, has become a standard procedure as part of fresh cow monitoring routines on many dairy farms.

For on-farm health monitoring, a variety of independently evaluated tests and devices are available. The use of hand-held devices for metabolic monitoring of dairy cows has attracted increasing interest from farmers and veterinarians in the last decade, and the use of these devices for ketosis monitoring has already become the standard (Overton et al., 2017). Many of these hand-held devices can be used even under harsh conditions on farms. They are easy to operate and provide reliable results in a very short time. Furthermore, displayed numerical measures are easy to interpret, in particular, compared with color change-based semi-quantitative tests. The availability of reliable test results during animal examination allows prompt treatment of diseased animals, which can contribute to improved animal welfare while securing the animals' productivity.

The use of sensor technologies in agriculture is considered as the next milestone in the development of the industry sector and is sometimes also referred to as the next industrial revolution (Zambon et al., 2019). In livestock, farmers increasingly use sensor-based PLF technologies for the monitoring of animal health, welfare, and performance. Due to the ongoing further development and cost reduction of the sensors, it can be assumed that their use in dairy farming will also continue to increase. Recently, several features of commercially available sensor systems have been validated to accurately classify distinct animal behaviors, for example, rumination, standing and lying times, as well as animal activities. Because specific behavioral patterns are associated with the disease, changes in behaviors detected by sensors are used to identify animals at risk. When PLF technologies are used as the sole test to detect specific diseases, the test performance is currently not sufficient for practical use. Nevertheless, sensor technologies can contribute to improved animal health by identifying animals at risk (e.g. subclinically diseased), which afterward can be examined with more specific tests. Without sensor technologies, these animals might be overlooked by farm personnel.

When introducing new technologies in practice, the question arises as to the economic benefit of this investment. Some models for the economic evaluation of different PLF technologies have already been developed (Steeneveld et al., 2015; Dolecheck et al., 2016). The economic value depends on numerous factors, which vary from farm to farm and situation to situation, for instance, on milk price and labor costs. Furthermore, the evaluation also depends on the number of features, which are offered for one system. With an increasing number of frequently used features, the economic benefit of a system will increase. One advantage of algorithm-based systems (e.g. CowManager Sensor; SMARTBOW®; SenseHub™ Dairy) is that the introduction of new features is usually not associated with a replacement of the entire sensor system, but, for example, 'only' with adoptions in the software. The optimization of algorithms during operation can also lead to an increase in economic efficiency. Due to these possibilities for expansion and improvement, a statement on the cost-benefit ratio is always only a snapshot. Cost-benefit calculations for the use of PLF technologies are currently under review. Furthermore, intensive research activities are focusing on the identification of additional biomarkers, for example, those used for the reliable assessment of individual animal's stress or welfare. Once these have been identified, it will then be necessary to develop reliable on-farm tests or to identify feature variables, which can be measured by sensor technologies. Linking the phenotype of an animal to measurements that can be derived from PLF technologies is also in the focus of actual research.

In the past, individual aspects of dairy farming, for example, animal nutrition, housing, climate, and animal performance, were often analyzed independently from each other, without considering potential interactions. Recent research focuses on integrating heterogeneous farm and animal data, generated by various technologies. In addition to facilitating the exchange of data between systems, integrated analyses of big data using, for example, artificial intelligence, enables new insights into the development of diseases as well as into the interaction of diseases among each other or with environmental factors, respectively. In this more 'holistic' approach, concepts of computer science, biostatistics, behavioral research, economics, animal sciences, and product development needs to interact across disciplines to gain additional information. For this, a close and trustworthy cooperation of, for example, engineering, animal science, veterinary medicine, economics, and social sciences is needed.

Tools that can also be used by laypersons in veterinary medicine, for example, farmers, are regularly criticized by some members of the veterinary profession, sometimes with the fear that they might give up a field of activity. For example, using ketometers or the automated and continuous monitoring of activity patterns by sensor technology are a tool, which not only are primarily used by farmers but also provides veterinarians with valuable information on animal and herd level. One of the advantages for veterinarians is that the identification of conspicuous animals does not have to be based solely on the farmer's skills to detect diseased animals or on single animal tests. Hence, sensor technologies have the potential of paying more attention to an individual animal within a herd.

With their expertise, veterinarians will remain a central element in herd health management, provided they are open-minded toward these technologies. For the veterinarian of the future, understanding the principles of data collection as well as their interpretation and use will be a self-evident part of the profession, just like today (and also in the future) using a stethoscope and rectal glove.

Despite all the optimism regarding the application of complex mathematical methods and the use of artificial intelligence to identify risk factors and for preventing diseases, there is still a need to maintain and sharpen our 'common sense' in animal husbandry. The potential of PLF and data integration should not be ignored by the veterinary profession and other stakeholders, as they will become valuable tools in daily practice. As a consequence of the general technical development, they are innovative tools in herd health management - nothing more, but nothing less either!

## 8 Where to look for further information

A comprehensive review of the evolution in the use of metabolic indicators for herd health management is found in 'A 100-year review: Metabolic health indicators and management of dairy cattle' presented by Overton et al. (2017).

Numerous recent publications are available on the development and use of sensor technologies in livestock farming:

- A general introduction and overview of biosensors are presented by Grieshaber et al. (2008).
- A structured review of 126 publications describing 139 sensor systems used for herd health management is provided in 'Invited review: Sensors to support health management on dairy farms' by Rutten et al. (2013).
- A comprehensive review of recent developments in the field of biosensors, including the detection of infectious agents, is presented in 'Recent advances in wearable sensors for animal health management' by Neethirajan (2017).
- An example of using artificial intelligence for the early detection of mastitis is presented by Hyde et al. (2020) in 'Automated prediction of mastitis infection patterns in dairy herds using machine learning'.

Journals and regular conferences providing recent research include:

- The *Journal of Dairy Science* is a renowned peer-reviewed dairy research journal.
- The European Association on Precision Livestock Farming (EA-PLF) holds regular meetings on PLF.
- The European Conference on Precision Livestock Farming (EC-PLF) presents innovations in PLF every 2 years.
- The World Buiatrics Congress (WBC) promotes knowledge transfer within the international veterinary community.

There are a number of current research projects on the use of modern management tools to support farmers and veterinarians in herd health management, including:

- EU-PLF focused on delivering a validated Blueprint for an animal and farm-centric approach to innovative livestock farming in Europe (www.eu-plf.eu).
- GENTORE will develop innovative genome-enabled selection and management tools to empower farmers to optimize cattle resilience and efficiency in different and changing environments (www.gentore.eu).

## 9 References

Abuelo, Á. and Alves-Nores, V. (2016). Point-of-care testing in cattle practice: reliability of cow-side diagnostic tests. *In Practice* 38(6): 293–302.

Adler, J. H., Roberts, S. J. and Steel, R. G. (1957). The relation between reactions to the Ross test on milk and urine and the degree of ketonemia in dairy cows. *The Cornell Veterinarian* 47(1): 101–111.

Afikim (2019). Afimilk's in-line milk lab and parlor automation solutions. In: Afimilk, K. A. (Ed). Israel: Kibbutz Afimilk.

Andersson, L. (1988). Subclinical ketosis in dairy cows. *Veterinary Clinics of North America: Food Animal Practice* 4(2): 233–251.

Arnold, J. E., Camus, M. S., Freeman, K. P., Giori, L., Hooijberg, E. H., Jeffery, U., Korchia, J., Meindel, M. J., Moore, A. R., Sisson, S. C., Vap, L. M. and Cook, J. R. (2019). ASVCP Guidelines: Principles of quality assurance and standards for veterinary clinical pathology (version 3.0): developed by the American Society for Veterinary Clinical Pathology's (ASVCP) Quality Assurance and Laboratory Standards (QALS) Committee. *Veterinary Clinical Pathology* 48: 542–618.

Bach, K. D., Heuwieser, W. and McArt, J. A. A. (2016). Technical note: Comparison of 4 electronic handheld meters for diagnosing hyperketonemia in dairy cows. *Journal of Dairy Science* 99: 9136–9142.

Banhazi, T. M., Lehr, H., Black, J. L., Crabtree, H., Schofield, P., Tscharke, M. and Berckmans, D. (2012). Precision Livestock Farming: an international review of scientific and commercial aspects. *International Journal of Agricultural and Biological Engineering* 5.

Banhazi, T. M., Rutley, D. L. and Pitchford, W. S. (2010). Validation and fine-tuning of a predictive model for air quality in livestock buildings. *Biosystems Engineering* 105(3): 395–401.

Bartens, M. C., Drillich, M., Rychli, K., Iwersen, M., Arnholdt, T., Meyer, L. and Klein-Jöbstl, D. (2016). Assessment of different methods to estimate bovine colostrum quality on farm. *New Zealand Veterinary Journal* 64(5): 263–267.

Berckmans, D. (2008). Precision livestock farming (PLF). *Computers and Electronics in Agriculture* 62(1): 1.

Berckmans, D. (2014). Precision livestock farming technologies for welfare management in intensive livestock systems. *Revue Scientifique et Technique* 33(1): 189–196.

Berckmans, D. and Guarino, M. (2008) Preface. *Computers and Electronics in Agriculture* 64(1): 1.

Bergman, E. N. (1971). Hyperketonemia-ketogenesis and ketone body metabolism. *Journal of Dairy Science* 54(6): 936–948.

Bertani, F., Mayfield, M. J., Mascilongo, V. and Nicols, S. (2019). Veterinary borderline products, medical devices and in vitro diagnostics: Global regulatory overview. *Regulatory Focus*. (accessed 21.04.2020).

Boivin, X., Lensink, J., Tallet, C. and Veissier, I. (2003). Stockmanship and farm animal welfare. *Animal Welfare* 12: 479–492.

Borchers, M. R. (2015). An evaluation of precision dairy farming technology adoption, perception, effectiveness, and use. Theses and Dissertations – Animal and Food Sciences, University of Kentucky.

Borchers, M. R., Chang, Y. M., Tsai, I. C., Wadsworth, B. A. and Bewley, J. M. (2016). A validation of technologies monitoring dairy cow feeding, ruminating, and lying behaviors. *Journal of Dairy Science* 99(9): 7458–7466.

Bradley, A. J. and Green, M. J. (2004). The importance of the nonlactating period in the epidemiology of intramammary infection and strategies for prevention. *Veterinary Clinics of North America: Food Animal Practice* 20(3): 547–568.

Bruss, M. L. (1989). Ketogenesis and ketosis. In: Kaneko, J. (Ed.) *Clinical Biochemistry of Domestic Animals* (4th edn.). Toronto: Academic Press, 86–105.

Burfeind, O., Schirmann, K., von Keyserlingk, M. A., Veira, D. M., Weary, D. M. and Heuwieser, W. (2011). Evaluation of a system for monitoring rumination in heifers and calves. *Journal of Dairy Science* 94(1): 426–430.

Burfeind, O., von Keyserlingk, M. A. G., Weary, D. M., Veira, D. M. and Heuwieser, W. (2010). Short communication: repeatability of measures of rectal temperature in dairy cows. *Journal of Dairy Science* 93(2): 624–627.

Cameron, R. E., Dyk, P. B., Herdt, T. H., Kaneene, J. B., Miller, R., Bucholtz, H. F., Liesman, J. S., Vandehaar, M. J. and Emery, R. S. (1998). Dry cow diet, management, and energy balance as risk factors for displaced abomasum in high producing dairy herds. *Journal of Dairy Science* 81(1): 132–139.

Carrier, J., Stewart, S., Godden, S., Fetrow, J. and Rapnicki, P. (2004). Evaluation and use of three cowside tests for detection of subclinical ketosis in early postpartum cows. *Journal of Dairy Science* 87(11): 3725–3735.

Chapinal, N., Carson, M., Duffield, T. F., Capel, M., Godden, S., Overton, M., Santos, J. E. and LeBlanc, S. J. (2011). The association of serum metabolites with clinical disease during the transition period. *Journal of Dairy Science* 94(10): 4897–4903.

Chapinal, N., LeBlanc, S. J., Carson, M. E., Leslie, K. E., Godden, S., Capel, M., Santos, J. E., Overton, M. W. and Duffield, T. F. (2012). Herd-level association of serum metabolites in the transition period with disease, milk production, and early lactation reproductive performance. *Journal of Dairy Science* 95(10): 5676–5682.

Clark Jr., L. C. and Lyons, C. (1962). Electrode systems for continuous monitoring in cardiovascular surgery. *Annals of the New York Academy of Sciences* 102: 29–45.

Cook, N., Oetzel, G. and Nordlund, K. (2006a). Modern techniques for monitoring high-producing dairy cows - 1. Principles of herd-level diagnoses. *In Practice* 28(9): 510–515.

Cook, N., Oetzel, G. and Nordlund, K. (2006b). Modern techniques for monitoring high-producing dairy cows - 2. Practical applications. *In Practice* 28(10): 598–603.

CowManager. Available at: https://www.cowmanager.com/en-us/ (accessed 29 December 2020).

Curtis, C. R., Erb, H. N., Sniffen, C. J., Smith, R. D. and Kronfeld, D. S. (1985). Path analysis of dry period nutrition, postpartum metabolic and reproductive disorders, and mastitis in Holstein cows. *Journal of Dairy Science* 68(9): 2347–2360.

Custer, E. M., Myers, J. L., Poffenbarger, P. L. and Schoen, I. (1983). The storage stability of 3-hydroxybutyrate in serum, plasma, and whole blood. *American Journal of Clinical Pathology* 80(3): 375–380.

Da Silva, J. C., Noordhuizen, J. P. T. M., Vagneur, M., Bexiga, R., Gelfert, C. C. and Baumgartner, W. (2006). Veterinary dairy herd health management in Europe: constraints and perspectives. *Veterinary Quarterly* 28(1): 23–32.

De Koning, C. (2010). Automatic milking – common practice on dairy farms. Proceedings of the 2nd North American Conference on Robotic Milking. Toronto, Canada: Precision Dairy Operators, V59–V63.

De Roos, A. P. W., Van Den Bijgaart, H. J. C. M., Hørlyk, J. and de Jong, G. (2007). Screening for subclinical ketosis in dairy cattle by Fourier transform infrared spectrometry. *Journal of Dairy Science* 90(4): 1761–1766.

Denis-Robichaud, J., Dubuc, J., Lefebvre, D. and DesCôteaux, L. (2014). Accuracy of milk ketone bodies from flow-injection analysis for the diagnosis of hyperketonemia in dairy cows. *Journal of Dairy Science* 97(6): 3364–3370.

Detilleux, J. C., Kehrli Jr., M. E., Stabel, J. R., Freeman, A. E. and Kelley, D. H. (1995). Study of immunological dysfunction in periparturient Holstein cattle selected for high and average milk production. *Veterinary Immunology and Immunopathology* 44(3–4): 251–267.

Dohoo, I. R. and Martin, S. W. (1984). Subclinical ketosis: prevalence and associations with production and disease. *Canadian Journal of Comparative Medicine (Revue Canadienne de Medecine Comparee)* 48(1): 1–5.

Dohoo, I. R., Martin, S. W. and Stryhn, H. (2009). *Veterinary Epidemiologic Research*. Charlottetown, PE, Canada: VER Incorporated.

Dolecheck, K. A., Heersche Jr., G. and Bewley, J. M. (2016). Retention payoff-based cost per day open regression equations: application in a user-friendly decision support tool for investment analysis of automated estrus detection technologies. *Journal of Dairy Science* 99(12): 10182–10193.

Drackley, J. K. (1999). ADSA Foundation Scholar Award. Biology of dairy cows during the transition period: the final frontier? *Journal of Dairy Science* 82(11): 2259–2273.

Drillich, M. and Wagener, K. (2018). Pathogenesis of uterine diseases in dairy cattle and implications for fertility. *Animal Reproduction* 15(Suppl. 1): 879–885.

Duffield, T. F. (1997). Effects of a monensin controlled release capsule on energy metabolism, health and production in lactating dairy cattle. Guelph, Canada: University of Guelph.

Duffield, T. (2000). Subclinical ketosis in lactating dairy cattle. *Veterinary Clinics of North America: Food Animal Practice* 16(2): 231-253.

Duffield, T. F., Kelton, D. F., Leslie, K. E., Lissemore, K. D. and Lumsden, J. H. (1997). Use of test day milk fat and milk protein to detect subclinical ketosis in dairy cattle in Ontario. *Canadian Veterinary Journal = la Revue Veterinaire Canadienne* 38(11): 713-718.

Duffield, T. F., Lissemore, K. D., McBride, B. W. and Leslie, K. E. (2009). Impact of hyperketonemia in early lactation dairy cows on health and production. *Journal of Dairy Science* 92(2): 571-580.

Eastwood, C. R., Chapman, D. F. and Paine, M. S. (2012). Networks of practice for co-construction of agricultural decision support systems: case studies of precision dairy farms in Australia. *Agricultural Systems* 108: 10-18.

Edwards, J. L. and Tozer, P. R. (2004). Using activity and milk yield as predictors of fresh cow disorders. *Journal of Dairy Science* 87(2): 524-531.

Egger-Danner, C., Fuerst-Waltl, B., Obritzhauser, W., Fuerst, C., Schwarzenbacher, H., Grassauer, B., Mayerhofer, M. and Koeck, A. (2012). Recording of direct health traits in Austria - Experience report with emphasis on aspects of availability for breeding purposes. *Journal of Dairy Science* 95(5): 2765-2777.

Erb, H. N. and Gröhn, Y. T. (1988). Epidemiology of metabolic disorders in the periparturient dairy cow. *Journal of Dairy Science* 71(9): 2557-2571.

Espadamala, A., Pallarés, P., Lago, A. and Silva-Del-Río, N. (2016). Fresh-cow handling practices and methods for identification of health disorders on 45 dairy farms in California. *Journal of Dairy Science* 99(11): 9319-9333.

Fetrow, J., McClary, D., Harman, R., Butcher, K., Weaver, L., Studer, E., Ehrlich, J., Etherington, W., Guterbock, W., Klingborg, D., Reneau, J. and Williamson, N. (1990). Calculating selected reproductive indices: recommendations of the American Association of Bovine Practitioners. *Journal of Dairy Science* 73(1): 78-90.

Fetrow, J., Nordlund, K. V. and Norman, H. D. (2006). Invited review: culling: nomenclature, definitions, and recommendations. *Journal of Dairy Science* 89(6): 1896-1905.

Fiore, F., Cocco, R., Musina, D. and Spissu, N. (2020). On-farm use of a water hardness test kit to assess total blood calcium level in dairy cattle. *Journal of Dairy Research* 87(1): 56-59.

Flatland, B., Freeman, K. P., Vap, L. M., Harr, K. E. and ASVCP (2013). ASVCP guidelines: quality assurance for point-of-care testing in veterinary medicine. *Veterinary Clinical Pathology* 42(4): 405-423.

Forshell, K. P. and Østerås, O. (2001). Trends in Norwegian udder health data during 1975 through 2000. Proceedings of the 2nd International Symposium on Mastitis and Milk Quality, Madison, WI: National Mastitis Council.

Galen, R. S. and Gambino, S. R. (1975). *Beyond Normality: The Predictive Value and Efficiency of Medical Diagnoses*. New Jersey, USA: John Wiley & Sons.

Gattani, A., Singh, S. V., Agrawal, A., Khan, M. H. and Singh, P. (2019). Recent progress in electrochemical biosensors as point of care diagnostics in livestock health. *Analytical Biochemistry* 579: 25-34.

Geishauser, T., Leslie, K., Kelton, D. and Duffield, T. (1998). Evaluation of five cowside tests for use with milk to detect subclinical ketosis in dairy cows. *Journal of Dairy Science* 81(2): 438-443.

Geishauser, T., Leslie, K., Tenhag, J. and Bashiri, A. (2000). Evaluation of eight cow-side ketone tests in milk for detection of subclinical ketosis in dairy cows. *Journal of Dairy Science* 83(2): 296–299.

George, J. W. (2001). The usefulness and limitations of hand-held refractometers in veterinary laboratory medicine: an historical and technical review. *Veterinary Clinical Pathology* 30(4): 201–210.

Godden, S. (2008). Colostrum management for dairy calves. *Veterinary Clinics of North America: Food Animal Practice* 24(1): 19–39.

Goldhawk, C., Chapinal, N., Veira, D. M., Weary, D. M. and von Keyserlingk, M. A. (2009). Prepartum feeding behavior is an early indicator of subclinical ketosis. *Journal of Dairy Science* 92(10): 4971–4977.

Grieshaber, D., MacKenzie, R., Vörös, J. and Reimhult, E. (2008). Electrochemical biosensors - sensor principles and architectures. *Sensors* 8(3): 1400–1458.

Grieve, D. G., Korver, S., Rijpkema, Y. S. and Hof, G. (1986). Relationship between milk composition and some nutritional parameters in early lactation. *Livestock Production Science* 14(3): 239–254.

Grummer, R. R. (1995). Impact of changes in organic nutrient metabolism on feeding the transition dairy cow. *Journal of Animal Science* 73(9): 2820–2833.

Grummer, R. R. (2008). Nutritional and management strategies for the prevention of fatty liver in dairy cattle. *The Veterinary Journal* 176(1): 10–20.

Grummer, R. R., Mashek, D. G. and Hayirli, A. (2004). Dry matter intake and energy balance in the transition period. *Veterinary Clinics of North America. Food Animal Practice* 20(3): 447–470.

Grummer, R. R. and Rastani, R. R. (2003). When should lactating cows reach positive energy balance? *The Professional Animal Scientist* 19(3): 197–203.

Guerci, B., Tubiana-Rufi, N., Bauduceau, B., Bresson, R., Cuperlier, A., Delcroix, C., Durain, D., Fermon, C., Le Floch, J. P., Le Devehat, C., Melki, V., Monnier, L., Mosnier-Pudar, H., Taboulet, P. and Hanaire-Broutin, H. (2005). Advantages to using capillary blood β-hydroxybutyrate determination for the detection and treatment of diabetic ketosis. *Diabetes and Metabolism* 31(4 Pt 1): 401–406.

Gussmann, M., Steeneveld, W., Kirkeby, C., Hogeveen, H., Farre, M. and Halasa, T. (2019). Economic and epidemiological impact of different intervention strategies for subclinical and clinical mastitis. *Preventive Veterinary Medicine* 166: 78–85.

Gusterer, E., Lidauer, L., Berger, A., Kickinger, F., Öhlschuster, M., Auer, W., Drillich, M. and Iwersen, M. (2019). Using the SMARTBOW system for monitoring animals suffering from periparturient disease. Proceedings of the 9th European Conference on Precision Livestock Farming, ECPLF 2019, 773–778.

Guterbock, W. M. (2004). Diagnosis and treatment programs for fresh cows. *Veterinary Clinics of North America: Food Animal Practice* 20(3): 605–626.

Hamann, J. and Krömker, V. (1997). Potential of specific milk composition variables for cow health management. *Livestock Production Science* 48(3): 201–208.

Hansen, P. W. (1999). Screening of dairy cows for ketosis by use of infrared spectroscopy and multivariate calibration. *Journal of Dairy Science* 82(9): 2005–2010.

Herdt, T. H. (2000). Ruminant adaptation to negative energy balance. Influences on the etiology of ketosis and fatty liver. *Veterinary Clinics of North America: Food Animal Practice* 16(2): 215–230.

Heuer, C., Schukken, Y. H. and Dobbelaar, P. (1999). Postpartum body condition score and results from the first test day milk as predictors of disease, fertility, yield, and culling in commercial dairy herds. *Journal of Dairy Science* 82(2): 295–304.

Heuwieser, W., Iwersen, M., Gossellin, J. and Drillich, M. (2010). Short communication: survey of fresh cow management practices of dairy cattle on small and large commercial farms. *Journal of Dairy Science* 93(3): 1065–1068.

Hyde, R. M., Down, P. M., Bradley, A. J., Breen, J. E., Hudson, C., Leach, K. A. and Green, M. J. (2020). Automated prediction of mastitis infection patterns in dairy herds using machine learning. *Scientific Reports* 10(1): 4289.

Ingvartsen, K. L. (2006). Feeding- and management-related diseases in the transition cow: physiological adaptations around calving and strategies to reduce feeding-related diseases. *Animal Feed Science and Technology* 126(3–4): 175–213.

Ingvartsen, K. L. and Andersen, J. B. (2000). Integration of metabolism and intake regulation: a review focusing on periparturient animals. *Journal of Dairy Science* 83(7): 1573–1597.

Itle, A. J., Huzzey, J. M., Weary, D. M. and von Keyserlingk, M. A. (2015). Clinical ketosis and standing behavior in transition cows. *Journal of Dairy Science* 98(1): 128–134.

Iwersen, M., Falkenberg, U., Voigtsberger, R., Forderung, D. and Heuwieser, W. (2009). Evaluation of an electronic cowside test to detect subclinical ketosis in dairy cows. *Journal of Dairy Science* 92(6): 2618–2624.

Iwersen, M., Klein-Jöbstl, D., Pichler, M., Roland, L., Fidlschuster, B., Schwendenwein, I. and Drillich, M. (2013). Comparison of 2 electronic cowside tests to detect subclinical ketosis in dairy cows and the influence of the temperature and type of blood sample on the test results. *Journal of Dairy Science* 96(12): 7719–7730.

Jenkins, N. T., Peña, G., Risco, C., Barbosa, C. C., Vieira-Neto, A. and Galvão, K. N. (2015). Utility of inline milk fat and protein ratio to diagnose subclinical ketosis and to assign propylene glycol treatment in lactating dairy cows. *Canadian Veterinary Journal = la Revue Veterinaire Canadienne* 56(8): 850–854.

Jeppesen, R., Enemark, J. M. and Enevoldsen, C. (2006). Ketone body measurement in dairy cows. Proceedings of the 24th World Buiatrics Congress, Nice, France. Vienna, Austria: World Association Buiatrics, OS 43-42.

Jones, G., Bork, O., Ferguson, S. A. and Bates, A. (2019). Comparison of an on-farm point-of-care diagnostic with conventional culture in analysing bovine mastitis samples. *Journal of Dairy Research* 86(2): 222–225.

Kanz, P., Drillich, M., Klein-Jöbstl, D., Mair, B., Borchardt, S., Meyer, L., Schwendenwein, I. and Iwersen, M. (2015). Suitability of capillary blood obtained by a minimally invasive lancet technique to detect subclinical ketosis in dairy cows by using 3 different electronic hand-held devices. *Journal of Dairy Science* 98(9): 6108–6118.

Khol, J. L., Freigassner, K., Stanitznig, A., Tichy, A. and Wittek, T. (2019). Evaluation of a handheld device for the measurement of beta-hydroxybutyrate in capillary blood obtained by the puncture of the vulva as well as in venous whole blood in cattle. *Polish Journal of Veterinary Sciences* 22: 557–564.

King, M. T. M., Duffield, T. F. and DeVries, T. J. (2019). Short communication: assessing the accuracy of inline milk fat-to-protein ratio data as an indicator of hyperketonemia in dairy cows in herds with automated milking systems. *Journal of Dairy Science* 102(9): 8417–8422.

Klein, D., Kern, A., Lapan, G., Benetka, V., Möstl, K., Hassl, A. and Baumgartner, W. (2009). Evaluation of rapid assays for the detection of bovine coronavirus, Rotavirus A and

*Cryptosporidium parvum* in faecal samples of calves. *Veterinary Journal* 182(3): 484–486.
Koenraadt, C. J. M., Balenghien, T., Carpenter, S., Ducheyne, E., Elbers, A. R., Fife, M., Garros, C., Ibáñez-Justicia, A., Kampen, H., Kormelink, R. J., Losson, B., van der Poel, W. H., De Regge, N., van Rijn, P. A., Sanders, C., Schaffner, F., Sloet van Oldruitenborgh-Oosterbaan, M. M., Takken, W., Werner, D. and Seelig, F. (2014). Bluetongue, Schmallenberg – what is next? Culicoides-borne viral diseases in the 21st century. *BMC Veterinary Research* 10: 77.
Lacasse, P., Vanacker, N., Ollier, S. and Ster, C. (2018). Innovative dairy cow management to improve resistance to metabolic and infectious diseases during the transition period. *Research in Veterinary Science* 116: 40–46.
Lago, A. and Godden, S. M. (2018). Use of rapid culture systems to guide clinical mastitis treatment decisions. *Veterinary Clinics of North America – Food Animal Practice* 34(3): 389–412.
Leal Yepes, F. A., Nydam, D. V., Heuwieser, W. and Mann, S. (2018). Technical note: evaluation of the diagnostic accuracy of 2 point-of-care β-hydroxybutyrate devices in stored bovine plasma at room temperature and at 37°C. *Journal of Dairy Science* 101(7): 6455–6461.
LeBlanc, S. J. (2006). *Monitoring Programs for Transition Dairy Cows. XXIV*. Nice, France: World Buiatrics Congress, 13.
LeBlanc, S. J. (2010). Monitoring metabolic health of dairy cattle in the transition period. *Journal of Reproduction and Development* 56 (Suppl.): 29–35.
LeBlanc, S. J., Leslie, K. E. and Duffield, T. F. (2005). Metabolic predictors of displaced abomasum in dairy cattle. *Journal of Dairy Science* 88(1): 159–170.
LeBlanc, S. J., Lissemore, K. D., Kelton, D. F., Duffield, T. F. and Leslie, K. E. (2006). Major advances in disease prevention in dairy cattle. *Journal of Dairy Science* 89(4): 1267–1279.
Liboreiro, D. N., Machado, K. S., Silva, P. R. B., Maturana, M. M., Nishimura, T. K., Brandão, A. P., Endres, M. I. and Chebel, R. C. (2015). Characterization of peripartum rumination and activity of cows diagnosed with metabolic and uterine diseases. *Journal of Dairy Science* 98(10): 6812–6827.
Lichtmannsperger, K., Hinney, B., Joachim, A. and Wittek, T. (2019). Molecular characterization of *Giardia intestinalis* and *Cryptosporidium parvum* from calves with diarrhoea in Austria and evaluation of point-of-care tests. *Comparative Immunology, Microbiology and Infectious Diseases* 66: 101333.
Luginbühl, A., Reitt, K., Metzler, A., Kollbrunner, M., Corboz, L. and Deplazes, P. (2005). Field study about prevalence and diagnostics of diarrhea causing agents in the new-born calf in a Swiss veterinary practice area. *Schweizer Archiv für Tierheilkunde* 147(6): 245–252.
Lumsden, J. H., Mullen, K. and Rowe, R. (1980). Hematology and biochemistry reference values for female Holstein cattle. *Canadian Journal of Comparative Medicine : Revue Canadienne de Medecine Comparee* 44(1): 24–31.
Macmillan, K., López Helguera, I., Behrouzi, A., Gobikrushanth, M., Hoff, B. and Colazo, M. G. (2017). Accuracy of a cow-side test for the diagnosis of hyperketonemia and hypoglycemia in lactating dairy cows. *Research in Veterinary Science* 115: 327–331.
Madoz, L. V., Prunner, I., Jaureguiberry, M., Gelfert, C. C., de la Sota, R. L., Giuliodori, M. J. and Drillich, M. (2017). Application of a bacteriological on-farm test to reduce

antimicrobial usage in dairy cows with purulent vaginal discharge. *Journal of Dairy Science* 100(5): 3875–3882.
Mahrt, A., Burfeind, O. and Heuwieser, W. (2014a). Effects of time and sampling location on concentrations of β-hydroxybutyric acid in dairy cows. *Journal of Dairy Science* 97(1): 291–298.
Mahrt, A., Burfeind, O. Voigtsberger, R., Müller, A. and Heuwieser, W. (2014b). Evaluation of a new electronic handheld meter for measurement of β-hydroxybutyric acid in dairy cows. *Tierärztliche Praxis Großtiere* 42(1): 5–10.
Matsas, D. J., Warnick, L. D., Mechor, G. D., Seib, L. N., Fatone, S., White, M. E. and Guard, C. L. (1999). Use of a water hardness test kit to measure serum calcium concentration in cattle. *Journal of the American Veterinary Medical Association* 214(6): 826–828.
Mazeris, F. (2010). DeLaval Herd Navigator: proactive herd management. Proceedings of the First North American Conference on Precision Dairy Management. Toronto, Canada, 26–27.
McArt, J. A. A., Nydam, D. V. and Oetzel, G. R. (2012). A field trial on the effect of propylene glycol on displaced abomasum, removal from herd, and reproduction in fresh cows diagnosed with subclinical ketosis. *Journal of Dairy Science* 95(5): 2505–2512.
McGuirk, S. M. and Collins, M. (2004). Managing the production, storage, and delivery of colostrum. *Veterinary Clinics of North America: Food Animal Practice* 20(3): 593–603.
McKenna, S. L. B. and Dohoo, I. R. (2006). Using and interpreting diagnostic tests. *Veterinary Clinics of North America: Food Animal Practice* 22(1): 195–205.
Miettinen, P. V. A. and Setala, J. J. (1993). Relationships between subclinical ketosis, milk production and fertility in Finnish dairy cattle. *Preventive Veterinary Medicine* 17(1–2): 1–8.
Müller, R. and Schrader, L. (2003). A new method to measure behavioural activity levels in dairy cows. *Applied Animal Behaviour Science* 83(4): 247–258.
Neethirajan, S. (2017). Recent advances in wearable sensors for animal health management. *Sensing and Bio-Sensing Research* 12: 15–29.
Negussie, M., Lidauer, M., Mäntysaari, E. A., Stranden, I., Pösö, J., Nielsen, U. S., Johansson, K., Eriksson, J. A. and Aamand, G. P. (2010). Combining test day SCS with clinical mastitis and udder type traits: a random regression model for joint genetic evaluation of udder health in Denmark, Finland and Sweden. *Interbull Bulletin* 42: 25–31.
Nelson, A. J. and Redlus, H. W. (1989). The key role of records in a production medicine practice. *Veterinary Clinics of North America: Food Animal Practice* 5(3): 517–552.
Nielen, M., Aarts, M. G., Jonkers, A. G., Wensing, T. and Schukken, Y. H. (1994). Evaluation of two cowside tests for the detection of subclinical ketosis in dairy cows. *Canadian Veterinary Journal = la Revue Veterinaire Canadienne* 35(4): 229–232.
Nielsen, N. I., Friggens, N. C., Chagunda, M. G. and Ingvartsen, K. L. (2005). Predicting risk of ketosis in dairy cows using in-line measurements of beta-hydroxybutyrate: a biological model. *Journal of Dairy Science* 88(7): 2441–2453.
Nielsen, U. S., Aamand, G. P. and Mark, T. (2000). National genetic evaluation of udder health and other health traits in Denmark. *Interbull Bulletin* 25: 143–150.
Nonnecke, B. J., Kimura, K., Goff, J. P. and Kehrli, M. E. (2003). Effects of the mammary gland on functional capacities of blood mononuclear leukocyte populations from periparturient cows. *Journal of Dairy Science* 86(7): 2359–2368.

Oetzel, G. R. (2004). Monitoring and testing dairy herds for metabolic disease. *Veterinary Clinics of North America: Food Animal Practice* 20(3): 651-674.

Osborne, T. M., Leslie, K. E., Duffield, T., Petersson, C. S., TenHag, J. and Okada, Y. (2002). Evaluation of Keto-Test in urine and milk for the detection of subclinical ketosis in periparturient Holstein dairy cattle. Proceedings of the 35th Annual Conference of the American Association of Bovine Practitioners. Madison, WI, 188–189.

Ospina, P. A., Nydam, D. V., Stokol, T. and Overton, T. R. (2010). Association between the proportion of sampled transition cows with increased nonesterified fatty acids and [beta]-hydroxybutyrate and disease incidence, pregnancy rate, and milk production at the herd level. *Journal of Dairy Science* 93(8): 3595–3601.

Overton, T. R., McArt, J. A. A. and Nydam, D. V. (2017). A 100-year review: metabolic health indicators and management of dairy cattle. *Journal of Dairy Science* 100(12): 10398-10417.

Pannwitz, G. (2015). Standardized analysis of German cattle mortality using national register data. *Preventive Veterinary Medicine* 118(4): 260–270.

Panousis, N., Kritsepi, M., Karagiannis, I., Kalaitzakis, E., Lafi, S. and Brozos, C. (2011). Evaluation of Precision Xceed® for on-site monitoring of blood β-hydroxybutyric acid and glucose in dairy cows. *Journal of the Hellenic Veterinary Medical Society* 62: 109–117.

Petersson, C. S., Steele, N. M., Swartz, T. H. and Dela Rue, B. T. (2017). Opportunities for identifying animal health and well-being disorders using precision technologies. In: Beede, D. K. (Ed.) *Large Dairy Herd Management* (3rd edn.). Champaign, IL: American Dairy Science Association, 1279–1292.

Philipsson, J. and Lindhé, B. (2003). Experiences of including reproduction and health traits in Scandinavian dairy cattle breeding programmes. *Livestock Production Science* 83(2–3): 99–112.

Pineda, A. and Cardoso, F. C. (2015). Technical note: Validation of a handheld meter for measuring β-hydroxybutyrate concentrations in plasma and serum from dairy cows. *Journal of Dairy Science* 98(12): 8818-8824.

Pothmann, H., Nechanitzky, K., Sturmlechner, F. and Drillich, M. (2014). Consultancy to dairy farmers relating to animal health and herd health management on small- and medium-sized farms. *Journal of Dairy Science* 97(2): 851-860.

Pralle, R. S. and White, H. M. (2020) Symposium review: Big data, big predictions: utilizing milk Fourier-transform infrared and genomics to improve hyperketonemia management. *Journal of Dairy Science* 103(4): 3867-3873.

Pyörälä, S. (2008). Mastitis in post-partum dairy cows. *Reproduction in Domestic Animals = Zuchthygiene* 43 (Suppl. 2): 252–259.

Raboisson, D., Mounié, M. and Maigné, E. (2014). Diseases, reproductive performance, and changes in milk production associated with subclinical ketosis in dairy cows: a meta-analysis and review. *Journal of Dairy Science* 97(12): 7547 -7563.

Reiter, S., Sattlecker, G., Lidauer, L., Kickinger, F., Öhlschuster, M., Auer, W., Schweinzer, V., Klein-Jöbstl, D., Drillich, M. and Iwersen, M. (2018). Evaluation of an ear-tag-based accelerometer for monitoring rumination in dairy cows. *Journal of Dairy Science* 101(4): 3398-3411.

Reneau, J. K. (1986). Effective use of Dairy Herd Improvement somatic cell counts in mastitis control. *Journal of Dairy Science* 69(6): 1708–1720.

Roberts, T., Chapinal, N., LeBlanc, S. J., Kelton, D. F., Dubuc, J. and Duffield, T. F. (2012). Metabolic parameters in transition cows as indicators for early-lactation culling risk. *Journal of Dairy Science* 95(6): 3057-3063.

Rodriguez, E. M., Aris, A. and Bach, A. (2017). Associations between subclinical hypocalcemia and postparturient diseases in dairy cows. *Journal of Dairy Science* 100(9): 7427-7434.

Rollin, F. (2006). Tool for a promt cowside diagnosis: what can be implemented by the bovine practitioner? Proceedings of the 24th World Buiatrics Congress, Nice, France. Vienna, Austria: World Association for Buiatrics.

Rothera, A. C. H. (1908). Note on the sodium nitro-prusside reaction for acetone. *The Journal of Physiology* 37(5-6): 491-494.

Rushen, J., de Passille, A. M., von Keyserlingk, M. A. G. and Weary, D. M. (2008). *The Welfare of Cattle*. The Netherlands: Springer.

Rutten, C. J., Velthuis, A. G. J., Steeneveld, W. and Hogeveen, H. (2013). Invited review: Sensors to support health management on dairy farms. *Journal of Dairy Science* 96(4): 1928-1952.

Schalm, O. W. and Noorlander, D. O. (1957). Experiments and observations leading to development of the *California mastitis* test. *Journal of the American Veterinary Medical Association* 130(5): 199-204.

Schirmann, K., von Keyserlingk, M. A. G., Weary, D. M., Veira, D. M. and Heuwieser, W. (2009). Technical note: Validation of a system for monitoring rumination in dairy cows. *Journal of Dairy Science* 92(12): 6052-6055.

SenseHub. Available at: https://www.allflex.global/product/sensehub-dairy/ (accessed 29 December 2020).

Sheldon, I. M., Molinari, P. C. C., Ormsby, T. J. R. and Bromfield, J. J. (2020). Preventing postpartum uterine disease in dairy cattle depends on avoiding, tolerating and resisting pathogonic bacteria. *Theriogenology* 150: 158-165.

SMARTBOW. Available at: www.smartbow.com.

Stangaferro, M. L., Wijma, R., Caixeta, L. S., Al-Abri, M. A. and Giordano, J. O. (2016a). Use of rumination and activity monitoring for the identification of dairy cows with health disorders: Part I. Metabolic and digestive disorders. *Journal of Dairy Science* 99(9): 7395-7410.

Stangaferro, M. L., Wijma, R., Caixeta, L. S., Al-Abri, M. A. and Giordano, J. O. (2016b). Use of rumination and activity monitoring for the identification of dairy cows with health disorders: Part II. Mastitis. *Journal of Dairy Science* 99(9): 7411-7421.

Stangaferro, M. L., Wijma, R., Caixeta, L. S., Al-Abri, M. A. and Giordano, J. O. (2016c). Use of rumination and activity monitoring for the identification of dairy cows with health disorders: Part III. Metritis. *Journal of Dairy Science* 99(9): 7422-7433.

Steeneveld, W., Hogeveen, H. and Oude Lansink, A. G. J. M. (2015). Economic consequences of investing in sensor systems on dairy farms. *Computers and Electronics in Agriculture* 119: 33-39.

Steeneveld, W., Tauer, L. W., Hogeveen, H. and Oude Lansink, A. G. (2012). Comparing technical efficiency of farms with an automatic milking system and a conventional milking system. *Journal of Dairy Science* 95(12): 7391-7398.

Stewart, S., Fetrow, J. and Eicker, S. (1994). Analysis of current performance on commercial dairies. *Compendium on Continuing Education for the Practicing Veterinarian* 16: 1099-1103.

Stokol, T. and Nydam, D. V. (2005). Effect of anticoagulant and storage conditions on bovine nonesterified fatty acid and beta-hydroxybutyrate concentrations in blood. *Journal of Dairy Science* 88(9): 3139–3144.

Sturm, V., Efrosinin, D., Öhlschuster, M., Gusterer, E., Drillich, M. and Iwersen, M. (2020). Combination of sensor data and health monitoring for early detection of subclinical Ketosis in dairy cows. *Sensors* 20(5).

Süss, D., Drillich, M., Klein-Jöbstl, D., Wagener, K., Krieger, S., Thiel, A., Meyer, L., Schwendenwein, I. and Iwersen, M. (2016). Measurement of β-hydroxybutyrate in capillary blood obtained from an ear to detect hyperketonemia in dairy cows by using an electronic handheld device. *Journal of Dairy Science* 99(9): 7362–7369.

Suthar, V. S., Canelas-Raposo, J., Deniz, A. and Heuwieser, W. (2013). Prevalence of subclinical ketosis and relationships with postpartum diseases in European dairy cows. *Journal of Dairy Science* 96(5): 2925–2938.

Titler, M., Maquivar, M. G., Bas, S., Gordon, E., Rajala-Schultz, P. J., McCullough, K. and Schuenemann, G. M. (2013). Effect of metritis on daily activity patterns in lactating Holstein dairy cows. *Journal of Dairy Science* 96(1): 647.

Tyopponen, J. and Kauppinen, K. (1980). The stability and automatic determination of ketone bodies in blood samples taken in field conditions. *Acta Veterinaria Scandinavica* 21(1): 55–61.

Van Asseldonk, M. A. P. M. (1999). *Economic Evaluation of Information Technology Applications on Dairy Farms*. S.l.: Van Asseldonk.

Van der Drift, S. G. A., Jorritsma, R., Schonewille, J. T., Knijn, H. M. and Stegeman, J. A. (2012). Routine detection of hyperketonemia in dairy cows using Fourier transform infrared spectroscopy analysis of β-hydroxybutyrate and acetone in milk in combination with test-day information. *Journal of Dairy Science* 95(9): 4886–4898.

Van Erp-Van der Kooj, E., Van de Brug, M. and Roelofs, J. (2016). Validation of the Nedap Smarttag Leg and Neck to assess behavioral activity level in dairy cattle. In: Kamphuis, C. and Steeneveld, W. (Eds) *Precision Dairy Farming 2016*. Wageningen Academic Publishers, 321–326.

Van Knegsel, A. T. M., Van der Drift, S. G. A., Horneman, M., de Roos, A. P., Kemp, B. and Graat, E. A. (2010). Short communication: ketone body concentration in milk determined by Fourier transform infrared spectroscopy: value for the detection of hyperketonemia in dairy cows. *Journal of Dairy Science* 93(7): 3065–3069.

Venjakob, P. L., Borchardt, S. and Heuwieser, W. (2017). Hypocalcemia – cow-level prevalence and preventive strategies in German dairy herds. *Journal of Dairy Science* 100(11): 9258–9266.

Villa-Godoy, A., Hughes, T. L., Emery, R. S., Chapin, L. T. and Fogwell, R. L. (1988). Association between energy balance and luteal function in lactating dairy cows. *Journal of Dairy Science* 71(4): 1063–1072.

Voyvoda, H. and Erdogan, H. (2010). Use of a hand-held meter for detecting subclinical ketosis in dairy cows. *Research in Veterinary Science* 89(3): 344–351.

Wagener, K., Gabler, C. and Drillich, M. (2017). A review of the ongoing discussion about definition, diagnosis and pathomechanism of subclinical endometritis in dairy cows. *Theriogenology* 94: 21–30.

Walsh, R. B., Walton, J. S., Kelton, D. F., LeBlanc, S. J., Leslie, K. E. and Duffield, T. F. (2007). The effect of subclinical ketosis in early lactation on reproductive performance of postpartum dairy cows. *Journal of Dairy Science* 90(6): 2788–2796.

Webster, J. (2016). Animal welfare: freedoms, dominions and 'A life worth living.' *Animals* 6(6): 35.

Wilson, A. J. and Mellor, P. S. (2009). Bluetongue in Europe: past, present and future. *Philosophical Transactions of the Royal Society of London. Series B, Biological Sciences* 364(1530): 2669-2681.

Wischenbart, M., Tomic, D., Iwersen, M., Schrefl, M. and Sturm, V. (2017). agriProKnow - process-oriented information management in precision dairy farming. In: (KTBL) AfTaSiA (Eds) 13th Conference on Contruction, Engineering and Environment in Livestock Farming. Stuttgart-Hohenheim.

Wolfger, B., Jones, B. W., Orsel, K. and Bewley, J. M. (2016). Technical note: Evaluation of an ear-attached real-time location-monitoring system. *Journal of Dairy Science* 100(3): 2219-2224.

Zakian, A., Tehrani-Sharif, M., Mokhber-Dezfouli, M. R., Nouri, M. and Constable, P. D. (2017). Evaluation of a point-of-care electrochemical meter to detect subclinical ketosis and hypoglycaemia in lactating dairy cows. *Australian Veterinary Journal* 95: 123-128.

Zambon, I., Cecchini, M., Egidi, G., Saporito, M. G. and Colantoni, A. (2019). Revolution 4.0: industry vs. agriculture in a future development for SMEs. *Processes* 7(1): 36.

# Chapter 4

# Data-driven decision support tools in dairy herd health

*Victor E. Cabrera, University of Wisconsin-Madison, USA*



## 1 Introduction

This chapter describes the development process of data-driven decision support tools for dairy herd management with an emphasis on real-time continuous data integration and its applications on dairy herd health. It includes concepts on big data analysis, expert systems, and artificial intelligence towards more sustainable dairy farm production systems.

Significant advancements in dairy herd genetics (Shook, 2006), nutrition (VandeHaar and St. Pierre, 2006), reproduction (Moore and Thatcher, 2006), health (LeBlanc et al., 2006), cropping systems (Tilman et al., 2002), and general management (Oltenacu and Broom, 2010) have increased dairy productivity by 40% during the last 50 years (Liang et al., 2018). However, real-time, continuous, integrated dairy farm data to project and optimize the whole production system through the use of decision support tools has not advanced at the same speed. Dairy farmers have embraced technological innovations that bring massive amounts of data. Today, animal scientists are considered to be part of the big data revolution (Madrigal, 2012) and precision livestock farming (Bewley and Russell, 2010). However, new challenges and opportunities of big data, precision livestock, and continuous data integration are not fully addressed (Madrigal, 2012).

http://dx.doi.org/10.19103/AS.2020.0086.19

Dairy producers and their consultants are using isolated expert systems to interpret and apply farm data recordings, which defeats the purpose of managing an integrated production system with data-rich pipelines. Even independently analysed, disparate data streams are informative and valuable. However, when integrated, they can generate new crucial insights (Cabrera et al., 2020). Knowledge from integrated data will promote a paradigm change in how dairy farms operate. Lack of data integration and subsequent absence of integrated data analysis and projections generate problems such as delayed optimal actions, increased risk of mistakes and failures, suboptimal use of on- and off-farm resources, narrow vision of opportunities for improvement, suboptimal profitability and therefore decreased sustainability and resilience (Cabrera et al., 2020). Dairy farm systems are highly integrated production systems in which every management change affects and is affected by the rest of factors in the system. It is recognized that expert systems related to the prevention or management of health require to be connected to production, nutrition, reproduction, genetics and other data sources inside the farm and weather, market, and other conditions outside the farm. Today, luckily, all or most of these data streams are available in modern dairy production systems. However, these data streams are not integrated.

There is a need for decision-support tools that account for biological, price, and weather uncertainties inherent to the production system (Mirando et al., 2012) using real-time, integrated big data analytical decision-making (Wolfert et al., 2017). There is an increasing need for the development of farm-specific models and decision support tools to improve decision-making in a permanent fashion (O'Grady and O'Hare, 2017). Machine-learning techniques and data mining could provide a much deeper understanding of big integrated data (Morota et al., 2018) after the data are transformed and quality controlled (Hashem et al., 2015).

It is also well documented that dairy farmers are making critical decisions relying mostly on intuition and experience (Groenendaal et al., 2004) without using all the data available and without using proper analyses and decision support tools. Dairy farmers use neither efficient simulation or optimization frameworks (Bewley et al., 2010) nor decision support tools available to them (Cabrera, 2018). Efficient decision support system frameworks are critical for the future of dairy farming management and decision-making (Meadows et al., 2005). Several research and extension groups have recognized this fact and are trying to cover this gap (Cabrera, 2018). One of these initiatives is the University of Wisconsin-Madison Dairy Management research and extension programme (https://DairyMGT.info) that provides the largest number of dairy farm decision support tools in one place. These tools aim to support improved decision-making of dairy farmers in distinct areas of management, including health. However, most of the DairyMGT.info tools are specific to a management area (e.g. reproduction, replacement, or rearing youngstock) and not integrated with other areas of management or sources of information on the farm. Also, most of the tools do not work automatically and/or do not use integrated real-time data connected with farm data pipelines.

The University of Wisconsin dairy management tools (DairyMGT.info -> tools) have proven to be very useful and demanded (Cabrera, 2018), but they require to keep-up with modern dairy farm practices. Part of the chapter is devoted to discussing the vision for a new generation of decision support tools that are automatic, use permanent or real-time integrated data and the vision of how the whole system will learn from itself becoming 'smarter' throughout time.

## 2 Big data and decision analysis

The University of Wisconsin (UW)-Dairy Brain[1] (https://DairyBrain.wisc.edu/) is a transdisciplinary research and extension project involving dairy, computer, and data scientists (>15 students, postdocs, faculty, and staff) engaged with community stakeholders (dairy farmers, industry suppliers, company representatives, consultants, etc.) - a *Coordinated Innovation Network* (CIN) of >60 members. The UW-Dairy Brain is addressing the issue of lack of real-time, continuous dairy farm data integration and its value-added opportunities for improved decision-making and operation efficiency. The UW-Dairy Brain is using the state-of-the-art database management system from the University of Wisconsin-Madison Center for High Throughput Computing (CTHC; http://chtc.cs.wisc.edu/) to develop an Agricultural Data Hub (AgDH) that connects and analyses cow and herd data on a permanent basis. This involves cleaning and normalizing the data as well as allowing data retrieval on demand (Cabrera et al., 2020). At the moment, five dairy farms are enrolled. These farms were purposefully selected to expose the team to the data available from the most widely used technologies and software available on dairy farms (e.g. milking robots, cow sensors, milking parlour systems, etc.). Once the proof of concept is demonstrated in these farms, the Dairy Brain team envisions to expand and scale up the developed technologies to the dairy industry at large.

The UW-Dairy Brain stores all data securely at the University of Wisconsin-Madison premises (HTCondor server, https://research.cs.wisc.edu/htcondor/). We installed in each collaborating farm a dedicated computer connected to the farm computers and software network. That computer is programmed to collect and place farm data files into specific server directories. Based on the farm configuration and vendor data source, specific data-parsing modules

1 The UW-Dairy Brain project is supported by the University of Wisconsin-Madison UW2020 initiative and by the USDA-NIFA-FACT grant 2019-68017-29935 (Washington, DC).

are set. This approach facilitates the addition or replacement of a vendor data source by simply replacing the module file reducing the amount of changes or code duplication of the whole process. Data will become big data as more farms with new technologies and software are added. Server computers prepare all received data via a programmatic interface through an automated process that detects the arrival of new data sets on which applies a specific set of programmes to extract and safely store the data. Then, data are cleaned by a process of data-mining and language-processing tool scripts, which detect, correct, and/or remove corrupt or inaccurate information. Data are then harmonized utilizing a Microsoft SQL relational database (Microsoft Corp., Redmond, WA) as a data warehouse (AgDH) with consistent data format and structure regardless of the data source. This process provides common key variables to query the data (Schuetz et al., 2018).

## 2.1 Data collection

The Dairy Brain project is collecting on-farm data according to their generation frequency and include: (1) herd management: records from routine and operational activities such as reproduction, calvings, health events, vaccinations, and treatments; (2) milking system: milk volume, flow, conductivity; (3) genetics and genomics: pedigree and DNA of tested animals; (4) dairy herd improvement control: monthly test-day visit variables such as milk volume, somatic cell count, milk fat, protein, and herd population; (5) feed: diets, consumption, and costs; and (6) processor: daily bulk tank milk volume, composition and somatic cell count. The Dairy Brain project also has access to off-farm data such as weather and market prices. These off-farm data are not collected or stored on University premises but are available to query on demand to integrate with on-farm data as needed. It is anticipated the amount of data and the number of farms will grow exponentially as the proof of concept is demonstrated (Cabrera et al., 2020).

## 2.2 Data warehouse (AgDH)

The Dairy Brain uses an extraction, transformation, and loading (ETL) artificial intelligence (AI) protocol to process the data and to prepare them for access via programmatic interfaces (Cabrera et al., 2020). This ETL process is managed by an automation framework that identifies the arrival of new data sets on which the pre-defined recipes are implemented to extract and integrate the information. Then, a process of transformation follows. Transformation of data involves cleaning using data-mining and language processing scripts, which detect, correct or remove corrupt or inaccurate information. Cleaned data are harmonized and then loaded into a Microsoft SQL relational database (Microsoft Corp., Redmond, WA) with a consistent format, structure, and key variables to query the data regardless of the source. The programme uses specific modules that contain particular data-parsing logic according to the software and farm configuration. This approach is flexible to facilitate the addition or replacement of a new module as new farms, and new technologies are implemented in farms.

### 2.3 Data analyses

Once the constant flow of dynamic data is set, descriptive, predictive, and/or prescriptive analyses can be applied on a permanent basis. This process is conceived as a real-time analytical engine and called the Dairy Brain. The Dairy Brain is capable of performing longitudinal historical analyses and forecasting future from past information in a closed loop (Cabrera et al., 2020). Analytical frameworks are then packaged in user-friendly visualization and decision support tools that have the unique characteristics of (1) automated data ETL, (2) permanent (real-time) data analyses utilizing combined data sets, and (3) integrated decision support tools at the fingertips of dairy farm decision makers.

### 2.4 New tools backing decision making

Better informed optimal decisions can be achieved by integrating data streams and applying descriptive, predictive, and prescriptive analytics within farm-specific decision-making tools. As pursued by the UW-Dairy Brain, these models and tools can demonstrate how integrated multiple data sources can leverage the agDH to generate more value within the data ecosystem. Models can be operationalized as data services to support dairy farmers' short-, medium- and long-term decision-making. An example of tools requiring permanent data integration is portrayed in Table 1.

Although relatively simple, critical for operational decision-making and performance monitoring are descriptive analytics such as daily calculation of daily feed efficiency (energy corrected milk produced by the unit of feed supplied) and daily milk income over feed cost (milk revenue minus feed costs), which at the moment are not readily available on dairy farms because they require data integration among systems. Farmers can effectively use this information to have early warnings of daily feed efficiency issues as well as detect management factors that could optimize margins.

Farmers also need tactical planning to support sustained efficiency, which, for example, can be achieved by improving feed nutritional accuracy by a continuous nutritional grouping of cows and by better managing clinical mastitis cases by anticipating their onset through early prediction of clinical mastitis (Table 1). The efficient use of integrated data can aid farmers in

**Table 1** Exemplars of data-integrated decision support tools

| Decision level | Tool | Benefit |
|---|---|---|
| **Operational Short-term** | Daily feed efficiency | Early warnings produced |
| | Daily milk income over feed cost | Margins optimized |
| **Tactical Mid-term** | Continuous nutritional grouping | Diets formulated more accurately |
| | Early prediction of clinical mastitis | Mastitis risk cows identified |
| **Strategic Long-term** | Dynamic net present value of a cow | Cow's long-term value assessed |
| | Lifetime expectation after an episode of clinical mastitis | Cow's fate after mastitis determined |
| | Selection of genetic traits to reduce clinical mastitis | Best replacements selected |

providing diets more accurate to the cow's requirements and select and better treat or prevent disease on cows at higher risk of mastitis.

Finally, strategic decisions, which are the most difficult to visualize and assess because the consequences of today's actions are only realized in the long-run and likely across several areas of management. As such, continuous data integration is paramount, along with more sophisticated analytics. Some exemplars in Table 1 are the evaluation of the cow's lifetime expectation after an episode of clinical mastitis or selecting cows by desired genetic traits such as those that reduce the risk of contracting clinical mastitis.

## 2.5 The evolving issues around decision making

Even though tools are based on state-of-the-art scientific knowledge, reliant on the latest computing technologies, and visually appealing, these facts do not guarantee that they will be adopted, remain useful, and endure throughout time. Adoption of scientific tools has not been optimal because of several challenges - among those, data overload, lack of data integration to run the models and tools, the complexity of using the tools and lack of clear messages when using the tools. Tools will only be adopted if they provide tangible value to the producer. As such, sustained adoption and application of tools for improved decision-making in health management or any other area of management depends critically on identification and communication on how these tools will make a difference in the industry. These tools have to clearly demonstrate the potential for continued impact by addressing the most pressing data-driven needs.

## 2.6 The growing importance of big data on dairy farms

Dairy farmers have embraced technological innovations that generate massive permanent data streams (Cabrera et al., 2020). The resulting system is complex, and integrating and interpreting all these data to improve whole-farm-based management and decision-making has proven challenging. There is a growing demand on big data use and application on dairy farms, and, at the moment, it is difficult for dairy farm managers and animal scientists to overcome the new challenge of taking full advantage of the opportunities that big data streams create (Madrigal, 2012).

## 3 Whole-dairy-farm systems simulation

Complex and highly interconnected dairy production systems require integrated simulation models to understand and project the short-, medium-, and long-term consequences of managerial decisions. In such vein, the Ruminant Farm Systems Model (RuFaS), under development, is conceptualized as the next-generation, open-source computational tool (Kebreab et al., 2019). It builds upon the long-standing Integrated Farm System Model (IFSM, Rotz et al., 2013). RuFaS is a process-based, daily time-step model that uses biophysical equations to represent farm processes. It is a whole-farm-dairy system modular model that integrates animal, manure, soil, crop, and feed storage modules; follows C, N, P, K, and water cycles through the dairy system; and monitors economics, energy, water, environment, and animal welfare systems within the farm boundaries. Briefly, the animal module follows the life cycle of each and all the animals in the herd; computes their production and nutrient balance according to facilities, management, and offered diets. The soil module uses weather data to simulate hydrology and soil C, N, P, and K dynamics with different chemical pools, whereas the crop modules represent the biological biomass growth according to weather conditions, soil nutrients, and management. Crops can be fertilized with cows' manure, and crops produced could become animal feeds. The feed storage module tracks C and N during harvesting and storage, whereas the water module follows the water cycle in all components of a dairy farm system monitoring water captured and loss out of the system. All modules will be integrated to streamline data input and output flows and will provide novel information about managing dairy production at a farm system level instead of optimizing separate farm operations (Kebreab et al., 2019). The RuFaS is being built in a collaborative environment with modular characteristics in which new developments can be easily incorporated as new information and knowledge are available. All the modules will be linked and organically integrated to ensure consistency and transparency and to streamline all the data input and output flows. Externalities, such as climate and weather conditions, government policy and incentive, connected water systems (e.g. precipitation, evapotranspiration, runoff, groundwater, etc.), energy systems (e.g. power grids that could accept renewable electricity generated from dairy manure), and food systems (e.g. agricultural operations, including conservation practices, of related farms) will be systematically integrated and accounted for throughout the 'cradle-to-grave' life cycle.

## 4 The University of Wisconsin-Madison Dairy Management website

The University of Wisconsin-Madison Dairy Management website (https://DairyMGT.info) is an online interactive informational resource that highlights computerized online decision support tools, all of which are scientifically documented through journal articles, extension publications, abstracts, and presentations. Decision support tools in DairyMGT.info are classified by management area, such as nutrition and feeding, youngstock, reproduction, production, replacement, health, financial, and environment. To date, there are more than 40 decision support tools openly available at DairyMGT.info. Cabrera (2012b, 2018) describes the general rationale for each management group of tools and the specifics for a large number of particular decision support tools available on DairyMGT.info.

Briefly, tools contained at DairyMGT.info support the decision-making of dairy farm managers with the goal of improving production performance and profitability. Users are able to customize the tools to represent desired farm conditions and perform analysis, projections, and optimizations. All tools allow users to enter their own input parameters. Some require users to manually define and enter their data, but many of the tools allow users to enter larger data sets as inputs through the use of spreadsheet templates. The protocol then involves opening the tool, enter the farm data, perform the analysis, study and/or retrieve the results, take action based on the outcomes, and repeat the protocol as desired and needed.

Following are some examples of usage of some common tools at DairyMGT.info. In the nutrition and feeding management area, FeedVal v6.0 tool calculates the relative value of feed ingredients according to their value per nutrient and their current market price, so it helps managers purchase the least cost feeds. The user needs to define reference nutrients and reference feed composition and their market price. The tool then ranks the best and worst feed purchases. In the reproduction management area, the DairyRepro$ tool projects the reproductive and economic performance of an actual and an alternative reproductive programme for adult cows (Giordano et al., 2012). The user defines the herd and economic conditions of the studied farm, along with the current and alternative reproduction parameters, including synchronization programmes, heat detection activities, service rates, and conception rates, among others. The tool projects the herd performance under current and alternative reproductive programmes and finds the differences in reproductive performances and economic returns. The tool is effective to virtually test the value of changes in reproductive management. In the replacement management area, the Economic Value of a Dairy Cow tool calculates the long-term projected net return of a cow compared with a potential replacement (Cabrera, 2012a). The user furnishes the farm, herd, and economic conditions together with a list of all adult cows and their current status (lactation, days after calving, days in pregnancy, and genetic make-up) and the possible genetic improvement of the herd replacements. The tool then calculates individual cow value and ranks the cows ascendingly. The tool is then effective for individual cow management, including replacement, treatment, or breeding decisions. In the health management area, the Improve Milk Bulk Tank SCC tool estimates the total milk bulk tank somatic cell count (SCC) and value according to a maximum cow's SCC threshold included in the tank and the milk price structure of premiums and penalties according to bulk tank SCC. The user defines milk price and SCC price structure, a list of lactating cows with their production and SCC, and a limit cow SCC threshold to discard milk out of the bulk tank. The tool then calculates the final bulk tank SCC, the estimated value of it and compares it with the situation of not discarding milk because of elevated SCC. The tool is effective for deciding which cow's milk should not be included in the bulk tank.

### *4.1 Limitations of decision support tools*

Decision support tools are valuable only if data required for its parameterization are readily available. As portrayed earlier, decision support tools contained at DairyMGT.info can be effective to inform improved decision-making, but they require active, and in most of the cases, manual, data input, which could be cumbersome, ineffective, and prone to errors. Indeed, it is believed that an important factor limiting a greater use of the DairyMGT.info tools is the lack of automation of data input/output of the tools. Following the examples above, FeedVal v6.0 requires the user extracts feed composition analysis data from the farm feed management software, from the laboratory analysis, or from vendors together with collecting local market prices for the feeds from vendors. All these data are available and could be systematically collected and organized to be input into the tool. An attempt to it is under course, FeedVal v7.0 that updates feed composition and prices from data and collaboration of a network of a feed laboratory company and field nutritionists. Even then, farm-level assessment needs additional data to fine-tune that could be automatized to the maximum extent possible. In the case of DairyRepro$ tool, most of the data required for the herd and current reproduction programme should be available from the management software but still requires careful extraction and organization that many times could be cumbersome and limited to experts (e.g. veterinarians with good knowledge of reproduction). Nonetheless, the tool could connect with the farm management software and extract all these data automatically, and so it would be readily available for the user on demand. Similarly, for the Economic Value of a Dairy Cow, the required data are normally available at the farm from the herd management software, DHIA, processing company, and feed management software. However, collecting the data and organizing it requires time and effort and needs to be repeated every time the analysis is required. All data could be extracted, cleaned, connected and made available on a permanent basis, so the tool can perform the analysis daily. The goal is that the farmer has a list of the value of all herd cows daily. In the case of the Improve Milk Bulk Tank SCC, the goal would be to connect DHIA control data automatically, or, if available, to milk parlour inline sensor data, that include individual cow SCC, together with daily cow-level milk production, so the manager can decide, daily as required, which cows' milk should not be included in the tank, by simply manipulating the threshold of maximum SCC. As seen, one limitation is the lack of automation of data extraction and furnishing to the tools, but a follow-up limitation is a fact that data from different systems at the farm are not connected and their integration is required for the right functioning of the tools.

Another current limitation is the lack of connection between and among tools. Most of the tools in DaiyMGT.info web portal deal with a specific management area, and although some of the most sophisticated tools require interaction with other areas of management (e.g. reproduction assessment interacts with productivity, herd dynamics, or replacement actions), these are not fully connected. For example, improved reproduction management is not connected with potential genetic improvement or feeding, and nutritional changes are not related to improved health of the herd.

Tools have not been developed with a modularity approach. Each tool is a stand-alone model that cannot be connected easily to other tools. Tools should share common variables and components so they could share information among then and could be connected. Tools can then have feedback among then, and, therefore, for example, health could be connected with reproduction performance; nutrition could be used in conjunction with replacement decisions, and replacement decision could affect reproduction and vice versa.

## 4.2 Opportunities for decision support tools' suites

Users benefit from information generated by decision support tools when they are able to custom-tailor the tools to their needs according to particular farm characteristics. As such, for example, the long-term vision of DairyMGT.info web portal is to allow users to interact with their live integrated data. Users would be able to run an integrated simulation and optimization models contained in decision support tools. The process would be set to retrieve data automatically and on demand, in other words, a combination of the Dairy Brain together with the RuFaS within an improved DairyMGT.info web portal. Currently, some software or vendor companies allow farmers' data retrieval, visualization, and dashboarding using cloud systems; however, the level of data integration between technologies in the farm is marginal (e.g. some management data with market prices) and the level of live data analytics is minimal (e.g. no predictive or prescriptive analytics or long-term simulation and life-cycle analysis as portrayed in the RuFaS). We have an opportunity to develop a system that could become the centre-piece for dairy farm management, informing farmers and farm managers the best actions for the short-, medium-, and long-term to take regarding critical areas of decision such as health, nutrition, breeding, treatments, replacement, crop systems, and environmental management.

## 5 Data-driven decision support tools: mastitis as a case example

This section illustrates the challenge and the opportunity with a practical example related to dairy herd health: clinical mastitis, the most important disease in dairy cattle. The vision is to use a series of integrated data pipelines in a unified framework to, first, alert the greater risk of early clinical mastitis of an animal, and, then, more specifically, alert the imminent onset of the disease for a cow. Dairy farmers could select animals and/or anticipate the events if they have the information handy and delivered as alerts in a timely fashion. A decision support tool would be connected with the dairy farm informational system using the DairyMGT.info platform as a bridge between systems and models. As discussed earlier, an important opportunity arises by combining the vision of the Dairy Brain and the development and delivery of decision support tools as portrayed in the University of Wisconsin-Madison Dairy Management tools (DairyMGT.info).

Clinical mastitis is the most expensive disease in dairy cattle that cost dairy farms between US$325 ±71 (Liang et al., 2017) and US$426±80 (Rollin et al., 2015) per case. Clinical mastitis not only has an important negative effect on dairy farm profitability (Delgado et al., 2017) but also shortens the productive life of dairy cattle, becoming one of the most important causes for earlier culling dairy cattle (Heikkilä et al., 2012; Rushen and de Passill, 2013). It is clear then that a decision support system that alerts farmers about cows having a high risk of ever have a clinical mastitis event during their first lactation and cows at imminent risk of contracting clinical mastitis at any time during any lactation would be of great value to producers and the dairy industry in general.

### 5.1 *Risk of mastitis during the first lactation*

Using machine-learning algorithms applied to integrated data, we demonstrated acceptable prediction power to quantify the risk of clinical mastitis on first lactation (Cabrera et al., 2020). These results can be used to tailor decision support tools to help farmers anticipate mastitis events consistent with their management style and available on-farm data (Delgado et al., 2019). The suggested procedure follows.

Historical records from animals between 15 and 300 days after first calving can be identified as healthy or sick if their culture tests were positive for either gram-negative, gram-positive bacteria, or other pathogens (explained variable). Animals that do not test positive in their culture or do not receive any further treatment can be labelled as free from clinical mastitis. These data can then be used to predict clinical mastitis risk in first lactation animals, according to a number of possible combined data streams available on the farm (explanatory variables). In our specific illustration, four groups of data streams were tested: (1) genetic records (Zoetis-Enlight), (2) dairy herd improvement, (3) production records, and (4) previous health events. These data streams were available in the Dairy Brain data sets, but the algorithms could and should be tested with more data as these become available. After testing a number of machine-learning algorithms, the Naïve Bayes Wrapper, the Random Forest algorithm with variables selected with the Gain Ratio Classifier and the Random Forest algorithm with variables selected with the Subset Classifier performed the best (Delgado et al., submitted 2020; Table 2).

As seen in Table 2, prediction precision (all correctly identified instances) can have an accuracy between 85% and 98% depending on the data available at the farm and the farm goals. If the farmer's main concern is to detect the true healthy animals, any of the algorithms, Naïve Bayes Wrapper, Random Forest Gain Ratio, or Random Forest Subset Classifier will work well with greater than 96% true healthy animals' detection. If the farmer's main objective is to detect true mastitis animals, Random Forest Subset Classifier will offer the best result, with over 80% of true mastitis cases detected. However, if the interest is early (3rd month or earlier in lactation) detection, any of the three algorithms will perform similarly. All three algorithms will select some genetic traits in the prediction, but each of them will combine these variables differently for better prediction. The Random Forest Subset Classifier will include in addition mature equivalent projected production at 305 days, the Random Forest Gain Ration will include in addition somatic cell count (SCC), whereas the Naïve Bayes Wrapper will include in addition fat corrected milk (FCM) production and previous health events such as ketosis, metritis, retained placenta, and abortion.

## *5.2 Continuous onset prediction of clinical mastitis*

The integration of different data sources could improve the performance of clinical mastitis prediction models compared with models using only variables pertaining to one data stream (Hogeveen et al., 2010). For example, the use of

**Table 2** Best performing machine-learning algorithms to predict mastitis in the first lactation[a]

| | Naïve Bayes Wrapper | Random Forest Gain Ratio | Random Forest Subset Classifier |
|---|---|---|---|
| True healthy, % | 96.6 | 99.9 | 99.5 |
| True mastitis, % | 38.8 | 76.7 | 80.4 |
| Weighted AUC-ROC[b], % | 85 | 98.7 | 97.9 |
| Somatic cell count-related variables[c] | - | SCC, logSCC | |
| Production-related variables[d] | FCM | - | ME305 |
| Genetic-related variables[e] | FAT, LIV, MAST, PL, PROT, SCS, UDC | CM$, DWP$, FM$, MAST, NM$, PL, SCS, TPI, UDC, WT$ | FM$, TPI |
| Other health-related variables | Ketosis, Metritis, Retained Placenta, Abortion | - | - |

[a] Performance improved as more data from control tests are included from tests 2 to 9. All performance indicators reported in the table are after 9 control tests.
[b] AUC-ROC = area under the curve of the receiver operating characteristic.
[c] SCC = somatic cell count.
[d] FCM = fat corrected milk, ME305 = mature equivalent milk production for 305 days lactation.
[e] Genetic traits: FAT = fat trait (kg), LIV = livability (%), MAST = mastitis trait (%), PL = productive life (month), PROT = protein trait (kg), SCS = somatic cell score (unitless), UDC = udder composite trait (points), CM$ = cheese merit ($), DWP$ = dairy wellness profit ($), FM$ = fluid merit ($), NM$ = net merit ($), TPI = total performance index (points), WT$ = wellness trait index ($). Source: Delgado et al. (2019).

integrated data from the milking system, together with management, which, though readily available on most modern dairy farm systems, is disjointed. These could include milk production at each milking (kg), milk conductivity (mS/cm), and/or milk flow speed (L/min) from the milking systems and previous health events and cow's lactation state from the herd management software. One of the limitations of the clinical mastitis studies is the low clinical mastitis prevalence (usually < 1%). However, there are different techniques to overcome this problem. For example, the synthetic minority over-sampling technique (SMOTE; Chawla et al., 2002) could be used to balance the data and improve predictability. After testing with different machine algorithms (Logistic Regression, Random Forest, Gradient Boosting) and stacking, an ensemble method, a combination of different machine-learning techniques into one predictive model (Dietterich, 2000), results showed that the Random Forest provided the best predictive capabilities in function to differences in milk production and milk conductivity between milkings from the milking system and lactation, days in milk, age at first calving and previous diseases events (metritis, retained placenta, abortion or ketosis) from the management software data. Good predictive assessments could be achieved seven milkings prior to the onset of clinical mastitis with a specificity of 88%, sensitivity of 93%, and an overall accuracy of 93%. Cows at a greater risk of becoming sick with clinical mastitis showed significantly greater conductivity (0.91 vs. 0.86 mS/cm; $P < 0.001$) and greater milk productivity (4.70 vs. 4.29 lbs; $P < 0.001$) than cows that did not contract clinical mastitis (Fadul et al., unpublished).

The model generates a herd-specific dynamic list of animals with the risk of clinical mastitis. Once the alert is emitted, a follow-up can be implemented for those animals at risk, according to the health protocol pre-established at the farm. The evolution of the mastitis risk forecast of the cow in the following milkings can inform the farmer of the best course of action. Having an alert from 2 to 4 days before the onset (depending on the number of milkings per day) could be a critical management tool to reduce the incidence and severity of mastitis on dairy farms. Early detection of (risk) of mastitis can help improve the welfare of cows by prompt intervention to alleviate pain and discomfort (Leslie and Petersson-Wolfe, 2012).

Moreover, farmers could adjust thresholds of predictions according to risk preferences to either having a greater detection rate (with more false alarms) or lower detection rate (with less false alarms but more true missing true events). This risk preference is associated with the clinical mastitis prevalence at the farm. For example, by ranging the threshold levels from 0.4 to 0.6, a farm with a low prevalence of clinical mastitis (i.e. 0.09%) could trade 9 percentual points (from 95% to 86%) of sensitivity (less detection rate) in exchange of 11 percentual points (from 89% to 78%) of specificity (less false alerts). For a farm with a greater prevalence of clinical mastitis (i.e. 1.23%), the trade could be 5

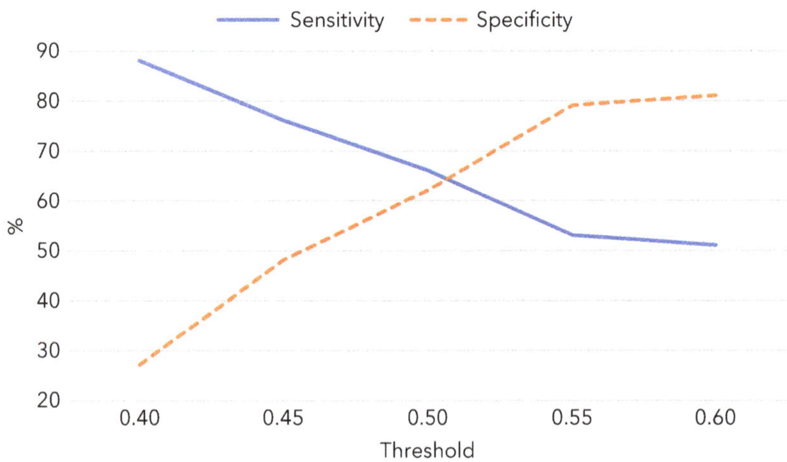


**Figure 1** Sensitivity and specificity of the Gradient Boost algorithm to predict the onset of clinical mastitis at different thresholds for a farm with a clinical mastitis prevalence of 1.1%. Machine learning prediction of 7 milkings before onset. Graphed with data from Fadul et al. (unpublished).

percentual points (from 98% to 93%) of sensitivity in exchange for 8 percentual points (from 90% to 98%) of specificity (Fig. 1).

### *5.3 Real time integrated decision support tools*

This process of integrating the data, applying machine algorithms, and generating lists of risk animals will be continuous, dynamic, and interactive. With more data, prediction accuracy will continue improving with the data selected. Nonetheless, the framework will be designed to test new data streams as they become available. The whole system will 'learn' as it goes, and therefore the projections and decisions will improve over time (Fig. 2).

## 6 Conclusion

Dairy herd health can be better managed by applying real-time, data-driven decision support tools based on integrated data coming from different sources. Artificial intelligence techniques add valuable insights to map complex connections that determine herd health. These techniques can successfully be incorporated on dairy farm management if data are properly collected, aggregated, integrated, and analysed. Highly sophisticated methods to manage larger amounts of data are available and should be used to help decision-makers take effective actions timely. The concept of developing a

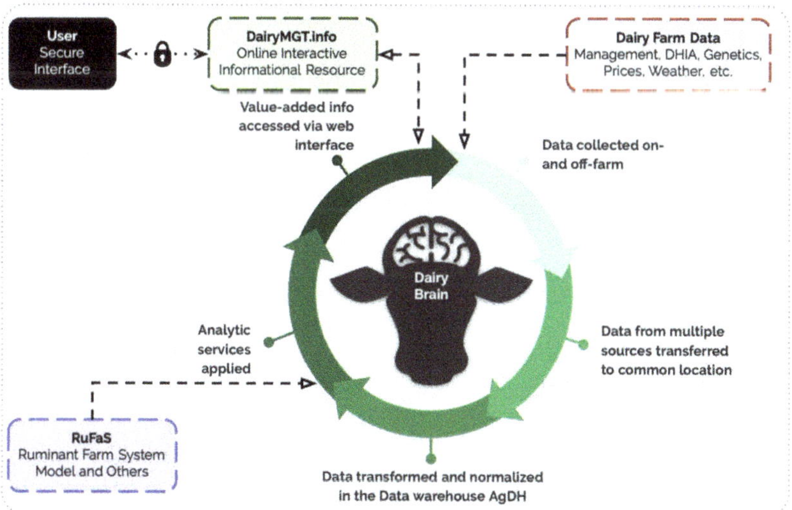


**Figure 2** Envisioned process of health decision support tools on the DairyMGT.info using integrated data from the Dairy Brain and predictive and prescriptive models. DHIA = dairy herd improvement association records; AgDH = Dairy Brain project anticipated agricultural data hub (data warehouse).

real-time, integrated, suite of data-driven descriptive and prescriptive decision support tools (e.g. the Dairy Brain), a whole farm system model (e.g. RuFaS), and an online platform to interact with users (e.g. the DairyMGT.info) are well developed, but these need to first, become fully functional, and second, be completely connected among themselves. Once these systems are totally operative and combined, their use and application will likely transform the way dairy farmers manage health-related issues for better. Their effective application will improve animal welfare, decrease health-related costs, increase overall herd profitability. The integrated structure could become the centre-piece for dairy farm management, informing farmers and farm managers the best actions for the short-, medium-, and long term regarding critical health management decisions and its interaction with others management decisions such as nutrition, breeding, treatments, replacement, crop systems, and environmental management.

## 7 Where to look for further information

Further live information can be found at the following live websites: the University of Wisconsin-Madison Dairy Management website (https://DairyMGT.info), the University of Wisconsin-Madison Dairy Brain project (https://DairyBrain.wisc.edu), and the US Ruminant Farm System Model (RuFaS) initiative (https://RuFaS.org).

## 8 References

Bewley, J. M. and Russell, R. A. 2010. Reasons for slow adoption rates of precision dairy farming technologies: evidence from a producer survey. In: *Proceedings of the First North American Conference on Precision Dairy Management, Toronto, Canada*.

Bewley, J. M., Boehlje, M. D., Gray, A. W., Hogeveen, H., Kenyon, S. J., Eicher, S. D. and Schultz, M. M. 2010. Stochastic simulation using @Risk for dairy business investment decisions. *Agricultural Finance Review* 70(1):97–125.

Cabrera, V. E. 2012a. A simple formulation and solution to the replacement problem: a practical tool to assess the economic cow value, the value of a new pregnancy, and the cost of a pregnancy loss. *Journal of Dairy Science* 95(8):4683–4698.

Cabrera, V. E. 2012b. DairyMGT: a suite of decision support systems in dairy farm management. In: *Decision Support Systems* Jao, C. (Ed.). INTECH, Rijeka, Croatia.

Cabrera, V. E. 2018. Invited review: Helping dairy farmers to improve economic performance utilizing data-driving decision support tools. *Animal* 12(1):134-144.

Cabrera, V. E., Barrientos-Blanco, J. A., Delgado, H. and Fadul-Pacheco, L. 2020. Symposium review: real-time continuous decision making using big data on dairy farms. *Journal of Dairy Science* 103(4):3856–3866.

Chawla, N. V., Bowyer, K.W., Hall, L. O. and Kegelmeyer, W. P. 2002. SMOTE: synthetic minority oversampling technique. *Journal of Artificial Intelligence Research* 16:321-357.

Delgado, H., Cue, R., Haine, D., Sewalem, A., Lacroix, R., Lefebvre, D., Dubuc, J., Bouchard, E. and Wade, K. M. 2017. Profitability measures as decision-making tools for Quebec dairy herds. *Canadian Journal of Animal Science*CJAS-2016-0202.

Delgado, H., Fadul-Pacheco, L. and Cabrera, V. E. 2019. The use of integrated data to identify first-lactation cows at high risk of clinical mastitis. *Journal of Dairy Science* 102(Suppl. 1):M134.

Dietterich, T. G. 2000. Ensemble methods in machine learning. In: *International Workshop on Multiple Classifier System*. Springer, Berlin, Heidelberg.

Giordano, J. O., Kalantari, A. S., Fricke, P. M., Wiltbank, M. C. and Cabrera, V. E. 2012. A daily herd Markov-chain model to study the reproductive and economic impact of reproductive programs combining timed artificial insemination and estrous detection. *Journal of Dairy Science* 95(9):5442–5460.

Groenendaal, H., Galligan, D. T. and Mulder, H. A. 2004. An economic spreadsheet model to determine optimal breeding and replacement decisions for dairy cattle. *Journal of Dairy Science* 87(7):2146–2157.

Hashem, I. A. T., Yaqoob, I., Anuar, N. B., Mokhtar, S., Gani, A. and Ullah Khan, S. 2015. The rise of "big data" on cloud computing: review and open research issues. *Information Systems* 47:98–115.

Heikkilä, A. M., Nousiainen, J. I. and Pyörälä, S. 2012. Costs of clinical mastitis with special reference to premature culling. *Journal of Dairy Science* 95(1):139–150.

Hogeveen, H., Kamphuis, C., Steeneveld, W. and Mollenhorst, H. 2010. Sensors and clinical mastitis–the quest for the perfect alert. *Sensors (Basel)* 10(9):7991–8009.

Kebreab, E., Reed, K. F., Cabrera, V. E., Vadas, P. A., Thoma, G. and Tricarico, J. M. 2019. A new modeling environment for integrated dairy system management. *Animal Frontiers: the Review Magazine of Animal Agriculture* 9(2):25–32.

LeBlanc, S. J., Lissemore, K. D., Kelton, D. F., Duffield, T. F. and Leslie, K. E. 2006. Major advances in disease prevention in dairy cattle. *Journal of Dairy Science* 89(4):1267–1279.

Leslie, K. E. and Petersson-Wolfe, C. S. 2012. Assessment and management of pain in dairy cows with clinical mastitis. *Veterinary Clinics of North America. Food Animal Practice* 28(2):289–305. doi:10.1016/j.cvfa.2012.04.002.

Liang, D., Arnold, L. M., Stowe, C. J., Harmon, R. J. and Bewley, J. M. 2017. Estimating US dairy clinical disease costs with a stochastic simulation model. *Journal of Dairy Science* 100(2):1472–1486.

Liang, D., Delgado, H. and Cabrera, V. E. 2018. A virtual dairy farm brain. In: *13th European International Farming System Association Symposium of the Farming and Rural Systems: Farming systems: Facing Uncertainties and Enhancing Opportunities*. Chania, Crete, Greece, 01–05 July 2018.

Madrigal, C. A. 2012. The Perfect Milk Machine - How big data transformed the Dairy Industry. *The Atlantic (online article)*. Available at: https://www.theatlantic.com/technology/archive/2012/05/the-perfect-milkmachine-how-big-data-transformed-the-dairy-industry/256423/. Accessed 20 October 2020.

Meadows, C., Rajala-Schultz, P. J. and Frazer, G. S. 2005. A Spreadsheet-based model demonstrating the nonuniform economic effects of varying reproductive performance in Ohio dairy herds. *Journal of Dairy Science* 88(3):1244–1254.

Mirando, M. A., Bewley, J. M., Blue, J., Amaral-Phillips, D. M., Corriher, V. A., Whittet, K. M., Arthur, N. and Patterson, D. J. 2012. Extension Education Symposium: reinventing extension as a resource--what does the future hold? *Journal of Animal Science*. 90(10):3677–3692.

Moore, K. and Thatcher, W. W. 2006. Major advances associated with reproduction in dairy cattle. *Journal of Dairy Science* 89(4):1254–1266.

Morota, G., Ventura, R. V., Silva, F. F., Koyama, M. and Fernando, S. C. 2018. Big Data Analytics and Precision Animal Agriculture Symposium: machine learning and data mining advance predictive big data analysis in precision animal agriculture. *Journal of Animal Science* 96(4):1540–1550.

O'Grady, M. J. and O'Hare, G. M. P. 2017. Modelling the smart farm. *Information Processing in Agriculture* 4(3):179–187.

Oltenacu, P. A. and Broom, D. M. 2010. The impact of genetic selection for increased milk yield on the welfare of dairy cows. *Animal Welfare* 19(1):39–49.

Rollin, E., Dhuyvetter, K. C. and Overton, M. W. 2015. The cost of clinical mastitis in the first 30 days of lactation: an economic modeling tool. *Preventive Veterinary Medicine* 122(3):257–264.

Rotz, C. A., Corson, M. S., Chianese, D. S., Montes, F., Hafner, S. D. and Coiner, C. U. 2013. Integrated farm system model: reference manual. USDA Agricultural Research Service, University Park (PA). [Accessed: February 3, 2020]. Available at: https://www.ars.usda.gov/SP2UserFiles/Place/19020500/Reference%20 Manual.pdf.

Rushen, J. and de Passill, A. M. 2013. The importance of improving cow longevity. In: Proceedings of the Cow Longevity Conference, Tumba, Sweden. DeLaval International, Tumba, Sweden, 28–29.

Shook, G. E. 2006. Major advances in determining appropriate selection goals. *Journal of Dairy Science* 89(4):1349–1361.

Schuetz, C. G., Schausberger, S. and Schrefl, M. 2018. Building an active semantic data warehouse for precision dairy farming. *Journal of Organizational Computing and Electronic Commerce* 28(2):122–141.

Tilman, D., Cassman, K. G., Matson, P. A., Naylor, R. and Polasky, S. 2002. Agricultural sustainability and intensive production practices. *Nature* 418(6898):671–677.

VandeHaar, M. J. and St. Pierre, N. 2006. Major advances in nutrition: relevance to the sustainability of the dairy industry. *Journal of Dairy Science* 89(4):1280–1291.

Wolfert, S., Ge, L., Verdouw, C. and Bogaardt, M. J. 2017. Big data in smart farming–a review. *Agricultural Systems* 153:69–80.

# Part 2

## Prerequisites

# Chapter 5

## Advances in understanding immune response in dairy cattle

*Bonnie Mallard, Mehdi Emam*, Shannon Cartwright, Tess Altvater-Hughes, Alexandra Livernois, Lauri Wagter-Lesperance, Douglas C. Hodgins and Heba Atalla, University of Guelph, Canada; Brad Hine, CSIRO Livestock & Aquaculture, Australia; Joshua Aleri, Murdoch University, Australia; and Andrew Fisher, University of Melbourne, Australia*



## 1 Introduction

### 1.1 A brief history of the role of cattle in immunology

The earliest accounts of immunology date back to 1549 with Dr. Wan Mizhai, a Chinese paediatrician, who used the practice of variolation (carried out by inserting or rubbing ground smallpox scabs or pus into scratches made in the skin) to inoculate children against smallpox (Plotkin, 2014). This method was also used in Africa and the Middle East before it was eventually introduced into England and North America in the 1720s. By 1770, physicians in England, Germany, France, and North America noticed that prior infection with cowpox, a similar but less virulent disease of cattle, rendered a person immune to smallpox. However, it was not until 1796 when Edward Jenner, a British physician, noted that milkmaids were generally immune to smallpox, and so scratched fluid from cowpox blisters into human skin, that the practice of vaccination began (Plotkin, 2014). In fact, he

* Bonnie Mallard and Mehdi Emam contributed equally to this work

http://dx.doi.org/10.19103/AS.2020.0086.06

coined the term 'vaccination' in one of his earliest publications on the topic from the Latin word 'vacca' (cow) or 'vaccinae', meaning, from the cow. In some areas the term 'calf lymph' was the name given to a type of smallpox vaccine in the 19th century, which was still manufactured in the USA up to the 1970s (Belongia and Naleway, 2003). Interestingly, although the practice of variolation has ceased, it has influenced the current day practices, such as 'Pox parties' where children are intentionally exposed to diseases like chickenpox, measles, and rubella in an attempt to gain immunity (Johnson, 2018). This practice is strongly discouraged by public health organizations but persists nonetheless.

Cattle have made other important contributions to the field of immunology, including the development of the first *Mycobacterium bovis* BCG vaccine for human tuberculosis in 1921 (Luca and Mihaescu, 2013). Also, in 1981 the first report of a biosynthesized polypeptide vaccine against foot and mouth disease virus (FMDV) using the VP3 protein expressed in *Escherichia coli* (*E. coli*) was made for cattle (Kleid *et al.*, 1981). Cattle and other farm animals also possess a substantial proportion of T cells expressing the γδ T-cell receptor, which helped to elucidate the role of these unique cells in host defence (Guzman and Montoya, 2018). More recently, Saini *et al.* (2003) discovered that cattle produce antibodies with ultra-long Complementarity Determining Region (CDR)-3 (Saini *et al.*, 2003). This seminal finding has allowed the production of bovine therapeutic broadly neutralizing antibodies (bnAb) with ultra-long CDRs to passively treat HIV-AIDS, rotavirus, and influenza (Sok *et al.*, 2017; Deiss *et al.*, 2019). No doubt the cow will continue to make important contributions to the field of immunology.

## 1.2 Cellular components of the immune system

The immune system of all mammals, including cattle, is a complex network of cells and molecules that have several key functions, including: 1 - surveillance and sensing of danger signals, 2 - protection against pathogenic infection and tumours, and 3 - maintenance of homeostasis. The two interconnected branches of the immune system, innate and adaptive host defence, differ in at least four ways: 1 - initiation of the response, 2 - clonal diversity, 3 - specificity of the response, and 4 - the magnitude of the memory or trained response in recurrent exposure to a foreign agent. Immune responses are classically initiated and directed via antigen presentation and the nature of the induced innate responses; specific adaptive immune responses follow if the foreign-infecting agent is not removed by innate mechanisms (Parkin and Cohen, 2001; Sattler, 2017). Granulocytes, monocytes, macrophages, dendritic cells (DCs), and natural killer cells constitute the cells of the innate system, and lymphocytes are the cells of the adaptive immune system. Cells of the innate immune system mainly recognize microbe-associated molecular patterns (MAMPs) via pathogen recognition receptors (PRRs). Toll-like receptors are a subgroup of PRRs that are classified into several types, each recognizing specific microbial structures. For instance, lipopolysaccharide (LPS) and lipoprotein are recognized by TLR-4 and TLR-2, respectively, on the surface of cells, whereas double-stranded RNA and single-stranded DNA are recognized in the endosome by TLR-3 and TLR-7, respectively. TLRs have been identified in cattle primarily by PCR (Fisher *et al.*, 2011); however, there are a limited number of reagents available for detecting these proteins, with the notable exception being an anti-bovine TLR2 antibody (Kwong *et al.*, 2011).

Macrophages and DCs become activated following the recognition of MAMPs and phagocytize the foreign agent or antigen. These cells are able to engulf microbes and present their antigens (Ag) on their surfaces via a highly polymorphic set of major histocompatibility complex (MHC) receptors. The Ag-MHC complex can then be recognized by B- and T-lymphocytes (Franken *et al.*, 2016). T-cells are further classified into three main sub-populations: helper, cytotoxic, and regulatory cells with many unique types of cells within each category. Cattle possess most of these cell types, but due to the lack of some reagents the complete list of T-cells is yet to be reported. Noteworthy is the fact that the proportion of various bovine T-cell subsets differ depending on bovine immune response genetics, particularly those expressing the γδ WC-1 T-cell marker (Hine *et al.*, 2012). Unique to cattle and other livestock species is that the γδ T-cells represent the major regulatory T-cell (Guzman *et al.*, 2014).

B-cells are generally classified into two broad categories: B-1 and B-2 cells. Following the activation, both types of B cells differentiate into plasma cells that produce antibodies (Ab). The self-renewing CD5+ B-1 cells produce the so-called natural antibodies that are present without previous exposure to antigen. In contrast, B-2 cells produce specific antibodies following antigen exposure. Both types of antibodies of various isotypes have been reported in cattle, but unlike in mice and humans it has not been possible to isolate and clone bovine B-1 cells (de Klerk *et al.*, 2018). Nonetheless, the proportion of bovine CD5+ B-cell differs based on the immune response phenotype (Hine *et al.*, 2012).

Although the immune system generally follows the above-mentioned pathways, the interaction between the host and the pathogen makes it difficult to predict the outcome of pathogenic infection. The multi-functionality of cells and molecules, numerous sub-populations and plasticity of cells, and redundancy in responses are a few examples that contribute to the complexity. Additionally, the genetics of the host, the site of infection, the virulence and dose of pathogen exposure, as well as internal and external stressors, influence the outcome of infection. These factors will be discussed in more detail with a focus on genetic and epigenetic regulation of the bovine immune system with the ultimate goal of selecting cattle for enhanced disease resistance.

## 1.3 Natural antibody (NAb)

Natural antibody (NAb) is often considered to represent a bridge between the innate and adaptive arms of the immune system (Ochsenbein and Zinkernagel, 2000). Antibodies are an essential component of the adaptive system, yet NAbs can respond quickly and without previous exogenous antigen stimulation or exposure (Baumgarth *et al.*, 2005). These antibodies are reportedly polyreactive, lack specificity, and often bind conserved epitopes such as carbohydrates, nucleic acids, and phospholipids (Briles *et al.*, 1981; Ehrenstein and Notley, 2010). NAbs play an essential role in complement activation, B-cell regulation, phagocytosis, antibody-dependent cellular cytotoxicity, and neutralization (Ehrenstein and Notley, 2010). In both mice and humans, these antibodies seem to be produced mainly from a CD5+ B1 cell lineage, although this has not been explicitly confirmed in cattle (Savage and Baumgarth, 2015). It is well accepted that NAbs are mainly isotype IgM, yet many studies explore the potential for NAb IgG and IgA, which are isotypes produced following class switch recombination (CSR). Researchers have shown that activation-induced cytidine deaminase (AID)-mediated expansion of the foetal bovine antibody repertoire was possible without previous exogenous antigen exposure (Liljavirta *et al.*, 2013).

NAb is often measured using an enzyme-linked immunosorbent assay (ELISA). Previous studies use antigens such as LPS, lipoteichoic acid (LTA), peptidoglycan (PTG), and keyhole limpet hemocyanin (KLH) to capture NAb. LPS, LTA, and PTG are surface molecules on gram-positive and gram-negative bacteria that are common in the environment. Due to the commonality of these antigens, cows may have been previously exposed, and antibodies measured may represent a specific antibody response. KLH has immunostimulatory properties and is a novel antigen from the giant keyhole limpet *Megathura crenulata* (Harris and Markl, 1999). Theoretically, this is an antigen that cattle would not have been exposed to within their lifetime and may allow a reliable measure of true NAb.

## 1.4 Broadly neutralizing antibody (bnAb)

Recently in cattle, bnAbs have been a growing topic of interest, especially in the context of viral and bacterial immunology. BnAbs are a group of antibodies that offer broad neutralization across different isolates of a virus (Sok *et al.*, 2017). The strength of these antibodies is measured by potency and breadth. Potency refers to the strength of neutralization, and breadth refers to the ability to neutralize across multiple strains and isolates (Walker and Burton, 2018). The ability to neutralize across pathogenic strains is a crucial feature of these antibodies and helps impede virus replication and mutation; similar characteristics are seen in antibodies against bacteria (Otto *et al.*, 2011; Sears *et al.*, 2017). These antibodies typically recognize a conserved epitope for broad neutralization to occur (Sok and Burton, 2018). Often the B cell is exposed to the antigen multiple times, driving rounds of somatic hypermutation in the CDR and framework regions (Kaushik *et al.*, 2009).

A major focus of the work using bnAbs from cattle has been with human immunodeficiency virus (HIV), and recently researchers were able to isolate a monoclonal bnAb against HIV (Sok *et al.*, 2017). One group focussed on immunizing cattle with an HIV trimer and then measured the antibody neutralization response (Sok *et al.*, 2017). Cattle that were immunized had a greater and quicker production of bnAbs (~48 days) than what had previously been seen in humans (~2 years) (Sok *et al.*, 2017). The production and potency of bnAbs in cattle are remarkable. In humans, it seems only certain individuals can produce these bnAbs and with lower efficiency and potency (Mascola and Haynes, 2013). Other studies on bovine bnAbs investigate the protection against the hepatitis C virus, influenza, and rotavirus.

Antibodies are comprised of a heavy (HC) and light chain (LC). The HC contains a variable region composed of three regions: CDR1, 2, and 3. Bovine CDR3 contains unique variable amino acid (aa) lengths compared to other species (humans 8-16 aa) (Shi *et al.*, 2014; Dong *et al.*, 2019). The most fascinating region is an ultra-long insert that is composed of 40–70 amino acids in cattle (Saini *et al.*, 1999; Haakenson, Huang, and Smider, 2018). This ultra-long insert creates a knob and stalk structure in the CDR3, where the knob will bind to the antigen (Haakenson *et al.*, 2018) (Fig. 1). Ultra-long CDR3s compose 10–15% of the total bovine antibody repertoire, and like bnAbs, it is suggested they are produced through AID-mediated mutations such as somatic hypermutation (Wang *et al.*, 2013; Deiss *et al.*, 2019). It remains unknown whether

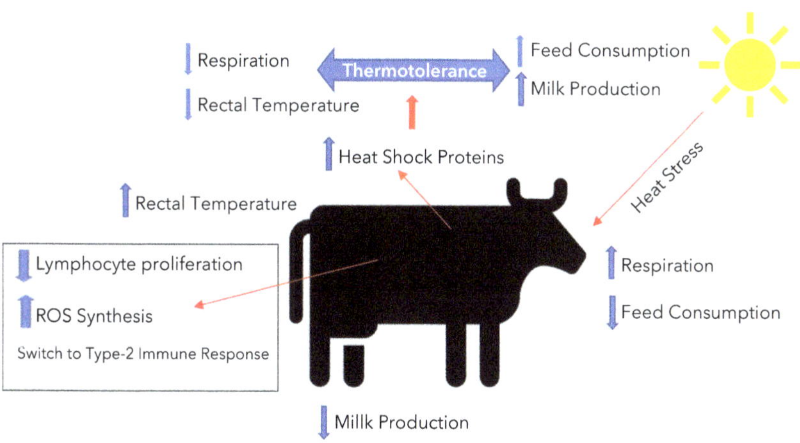


**Figure 1** Effects of heat stress and climate change on bovine immunity (Abdelnour *et al.*, 2019).

these antibodies are found equally across all isotypes, but one of the first studies on bovine CDR3 identified antibodies with ultra-long CDR3 that were restricted to IgM (Kaushik *et al.*, 2009). The ultra-long CDR3 plays an important role in antibody flexibility, neutralization, and reaching hidden epitopes that are not easily accessible (Sok *et al.*, 2017). These bovine antibodies are anticipated to have a great deal of across species therapeutic potential.

## 2 Genetics for dairy health

There is considerable variation in cattle in their diseases that provides scope for the effective selection of healthier and more productive animals. Identifying the most appropriate measure of disease resistance and its heritability, along with the precise selection goal, is critical to ensure the desired health outcome (Stear *et al.*, 2017). Resistance to infectious diseases is generally considered as complex quantitative traits. In fact, the immune system is regulated by about 5500 genes of the total ~23 000 genes within the mammalian genome. Improving resistance to infectious disease of cattle has been tackled genetically using three different approaches: 1 - by selecting for resistance to important diseases, using the information on the occurrence of disease (e.g. clinical mastitis) generally collected by the producer and/or veterinarian or indirectly using a substitute measurement (e.g. SSC of milk); 2 - by selecting for animals with strong innate and/or adaptive immune responses to a challenge with a vaccine, live organisms, or immunization with specified test antigens; 3 - by selecting for animals that perform well in environments where the disease is endemic (e.g. Johnes' disease). This later method yields animals that are disease tolerant but not necessarily disease resistant. For example, tolerant animals may themselves be healthy, even though they are infected and shedding the pathogen. Therefore, resistance is generally the preferred breeding goal. High-density SNP chips and improved sequencing methodologies lead to rapid improvement in the technical ability to conduct genome-wide association studies (GWAS) to the extent that the collection of phenotypic data is becoming the primary limitation. GWAS can be used directly in genomic selection and can also identify candidate genes for a detailed examination of their role in immunity.

Since the preferred breeding goal is most often disease resistance rather than disease tolerance, strategies to improve immune responses are ideal, given their moderate to high heritability, the accuracy of measurements, and the role they play directly in health (Groves *et al.*, 1993; Thompson-Crispi *et al.*, 2012b). Research in pigs, chickens, and cattle has clearly shown that immune responses can be improved by selective breeding (Groves *et al.*, 1993; Thompson-Crispi *et al.*, 2012a; Zhao, Honaker, and Siegel, 2012). Animals with inherently higher immune responses are generally healthier and more productive. This approach has been used in the patented High Immune Response (HIR™) technology and does not target a particular pathogen but is designed to facilitate improvements in broad-based disease resistance. Since antibody-mediated immune responses (AMIRs) and cell-mediated immune responses (CMIRs) are effective against different types of pathogens, it is advantageous to select for a balanced and robust AMIR and CMIR. The HIR method accomplishes balanced selection by estimating breeding values (EBVs) for these two traits and selecting animals that have the highest combined EBVs for both these traits. Identification of individuals with the most robust AMIR and CMIR also includes the ability to capture additional benefits, such as immunological specificity and memory, which can generate a quicker and greater response upon second and subsequent exposure to the foreign agent. Additionally, innate host defence mechanisms required for pathogen recognition and appropriate initiation of adaptive immunity are included in this type of selection. These immune response traits in cattle are moderately to highly heritable (Thompson-Crispi *et al.*, 2012a; de Klerk *et al.*, 2018). This is in line with other milk production traits, where genetic progress has been consistently achieved. Dairy cattle with the highest EBVs for AMIR and CMIR consistently have the lowest occurrence of disease in both research and commercial applications (Thompson-Crispi *et al.*, 2012b; Larmer and Mallard, 2016). Breeding for genetically enhanced immunity ensures the transmission of beneficial immune response genes to future generations. This method has a sound theoretical basis given the role of the immune system in maintaining health and the heritability of these traits, and it eliminates genes that cause suboptimal immune responses. The HIR technology has been marketed by the Semex Alliance under the tradename Immunity+ since 2013 and is well accepted by the global dairy industry to improve animal health (Larmer and Mallard, 2016).

## 2.1 Selection for immuno-competence in pasture-based production: the Australian experience

The dairy cattle industry in Australia comprises about 2.6 million cows, with approximately two-thirds of the population being Holstein Friesians. Although the majority of Australian dairy herds are housed outdoors in extensive pasture-based systems, an environment very different from the North-American intensively housed system, cows face similar disease challenges, especially during the periparturient period. A long-term genetic strategy aimed at reducing the incidence of diseases in livestock production systems is to breed animals with an enhanced ability to respond to disease challenges posed by their production environment. Placing selection emphasis on health and fitness traits, in conjunction with important productivity-related traits, has the potential to simultaneously improve the health, welfare, and productivity of livestock. The first step in developing a genetic strategy to improve any trait is to define the phenotype associated with the trait and to develop methodologies to assess an animal's phenotype.

Following the success of the HIR technology in North America to identify animals with enhanced general disease resistance (Mallard *et al.*, 2015), a modified immune competence testing protocol was developed, which was practical to apply in commercial Australian dairy herds. The use of a modified system was necessary to avoid restrictions imposed by using unregistered products in food-producing animals in Australia. To date, research undertaken in the Australian context has focused on investigating the associations which exist between immune competence phenotype, as determined using this modified methodology and stress-coping ability, feed efficiency, productivity, and disease incidence.

To validate the suitability of the modified immune testing protocol, we assessed the repeatability of secondary and tertiary immune responses measured in calves at 5–6 months of age (secondary response) and again at 12–13 months of age (tertiary response) (Aleri *et al.*, 2015). Repeated measures of AMIR were strongly positively correlated ($r = 0.651$, $P < 0.0001$); however, repeated measures of CMIRs were not correlated ($r = 0.078$, $P = 0.450$). Based on test results, animals were ranked for AMIR, CMIR, or combined AMIR/CMIR. The Cohen Kappa ($\kappa$) test of the agreement was used to test the consistency of the immune response category ranking of individual animals. The level of agreement for an animal to be identified as either a high or low responder across the two test periods was moderate to high ($\kappa = 0.445$) for AMIR, low for CMIR ($\kappa = -0.055$), and moderate for overall immune competence (combined AMIR/CMIR, $\kappa = 0.395$). Although the influence of age at the time of testing on the repeatability of immune responses observed could not be determined in the study, age has been shown to influence both the magnitude and the type of immune responses induced by vaccination (Hine *et al.*, 2011). Results highlight the need for animals to be of a similar age and with an identical vaccination history when using this methodology to assess the immune competence phenotype of cattle.

Investigations have been undertaken to assess the relationship between immune competence phenotype and traits related to health (e.g. somatic cell count (SCC), odds of being culled for mastitis and worm egg count (WEC)), stress-coping ability (cortisol responses to handling), growth (average daily gain (ADG)), productivity (milk production and composition), and feed efficiency (residual feed intake) in Australian pasture-based production system (Aleri *et al.*, 2015, 2016, 2017, 2019). The incidence of disease increases dramatically in many production systems at times when animals are exposed to management-induced stressors. Examples of such times include the periparturient period (Aleri *et al.*, 2016) and the period around feedlot induction (Taylor *et al.*, 2010).

Therefore, investigations were undertaken to investigate the association between immune competence and stress-responsiveness, as assessed by changes in serum cortisol concentrations induced by mustering, yarding, and handling (Aleri *et al.*, 2019). Animals with below-average AMIR and CMIR had higher serum cortisol concentrations following handling when compared to their counterparts, suggesting that they had a reduced ability to cope with management-induced stress. Furthermore, a significant negative (favourable) correlation was observed between antibody production in response to vaccination and stress-responsiveness. Favourable associations between stress responses induced by handling and temperament have been reported previously in beef cattle (Fell *et al.*, 1999), suggesting that immune competence and temperament may also be favourably correlated.

Associations between immune competence and feed efficiency have also been investigated using animals representing the extremes of high and low feed conversion efficiency genotypes (Aleri *et al.*, 2017). No significant differences in AMIR or CMIR were observed in extreme high and low feed conversion efficiency genotype animals. Associations between immune competence and productivity have also been investigated in first lactation dairy heifers by assessing milk yield, milk fat, and milk protein for 305 days (Aleri *et al.*, 2019). Although milk yield and AMIR and overall immune competence were not correlated, the correlation between milk yield and CMIR trended towards favourable. No significant correlations were observed between overall immune competence and milk yield, milk fat, and milk protein content, confirming the previous finding by the Mallard group that AMIR and CMIR are not strongly correlated with milk production traits. Associations between growth rate and immune competence phenotype in dairy calves have also been investigated, with results showing that calves categorized as having a high AMIR phenotype had a greater ADGs than did their low-responder counterparts. However, ADGs in immune competence phenotype groups based on CMIR or overall immune competence were not significantly different. The increased growth associated with high immune responders was also evident in earlier studies in pigs selected for high and low AMIR and CMIR.

It is well documented that the HIR phenotyping methodology used to assess immune competence can accurately predict health outcomes (Thompson-Crispi *et al.*, 2012a). In an effort to validate the ability of the modified phenotyping method developed to predict health outcomes in Australian pasture-based dairy systems, the relationship between immune competence phenotype, WEC, and SCC was investigated. Internal parasites pose a significant threat to livestock grazing on pasture, and the level of infection is commonly monitored by quantifying numbers of worm eggs in fresh faecal samples. While SCC is commonly used as an indicator of general dairy cow herd health, results from these studies demonstrated that AMIR and WEC were negatively (favourably) correlated. However, no correlation was observed with CMIR or overall immune competence and WEC or between immune competence phenotype and SCC (Aleri *et al.*, 2019).

Clearly more work will be required to validate the ability of the modified immune competence phenotyping methodology to predict health outcomes in Australian commercial pasture-based dairy production systems. However, results to date suggest that favourable correlations exist between immune competence, the ability to cope with management-induced stress, ADG, and WEC, highlighting the potential benefits of selecting animals based on their immune responses within the Australian pasture-based production system.

## 2.2 Genomic selection for health

Genomic selection has also proven useful for rapidly improving production, as well as immune response traits. It is possible to identify unique SNPs associated with high and low AMIR and CMIR using the Illumina 50K SNP panel (Thompson-Crispi *et al.*, 2014). A genomics test for Immunity+ has been available since 2018 and is offered commercially for Holsteins as part of the Semex Elevate platform. Several other companies have introduced genomic tests for disease resistance of cattle that are based on disease records rather than immune response traits. The best known is the Zoetis Wellness Index, which is part of their Clarified programme available in North America (Dairy Business, June 12, 2018, www.dairybusiness.com). There is also a genomics test to reduce tuberculosis in the UK, which became available for Holsteins following research at the Roslin Institute, in conjunction with Genus ABS and the UK Dairy Genetics Advisory Forum (Agriculture and Horticulture Development Board, January 8, 2016, www.dairy.ahdb.org.uk). This genomics test, known as the TB Advantage, shows genetic variation between animals in response to the tuberculin skin test. Like some other indexes the scale ranges from −3 to +3, with a positive number predicting daughters with greater resistance to tuberculosis. For every +1 point, 1% fewer daughters are expected to become infected. The Canadian Dairy Network, now known as Lactanet, is also expected to launch a genomics test for disease resistance in Holsteins. The main drawback of the above tests is that the heritability of the targeted traits is low, generally 10% or less, and the disease information is obtained from on-farm records that are not necessarily scored using the same criteria by each producer or veterinarian. This approach may work best in Scandinavian countries where all disease treatments are given and recorded by trained veterinarians. However, in Canada, this practice differs depending on the province. For example, in Quebec, disease records are transmitted by the veterinarian and coding is differentiated if a disease is recorded by a veterinarian or another person. Ideally, this will eventually be extended to other provinces as well.

## 2.3 Reductionist models in understanding genetic regulation of disease resistance

Resistance to infectious diseases is the outcome of interactions between the host immune system, pathogens, the environment, and the host microbiome. In addition to the interactions between these four main elements, there is notable complexity within each element (Goh and Knight, 2017). For instance, the response of the immune system to the pathogen is generated through an intricate network of cells and molecules. These responses in cattle are controlled by more than 5000 genes (ENSEMBL, Release 98). Moreover, redundancy in responses is one of the features of the immune system. The redundancy in response is a strategy of the host to maximize the chance of mounting a protective response in case of failure of a minor pathway or molecule (due to genetic mutation in the host or immune evasion mechanisms of pathogens) (Fischer and Rausell, 2018). Therefore, when the phenotype is defined based on the outcome of infection in the host (or, in other words, when all the above-mentioned elements are contributing to the phenotype), the detection power in identifying the association between the genetic polymorphism and the phenotype is strongly compromised. The low heritability of health phenotypes, despite the sample size and all efforts to remove non-genetic effects, indicates that the majority of the genetic contribution to the phenotype is still missing (Emam *et al.*, 2019a).

Along with technological advances, linking quantitative genetics/genomic methods with cell-based functional immuno-phenotyping experiments has recently been introduced as an alternative approach (Ko and Jaslow, 2014). Due to the nature of the cell-based functional experiments, the results of these experiments are more accurately recorded, less subjective and less affected by environments. In addition, the cell-based phenotypes are relatively simpler than whole-host-based phenotypes and controlled by a limited number of genes with large effects (Miller and Chaudhary, 2016).

This strategy has been employed to study the genetic regulation of resistance to *Salmonella typhi* (*S. typhi*) infection and sepsis in humans. Alvarez and her colleague reported the association between a genetic polymorphism that affects the expression of *Vac14* genes and subsequently the cholesterol level on the cell surfaces, which can facilitate *Salmonella* docking and invasion. In this study, they used lymphoblastoid cell lines (LCLs) and the percentage of infected cells was used as the phenotype. The heritability of this cellular trait was estimated to be 0.47. The association of findings of this cellular trait with human typhoid fever was also confirmed using an independent population from Vietnam (Alvarez *et al.*, 2017). An independent study has also confirmed these findings in African children (Gilchrist *et al.*, 2018). Genome-wide association of quantitative trait loci with the level of cytokine expression by different cell types in response to various pathogens has also been investigated (Li *et al.*, 2016). In this study, the heritability of Interleukin (IL)-1b expression by blood mononuclear cells (BMCs), 24 h after stimulation with *E. coli*, has been estimated to be 0.70. Moreover, the heritability of expression of IFN-γ and IL6 by BMCs in response to *Borrelia burgdorferi*, *Candida albicans*, CpG, *Mycobacterium tuberculosis*, and *Staphylococcus aureus* was estimated to be 0.999. These results indicate the robustness of the reductionist approach in identifying the genetic effects on immune phenotypes.

In dairy cattle a similar approach has been explored (Emam *et al.*, 2018). The authors estimated the heritability of respiratory burst (as measured by the concentration of nitric oxide ($NO^-$) in the supernatant of monocyte-derived macrophage (MDM) culture in response to *E. coli* to be between 0.77 and 0.99 (Emam *et al.*, 2019b). In addition, there was a strong, significant, and positive correlation between respiratory burst and phagocytosis ability of bovine MDMs. A positive correlation has also been reported between the respiratory burst in response to *E. coli* and *S. aureus*, indicating that the genetic effect on respiratory burst has a broad-range effect. Further study by this group revealed a genome-wide effect of host genetics on the expression of more than 500 genes in MDMs. Among the differentially expressed genes, some key transcription factors such as IRF1 and STAT1, HIF1-α have been identified. The transcriptomic expression of MDMs, which had been classified as high and low responders based on $NO^-$ in response to *E. coli*, revealed that the effect of host genetics on macrophages extends beyond the respiratory burst and includes the overall inflammatory capability of MDMs (Emam *et al.*, 2020). A GWAS revealed 52 loci that are significantly associated with the macrophage's ability to mount oxidative burst in response to *E. coli*. Some of the loci described more than 30% of the phenotypic variation, individually. Functional annotation analysis revealed that the response to hypoxia following phagocytosis is the candidate metabolic pathway that shapes the cellular phenotype (Emam et al., 2020). Preliminary results from a cross-validation study demonstrated a significant association between this cell-based phenotype and the incidence of clinical mastitis in dairy cattle (unpublished results).

Despite the complexity of the immune system, reductionist approaches have shown promising results for identifying the genetic effects not only at the cellular level but also on the outcome of disease in the host. This notable effect might have been associated with the fundamental roles of the pathways that can be detected using the cellular immuno-genomic approach. This method is likely to be more widely applied in the future.

## 3 Epigenetics

### 3.1 Introduction to epigenetic regulation of the immune system

C. H. Waddington first described 'epigenetics' in 1942 as phenotypic changes without genotypic change (Waddington, 2012). The definition has been extended to include transient and stable 'structural adaptation of chromatin regions so as to register, signal or perpetuate altered activity states' (Bird, 2007). Epigenetic modification of the genome encompasses covalent and noncovalent modifications to DNA and its associated proteins that make up the chromatin fibre. DNA is wound around nucleosome protein complexes, which are each made up of an octamer of four core histone proteins: two each of H3, H2A, H4, and H2B. The nucleosomes are further wrapped into a solenoid, and H1 linker proteins are added to maintain the chromatin structure. The tightness and configuration of how DNA is wound around nucleosomes, and the density of nucleosomes, determine the DNA accessibility to RNA polymerase II (pol II), thus regulating transcription.

The plasticity of the epigenome allows for changes in the chromatin structure in response to external stimuli, thereby playing a fundamental role in biological and pathological processes. The structure and organization of chromatin are dynamic and thereby regulate fundamental nuclear processes, such as DNA replication and repair, cell division and growth, and gene transcription. Chromatin can exist in two major categories of higher-order structure: euchromatin and heterochromatin. Euchromatin is open and transcriptionally accessible and contains genes that are constitutively required, such as housekeeping genes. In contrast, heterochromatin is compact and inaccessible to transcription. Heterochromatin can be further subdivided into constitutive and facultative heterochromatin. Constitutive heterochromatin is found at centromeres and telomeres and where repeat sequences and transposons are located in eukaryotic cells. Facultative heterochromatin represses transcription in higher eukaryotes but, in contrast to constitutive heterochromatin, can lose its condensed structure and become transcriptionally active under specific developmental or environmental cues (Oberdoerffer and Sinclair, 2007).

The epigenome is unique for each cell type and is established and maintained through development and differentiation. The most well-studied epigenetic modification in mammals is DNA methylation, which is classically associated with genomic imprinting and X chromosome inactivation but is integral to transcriptional regulation across the entire genome. Reversible chemical modification of the tails of histone proteins affects gene expression positively or negatively by changing how the histones interact with or bind to the DNA. A cell's epigenome also includes a collection of non-coding RNAs (ncRNAs), including long non-coding RNAs (lncRNAs) and microRNAs (miRNAs) that play an indispensable role in regulating cellular processes. Perhaps the most well-known ncRNA is Xist, the lncNC RNA-regulating X chromosome inactive in placental mammals.

The immune system is prepared to identify, respond to, and eradicate invading pathogens from the body. It is now clear that epigenetic mechanisms play an integral role in the complex signalling pathways by sensing pathogens and activating infection-specific transcription patterns to support the immune response and inflammation. The innate immune system responds quickly to invading pathogens by maintaining high expression levels of positive signalling transducers and regulators and also maintains homeostasis by expressing negative signalling regulators (reviewed in Zhang and Cao, 2019).

The chromatin landscape is integral to providing a memory of past transcriptional events. Immunological memory is the ability of the immune system to respond faster to a pathogen that has been previously encountered. Cells of the adaptive immune system - B and T cells - circulate through the body until those specific to an invading pathogen are stimulated to proliferate, differentiate, and respond. B and T cells for a specific pathogen are retained as memory B and T cells, which are 'primed' to respond with a more rapid and effective transcriptional response upon a second exposure to the pathogen (reviewed in Holloway and Woodworth, 2017).

Epigenetic mechanisms are also integral to immune cell lineage specification, development, and immune effector function of mature cell types. Indeed, immune cells develop from a common multipotent progenitor - the haematopoietic stem cell (HSC) - in the bone marrow (reviewed in Kondilis-Mangum and Wade, 2013). HSCs differentiate into three main lineages: red blood cells, innate host defence cells, and cells of the adaptive immune system (Seita and Weissman, 2010). Epigenetic mechanisms play a role in the stepwise development of each lineage by tuning the expression of genes for the development of the appropriate lineage. Knockout (KO) studies in mice have revealed genes that are integral to HSC development and function, including DNA methylating enzymes, histone modifiers, and genes involved in microRNA synthesis (reviewed in Cullen *et al.*, 2014). These epigenetic mechanisms also underlie bovine immune responses.

## 3.2 Chromatin remodelling in cells of the bovine immune system

Immunity is the result of coordinated responses among different cells of the immune system (Emam *et al.*, 2019a). Although these cells all originated from HSCs, they exhibit diverse phenotypic and functional characteristics (Sawai *et al.*, 2016). In addition, it is well documented that these characteristics change after exposure to pathogens when the cells of the immune system are actively responding to pathogens. For instance, monocytes transform into macrophages as they leave blood vessels and migrate to tissues during inflammation (Glass and Natoli, 2015). These phenotypic and functional transformations require the meticulous regulation of gene expression during the maturation process, quiescent phase of steady-state, active responding to pathogens, and healing process (Bossel Ben-Moshe *et al.*, 2019). The regulation of gene expression is coordinated via the accessibility of chromatin, also known as chromatin state (Thurman *et al.*, 2012). Specific epigenetic marks, or a combination of marks, are associated with active chromatin and others with silent chromatin. For example, a combination of H3K4me1 and K3K27ac are found on active enhancers, whereas the combination of H3K4me1 and H3K9me3 is found on inactive enhancers. The combination of the following six markers can represent 18 different chromatin states: H3K4me1, H3K4me3, H3K27ac, H3K27me3, H3K36me3, H3K9me3 (Ernst and Kellis, 2017). Changes in chromatin state result in phenotypic alterations to the cells of the immune system and consequently to livestock resistance or susceptibility to diseases. Improved understanding in this area will allow more precise management (identifying novel targets for medical intervention) and breeding of livestock more accurately (revealing the source of missing heritability in complex traits, such as health).

The epigenome is specific to the cell type and the physiological state of the cells. Therefore, investigating the epigenomic regulation of an immune response is complex and requires coordinated international efforts to put pieces of the puzzle together to illustrate the blueprint of epigenomic regulation. Such efforts in human were initiated by the ENCODE project (Encyclopedia of DNA Elements), and international consortiums such as FANTOM (coordinated by RIKEN, Japan) are currently leading the investigation of this field (ENCODE Project Consortium, 2004; de Hoon *et al.*, 2015). Considerably less is known about epigenomics in non-model and livestock species, but as more genomes are becoming publicly available, the area of epigenomics in animal science is gaining attention. AgENCODE in France and FAANG (Functional Annotation of Animal Genomes) are the current leaders of the epigenomic regulation of complex traits, including animal health (Giuffra and Tuggle, 2019), making epigenomic data available for public use. For example, the chromatin accessibility of bovine T cells is available via AgENCODE and FAANG. These data were generated using Assay for Transposase-Accessible Chromatin using sequencing (ATAC-seq). Recently, ChIP-seq and whole-genome bisulphite sequencing (WGBS) were used to elucidate changes in the chromatin state of bovine alveolar macrophages (BAMs) in response to *Mycobacterium bovis* (*M. bovis*.) infection by MacHugh's Laboratory at the University College Dublin, Ireland (Hall *et al.*, 2017; O'Doherty *et al.*, 2019). This same group also demonstrated that immune response genes in macrophages are intermediately methylated in response to *M. bovis*. Thomas Hall and his colleagues reported the presence of H3K4me3 and RNA polymerase II at the transcription start site of *Arginase 2* (*Arg2*) in infected macrophages, indicating transcriptional upregulation. Indeed, the expression of *Arg2* has been shown to be associated with M2-like phenotype of macrophages (Hall *et al.*, 2017).

A multi-omics approach (e.g. RNA-Seq, ATAC-seq, and ChIP-seq) will be instrumental in identifying the chromatin remodelling of macrophage response to infection. Further linking these data with genetic markers identified in GWAS will help elucidate the molecular mechanisms of disease resistance, which could be used in breeding programmes to improve health in cattle.

The spatial shape (three-dimentional (3D) shape) of the chromatin can also alter the gene expression and function of the cells of the immune system. This level of regulatory mechanisms is challenging to elucidate because the analysis of the 3D chromatin state at the genome-wide scale is required. Chromosome conformation capture (also known as 3C) techniques are the methods of choice at this time to study trans-regulatory mechanisms (Sati and Cavalli, 2017). The development of 3C methods to Hi-C, which can analyse the 3D chromatin state at a genome-wide scale, has recently been used in livestock species, such as chicken, goat, pig, and cattle, to study regulation in liver cells (Giuffra and Tuggle, 2019). As such, cells of the immune system are yet to be studied.

The effect of genetics on epigenomic regulation is also a novel concept in the immuno-epigenomic field. In 2018, quantitative trait loci (QTL) associated with chromatin accessibility (caQTL) in human macrophages in response to IFNγ and *Salmonella enterica* serovar *typhimurium* were reported. These findings indicate that chromatin accessibility and expression of some transcription factors, such as PU.1, are associated with SNPs in the human genome (Alasoo *et al.*, 2018). Considerably less omics data are available for livestock species. However, the mRNA expression of IRF1, STAT1, and HIF1α, key transcription factors in regulating macrophage inflammatory responses, has been shown to be associated with host genetics in Holsteins (Emam *et al.*, 2020). Additionally, a candidate gene approach study in Holsteins demonstrated an association between host genetics and methylation of CpG islands in the promoters of *IFN-gamma* and *IL-4* (Paibomesai, 2017).

### 3.3 MicroRNAs and bovine immunity

Small ncRNAs are integral to the precise control of gene expression that is essential to the immune system. MicroRNAs (miRNAs) are a class of ~22 base pair small ncRNA molecules that regulate the gene expression of diverse biological processes through both translational repression and mRNA degradation (Huntzinger and Izaurralde, 2011). The location and number of target sites within the 3′UTR of the target mRNA influence the miRNA's function (Bartel, 2004). Most of our understanding of the role of miRNAs in immune system regulation comes from studies in humans and mice. For example, miRNA-155 (miR-155) was shown to play a role in adaptive immunity, including the regulation of B and T cell functions and antibody production (O'Connell *et al.*, 2012). MiRNAs are also important for regulating innate immune system activation; human miR-146a targets tumour necrosis factor receptor-associated factor 6 (TRAF6) and interleukin-1 (IL1) receptor-associated kinase (IRAK1), which control innate immune signalling in response to LPS (Taganov *et al.*, 2006; Gantier, 2010).

miRNAs are also integral players in regulatory networks of HPC fate decisions and functional responses (Ivey and Srivastava, 2010; Ebert and Sharp, 2012) and can regulate several targets in a network in a cooperative or antagonistic fashion (Na and Kim, 2013). For example, miR125b, miR126, and miR-196b have been shown to influence HPC differentiation into myeloid and lymphoid lineages (O'Connell *et al.*, 2010; Rossi *et al.*, 2011). Readers are directed to Arnav *et al.* (Mehta and Baltimore, 2016) for an in-depth review of the role of miRNAs in regulating HPC fate decisions.

Evolutionary conservation of miRNAs implies their integral role in the genome across species (Siomi and Siomi, 2010), and many of the miRNAs identified in the human genome have orthologues in cattle (Lawless *et al.*, 2014). As such, the prediction of miRNA function in bovine disease can be inferred from human and mouse studies (Lawless *et al.*, 2014). For example, human miRNAs 155 (hsa-miR-155) is an identical homologue of bovine miRNAs 155 (bsa-miR-155). However, the identification of novel miRNAs will provide insight into the bovine immune system from their miRNAs repertoire because certain miRNA families have expanded or contracted based on pathogen-specific response profile (Lawless *et al.*, 2014).

The first group to use high-throughput sequencing to profile the expression of miRNAs in BAMs from Holstein-Friesian calves was Vegh *et al.* (2013). This work provided an important reference for miRNA expression in unchallenged BAMs, which are the primary target for infection by *M. bovis*, the causative agent of bovine tuberculosis (Pollock *et al.*, 2006). MiR-21 was the most highly expressed miRNA in this study, which is known to be an important regulator of antimicrobial peptides in human monocytes infected with *Mycobacterium leprae* (Vegh *et al.*, 2013).

Differential expression of miRNAs in bovine mammary epithelial cells (BMEs) in response to challenge with bovine mastitis-causing pathogens, such as *E. coli*, *S. aureus*, and *S. uberis*, indicated that miRNAs potentially act as central regulators of mobilizing the innate immune system (Lawless *et al.*, 2013). Several of the most highly expressed miRNAs have a role in immunity. For example, MiR-27b is known to negatively regulate the inflammatory response by destabilizing mRNA of the transcriptional regulator peroxisome proliferator-activated receptor gamma (PPAR-γ) (Jennewein *et al.*, 2010).

Cells of the immune system can communicate by the specific transfer of miRNAs from one cell to another via extracellular vesicles to prevent their degradation (De Candia *et al.*, 2016). MiRNAs have been shown to be abundant in bovine colostrum and milk, either free or enclosed within exosomes (Chen *et al.*, 2010; Izumi *et al.*, 2012, 2015). Immune-related miRNAs are highly expressed in milk, particularly in colostrum, suggesting that they are crucial for mammary gland immune regulation as well as promoting the development of the calf gut mucosal immune system (Liang *et al.*, 2015). The Mallard lab studied the bioactivity of bovine colostrum and milk exosomes containing miRNA affected to predict the effect on calf gut health. Metabolic activity after co-culture of the human intestinal epithelial cell line (Caco-2 cells) with colostrum and milk exosomes from high-immune-responder cows (based on estimated breeding values using the patented HIR technology) was significantly greater than those with exosomes isolated from low-responder cows ($P = 0.0198$). Viability of Caco-2 cells co-cultured with either colostrum or milk exosomes from high immune responders was significantly greater ($P < 0.0024$ and $P < 0.0048$) than from low responders at 72 h in culture (Ross, 2017). A similar observation was previously reported in porcine milk exosomes (Chen *et al.*, 2016).

Functional and pathway analysis using ingenuity pathway analysis (IPA) of miRNA high-throughput expression analysis from a recent miRNA study revealed that immune-related pathways were among the top-10 enriched pathways. The expression of important immune-related miRNAs (including miR-148a, miR-155, miR-21, miR-26a, and miR-29b) was validated with qRT-PCR in colostrum and milk exosomes, showing significantly ($P < 0.05$) higher expression of miR-155 in colostrum compared to milk exosomes. MiR-155 is known to play a role in regulating adaptive immunity in humans, including antibody production (O'Connell, Rao, and Baltimore, 2012). These results demonstrate that immune-related miRNAs are present in milk and colostrum exosomes in healthy cows and provide a foundation for future investigation of novel miRNAs that could be used as biomarkers for milk and colostrum quality control or disease surveillance.

## 4 Heat stress, climate change and immunity

One of the most significant stressors that dairy cattle currently encounter is environmental stress. Climate change is a serious issue that results due to increases in greenhouse gases that continue to enter the atmosphere. This leads to an overall warming of the atmosphere, causing extremes in temperature swings and an increase in the frequency of extreme weather events (Solomon *et al.*, 2007), making it difficult for dairy cattle to adapt to the constantly changing and warming environment. The overall increase in atmospheric temperature has led to increases in the heat stress in dairy cattle, which in turn can lead to increased disease and mortality (Bishop-Williams *et al.*, 2015). This increase in disease and mortality during heat stress is likely due, at least in part, to the impact heat stress has on the immune system (Fig. 1).

Heat stress can impact both the innate and adaptive immune systems. One of the effects of heat stress on the adaptive immune response is the disruption of the balance between T-helper 1 and T-helper 2 responses, causing a shift towards a T-helper 2 response (Salak-Johnson and McGlone, 2007). The glucocorticoid produced in response to heat stress is cortisol, which can bind to DNA and cause transcriptional inhibition of genes involved in the activation of T-helper 1 cells and pro-inflammatory cytokine production (Bagath *et al.*, 2019). This results in an impaired cell-mediated immune response, which is important for defence against intra-cellular pathogens, such as viruses. Similarly, it has also been shown that heat stress is involved in the down-regulation of pro-inflammatory or T-helper 1 cytokines and the upregulation of T-helper 2 or regulatory cytokines (Salak-Johnson and McGlone, 2007; Bagath *et al.*, 2019). This suggests a shift towards a T-helper 2 response or potentially a suppressive immune response.

Studies have shown that heat stress in dairy cattle leads to reduced lymphocyte proliferation (Bhanuprakash *et al.*, 2016; Bagath *et al.*, 2019). Lymphocytes that become activated when exposed to pathogens will rapidly proliferate in order to fight infection. Accordingly, reduced lymphocyte proliferation hinders defence against invading pathogens. Similarly, it has also been suggested that heat stress can cause impaired function in neutrophils (Evans *et al.*, 2016). The impaired function of cells could be attributed to cell damage or death because heat stress increases the production of reactive oxygen species and reduces antioxidant defence, leading to oxidative cell damage (Abdelnour *et al.*, 2019).

Another component of the innate immune system that is impacted by heat stress is complement activation. The genes that encode the components of the complement cascade, as well as factors B and H, become down-regulated during heat stress in dairy cattle (Min *et al.*, 2016). The complement system helps enhance antibody response and phagocytic cell activity; therefore, the impairment of this system makes it more difficult for cattle to clear invading pathogens.

Heat stress has been shown to have a negative impact on the immune system during late gestation in dairy cattle. Late-gestation dairy cows under heat stress have been shown to have an increased concentration of circulating prolactin, which causes a decrease in the expression of prolactin receptors. Prolactin is an important metabolite for lactogenesis. The negative feedback loop resulting from increased circulating prolactin has been associated with reduced lymphocyte proliferation and reduced milk production (do Amaral *et al.*, 2011). It has also been shown that relative to cooled cows, cells from heat-stressed cows during late gestation exhibit reduced oxidative burst and phagocytosis, which leaves these animals susceptible to invading pathogens (do Amaral *et al.*, 2011). Furthermore, heat stress also reduces specific IgG responses in late gestation compared to cows that have been cooled (do Amaral *et al.*, 2011), indicating that animals may not be able to respond as well to invading extracellular pathogens. Evidently, heat stress introduces additional strain to an already-challenged immune system in the pregnant cow, leaving them more susceptible to disease.

Calves are born with minimal immunity and therefore rely on immunoglobulins (Ig) in the maternal colostrum, transferred through the gut, for protection until their own immune system matures. Failure of passive transfer (FPT) means the calves have not been able to transfer an adequate amount of Ig from colostrum, and this leaves them susceptible to undesirable infection. Heat stress during late gestation has been shown to have an impact on the immune system of the calf in-utero. Calves born from heat-stressed dams have an increased FPT (Dahl *et al.*, 2016). It has been suggested that the increase in FPT could be due to lower IgG in the heat-stressed dam's colostrum. However, even when those calves were fed colostrum with high concentrations of IgG, they still exhibited low levels of IgG in their serum (Monteiro *et al.*, 2014). Therefore, it seems that FPT is a result of reduced absorption in the gut of the calf, and not from reduced Ig content in the heat-stressed dam's colostrum. Additionally, calves born to heat stress dams have reduced lymphocyte proliferation, which impairs their immune system and therefore their health and survival (Monteiro *et al.*, 2014; Dahl *et al.*, 2016).

Considering the negative impacts of heat stress on the immune system, it is not surprising that dairy cattle experience increased incidence of disease and mortality, which can even extend to their offspring as the temperature humidity index (THI) rises. Considering the current and future environmental challenges, more research is urgently needed to identify animals that are more resilient to the impacts of heat stress. One avenue for the investigation of increased heat stress resilience is the examination of heat shock proteins (HSPs), which have been identified as one of the major players in response to heat stress (Abdelnour *et al.*, 2019). HSPs aid in the repair and protection of cells during heat stress and can also help in the initiation of many immune responses (Moseley, 2000). Some studies have suggested that increased expression of HSP (specifically HSP70) may be associated with thermo-tolerance, or heat stress resilience, in cattle (Bhanuprakash *et al.*, 2016; Abdelnour *et al.*, 2019). Preliminary data from Holstein cows suggest that high immune responders, identified based on their EBVs for immune responsiveness, produce more HSP70 under heat stress than average or low immune responders (Cartwright *et al.*, 2019). Further studies are being conducted to determine if certain immune phenotypes in dairy cattle are associated with resilience to climate change, with further investigation of the production of HSP70 being one of the variables evaluated (Cartwright *et al.*, 2019).

## 5 Crossbreeding and immunity in dairy cattle

Inbreeding in the dairy cattle population continues to increase. In 2018 inbreeding in Canadian dairy cattle ranged from 2.95% for Milking Shorthorns to 8.45% for Canadiennes, with Holsteins sitting at 7.85% (Van Doormaal, 2019). Inbreeding depression causes many issues, but the greatest negative effects are seen with functional traits like health and survival (Hansen, 2000). Studies have shown an increased incidence of mastitis in more inbred individuals as well as reduced survival (Sewalem *et al.*, 2006; Sørensen *et al.*, 2006). Therefore, one solution to these issues might be crossbreeding. Studies have shown crossbred cattle have decreased disease occurrence and increased survival compared to purebreds (Touchberry, 1992; Sørensen *et al.*, 2008). This could be in part due to increased immune responsiveness observed in crossbred compared to purebred cattle. Second lactation crossbred cattle have been shown to have increased specific primary IgG response and reduced incidence of mastitis compared to purebred animals (Begley *et al.*, 2009a). Similarly, it was observed that crossbred calves had a greater specific IgM and IgG antibody response, as well as greater survival to calving, compared to purebreds (Begley *et al.*, 2009b; Cartwright *et al.*, 2011). These calves were later re-evaluated as first calf heifers, and similar results were found. Crossbred heifers had greater specific primary and tertiary IgG antibody response as well as a greater delayed-type hypersensitivity, indicating an enhanced cell-mediated immune response, compared to purebred heifers (Cartwright *et al.*, 2012). These studies provide evidence of enhanced immune responsiveness in F1 crosses compared to purebred animals.

One issue with crossbreeding is that heterosis is fully expressed in the F1 generation and therefore is not well maintained in subsequent generations unless new breeds are introduced (Sørensen *et al.*, 2008). This can make breeding decisions more complicated and could result in reduced performance of the F2 generation compared to the F1 if a new breed is not introduced (McDowell, 1985).

Overall crossbreeding may be a viable solution for improving immune responsiveness and overall health in dairy cattle even though it will require careful planning of rotational crosses. Only time will tell if the majority of dairy producers will adopt this approach in the future.

## 6 Colostrum and calf health

In cattle the syndesmochorial placenta does not mediate transfer of Ig from the blood of the dam to the foetus; calves are essentially agammaglobulinemic at birth. High concentrations of Ig, especially IgG1, are present in the dam's first milk (colostrum) at the time of calving. Calves which ingest colostrum within the first 24 h after birth can absorb colostral Ig intact into the blood circulation. These Igs ('maternal antibodies') potentially can protect calves against a spectrum of pathogens but decline in concentration in the blood over weeks or months, with a nominal half-life of 20 days (Murphy *et al.*, 2014).

## 6.1 Studies of colostrum management and failure of passive transfer

The importance of early ingestion of colostrum for calf survival, and the potential use of serum from mature cattle (administered orally and/or intravenously) to enhance the survival of colostrum-deprived calves, was noted as early as 1922 (Smith and Little, 1922a,b). Efforts to refine and communicate best management practices for administering colostrum to newborn calves continue to the present day; a recent, thorough review by Godden *et al.* (2019) summarizes current best practices (Godden *et al.*, 2019). Administration of a sufficient quantity (volume or mass) of colostrum of high quality in a timely fashion has been a consistent message to dairy producers for decades (Stott *et al.*, 1979a,b), with a more recent emphasis on colostral cleanliness, including low bacterial (Stewart *et al.*, 2005) and viral contents. Calves with serum IgG concentrations of less than 10 g/litre are considered historically to have FPT of maternal Ig. It has been estimated that FPT occurred in 19.2% of dairy calves in the USA in 2007 (National Animal Health Monitoring System, 2010); more recently FPT was estimated to occur in 12.1% of dairy calves in 2014 (Shivley *et al.*, 2018, data from National Animal Health Monitoring System, 2018). One factor that could contribute to a reduction in the failure of FPT on a national scale was the increased use of routine monitoring of 'total serum protein' (TS) concentrations by refractometry in calves 1-7 days of age. In 2007, 2.1% of dairy operations surveyed and routinely monitored TS, but this had increased to 6.2% by 2014. More importantly, in 2007, 14.5% of large dairy operations (500 cows or more) routinely monitored calves, but by 2014 this had risen to 38.3% of large operations. In the context of large dairy operations, 53.9% of heifer calves were raised in operations that routinely monitored TS in calf sera. Thus larger dairy operations (which in recent years constitute an increasing proportion of the national dairy herd), through routine testing, are better able to detect and remedy weaknesses in their protocols for colostrum feeding to newborn calves.

The volumes of colostrum recommended for feeding newborn calves have been increasing in recent years. It should be noted that there is a physiological/anatomical limit to the amount of colostrum that can be fed, by bottle or by oesophageal feeder, to newborn calves. Anderson (2011) has suggested that feeding excessive volumes of colostrum at one time can predispose to rumenitis, systemic acidosis, and proliferation of *Clostridium perfringens* (Anderson, 2011). Feeding colostrum containing higher concentrations of IgG would alleviate the need to feed large volumes at each feeding.

## 6.2 Genetics as a means of increasing colostrum quality and absorption

A supplementary approach to increase the concentration of IgG in colostrum is the use of selective breeding strategies. Fleming *et al.* (2016) have reported significantly higher concentrations of IgG in colostrum from cows with higher EBVs for serum antibody responses to a type-2 antigen (Fleming *et al.*, 2016). Kiser *et al.* (2019) used GWAS in a herd of 345 Jersey cows to identify genes associated with the concentration of IgG in colostrum and the volume of colostrum at the time of first milking (Kiser *et al.*, 2019). Seven loci were identified that were associated with the quantity of colostrum produced, and one locus was identified that was associated with the concentration of IgG in colostrum.

Immunohistological evidence suggests that in cattle the Fc receptor neonatal (FcRn) is responsible for the transport of bovine IgG1 from the blood into colostrum in the final weeks before calving (Cervenak and Kacskovics, 2009). Polymorphisms in the FCGRT gene encoding FcRn have been associated with variation in the concentration of IgG in colostrum of Holsteins (Zhang *et al.*, 2009) and beef breeds (Laegreid *et al.*, 2002). The heavy chain of the FcRn receptor forms a heterodimer with beta-2-microglobulin. Polymorphisms of B2M, the gene encoding beta-2-microglobulin, have been associated with variation in the absorption of IgG from colostrum into the bloodstream of newborn beef calves (Clawson *et al.*, 2004). These findings suggest that there are opportunities to improve the efficiency of passive transfer by means of selective breeding.

## 6.3 Periparturient vaccination to improve colostrum quality

Considerable efforts have been invested in improving protocols for administering colostrum to newborn dairy calves. A complementary strategy is to vaccinate the dams before calving in order to increase antibody titres (in their colostrum) to specific pathogens (Hodgins and Shewen, 1994; Reppert *et al.*, 2019). The goal is to provide immediate passive protection against pathogens responsible for early calf-hood diseases, even in calves that receive colostrum of suboptimal quality or in suboptimal quantity.

For this strategy to be of practical benefit: 1 – there must be commercial vaccines that target an infectious disease that is common during the first months of life; 2 – commercial vaccines must address relevant field strains of the target pathogens; 3 - vaccination must be capable of increasing antibody titres in the dams sufficiently that serum antibody titres in calves receiving suboptimal volumes or quality of colostrum will still attain protective (passive) antibody titres; 4 - protection against the target pathogen must be antibody-mediated. There are examples in the literature of vaccination to induce passive protection against experimental challenge (Snodgrass *et al.*, 1982; Rabinovitz *et al.*, 2012), but the demonstration of the principle in field studies is more elusive (Kohara *et al.*, 1997). There is also evidence that leukocytes present in fresh versus frozen colostrum contributes to the level of calf protection (Novo *et al.*, 2017).

## 6.4 Colostrum contains more than IgG

In addition to Ig, bovine colostrum contains cytokines, defensins, cathelicidins, lactoferrin (Stelwagen *et al.*, 2009), lysozyme, lactoperoxidase (Sordillo *et al.*, 1997), oligosaccharides, glycoconjugates (Gopal and Gill, 2000), and multiple components of the complement system (Zhang *et al.*, 2015), to name a few, all of which can mediate or modify immune defence. An analysis of the proteome of bovine colostrum has identified and quantified over 200 proteins; their roles in passive protection of calves and in directing immune system development await further investigation (Zhang *et al.*, 2015).

## 6.5 Cellular components of colostrum

Intermittent studies over the last three decades have documented the ability of live maternal lymphocytes in the colostrum of the dam to pass through the intestinal epithelium of their offspring and to modulate immune function, at least in the short term. The importance of these cells in immune protection of neonates against common pathogens remains unclear (see Godden *et al.* 2019, for a recent synopsis). Nonetheless, recent studies in humans have shown that maternal leukocytes from breast milk establish lifelong maternal chimerism, which appears pivotal to immune system development in infants (Molès *et al.*, 2018).

## 6.6 Pasteurization and heat-labile components of colostrum

In recent years pasteurization of colostrum has been advocated as a means of reducing bacterial contamination of colostrum (to enhance the uptake of colostral IgG) and to minimize the transmission of *Mycobacterium avium* subspecies paratuberculosis (MAP, the pathogen causing Johne's disease) and bovine leukaemia virus. It should be noted that pasteurization, although enhancing uptake of IgG and minimizing the transmission of pathogens, can also inactivate some non-immunoglobulin components that contribute to host defence, including maternal leukocytes.

## 6.7 Natural antibody (NAb) in colostrum

NAb has been detected in the serum, colostrum, and milk of cows (Mayasari *et al.*, 2015; Denholm *et al.*, 2018). In Holstein dairy cows, increased NAb IgM has been associated with decreased incidence of mastitis, which is often caused by bacteria such as *E. coli* and *S. aureus* (Thompson-Crispi *et al.*, 2013). A strong positive correlation (0.69–0.91) between serum and milk NAb in Holstein dairy cows was also noted (Denholm *et al.*, 2018). NAb in milk tends to increase with parity, but in serum, it remains constant (de Klerk *et al.*, 2015). Guidelines for ideal NAb concentrations in colostrum and milk remain undetermined, yet these can be transferred to the calf following colostrum and milk consumption (Mayasari *et al.*, 2015). Researchers noted that NAb concentrations in colostrum were positively correlated with NAb concentrations in the calf in the first 2 weeks of age (Mayasari *et al.*, 2015).

Using a GWAS approach, researchers found that bovine serum NAb IgG responses were associated with 23 SNPs located on chromosome 1, 20, and 21 (de Klerk *et al.*, 2018). There were no significant SNP associations with NAb IgM (de Klerk *et al.*, 2018). In another study, chromosome 21 was also associated with milk NAb IgG, with the use of a different antigen, PTG (Cordero-Solorzano *et al.*, 2019). Both NAb IgM and IgG have a relatively high heritability of 0.31 (± 0.065) and 0.27 (± 0.064), respectively (de Klerk *et al.*, 2018). Incorporating this information into breeding decisions could improve the breadth and time of response to pathogens.

## 6.8 Broadly neutralizing antibody (bnAb) in colostrum

A valuable feature of bnAbs is their presence in colostrum, although there is not extensive research in this area. In one study, researchers immunized cows with a veterinary rotavirus vaccine, which resulted in antibodies that showed neutralization across multiple strains of bovine rotavirus and even a human strain (Civra *et al.*, 2019). The use of hyperimmune bovine colostrum has been investigated as a prophylactic or therapeutic in studies using rodents, calves, and humans (Fayer *et al.*, 1989; Otto *et al.*, 2011; Sears *et al.*, 2017). Hyperimmune bovine colostrum contains increased concentrations of antibodies, and often they are pathogen-specific. Antibody concentrations generally increase through hyper-immunization and often neutralize a viral pathogen or agglutinate a bacterial pathogen. Hyperimmune bovine colostrum could be important for calf health in decreasing on-farm morbidity and mortality, especially with prevalent pathogens. Evaluation of bovine antibodies in hyperimmune colostrum following immunization is necessary to identify the presence of bnAbs and whether or not there are antibodies with UL-CDR3s.

Previous studies have reported that cows with high immune response have inherently greater concentrations of serum antibody and also have higher specific antibody and total IgG in their colostrum and milk. Therefore, the more specific antibody a dam can produce in response to a prevalent environmental pathogen, the better her calf can resist these diseases after the consumption of pathogen-specific antibody-rich colostrum. This is known as passive immunity. Additionally, milk from high-immune-responder cows, based on the HIR technology, up to 6 weeks postpartum, have more antibodies compared to the colostrum from average- or low-immune-response cows (Wagter *et al.*, 2000). Thus, colostrum from HIR cows could be banked and given to calves born of low or average immune responders, thereby improving calf health. Conversely, cows with a low or average adaptive immune response phenotype are not capable of producing high levels of specific antibody and thus cannot provide the same degree of passive immunity. These individuals cost the producer more money in managing disease costs. In a recent study, calves given colostrum with high concentrations of antibodies from their dams were also demonstrated to have higher concentrations of specific antibodies in their blood at 2 days of age (Wagter-Lesperance, 2017). Therefore, HIR cows provide enhanced passive immunity to their calves and are more able to protect their calves from the disease than animals that have a low or average immune response. This improved passive immunity is another advantage of genetic selection for improved overall immuno-competence.

## 6.9 Colostral defensins

Defensins are small cationic peptides (28–42 amino acids; ~2–10 kDa) produced by epithelial cells, macrophages, and neutrophils and are characterized by six cysteine residues that form three disulphide bonds. While defensins across species can be organized into three different sub-families of peptides according to their structure (α, β, θ), bovine defensin genes belong to the β subfamily (Meade *et al.*, 2014). The first β-defensin identified in cattle was TAP (tracheal antimicrobial peptide) (Diamond *et al.*, 1991), and 18 complete and partial bovine β-defensin (bBD) sequences have been identified since then through genomic sequence analysis and direct sequencing of isolated purified proteins from neutrophils (Roosen *et al.*, 2004). Antimicrobial mechanisms of β-defensins include: aggregation, pore formation, interference with cell wall synthesis, and prokaryotic membrane depolarization (Sahl *et al.*, 2005). Bovine β-defensins have been reported to have a direct anti-microbicidal effect against a broad spectrum of bacteria and have recently been described as 'natural antibiotics' useful for maintaining homeostasis and health by managing the cross-talk between host and resident mucosal bacteria at mucosal surfaces. Mucosal surfaces of the cow include the oral cavity, respiratory tract, digestive tract, reproductive tract, and the mammary gland. β-Defensins are both constitutive and inducible and can recruit immature dendritic cells and memory T cells to infection sites (Machado and Ottolini, 2015), thus providing a link between innate and adaptive immune responses. However, disturbances in the ability of defensins to provide a regulatory function may lead to susceptibility to bovine disease (Meade and O'Farrelly, 2019). The healthy mammary gland has been shown to produce β-defensins including TAP and LAP (lingual antimicrobial peptide), as well as DEFB1, DEFB4, and DEFB5, and their presence may help regulate the microflora and be beneficial for the ruminal and intestinal development of the calf (Meade and O'Farrelly, 2019). Microbial analysis of milk has shown the presence of bacterial populations, and changes in diversity of these microbe populations have been associated with the disease. A study (Wang *et al.*, 2015) has shown an abundant concentration of DEFB1 and DEFB2 defensins in bovine colostrum and milk when measured by ELISA. Additionally, DEFB2 was shown to have a higher concentration in fresh bovine milk compared to DEFB1, but further data and validation of defensin proteins, as measured by ELISA with other molecular methods, are required.

A total of 58 β-defensin genes within four clusters have been identified on chromosomes 8, 13, 23, and 27. Most β-defensin genes have a two-exon structure: the first encodes a prepropeptide, and the second exon encodes the mature peptide. It is important to note that the genes encoding β-defensins exhibit copy number variation (CNV) as a result of gene duplication events (Machado and Ottolini, 2015), and this has led to a large family of β-defensin peptides with diverse amino acid sequences. Genes associated with resistance against mastitis have been identified on chromosome 27 (Gurao, Kashyap, and Singh, 2017). β-Defensin gene CNV has a documented association with disease resistance, and altering gene-copy number may contribute to host defence. Defensins and other bioactive molecules found in bovine serum, colostrum, and milk warrant further investigation.

## 7 Conclusion

Cattle have played a critical role in the field of immunology, from the development of the first vaccines to modern-day use of bovine bnAb with UL-CDRs for the treatment of viral infections. They have also made important contributions to the basic understanding of immunological mechanisms, such as establishing the role of gamma-delta T cells in immune regulation and protection. Improvements to herd health have also benefits from continued insights stemming from the field of immunology. Examples of this include improved vaccines for mastitis and other infectious diseases of cattle, diagnostic tests based on the presence of antibodies such as the IDEXX test for Johne's Disease, and treatments to boost the bovine immune system, including Imrestor (Elanco), just to name a few.

High-producing dairy cattle are particularly prone to periparturient stress; however, the development of new vaccines, immune modulators, and breeding strategies are all contributing to maintaining animal health without compromising important production traits. Dairy cattle, particularly Holsteins, are also known to respond poorly to heat stress. With the advent of global warming, this is a mounting problem, particularly in the northern hemisphere where it has been reported that warming is occurring two times faster than in more southerly latitudes. Heat stress can impact both the innate and adaptive immune systems. One of the effects of heat stress on the adaptive immune response is the disruption of the balance between T-helper 1 and T-helper 2 responses, causing a shift towards a T-helper 2 response. Heat stress is associated with cortisol production, which can bind to DNA and cause transcriptional inhibition of genes involved in the activation of T-helper 1 cells and pro-inflammatory cytokine production. This results in an impaired cell-mediated immune response, which is important for defence against intra-cellular pathogens, such as viruses. Similarly, it has been shown that heat stress is involved in the down-regulation of pro-inflammatory or T-helper 1 cytokines and the upregulation of T-helper 2 or regulatory cytokines. Collectively these data indicate a shift towards type 2 immune responses, which can be associated with immuno-suppression. Heat stress needs to be carefully monitored during the peripartum period when immunity is already suboptimal. Currently, there is a window of opportunity to improve on-farm cooling management systems, as well as to select cattle that carry genes associated with heat tolerance. Studies have shown that heat tolerance is a heritable trait in dairy, but caution must be exercised as to not compromise the hard-earned gains made in health and production during this selection process. Of note is that recent studies demonstrate an interaction between particular bovine immune response genes associated with increased antibody production and better ability to regulate body temperature when the THI is above optimal. More study is warranted in this area to identify genes associated with overall resilience. In other words, to locate the genes associated with overall fitness, which includes immunity as well as adaptability to climate change. Genetic improvement takes time, but the gains made are permanent and passed on to future generations of offspring.

Healthy calves make the healthiest cows, and therefore it is critical to ensure that calves get the best colostrum. This is especially relevant to dairy calves, who, unlike beef calves, are not left with their mothers to suckle. Advances in colostrum management and a better understanding of FPT are supporting improved dairy calf health. Additionally, dairy cows classified as high immune responders using the HIR™/Immunity+ technology are known to have superior health genetics and produce colostrum with greater amounts of antibody compared to their herd-mates. High responders also meet the minimal criteria for optimal colostrum quality more often that their herd-mates. This gives dairy producers a new tool to identify the cows in their herds that have high immune response genetics and thus superior colostrum quality. Furthermore, these superior genes for immunity can be enriched in their herds through selective breeding, ensuring that these genes are passed to future generations of dairy cows.

## 8 Where to look for further information

The main consortiums on epigenomic investigations on livestock are FR-AgENCODE and FAANG. They maintain a database of the publicly available resources for further studies. They can be accessed from:

- http://www.fragencode.org/.
- https://www.faang.org/.

The BovReg is a project under the FAANG consortium. It is composed of international research groups aimed to expand the knowledge on the bovine genome at the functional level. The BovReg project can be accessed from:

- https://www.bovreg.eu/.

Every five to seven years, the United States Department of Agriculture provides an update report on animal health in the USA. It can be accessed from the National Animal Health Monitoring System (NAHMS), which offers a valuable source for strategizing animal health research.

## 9 References

Abdelnour, S. A., Abd El-Hack, M. E., Khafaga, A. F., Arif, M., Taha, A. E. and Noreldin, A. E. (2019). Stress biomarkers and proteomics alteration to thermal stress in ruminants: a review, *Journal of Thermal Biology* 79, 120–134. doi: 10.1016/j.jtherbio.2018.12.013.

Alasoo, K., Rodrigues, J., Mukhopadhyay, S., Knights, A. J., Mann, A. L., Kundu, K., HIPSCI Consortium, Hale, C., Dougan, G. and Gaffney, D. J. (2018). Shared genetic effects on chromatin and gene expression indicate a role for enhancer priming in immune response, *Nature Genetics*. Nature Publishing Group 50(3), 424–431. doi: 10.1038/s41588-018-0046-7.

Aleri, J. W., Hine, B. C., Pyman, M. F., Mansell, P. D., Wales, W. J., Mallard, B. and Fisher, A. D. (2015). Assessing adaptive immune response phenotypes in Australian Holstein-Friesian heifers in a pasture-based production system1, *Journal of Animal Science* 93(7), 3713–3721. doi: 10.2527/jas.2015-9078.

Aleri, J. W., Hine, B. C., Pyman, M. F., Mansell, P. D., Wales, W. J., Mallard, B. and Fisher, A. D. (2016). Periparturient immunosuppression and strategies to improve dairy cow health during the periparturient period, *Research in Veterinary Science*. Elsevier B.V. 108, 8–17. doi: 10.1016/j.rvsc.2016.07.007.

Aleri, J. W., Hine, B. C., Pyman, M. F., Mansell, P. D., Wales, W. J., Mallard, B. and Fisher, A. D. (2017). An assessment of immune and stress responsiveness in Holstein-Friesian cows selected for high and low feed conversion efficiency, *Animal Production Science* 57(2), 244. doi: 10.1071/AN15406.

Aleri, J. W., Hine, B. C., Pyman, M. F., Mansell, P. D., Wales, W. J., Mallard, B., Stevenson, M. A. and Fisher, A. D. (2019). Associations between immune competence, stress responsiveness, and production in Holstein-Friesian and Holstein-Friesian × Jersey heifers reared in a pasture-based production system in Australia, *Journal of Dairy Science* 102(4), 3282–3294. doi: 10.3168/jds.2018-14578.

Alvarez, M. I., Glover, L. C., Luo, P., Wang, L., Theusch, E., Oehlers, S. H., Walton, E. M., Tram, T. T. B., Kuang, Y. L., Rotter, J. I., McClean, C. M., Chinh, N. T., Medina, M. W., Tobin, D. M., Dunstan, S. J. and Ko, D. C. (2017). Human genetic variation in VAC14 regulates Salmonella invasion and typhoid fever through modulation of cholesterol, *Proceedings of the National Academy of Sciences of the United States of America*. National Academy of Sciences 114(37), E7746–E7755. doi: 10.1073/pnas.1706070114.

Anderson, N. (2011). Accelerated colostrum feeding to dairy calves, *Ceptor Animal Health News*, 13–16.

Bagath, M., Krishnan, G., Devaraj, C., Rashamol, V. P., Pragna, P., Lees, A. M. and Sejian, V. (2019). The impact of heat stress on the immune system in dairy cattle: a review, *Research in Veterinary Science*. Elsevier 126(March), 94–102. doi: 10.1016/j.rvsc.2019.08.011.

Bartel, D. P. (2004). MicroRNAs: genomics, biogenesis, mechanism, and function, *Cell* 116(2), 281–297. doi: 10.1016/S0092-8674(04)00045-5.

Baumgarth, N., Tung, J. W. and Herzenberg, L. A. (2005). Inherent specificities in natural antibodies: a key to immune defense against pathogen invasion, *Springer Seminars in Immunopathology*. Springer-Verlag 26(4), 347–362. doi: 10.1007/s00281-004-0182-2.

Begley, N., Buckley, F., Pierce, K. M., Fahey, A. G. and Mallard, B. A. (2009a). Differences in udder health and immune response traits of Holstein-Friesians, Norwegian Reds, and their crosses in second lactation, *Journal of Dairy Science* 92(2), 749–757. doi: 10.3168/jds.2008-1356.

Begley, N., Buckley, F., Burnside, E. B., Schaeffer, L., Pierce, K. and Mallard, B. A. (2009b). Immune responses of Holstein and Norwegian Red x Holstein calves on Canadian dairy farms, *Journal of Dairy Science* 92(2), 518–525. doi: 10.3168/jds.2008-1300.

Belongia, E. A. and Naleway, A. L. (2003). Smallpox vaccine: the good, the bad, and the ugly, *Clinical Medicine and Research* 1(2), 87–92. doi: 10.3121/cmr.1.2.87.

Bhanuprakash, V., Singh, U., Sengar, G., Sajjanar, B., Bhusan, B., Raja, T. V., Alex, R., Kumar, S., Singh, R., Ashish Kumar, Alyethodi, R. R., Kumar, S. and Deb, R. (2016). Differential effect of thermal stress on HSP70 expression, nitric oxide production and cell proliferation among native and crossbred dairy cattle, *Journal of Thermal Biology*. Elsevier 59, 18–25. doi: 10.1016/j.jtherbio.2016.04.012.

Bird, A. (2007). Perceptions of epigenetics, *Nature* 447(7143), 396–398. doi: 10.1038/nature05913.

Bishop-Williams, K. E., Berke, O., Pearl, D. L., Hand, K. and Kelton, D. F. (2015). Heat stress related dairy cow mortality during heat waves and control periods in rural Southern

Ontario from 2010–2012, *BMC Veterinary Research* 11(1), 1–10. doi: 10.1186/s12917-015-0607-2.

Bossel Ben-Moshe, N., Hen-Avivi, S., Levitin, N., Yehezkel, D., Oosting, M., Joosten, L. A. B., Netea, M. G. and Avraham, R. (2019). Predicting bacterial infection outcomes using single cell RNA-sequencing analysis of human immune cells, *Nature Communications*. Nature Publishing Group 10(1), 3266. doi: 10.1038/s41467-019-11257-y.

Briles, D. E., Nahm, M., Schroer, K., Davie, J., Baker, P., Kearney, J. and Barletta, R. (1981). Antiphosphocholine antibodies found in normal mouse serum are protective against intravenous infection with type 3 streptococcus pneumoniae, *Journal of Experimental Medicine* 153(3), 694–705.

Cartwright, S. L., Begley, N., Schaeffer, L. R., Burnside, E. B. and Mallard, B. A. (2011). Antibody and cell-mediated immune responses and survival between Holstein and Norwegian Red × Holstein Canadian calves, *Journal of Dairy Science*. Elsevier 94(3), 1576–1585. doi: 10.3168/jds.2010-3502.

Cartwright, S. L., Schaeffer, L. R., Burnside, E. B. and Mallard, B. A. (2012). Adaptive immune response, survival, and somatic cell score between postpartum Holstein and Norwegian Red × Holstein first-calf heifers, *Journal of Animal Science* 90(9), 2970–2978. doi: 10.2527/jas.2011-4233.

Cartwright, S. L. Livernois, A., McKenchnie, M., Schmied, J. and Mallard, B. (2019). Effect of heat stress on peripheral blood mononuclear cell function in dairy cattle ranked as high, average and low immune responders. In: *International Veterinary Immunology Symposium*. Seattle, Washington.

Cervenak, J. and Kacskovics, I. (2009). The neonatal Fc receptor plays a crucial role in the metabolism of IgG in livestock animals, *Veterinary Immunology and Immunopathology* 128(1–3), 171–177. doi: 10.1016/j.vetimm.2008.10.300.

Chen, T., Xie, M. Y., Sun, J. J., Ye, R. S., Cheng, X., Sun, R. P., Wei, L. M., Li, M., Lin, D. L., Jiang, Q. Y., Xi, Q. Y. and Zhang, Y. L. (2016). Porcine milk-derived exosomes promote proliferation of intestinal epithelial cells, *Scientific Reports*. Nature Publishing Group 6, 33862. doi: 10.1038/srep33862.

Chen, X., Gao, C., Li, H., Huang, L., Sun, Q., Dong, Y., Tian, C., Gao, S., Dong, H., Guan, D., Hu, X., Zhao, S., Li, L., Zhu, L., Yan, Q., Zhang, J., Zen, K. and Zhang, C. Y. (2010). Identification and characterization of microRNAs in raw milk during different periods of lactation, commercial fluid, and powdered milk products, *Cell Research*. Nature Publishing Group 20(10), 1128–1137. doi: 10.1038/cr.2010.80.

Civra, A., Altomare, A., Francese, R., Donalisio, M., Aldini, G. and Lembo, D. (2019). Colostrum from cows immunized with a veterinary vaccine against bovine rotavirus displays enhanced in vitro anti-human rotavirus activity, *Journal of Dairy Science*. Elsevier Inc. 102(6), 4857–4869. doi: 10.3168/jds.2018-16016.

Clawson, M. L., Heaton, M. P., Chitko-McKown, C. G., Fox, J. M., Smith, T. P., Snelling, W. M., Keele, J. W. and Laegreid, W. W. (2004). Beta-2-microglobulin haplotypes in U.S. beef cattle and association with failure of passive transfer in newborn calves, *Mammalian Genome: Official Journal of the International Mammalian Genome Society* 15(3), 227–236. doi: 10.1007/s00335-003-2320-x.

Cordero-Solorzano, J., Parmentier, H. K., Arts, J. A. J., van der Poel, J., de Koning, D. J. and Bovenhuis, H. (2019). Genome-wide association study identifies loci influencing natural antibody titers in milk of Dutch Holstein-Friesian cattle, *Journal of Dairy Science*. American Dairy Science Association 102(12), 11092–11103. doi: 10.3168/jds.2019-16627.

Cullen, S. M., Mayle, A., Rossi, L. and Goodell, M. A. (2014). Hematopoietic stem cell development: an epigenetic journey, *Current Topics in Developmental Biology* 107, 39–75. doi: 10.1016/B978-0-12-416022-4.00002-0.

Dahl, G. E., Tao, S. and Monteiro, A. P. A. (2016). Effects of late-gestation heat stress on immunity and performance of calves, *Journal of Dairy Science*. American Dairy Science Association 99(4), 3193–3198. doi: 10.3168/jds.2015-9990.

De Candia, P., De Rosa, V., Casiraghi, M. and Matarese, G. (2016). Extracellular RNAs: a secret arm of immune system regulation, *Journal of Biological Chemistry* 291(14), 7221–7228. doi: 10.1074/jbc.R115.708842.

de Hoon, M., Shin, J. W. and Carninci, P. (2015). Paradigm shifts in genomics through the FANTOM projects, *Mammalian Genome: Official Journal of the International Mammalian Genome Society*. Springer New York LLC 26(9–10), 391–402. doi: 10.1007/s00335-015-9593-8.

Deiss, T. C., Vadnais, M., Wang, F., Chen, P. L., Torkamani, A., Mwangi, W., Lefranc, M. P., Criscitiello, M. F. and Smider, V. V. (2019). Immunogenetic factors driving formation of ultralong VH CDR3 in *Bos taurus* antibodies, *Cellular and Molecular Immunology* 16(1), 53–64. doi: 10.1038/cmi.2017.117.

de Klerk, B., Ducro, B. J., Heuven, H. C., den Uyl, I., van Arendonk, J. A., Parmentier, H. K. and van der Poel, J. J. (2015). Phenotypic and genetic relationships of bovine natural antibodies binding keyhole limpet hemocyanin in plasma and milk, *Journal of Dairy Science* 98(4), 2746–2752. doi: 10.3168/jds.2014-8818.

de Klerk, B., Emam, M., Thompson-Crispi, K. A., Sargolzaei, M., van der Poel, J. J. and Mallard, B. A. (2018). A genome-wide association study for natural antibodies measured in blood of Canadian Holstein cows, *BMC Genomics* 19(1), 694. doi: 10.1186/s12864-018-5062-6.

Denholm, S. J., McNeilly, T. N., Banos, G., Coffey, M. P., Russell, G. C., Bagnall, A., Mitchell, M. C. and Wall, E. (2018). Immune-associated traits measured in milk of Holstein-Friesian cows as proxies for blood serum measurements, *Journal of Dairy Science* 101(11), 10248–10258. doi: 10.3168/jds.2018-14825.

Diamond, G., Zasloff, M., Eck, H., Brasseur, M., Maloy, W. L. and Bevins, C. L. (1991). Tracheal antimicrobial peptide, a cysteine-rich peptide from mammalian tracheal mucosa: peptide isolation and cloning of a cDNA, *Proceedings of the National Academy of Sciences of the United States of America* 88(9), 3952–3956. doi: 10.1073/pnas.88.9.3952.

do Amaral, B. C., Connor, E. E., Tao, S., Hayen, M. J., Bubolz, J. W. and Dahl, G. E. (2011). Heat stress abatement during the dry period influences metabolic gene expression and improves immune status in the transition period of dairy cows, *Journal of Dairy Science*. Elsevier 94(1), 86–96. doi: 10.3168/jds.2009-3004.

Dong, J., Finn, J. A., Larsen, P. A., Smith, T. P. and Crowe Jr, J. E. (2019). Structural diversity of ultralong CDRH3s in seven bovine antibody heavy chains, *Frontiers in Immunology* 10(Mar), 1–9. doi: 10.3389/fimmu.2019.00558.

Ebert, M. S. and Sharp, P. A. (2012). Roles for microRNAs in conferring robustness to biological processes, *Cell* 149(3), 515–524. doi: 10.1016/j.cell.2012.04.005.

Ehrenstein, M. R. and Notley, C. A. (2010). The importance of natural IgM: scavenger, protector and regulator, *Nature Reviews Immunology*. Nature Publishing Group 10(11), 778–786. doi: 10.1038/nri2849.

Emam, M. et al. (2018). Cellular immuno-genomics: a novel approach to examine genetic regulation of disease resistance in cattle. *Proceedings of the World Congress on*

*Genetics Applied to Livestock Production*. World Congress on Genetics Applied to Livestock Production, 738. Available at: http://www.wcgalp.org/proceedings/2018/cellular-immuno-genomics-novel-approach-examine-genetic-regulation-disease (Accessed: 19 March 2019).

Emam, M., Livernois, A., Paibomesai, M., Atalla, H. and Mallard, B. (2019a). Genetic and epigenetic regulation of immune response and resistance to infectious diseases in domestic ruminants, *Veterinary Clinics of North America. Food Animal Practice* 35(3), 405–429. doi: 10.1016/j.cvfa.2019.07.002.

Emam, M., Tabatabaei, S., Sargolzaei, M., Sharif, S., Schenkel, F. and Mallard, B. (2019b). The effect of host genetics on in vitro performance of bovine monocyte-derived macrophages, *Journal of Dairy Science*. Elsevier 102(10), 9107–9116. doi: 10.3168/JDS.2018-15960.

Emam, M., Cánovas, A., Islas-Trejo, A. D., Fonseca, P. A. S., Medrano, J. F. and Mallard, B. (2020). Transcriptomic profiles of monocyte-derived macrophages in response to Escherichia coli is associated with the host genetics, *Scientific Reports* 10(1), 271. doi: 10.1038/s41598-019-57089-0.

ENCODE Project Consortium (2004). The ENCODE (ENCyclopedia of DNA Eelements) project, *Science* 306(5696), 636–640. doi: 10.1126/science.1105136.

Ernst, J. and Kellis, M. (2017). Chromatin-state discovery and genome annotation with ChromHMM, *Nature Protocols*. Nature Publishing Group 12(12), 2478–2492. doi: 10.1038/nprot.2017.124.

Evans, S. S., Repasky, E. A. and Fisher, D. T. (2016). System feels the heat, *Nature Reviews Immunology* 15(6), 335–349. doi: 10.1038/nri3843.Fever.

Fayer, R., Perryman, L. E. and Riggs, M. W. (1989). Hyperimmune bovine colostrum neutralizes Cryptosporidium Sporozoites and protects mice against oocyst challenge, *The Journal of Parasitology* 75(1), 151–153. doi: 10.2307/3282956.

Fell, L. R., Colditz, I. G., Walker, K. H. and Watson, D. L. (1999). Associations between temperament, performance and immune function in cattle entering a commercial feedlot, *Australian Journal of Experimental Agriculture* 39(7), 795–802. doi: 10.1071/EA99027.

Fischer, A. and Rausell, A. (2018). What do primary immunodeficiencies tell us about the essentiality/redundancy of immune responses?, *Seminars in Immunology*. Academic Press 36, 13–16. doi: 10.1016/j.smim.2017.12.001.

Fisher, C. A., Bhattarai, E. K., Osterstock, J. B., Dowd, S. E., Seabury, P. M., Vikram, M., Whitlock, R. H., Schukken, Y. H., Schnabel, R. D., Taylor, J. F., Womack, J. E. and Seabury, C. M. (2011). Evolution of the bovine TLR gene family and member associations with *Mycobacterium avium* subspecies paratuberculosis infection, *PLoS ONE* 6(11), e27744. doi: 10.1371/journal.pone.0027744.

Fleming, K., Thompson-Crispi, K. A., Hodgins, D. C., Miglior, F., Corredig, M. and Mallard, B. A. (2016). Short communication: variation of total immunoglobulin G and β-lactoglobulin concentrations in colostrum and milk from Canadian Holsteins classified as high, average, or low immune responders, *Journal of Dairy Science* 99(3), 2358–2363. doi: 10.3168/jds.2015-9707.

Franken, L., Schiwon, M. and Kurts, C. (2016). Macrophages: sentinels and regulators of the immune system, *Cellular Microbiology* 18(4), 475–487. doi: 10.1111/cmi.12580.

Gantier, M. P. (2010). New perspectives in microRNA regulation of innate immunity, *Journal of Interferon and Cytokine Research: The Official Journal of the International Society for Interferon and Cytokine Research* 30(5), 283–289. doi: 10.1089/jir.2010.0037.

Gilchrist, J. J., Mentzer, A. J., Rautanen, A., Pirinen, M., Mwarumba, S., Njuguna, P., Mturi, N., Wellcome Trust Case-Control Consortium 2, Kenyan Bacteraemia Study Group, Williams, T. N., Scott, J. A. G. and Hill, A. V. S. (2018). Genetic variation in VAC14 is associated with bacteremia secondary to diverse pathogens in African children, *Proceedings of the National Academy of Sciences of the United States of America*. National Academy of Sciences 115(16), E3601–E3603. doi: 10.1073/pnas.1802071115.

Giuffra, E., Tuggle, C. K. and FAANG Consortium (2019). Functional annotation of animal genomes (FAANG): current achievements and roadmap, *Annual Review of Animal Biosciences* 7(1), 65–88. doi: 10.1146/annurev-animal-020518-114913.

Glass, C. K. and Natoli, G. (2015). Molecular control of activation and priming in macrophages, *Nature Immunology* 17(1), 26–33. doi: 10.1038/ni.3306.

Godden, S. M., Lombard, J. E. and Woolums, A. R. (2019). Colostrum management for dairy calves, *Veterinary Clinics of North America. Food Animal Practice* 35(3), 535–556.

Goh, C. and Knight, J. C. (2017). Enhanced understanding of the host-pathogen interaction in sepsis: new opportunities for omic approaches, *The Lancet. Respiratory Medicine* 5(3), 212–223. doi: 10.1016/S2213-2600(17)30045-0.

Gopal, P. K. and Gill, H. S. (2000). Oligosaccharides and glycoconjugates in bovine milk and colostrum, *British Journal of Nutrition* 84(Suppl. 1), S69–S74. doi: 10.1017/s0007114500002270.

Groves, T. C., Wilkie, B. N., Kennedy, B. W. and Mallard, B. A. (1993). Effect of selection of swine for high and low immune responsiveness on monocyte superoxide anion production and Class II MHC antigen expression, *Veterinary Immunology and Immunopathology* 36(4), 347–358. doi: 10.1016/0165-2427(93)90030-8.

Gurao, A., Kashyap, S. K. and Singh, R. (2017). β-defensins: an innate defense for bovine mastitis, *Veterinary World* 10(8), 990–998. doi: 10.14202/vetworld.2017.990-998.

Guzman, E., Hope, J., Taylor, G., Smith, A. L., Cubillos-Zapata, C. and Charleston, B. (2014). Bovine γδ T cells are a major regulatory T cell subset, *Journal of Immunology* 193(1), 208–222. doi: 10.4049/jimmunol.1303398.

Guzman, E. and Montoya, M. (2018). Contributions of farm animals to immunology, *Frontiers in Veterinary Science* 5, 307. doi: 10.3389/fvets.2018.00307.

Haakenson, J. K., Huang, R. and Smider, V. V. (2018). Diversity in cow ultralong CDR H3 antibody repetoire. *Frontiers in Immunology*, 1262. doi: 10.3389/fimmu.2018.01262.

Hall, T. J., Vernimmen, D., Browne, J. A., Mullen, M. P., Gordon, S. V., MacHugh, D. E. and O'Doherty, A. M. (2020). Alveolar macrophage chromatin is uniquely modified to orchestrate host response to Mycobacterium bovis infection, *Front. Genet.* 10:1386. doi: 10.3389/fgene.2019.01386.

Hansen, L. B. (2000). Consequences of selection for milk yield from a geneticist's viewpoint, *Journal of Dairy Science* 83(5), 1145–1150. doi: 10.3168/jds.S0022-0302(00)74980-0.

Harris, J. R. and Markl, J. (1999). Keyhole limpet hemocyanin (KLH): A biomedical review, *Micron* 30(6), 597–623. doi: 10.1016/S0968-4328(99)00036-0.

Hine, B. C., Cartwright, S. L. and Mallard, B. A. (2011). Effect of age and pregnancy status on adaptive immune responses of Canadian Holstein replacement heifers, *Journal of Dairy Science* 94(2), 981–991. doi: 10.3168/jds.2010-3329.

Hine, B. C., Cartwright, S. L. and Mallard, B. A. (2012). Analysis of leukocyte populations in Canadian Holsteins classified as high or low immune responders for antibody- or cell-mediated immune response, *Canadian Journal of Veterinary Research = Revue Canadienne de Recherche Veterinaire* 76(2), 149–156. Available at: http://www.ncbi.nlm.nih.gov/pubmed/23024458 (Accessed: 1 January 2020).

Hodgins, D. C. and Shewen, P. E. (1994). Passive immunity to Pasteurella haemolytica A1 in dairy calves: effects of preparturient vaccination of the dams, *Canadian Journal of Veterinary Research = Revue Canadienne de Recherche Veterinaire* 58(1), 31–35.

Holloway, A. F. and Woodworth, A. M. (2017). Chapter three - the role of epigenetic regulation in transcriptional memory in the immune system, *Advances in Protein Chemistry and Structural Biology* 106, 43–69.

Huntzinger, E. and Izaurralde, E. (2011). Gene silencing by microRNAs: contributions of translational repression and mRNA decay, *Nature Reviews Genetics* 12(2), 99–110. doi: 10.1038/nrg2936.

Ivey, K. N. and Srivastava, D. (2010). MicroRNAs as regulators of differentiation and cell fate decisions, *Cell Stem Cell* 7(1), 36–41. doi: 10.1016/j.stem.2010.06.012.

Izumi, H., Kosaka, N., Shimizu, T., Sekine, K., Ochiya, T. and Takase, M. (2012). Bovine milk contains microRNA and messenger RNA that are stable under degradative conditions, *Journal of Dairy Science*. Elsevier 95(9), 4831–4841. doi: 10.3168/jds.2012-5489.

Izumi, H., Tsuda, M., Sato, Y., Kosaka, N., Ochiya, T., Iwamoto, H., Namba, K. and Takeda, Y. (2015). Bovine milk exosomes contain microRNA and mRNA and are taken up by human macrophages, *Journal of Dairy Science*. Elsevier 98(5), 2920–2933. doi: 10.3168/jds.2014-9076.

Jennewein, C., von Knethen, A., Schmid, T. and Brüne, B. (2010). MicroRNA-27b contributes to lipopolysaccharide-mediated peroxisome proliferator-activated receptor γ (PPARγ) mRNA destabilization, *Journal of Biological Chemistry* 285(16), 11846–11853. doi: 10.1074/jbc.M109.066399.

Johnson, J. (2018). What are the risks and benefits of pox parties? Available at: https://www.medicalnewstoday.com/articles/323728.php (Accessed: 19 November 2018).

Kaushik, A. K., Kehrli, M. E., Kurtz, A., Ng, S., Koti, M., Shojaei, F. and Saini, S. S. (2009). Somatic hypermutations and isotype restricted exceptionally long CDR3H contribute to antibody diversification in cattle, *Veterinary Immunology and Immunopathology* 127(1–2), 106–113. doi: 10.1016/j.vetimm.2008.09.024.

Kiser, J. N., Cornmesser, M. A., Gavin, K., Hoffman, A., Moore, D. A. and Neibergs, H. L. (2019). Rapid Communication: genome-wide association analyses identify loci associated with colostrum production in Jersey cattle, *Journal of Animal Science* 97(3), 1117–1123. doi: 10.1093/jas/sky482.

Kleid, D. G., Yansura, D., Small, B., Dowbenko, D., Moore, D. M., Grubman, M. J., McKercher, P. D., Morgan, D. O., Robertson, B. H. and Bachrach, H. L. (1981). Cloned viral protein vaccine for foot-and-mouth disease: responses in cattle and swine, *Science (New York, N.Y.)* 214(4525), 1125–1129. doi: 10.1126/science.6272395.

Ko, D. C. and Jaslow, S. L. (2014). The marriage of quantitative genetics and cell biology: a novel screening approach reveals people have genetically encoded variation in microtubule stability, *Bioarchitecture*. Taylor & Francis 4(2), 58–61. doi: 10.4161/bioa.28481.

Kohara, J., Hirai, T., Mori, K., Ishizaki, H. and Tsunemitsu, H. (1997). Enhancement of passive immunity with maternal vaccine against newborn calf diarrhea, *Journal of Veterinary Medical Science* 59(11), 1023–1025. doi: 10.1292/jvms.59.1023.

Kondilis-Mangum, H. D. and Wade, P. A. (2013). Epigenetics and the adaptive immune response, *Molecular Aspects of Medicine* 34(4), 813–825. doi: 10.1016/j.mam.2012.06.008.

Kwong, L. S., Parsons, R., Patterson, R., Coffey, T. J., Thonur, L., Chang, J. S., Russell, G., Haig, D., Werling, D. and Hope, J. C. (2011). Characterisation of antibodies to bovine toll-like receptor (TLR)-2 and cross-reactivity with ovine TLR2, *Veterinary Immunology and Immunopathology* 139(2–4), 313–318. doi: 10.1016/j.vetimm.2010.10.014.

Laegreid, W. W., Heaton, M. P., Keen, J. E., Grosse, W. M., Chitko-McKown, C. G., Smith, T. P., Keele, J. W., Bennett, G. L. and Besser, T. E. (2002). Association of bovine neonatal Fc receptor α-chain gene (FCGRT) haplotypes with serum IgG concentration in newborn calves, *Mammalian Genome: Official Journal of the International Mammalian Genome Society* 13(12), 704–710. doi: 10.1007/s00335-002-2219-y.

Larmer, S. and Mallard, B. (2016). High immune response sires reduce disease incidence in North American large commercial dairy populations, *Cattle Practice* 25(2), 74–81.

Lawless, N., Foroushani, A. B., McCabe, M. S., O'Farrelly, C. and Lynn, D. J. (2013). Next generation sequencing reveals the expression of a unique miRNA profile in response to a Gram-positive bacterial infection, *PLoS ONE* 8(3), e57543. doi: 10.1371/journal.pone.0057543.

Lawless, N., Vegh, P., O'Farrelly, C. and Lynn, D. J. (2014). The role of microRNAs in bovine infection and immunity, *Frontiers in Immunology* 5(V), 611. doi: 10.3389/fimmu.2014.00611.

Li, Y., Oosting, M., Deelen, P., Ricaño-Ponce, I., Smeekens, S., Jaeger, M., Matzaraki, V., Swertz, M. A., Xavier, R. J., Franke, L., Wijmenga, C., Joosten, L. A., Kumar, V. and Netea, M. G. (2016). Inter-individual variability and genetic influences on cytokine responses to bacteria and fungi, *Nature Medicine* 22(8), 952–960. doi: 10.1038/nm.4139.

Liang, G., Malmuthuge, N., Guan, le L. and Griebel, P. (2015). Model systems to analyze the role of miRNAs and commensal microflora in bovine mucosal immune system development, *Molecular Immunology*. Elsevier Ltd 66(1), 57–67. doi: 10.1016/j.molimm.2014.10.014.

Liljavirta, J., Ekman, A., Knight, J. S., Pernthaner, A., Iivanainen, A. and Niku, M. (2013). Activation-induced cytidine deaminase (AID) is strongly expressed in the fetal bovine ileal Peyer's patch and spleen and is associated with expansion of the primary antibody repertoire in the absence of exogenous antigens, *Mucosal Immunology*. Nature Publishing Group 6(5), 942–949. doi: 10.1038/mi.2012.132.

Luca, S. and Mihaescu, T. (2013). History of BCG vaccine, *Maedica* 8(1), 53–58. Available at: http://www.ncbi.nlm.nih.gov/pubmed/24023600 (Accessed: 1 January 2020).

Machado, L. R. and Ottolini, B. (2015). An evolutionary history of defensins: A role for copy number variation in maximizing host innate and adaptive immune responses, *Frontiers in Immunology*. Frontiers Media S.A. 6, 115. doi: 10.3389/fimmu.2015.00115.

Mallard, B. A., Emam, M., Paibomesai, M., Thompson-Crispi, K. and Wagter-Lesperance, L. (2015). Genetic selection of cattle for improved immunity and health, *The Japanese Journal of Veterinary Research* 63(Suppl. 1), S37–S44. Available at: http://www.ncbi.nlm.nih.gov/pubmed/25872325 (Accessed: 19 July 2015).

Mascola, J. R. and Haynes, B. F. (2013). HIV-1 neutralizing antibodies: understanding nature's pathways, *Immunological Reviews* 254(1), 225–244. doi: 10.1111/imr.12075.

Mayasari, N., de Vries Reilingh, G., Nieuwland, M. G., Remmelink, G. J., Parmentier, H. K., Kemp, B. and van Knegsel, A. T. (2015). Effect of maternal dry period length on colostrum immunoglobulin content and on natural and specific antibody titers in calves, *Journal of Dairy Science*. Elsevier 98(6), 3969–3979. doi: 10.3168/jds.2014-8753.

McDowell, R. E. (1985). Crossbreeding in tropical areas with emphasis on milk, health, and fitness, *Journal of Dairy Science* 68(9), 2418–2435. doi: 10.3168/jds.S0022-0302(85)81118-8.

Meade, K. G., Cormican, P., Narciandi, F., Lloyd, A. and O'Farrelly, C. (2014). Bovine β-defensin gene family: opportunities to improve animal health?, *Physiological Genomics* 46(1), 17–28. doi: 10.1152/physiolgenomics.00085.2013.

Meade, K. G. and O'Farrelly, C. (2019). Β-Defensins: farming the microbiome for homeostasis and health, *Frontiers in Immunology* 10(Jan), 1–20. doi: 10.3389/fimmu.2018.03072.

Mehta, A. and Baltimore, D. (2016). MicroRNAs as regulatory elements in immune system logic, *Nature Reviews Immunology* 16(5), 279–294. doi: 10.1038/nri.2016.40.

Miller, S. and Chaudhary, A. (2016). A cellular GWAS approach to define human variation in cellular pathways important to inflammation, *Pathogens*. Multidisciplinary Digital Publishing Institute 5(2), 39. doi: 10.3390/pathogens5020039.

Min, L., Zheng, N., Zhao, S., Cheng, J., Yang, Y., Zhang, Y., Yang, H. and Wang, J. (2016). Long-term heat stress induces the inflammatory response in dairy cows revealed by plasma proteome analysis, *Biochemical and Biophysical Research Communications* 471(2), 296–302. doi: 10.1016/j.bbrc.2016.01.185.

Molès, J. P., Tuaillon, E., Kankasa, C., Bedin, A. S., Nagot, N., Marchant, A., McDermid, J. M. and Van de Perre, P. (2018). Breastmilk cell trafficking induces microchimerism-mediated immune system maturation in the infant, *Pediatric Allergy and Immunology: Official Publication of the European Society of Pediatric Allergy and Immunology* 29(2), 133–143. doi: 10.1111/pai.12841.

Monteiro, A. P. A., Tao, S., Thompson, I. M. and Dahl, G. E. (2014). Effect of heat stress during late gestation on immune function and growth performance of calves: isolation of altered colostral and calf factors, *Journal of Dairy Science*. Elsevier 97(10), 6426–6439. doi: 10.3168/jds.2013-7891.

Moseley, P. (2000). Stress proteins and the immune response, *Immunopharmacology* 48(3), 299–302. doi: 10.1016/S0162-3109(00)00227-7.

Murphy, J. M., Hagey, J. V. and Chigerwe, M. (2014). Comparison of serum immunoglobulin G half-life in dairy calves fed colostrum, colostrum replacer or administered with intravenous bovine plasma, *Veterinary Immunology and Immunopathology*. Elsevier B.V. 158(3–4), 233–237. doi: 10.1016/j.vetimm.2014.01.008.

Na, Y. J. and Kim, J. H. (2013). Understanding cooperativity of microRNAs via microRNA association networks, *BMC Genomics* 14(Suppl. 5), S17. doi: 10.1186/1471-2164-14-S5-S17.

National Animal Health Monitoring System (2010). *Dairy 2007: Heifer Calf Health and Management Practices on U.S. Dairy Operations, 2007*. U.S. Department of Agriculture.

National Animal Health Monitoring System (2018). *Dairy 2014: Health and Management Practices on U.S. Dairy Operations, 2014*. U.S. Department of Agriculture

Novo, S. M. F., Costa, J. F. D. R., Baccili, C. C., Sobreira, N. M., Silva, B. T., de Oliveira, P. L., Hurley, D. J. and Gomes, V. (2017). Effect of maternal cells transferred with colostrum on the health of neonate calves, *Research in Veterinary Science*. Elsevier B.V. 112, 97–104. doi: 10.1016/j.rvsc.2017.01.025.

O'Connell, R. M., Rao, D. S., Chaudhuri, A. A. and Baltimore, D. (2010). Physiological and pathological roles for microRNAs in the immune system, *Nature Reviews Immunology* 10(2), 111–122. doi: 10.1038/nri2708.

O'Connell, R. M., Rao, D. S. and Baltimore, D. (2012). MicroRNA regulation of inflammatory responses, *Annual Review of Immunology* 30, 295–312. doi: 10.1146/annurev-immunol-020711-075013.

O'Doherty, A. M., Rue-Albrecht, K. C., Magee, D. A., Ahting, S., Irwin, R. E., Hall, T. J., Browne, J. A., Nalpas, N. C., Walsh, C. P., Gordon, S. V., Wojewodzic, M. W. and MacHugh, D. E. (2019). The bovine alveolar macrophage DNA methylome is resilient to infection with Mycobacterium bovis, *Scientific Reports*. Nature Publishing Group, 9(1). doi: 10.1038/s41598-018-37618-z.

Oberdoerffer, P. and Sinclair, D. A. (2007). The role of nuclear architecture in genomic instability and ageing, *Nature Reviews Molecular Cell Biology* 8(9), 692–702. doi: 10.1038/nrm2238.

Ochsenbein, A. F. and Zinkernagel, R. M. (2000). Natural antibodies and complement link innate and acquired immunity, *Immunology Today*. Elsevier Current Trends 21(12), 624–630. doi: 10.1016/S0167-5699(00)01754-0.

Otto, W., Najnigier, B., Stelmasiak, T. and Robins-Browne, R. M. (2011). Randomized control trials using a tablet formulation of hyperimmune bovine colostrum to prevent diarrhea caused by enterotoxigenic Escherichia coli in volunteers, *Scandinavian Journal of Gastroenterology* 46(7–8), 862–868. doi: 10.3109/00365521.2011.574726.

Paibomesai, M. (2017). *Epigenetic Influences on Bovine T-helper 1 and T-helper 2 Cytokines (Interferon-Gamma and Interleukin-4) in High and Low Immune Responders around the Peripartum Period*. University of Guelph.

Parkin, J. and Cohen, B. (2001). An overview of the immune system, *Lancet*. Elsevier Limited 357(9270), 1777–1789. doi: 10.1016/S0140-6736(00)04904-7.

Plotkin, S. (2014). History of vaccination, *Proceedings of the National Academy of Sciences of the United States of America*. National Academy of Sciences 111(34), 12283–12287. doi: 10.1073/pnas.1400472111.

Pollock, J. M., Rodgers, J. D., Welsh, M. D. and McNair, J. (2006). Pathogenesis of bovine tuberculosis: the role of experimental models of infection, *Veterinary Microbiology* 112(2–4), 141–150. doi: 10.1016/j.vetmic.2005.11.032.

Rabinovitz, B. C., Gerhardt, E., Tironi Farinati, C., Abdala, A., Galarza, R., Vilte, D. A., Ibarra, C., Cataldi, A. and Mercado, E. C. (2012). Vaccination of pregnant cows with EspA, EspB, γ-intimin, and Shiga toxin 2 proteins from Escherichia coli O157:H7 induces high levels of specific colostral antibodies that are transferred to newborn calves, *Journal of Dairy Science* 95(6), 3318–3326. doi: 10.3168/jds.2011-5093.

Reppert, E. J., Chamorro, M. F., Robinson, L., Cernicchiaro, N., Wick, J., Weaber, R. L. and Haines, D. M. (2019). Effect of vaccination of pregnant beef heifers on the concentrations of serum igg and specific antibodies to bovine herpesvirus 1, bovine viral diarrhea virus 1, and bovine viral diarrhea virus 2 in heifers and calves, *Canadian Journal of Veterinary Research = Revue Canadienne de Recherche Veterinaire* 83(4), 313–316.

Roosen, S., Exner, K., Paul, S., Schröder, J. M., Kalm, E. and Looft, C. (2004). Bovine β-defensins: identification and characterization of novel bovine β-defensin genes and their expression in mammary gland tissue, *Mammalian Genome: Official Journal of the International Mammalian Genome Society* 15(10), 834–842. doi: 10.1007/s00335-004-2387-z.

Ross, M. (2017). Bioactivity of bovine colostrum and milk exosomes from high, average, and low immune responders on human intestinal epithelial cells. University of Guelph. Available at: http://hdl.handle.net/10214/11538.

Rossi, R. L., Rossetti, G., Wenandy, L., Curti, S., Ripamonti, A., Bonnal, R. J., Birolo, R. S., Moro, M., Crosti, M. C., Gruarin, P., Maglie, S., Marabita, F., Mascheroni, D., Parente, V., Comelli, M., Trabucchi, E., De Francesco, R., Geginat, J., Abrignani, S. and Pagani, M. (2011). Distinct microRNA signatures in human lymphocyte subsets and enforcement of the naive state in CD4+ T cells by the microRNA miR-125b, *Nature Immunology* 12(8), 796–803. doi: 10.1038/ni.2057.

Sahl, H. G., Pag, U., Bonness, S., Wagner, S., Antcheva, N. and Tossi, A. (2005). Mammalian defensins: structures and mechanism of antibiotic activity, *Journal of Leukocyte Biology* 77(4), 466–475. doi: 10.1189/jlb.0804452.

Saini, S. S., Allore, B., Jacobs, R. M. and Kaushik, A. (1999). Exceptionally long CDR3H region with multiple cysteine residues in functional bovine IgM antibodies, *European Journal of Immunology* 29(8), 2420–2426. doi: 10.1002/(SICI)1521-4141(199908)29:08<2420::AID-IMMU2420>3.0.CO;2-A.

Saini, S. S., Farrugia, W., Ramsland, P. A. and Kaushik, A. K. (2003). Bovine IgM antibodies with exceptionally long complementarity-determining region 3 of the heavy chain share unique structural properties conferring restricted VH + Vlambda pairings, *International Immunology* 15(7), 845–853. doi: 10.1093/intimm/dxg083.

Salak-Johnson, J. L. and McGlone, J. J. (2007). Making sense of apparently conflicting data: stress and immunity in swine and cattle, *Journal of Animal Science* 85(13 Suppl), E81–E88. doi: 10.2527/jas.2006-538.

Sati, S. and Cavalli, G. (2017). Chromosome conformation capture technologies and their impact in understanding genome function, *Chromosoma*. Springer Science and Business Media Deutschland GmbH 126(1), 33–44. doi: 10.1007/s00412-016-0593-6.

Sattler, S. (2017). The role of the immune system beyond the fight against infection, *Advances in Experimental Medicine and Biology*. Springer New York LLC, 3–14. doi: 10.1007/978-3-319-57613-8_1.

Savage, H. P. and Baumgarth, N. (2015). Characteristics of natural antibody-secreting cells, *Annals of the New York Academy of Sciences* 1362, 132–142. doi: 10.1111/nyas.12799.

Sawai, C. M., Babovic, S., Upadhaya, S., Knapp, D. J. H. F., Lavin, Y., Lau, C. M., Goloborodko, A., Feng, J., Fujisaki, J., Ding, L., Mirny, L. A., Merad, M., Eaves, C. J. and Reizis, B. (2016). Hematopoietic stem cells are the major source of multilineage hematopoiesis in adult animals, *Immunity*. Cell Press 45(3), 597–609. doi: 10.1016/j.immuni.2016.08.007.

Sears, K. T., Tennant, S. M., Reymann, M. K., Simon, R., Konstantopoulos, N., Blackwelder, W. C., Barry, E. M. and Pasetti, M. F. (2017). Bioactive immune components of anti-diarrheagenic enterotoxigenic Escherichia coli hyperimmune bovine colostrum products, *Clinical and Vaccine Immunology*. American Society for Microbiology 24(8). doi: 10.1128/CVI.00186-16.

Seita, J. and Weissman, I. L. (2010). Hematopoietic stem cell: self-renewal versus differentiation, *Wiley Interdisciplinary Reviews: Systems Biology and Medicine* 2(6), 640–653. doi: 10.1002/wsbm.86.

Sewalem, A., Kistemaker, G. J., Miglior, F. and Van Doormaal, B. J. (2006). Analysis of inbreeding and its relationship with functional longevity in Canadian dairy cattle, *Journal of Dairy Science* 89(6), 2210–2216. doi: 10.3168/jds.S0022-0302(06)72291-3.

Shi, B., Ma, L., He, X., Wang, X., Wang, P., Zhou, L. and Yao, X. (2014). Comparative analysis of human and mouse immunoglobulin variable heavy regions from IMGT/LIGM-DB with IMGT/HighV-QUEST, *Theoretical Biology and Medical Modelling* 11(1), 30. doi: 10.1186/1742-4682-11-30.

Shivley, C. B., Lombard, J. E., Urie, N. J., Haines, D. M., Sargent, R., Kopral, C. A., Earleywine, T. J., Olson, J. D. and Garry, F. B. (2018). Preweaned heifer management on US dairy operations: Part II. Factors associated with colostrum quality and passive transfer status of dairy heifer calves, *Journal of Dairy Science*. American Dairy Science Association 101(10), 9185–9198. doi: 10.3168/jds.2017-14008.

Siomi, H. and Siomi, M. C. (2010). Posttranscriptional regulation of microRNA biogenesis in animals, *Molecular Cell* 38(3), 323–332. doi: 10.1016/j.molcel.2010.03.013.

Smith, T. and Little, R. (1922a). Cow serum as a substitution for colostrum in new-born calves, *Journal of Experimental Medicine* 34(4), 453–468.

Smith, T. and Little, R. B. (1922b). The significance of colostrum to the new-born calf, *Journal of Experimental Medicine* 36(2), 181–198. doi: 10.1084/jem.36.2.181.

Snodgrass, D. R., Nagy, L. K., Sherwood, D. and Campbell, I. (1982). Passive immunity in calf diarrhea: vaccination with K99 antigen of enterotoxigenic Escherichia coli and rotavirus, *Infection and Immunity* 37(2), 586–591.

Sok, D., Le, K. M., Vadnais, M., Saye-Francisco, K. L., Jardine, J. G., Torres, J. L., Berndsen, Z. T., Kong, L., Stanfield, R., Ruiz, J., Ramos, A., Liang, C. H., Chen, P. L., Criscitiello, M. F., Mwangi, W., Wilson, I. A., Ward, A. B., Smider, V. V. and Burton, D. R. (2017). Rapid elicitation of broadly neutralizing antibodies to HIV by immunization in cows, *Nature* 548(7665), 108–111. doi: 10.1038/nature23301.

Sok, D. and Burton, D. R. (2018). Recent progress in broadly neutralizing antibodies to HIV, *Nature Immunology*. Springer US 19(11), 1179–1188. doi: 10.1038/s41590-018-0235-7.

Solomon, S. Qin, D., Manning, M., Marquis, M., Averyt, K., Tignor, M., Miller, H. and Chen, Z. (2007). *Contribution of Working Group I to the Forth Assessment Report of the Intergovernmental Panel on Climate Change*. Cambridge University Press, New York.

Sordillo, L. M., Shafer-Weaver, K. and DeRosa, D. (1997). Immunobiology of the mammary gland. *Journal of Dairy Science*. American Dairy Science Association 80(8), 1851–1865. doi: 10.3168/jds.S0022-0302(97)76121-6.

Sørensen, A. C., Madsen, P., Sørensen, M. K. and Berg, P. (2006). Udder health shows inbreeding depression in Danish Holsteins, *Journal of Dairy Science* 89(10), 4077–4082. doi: 10.3168/jds.S0022-0302(06)72452-3.

Sørensen, M. K., Norberg, E., Pedersen, J. and Christensen, L. G. (2008). Invited review: crossbreeding in dairy cattle: a Danish perspective, *Journal of Dairy Science* 91(11), 4116–4128. doi: 10.3168/jds.2008-1273.

Stear, M., Fairlie-Clarke, K., Jonsson, N., Mallard, B. and Groth, D. (2017). Genetic variation in immunity and disease resistance in dairy cows and other livestock. In: *Burleigh Dodds Series in Agricultural Science*, 509–532. Burleigh Dodds Science Publishing, Cambridge, UK. doi: 10.19103/AS.2016.0006.25.

Stelwagen, K., Carpenter, E., Haigh, B., Hodgkinson, A. and Wheeler, T. T. (2009). Immune components of bovine colostrum and milk, *Journal of Animal Science* 87(13 Suppl), 3–9. doi: 10.2527/jas.2008-1377.

Stewart, S., Godden, S., Bey, R., Rapnicki, P., Fetrow, J., Farnsworth, R., Scanlon, M., Arnold, Y., Clow, L., Mueller, K. and Ferrouillet, C. (2005). Preventing bacterial contamination and proliferation during the harvest, storage, and feeding of fresh bovine colostrum, *Journal of Dairy Science*. Elsevier 88(7), 2571–2578. doi: 10.3168/jds.S0022-0302(05)72933-7.

Stott, G. H., Marx, D. B., Menefee, B. E. and Nightengale, G. T. (1979a). Colostral immunoglobulin transfer in calves I. Period of absorption, *Journal of Dairy Science* 62(10), 1632–1638. doi: 10.3168/jds.S0022-0302(79)83472-4.

Stott, G. H., Marx, D. B., Menefee, B. E. and Nightengale, G. T. (1979b). Colostral immunoglobulin transfer in calves II. The rate of absorption, *Journal of Dairy Science* 62(11), 1766–1773. doi: 10.3168/jds.S0022-0302(79)83495-5.

Taganov, K. D., Boldin, M. P., Chang, K. J. and Baltimore, D. (2006). NF-κB-dependent induction of microRNA miR-146, an inhibitor targeted to signaling proteins of innate immune responses, *Proceedings of the National Academy of Sciences of the United States of America* 103(33), 12481–12486. doi: 10.1073/pnas.0605298103.

Taylor, J. D., Fulton, R. W., Lehenbauer, T. W., Step, D. L. and Confer, A. W. (2010). The epidemiology of bovine respiratory disease: what is the evidence for predisposing factors?, *Canadian Veterinary Journal = la Revue Veterinaire Canadienne* 51(10), 1095–1102.

Thompson-Crispi, K. A., Sewalem, A., Miglior, F. and Mallard, B. A. (2012a). Genetic parameters of adaptive immune response traits in Canadian Holsteins, *Journal of Dairy Science*. Elsevier 95(1), 401–409. doi: 10.3168/jds.2011-4452.

Thompson-Crispi, K. A., Hine, B., Quinton, M., Miglior, F. and Mallard, B. A. (2012b). Short communication: association of disease incidence and adaptive immune response in Holstein dairy cows, *Journal of Dairy Science* 95(7), 3888–3893. doi: 10.3168/jds.2011-5201.

Thompson-Crispi, K. A., Miglior, F. and Mallard, B. A. (2013). Incidence rates of clinical mastitis among Canadian Holsteins classified as high, average, or low immune responders, *Clinical and Vaccine Immunology*. American Society for Microbiology 20(1), 106–112. doi: 10.1128/CVI.00494-12.

Thompson-Crispi, K. A., Sargolzaei, M., Ventura, R., Abo-Ismail, M., Miglior, F., Schenkel, F. and Mallard, B. A. (2014). A genome-wide association study of immune response traits in Canadian Holstein cattle, *BMC Genomics* 15(1), 559. doi: 10.1186/1471-2164-15-559.

Thurman, R. E., Rynes, E., Humbert, R., Vierstra, J., Maurano, M. T., Haugen, E., Sheffield, N. C., Stergachis, A. B., Wang, H., Vernot, B., Garg, K., John, S., Sandstrom, R., Bates, D., Boatman, L., Canfield, T. K., Diegel, M., Dunn, D., Ebersol, A. K., Frum, T., Giste, E., Johnson, A. K., Johnson, E. M., Kutyavin, T., Lajoie, B., Lee, B. K., Lee, K., London, D., Lotakis, D., Neph, S., Neri, F., Nguyen, E. D., Qu, H., Reynolds, A. P., Roach, V., Safi, A., Sanchez, M. E., Sanyal, A., Shafer, A., Simon, J. M., Song, L., Vong, S., Weaver, M., Yan, Y., Zhang, Z., Zhang, Z., Lenhard, B., Tewari, M., Dorschner, M. O., Hansen, R. S., Navas, P. A., Stamatoyannopoulos, G., Iyer, V. R., Lieb, J. D., Sunyaev, S. R., Akey, J. M., Sabo, P. J., Kaul, R., Furey, T. S., Dekker, J., Crawford, G. E. and Stamatoyannopoulos, J. A. (2012). The accessible chromatin landscape of the human genome, *Nature* 489(7414), 75–82. doi: 10.1038/nature11232.

Touchberry, R. W. (1992). Crossbreeding effects in dairy cattle: the Illinois experiment, 1949 to 1969, *Journal of Dairy Science* 75(2), 640–667. doi: 10.3168/jds.S0022-0302(92)77801-1.

Van Doormaal, B. (2019). *Inbreeding Update - August 2017*. Canadian Dairy Network.

Vegh, P., Foroushani, A. B., Magee, D. A., McCabe, M. S., Browne, J. A., Nalpas, N. C., Conlon, K. M., Gordon, S. V., Bradley, D. G., MacHugh, D. E. and Lynn, D. J. (2013).

Profiling microRNA expression in bovine alveolar macrophages using RNA-seq, *Veterinary Immunology and Immunopathology*. Elsevier B.V. 155(4), 238–244. doi: 10.1016/j.vetimm.2013.08.004.

Waddington, C. H. (2012). The epigenotype 1942, *International Journal of Epidemiology* 41(1), 10–13. doi: 10.1093/ije/dyr184.

Wagter, L. C., Mallard, B. A., Wilkie, B. N., Leslie, K. E., Boettcher, P. J. and Dekkers, J. C. (2000). A quantitative approach to classifying Holstein cows based on antibody responsiveness and its relationship to peripartum mastitis occurrence, *Journal of Dairy Science* 83(3), 488–498. doi: 10.3168/jds.S0022-0302(00)74908-3.

Wagter-Lesperance, L. (2017). Differences in TLR2 pattern recognition receptor expression on blood mononuclear cells with and without ligand stimulation among dairy cattle classified by estimated breeding value (EBV) of adaptive immune response. University of Guelph. Available at: http://hdl.handle.net/10214/10438.

Walker, L. M. and Burton, D. R. (2018). Passive immunotherapy of viral infections: "super-antibodies" enter the fray, *Nature Reviews Immunology*. Nature Publishing Group 18(5), 297–308. doi: 10.1038/nri.2017.148.

Wang, F., Ekiert, D. C., Ahmad, I., Yu, W., Zhang, Y., Bazirgan, O., Torkamani, A., Raudsepp, T., Mwangi, W., Criscitiello, M. F., Wilson, I. A., Schultz, P. G. and Smider, V. V. (2013). XReshaping antibody diversity, *Cell*. Elsevier Inc. 153(6), 1379–1393. doi: 10.1016/j.cell.2013.04.049.

Wang, X. F., Cao, R., Li, J., Wu, S. and Chen, T. (2015). Concentration characteristics of bovine β-defensin 1 and 2 in fresh bovine milk and infant formula, *International Journal of Dairy Technology* 68(2), 299–301. doi: 10.1111/1471-0307.12214.

Zhang, L., Boeren, S., Hageman, J. A., van Hooijdonk, T., Vervoort, J. and Hettinga, K. (2015). Bovine milk proteome in the first 9 days: protein interactions in maturation of the immune and digestive system of the newborn, *PLoS ONE* 10(2), e0116710. doi: 10.1371/journal.pone.0116710.

Zhang, Q. and Cao, X. (2019). Epigenetic regulation of the innate immune response to infection, *Nature Reviews Immunology* 19(7), 417–432. doi: 10.1038/s41577-019-0151-6.

Zhang, R., Zhao, Z., Zhao, Y., Kacskovics, I., Eijk, Mv, Groot, Nd, Li, N. and Hammarstrom, L. (2009). Association of FcRn Heavy Chain encoding gene (FCGRT) POLYMORPHISMS with IgG content in bovine colostrum, *Animal Biotechnology* 20(4), 242–246. doi: 10.1080/10495390903196448.

Zhao, X. L., Honaker, C. F. and Siegel, P. B. (2012). Phenotypic responses of chickens to long-term selection for high or low antibody titers to sheep red blood cells, *Poultry Science* 91(5), 1047–1056. doi: 10.3382/ps.2011-01892.

# Chapter 6

## Dairy cattle welfare and health: an intimate partnership

*Clive Phillips, Curtin University Sustainable Policy (CUSP) Institute, Australia*



### 1 Introduction

The welfare of dairy cows refers to 'their state with respect to the quality and quantity of their experiences' (Phillips, 2009, p. 4), from which they derive their feelings, particularly in response to their environment. Health, according to the World Health Organisation (WHO) (2020), is broadly defined as 'a state of complete physical, mental and social well-being, and not merely the absence of disease and infirmity'. Thus welfare is more directly focussed on the animals' feelings, and health is dependent on good welfare, which is not necessarily conferred just by the absence of disease. The World Animal Health Organisation (OIE) defines health even more narrowly as just one of the pillars of welfare: 'An animal experiences good welfare if the animal is healthy, comfortable, well nourished, safe, is not suffering from unpleasant states such as pain, fear and distress, and is able to express behaviours that are important for its physical and mental state' (OIE, 2019). For non-human animals, therefore, welfare is a broader concept than health. This narrower perspective of health is appropriate for this review on dairy cows as we currently have little understanding of cows' mental life and social requirements for good welfare.

http://dx.doi.org/10.19103/AS.2020.0086.08

Health is clearly an important part of welfare, but can the other pillars expounded by the OIE, comfort, good nutrition, safety and not experiencing pain, fear or distress, also contribute to the good health of an animal? The unequivocal viewpoint of most specialists involved in animal welfare is that they can, and therefore a holistic approach to animal welfare is necessary. Considering the dairy cow, her comfort in a free stall will depend on suitable bedding, which will also impact on her udder health, in particular. As an example of this relationship between comfort and health, in research to investigate the relationships between cow lying behaviour and their hygiene, DeVries et al. (2012) found that cows that had more lying bouts in their stalls, demonstrating greater comfort, also had better leg hygiene. The relationship between nutrition and health is even more clearly established. Recently reviewed by Bertoni et al. (2016), the connection between nutrition and welfare is demonstrated to be closely related to animals' health status. Immune function is dependent on good protein nutrition, in particular, with a response to parasitization being particularly dependent on this aspect of nutrition. The inflammatory response, too, has important nutritional correlates, in particular amino and fatty acids and micronutrient status, with both direct and indirect (antioxidant) effects. Finally, safety for dairy cows is also closely related to their health, being principally concerned with mitigating any biosecurity risks and providing an environment in which the animal can coexist with conspecifics peacefully and without risk of injury. Biosecurity risks should focus on listed diseases, infection and infestation.

Two key components of the contribution of health to welfare are the duration and severity of diseases, which are related to milk production or the cows' environment, or they are of congenital or nutritional origin. Dairy cow production systems are intensifying rapidly in countries that are moving from a traditional Asian diet to a Western diet, and we need to understand how the new intensive production systems impact on cows' stress levels and their immune function.

According to the WHO definition of health, mental health is a very important component, but we do not understand it well in cattle. Cattle do not readily display their mental state; they evolved as prey animals, which rarely give away their secrets of vulnerability for fear of predation. There are other aspects of diseases of dairy cows that we do not understand. Is there evidence for referred pain, for example, or any impact of reproductive state and subclinical disease on welfare? Many of these could be easily researched, but our primary reason for addressing cattle diseases has been to maintain productivity and hence profit for the farmer. Now that the public is evidently concerned about the quality of lives that cows have, short though they may be, the welfare impact of disease needs to be much better understood. When we have that knowledge, we need to use it to improve welfare through incentive schemes for farmers, perhaps involving accreditation by retailers or setting of legislated standards by governments.

## 2 The welfare implications of common dairy cow diseases

The impact of a disease on cows' welfare relates to how severe the disease state is and its duration. The two are interrelated; severity may be low at the start of a disease bout, increase to a maximum and decline during the later stages, as the disease is overcome and the cow learns how to cope with the challenges to her welfare. If she does not cope and dies, it may be in response to a particularly high challenge from the environment, or if in response to an infectious disease, the pathogens may not survive their occupation of the animal and may have difficulty in passing to other animals.

The durations of even the common diseases have received little attention despite their importance to welfare. Most research on severity has focussed on minimising production losses and correcting the health of the cow. However, the most common ailments affecting dairy cows in modern dairy systems are now considered for their impact on welfare.

### *2.1 Lameness*

Lameness is a very common welfare problem in most dairy farming systems, with multiple causes and impacts on the cows' health (Fig. 1). The most common causes are nutritional and environmental, specifically divergence

**Figure 1** Lameness is one of the major health problems of cows in intensive dairy farming systems.

from a natural diet (cows need very high energy and protein supply for high milk yields) and natural environment (cows are now mostly kept indoors, on concrete floors for the ease of cleaning and longevity). The biggest influence of lameness on welfare is due to the pain experienced, particularly during walking, but an inability to stand to feed, to walk comfortably to the milking parlour and to interact with conspecifics will all cause considerable distress. In extreme cases lameness increases heart and respiratory rates and rectal temperature (Tadich et al., 2013). Lameness in a typical British dairy herd has been recorded to last, on average, 1.88 months (Phillips, 1990), but this clearly varies between the different types of lameness. Those that are lame because of interdigital phlegmon or digital dermatitis may require 28 days to heal, but sole ulcers, white line disease and heel erosion all take longer (Whay et al., 1998). However, Eriksson et al. (2020) recorded that lameness in cows' dry period lasts for shorter periods, most commonly only 2 weeks for mild lameness and 4 weeks for severe lameness. Interventions reduce the duration of lameness and make the recording of the natural duration difficult.

### 2.2 Mastitis

Mastitis is the most common infectious disease caused by bacteria in dairy cows. There has been considerable research to reduce the duration of the infection by the use of intra-mammary antimicrobials (Ruegg, 2017). Although the recording of milk samples is not usually frequent enough to give a precise duration on a national basis, the fact that approximately 8–10% of recorded cows in Sweden have elevated SCC (>200 000/mL) suggests an average duration of infection of about 1 month (Frossling et al., 2017). Qualitative behaviour analysis suggests that infected cows are dejected and lethargic (des Roches et al., 2018), and it is therefore highly likely that mastitis causes pain, as well as the cows feeling the adverse effects of pyrexia, particularly during coliform mastitis. The pain derives from udder inflammation and sensitivity to touch, with swelling of the udder or individual glands, as well as pressure on the swollen udder from other body parts, such as limbs. Pain is also evident from the cows' behaviour, making them vocalize, flinch on touch and leg raise, and also from increased stress-related hormones, in particular cortisol and acute-phase proteins. In addition, mastitis increases cows' heart rate and stress responses associated with cortisol release (des Roches et al., 2018).

### 2.3 Metritis

Metritis, or uterine infection, is also common in dairy cows and is associated with pain, low feed intake and reduced competitiveness at a feeding bunk (Stojkov et al., 2015). The vulval lips become painful to touch, the vagina inflamed and the passage of urine is difficult and painful. The palpation of the uterus is associated with further pain and should be avoided. Cows become withdrawn, engage in bruxism, in which they clench their teeth to try to control the pain and are reluctant to get involved in normal agonistic interactions (Sahar et al., 2020). Rumination is suppressed, and pyrexia often ensues.

### 2.4 Acidosis

Acidosis is a digestive disorder occurring when cows consume large amounts of highly fermentable carbohydrates. It occurs in about 20% of early- to mid-lactation cows, making it one of the most common health problems in cows. It is associated with a reduction in rumen pH to below 5.6 for at least 3 h/d, as a result of excessive accumulation of the end products of microbial digestion - volatile fatty acids and lactic acid - and also perhaps the lysis of gram-negative rumen bacteria because of the low pH conditions. Free lipopolysaccharide following the cell lysis may damage the rumen wall (rumenitis) through the release of inflammatory cytokines (Zhao et al., 2018), and it is notable that inflammatory markers increase in the blood during sub-acute ruminal acidosis (SARA) (Gozho et al., 2005). SARA is often associated with sub-acute and acute laminitis, a painful form of lameness that is thought to derive from toxins entering the blood at the damage sites in the rumen. High production, intensively managed herds have a high prevalence of subclinical laminitis, which predisposes cows to other hoof disorders, such as sole ulcer, white line disease and toe necrosis.

### 2.5 Ketosis

Ketosis or acetonaemia is a disease of high-yielding cows, which stems from an accumulation of ketone bodies in the blood and is associated with a shortage of glucose precursors. Cows characteristically have a reduced feeding time; become withdrawn and lethargic; and do not engage in agonistic interactions (Sahar et al., 2020). It is often associated with liver lipidosis.

### 2.6 Other production-related diseases

Other production-related diseases mainly relate to mineral nutrition management, in particular, hypocalcaemia and hypomagnesaemia (colloquially known as milk fever and grass staggers, respectively). The former is usually caused by the major increase in calcium secretion into the mammary gland at the start of lactation and is best avoided by encouraging cows to mobilize calcium from their bones during their transition to lactation by feeding a low calcium diet. Cows quickly become weak as a result of muscle paralysis, stop ruminating and defaecating and have a reduced temperature. Death may ensue when a cow rolls onto her side and rumen gases cannot escape, causing excessive pressure on the heart. Hypomagnesaemia is common if excessive potassium is used to fertilize pasture grasses, to which they respond with faster growth, but the high potassium and low sodium contents create an imbalance of divalent cations in the rumen that retards magnesium absorption. Cows become excitable and may go into tetany with their legs paddling violently. Death may ensue as a result of heart failure. Replacement of some of the potassium fertilizer with a sodium fertilizer should solve the problem.

### 2.7 Respiratory diseases

As well as nutritional disorders, some of which involve micro-organisms, such as Johne's disease, respiratory diseases are among the most common infectious diseases in dairy cattle. They are particularly likely to infect cattle that are stressed, such as transported cattle or weaned dairy calves that are housed. Respiratory infections are characterized by initial viral pathogens that kill alveolar and intravascular macrophages, followed by a secondary bacterial infection. The resultant hypoxia, hypercapnia and consequent respiratory failure is likely to be very stressful. Mentally, the hypoxia can lead to reduced alertness and eventually coma and death. Oxygen requirements are reduced by minimizing activity, including moving and eating. Cardiac output increases in an attempt to compensate, mainly by elevated heart rate but also stroke volume. Hypercapnia causes anxiety, accompanied by respiratory acidosis, but renal mechanisms attempt to return arterial pH to normal. Both hypoxia and hypercapnia stimulate increased respiratory rate, leading eventually to gasping, dyspnea, apnoea and, ultimately, death.

### 2.8 Infectious diseases

Many infectious diseases are controlled in modern, intensive dairy farms, through vaccination and culling of infected animals. However, in developing countries, which have the largest number of cows worldwide, they are common. One of the most common that has a large welfare impact is foot and mouth disease. Initially it causes extremely painful lesions on the tongue and in the mouth that prevent the cow from feeding. Later lesions form in the hooves, typically in the interdigital space, causing discomfort in the feet, and stepping, kicking and shaking of the limbs. Other important infectious diseases with a significant impact on cattle welfare include fascioliasis, trypanosomiasis, tuberculosis, brucellosis, pasteurellosis, Johne's disease, bovine viral diarrhoea (BVD) and pneumonia.

## 3 Subclinical diseases

The impact of subclinical diseases on welfare is understood even less than that of clinical disease. However, there is evidence of cows experiencing pain during subclinical mastitis (Peters et al., 2015) before they experience the swelling of the udder, fever and loss of appetite typical of cows with clinical mastitis. Mild locomotion disorders, suggesting subclinical lameness, have produced increases in haptoglobin concentrations in blood, demonstrating an inflammatory response (Tadich et al., 2013). These two examples suggest that subclinical disease can adversely affect cows' welfare, but, as with clinical disease, the impacts need to be better understood. Detection of subclinical disease is still in its infancy, and many cases will self-heal, meaning that treating subclinical disease may not be necessary. Thus, recognition of subclinical mastitis is theoretically possible (e.g. Polat et al., 2010; des Roches et al., 2018), but the question remains as to whether cases should be treated, and if so how. The use of antibiotics renders milk unsaleable for several days and may contribute to antibiotic resistance. Separating cows with subclinical mastitis from their herdmates may stress them and possibly even exacerbate the disease but does help to limit the contiguous spread of the responsible organism to other cows. Milking affected cows last, as well as paying careful attention to the cleanliness of these cows, is a good strategy to limit the transmission of the organism.

## 4 Stress and immune function

The short lives of dairy cows in industrial farming systems (De Vries, 2013) and high culling rate as a result of diseases such as lameness and mastitis suggest that cows in such systems are exposed to stressful conditions. The development of cows with genetic potential for high milk yields has been associated with a steep decline in fertility, demonstrating that the physiological demands on cows are too great for them to produce milk and calves at the rate expected over several years (Grandin, 2013). Other problems include difficulty in maintaining body condition in early lactation cows and loss of disease resistance. Although restrictions on output (growth) rate have been introduced into the broiler industry by some progressive farming system-accrediting bodies (RSPCA, undated), no such move has been taken in relation to the ever-increasing milk output of cows in the dairy industry.

Cortisol can indicate stress responses to diseases such as mastitis, lameness, heat stress and production diseases (e.g. Gellrich et al., 2015). Cortisol concentration in hair is a useful long-term stress measure that detects impacts on welfare (Sharma et al., 2019a) since it has been correlated with body lesions and other health problems in dairy cows. Even small problems, which we might consider relatively harmless to cows, such as faeces in their passageways and lack of access to outdoors, can cause long-term secretion of cortisol that is detected in hair concentrations (Sharma et al., 2019a). Care in interpreting results must also be taken because cortisol secretion is reduced during energy deficiencies, and it has a role in glucose homeostasis

(Gross et al., 2015). There are reduced anti-inflammatory effects of cortisol in cows in a negative energy balance in early lactation (Burton and Erskine, 2003). Stress may change a normal inflammatory response to a $T_H2$-dominated antibody response of B cell production of IgM and $IgG_1$, leading to limited neutrophil recruitment and proliferation of the pathogen (Burton and Erskine, 2003). Furthermore, cortisol that is required to be released around parturition for the expulsion of the foetus and establishment of milk production is also suppressing the inflammatory responses and rendering cows susceptible to mastitis. Glucocorticosteroids released at parturition suppress CD62L gene expression in neutrophils, preventing them from performing their surveillance function in the blood. In future a better understanding of genetic variation in mammary defence responses may be able to be used to breed cows with reduced susceptibility to mastitis after calving.

A cows' stress response depends on the context, a settled cow in a familiar environment will respond less than a nervous cow in a novel environment. Multiple stressors are likely to have cumulative effects. At all times it must be remembered that cows are gregarious animals, with complex social groups. Isolation stresses cattle.

## 5 Mental health

Most animal welfare scientists accept that animals' experiences, and their feelings in response to these experiences, are the key concepts of animal welfare. The severity of a disease could impact on the cows' feelings through the levels of pain, fatigue, weakness, irritability and a feeling of being anxious or stressed (Desire et al., 2002). All these are common effects of diseases. Less common are feelings of being frightened, angry or unfriendly to conspecifics or humans, but these could be secondary emotional responses to disease and dependant on an animal experiencing one of the primary emotional responses. Some responses may be seen only in the early stages, such as when animals are restless and easily panic. Some may be more likely in the later stages, such as when animals become frustrated and withdrawn. Cows' initial emotional response to mastitis appears to be one of suffering, dejection and lethargy, which later change to make the cows appear more calm, confident and relaxed (des Roches et al., 2018). From all the negative emotions listed above, pain is the one that is of greatest concern.

### *5.1 Pain*

Although pain is one of the most common and obvious welfare impacts, it is hard to recognize in cattle, as overt displays of responses to pain would, during the course of their evolution, attract predators. Nevertheless, it is believed that there are many procedures that produce pain in dairy cattle, including disbudding/dehorning, branding, supernumerary teat removal, caesarean section and claw amputation. Many disease conditions will also cause pain, especially lameness, mastitis, pneumonia, joint ill, dystocia, prolapsed uterus and peritonitis (Cockroft, 2015 pp. 239). Topical analgesia and anaesthesia can control pain but are often underused, with inadequate recognition by farmers and veterinarians of the pain that cattle experience (Becker et al., 2014).

A pain scale for cattle has been developed, which uses many of the more subtle measures, including their attention towards the surroundings, head, back and ear positions; facial expression; and response to approach by a human (Gleerup et al., 2015). Subtle indicators can be detected by good stockpeople, in particular, backward-facing and drooping ears, dilated nostrils, an open mouth and raised eyebrows (Muller et al., 2019). Other associated behaviour changes include decreased interaction with conspecifics and reduced responsiveness to stimuli. Many responses depend on the source of the pain; for example, reduced locomotion and kicking when touched are indicative of leg or hoof pain, inspiratory grunting of thoracic pain or pneumonia. Withdrawal of a body part is a standard response to palpation of any area producing pain.

For research purposes, an algometer, measuring the maximum force that can be applied to a point on the animal's body before an avoidance behaviour is evident, can be used to evaluate the pain caused by mastitis and lameness (Fitzpatrick et al., 2013). However, animals cannot tell you when they are in pain and just relying on a movement response by the animal may lead to erroneous conclusions, since they may relate to other factors, such as the proximity or known personality of the person.

The use of analgesics is clearly warranted from an animal welfare perspective, but it is less clear whether farmers will benefit financially. Farmers are generally aware that cows experience pain from the most common diseases, particularly *E. coli* mastitis, digital dermatitis and interdigital necrobacillosis, but they also believe that other forms of mastitis produce only low levels of pain (Thomsen et al., 2012). Worryingly, in Thomsen et al.'s research veterinarians were less aware than farmers of pain suffered by cows when afflicted with these most painful diseases, but they were more likely to advocate the use of analgesics. There are relationships between dairy farmers' attitudes towards their cows and the cows' productivity. Dairy farmers with high levels of empathy for their cows are reported to have lower yields from their cows (Kielland et al., 2010), perhaps because they do not want to push their cows to high yields requiring high energy feeds, with all the associated disease risks.

Pain can be referred to other areas in the body than the one in which the damage has occurred. For example, cows with mastitis have a reduced pain threshold when a thermal stimulus is applied to their leg, demonstrating systemically increased nociception (Peters et al., 2015). This may derive from inflammation spreading from the mammary gland to the leg. The disease may cause allodynia or central pain sensitization, which may progress to hyperalgesia in extreme cases. Cows with even moderately severe lameness have reduced nociceptive thresholds (Tadich et al., 2013). There has been little exploration of allodynia and hyperalgesia in cows since pain is still little understood in dairy cows.

## 5.2 Anxiety, fear and depression

Anxiety is associated with an increased expectation of negative events. Depression is associated with a reduced expectation of positive events and an inability to feel pleasure in activities that cattle usually enjoy (anhedonia). Presenting calves with ambiguous cues in a judgement bias test has provided evidence that disbudding creates both a state of anxiety (Neave et al., 2013) and depression (Lecorps et al., 2019). The separation of the dam from her calf makes her decisions more pessimistic (Daros et al., 2014), which could also indicate depression. Evidence of depression in cows is further provided by their tendency to have their heads turned downwards during stressful events, such as in heat stress (Idrus, 2020) or on entering an abattoir (Hemsworth et al., 2011), in the same way that depressed people look down to avoid eye contact.

Observations of lateralized behaviours provide another indicator of the mental impact of different situations for cows. These behaviours include passing a person on the right or left to view them with their left or right eye, respectively, cows' acceptance of a person in close proximity of their left or right side or using their left or right limbs when they stamp in response to a stressor, which have all provided evidence that some stressors, particularly those involving people, invoke an anxiety response in cattle (Goma et al., 2018; Uddin et al., 2019). We have also recently found that groups of cattle accept people more readily when they are positioned in their left field of view (Robins et al., 2018) and that subordinate cows are more likely to view dominant cows and unfamiliar humans with their left than right eye (Robins and Phillips, 2010; Phillips et al., 2015). Such lateralized behaviour appears more pronounced in intensive dairy herds than in extensive partially suckled herds (Phillips et al., 2003). There is also evidence that feeding cows with a truck that arrives in cows' left eye field of view improves their productivity (Rizhova and Kokorina, 2005). The underlying physiological basis for this new evidence of anxiety and fear arising in response to everyday events in dairy cows' lives is that cows' brains are specialized to process different types of signals in their two hemispheres. Avoiding duplication in this way increases the efficiency of signal processing, particularly in gregarious ungulates, which can then more rapidly synchronize their escape movements with conspecifics when confronted by a predator (Robins and Phillips, 2010). Thus the right brain hemisphere, connected to the left eye, processes flight/fight and spatial responses, and it has a diffuse focus. Conversely the left brain hemisphere, connected to the right eye, has a narrow focus and processes communication with conspecifics and temporal responses. The sympathetic nervous system, responsible for controlling involuntary responses to stressful situations, is primarily organized by the right brain hemisphere.

### 5.3 Contentment

It is harder to find indicators of contentment, or positive welfare, in dairy cows. However, rumination may be one of the best subtle indicators. Cows that are in pain or suffering in other ways, for example, during mastitis (Fitzpatrick et al., 2013) spend less time ruminating. Other indicators of contentment include a downward ear position and eyes half closed, revealing limited sclera (Battini et al., 2019).

## 6 Case study: the health and welfare of cows in Indian shelters

Much is known about the relationships between health and welfare in dairy cows in the intensive production systems favoured in developed countries. However, in the country with the most dairy cows, and therefore the greatest risk to global cow welfare, the relationship is little understood. India has at least 190 million dairy cows, more than any other country in the world, and respect for their welfare and ethical treatment has led to the development of a network of shelters across India to which they can be retired after they have served their useful, productive life (Kennedy et al., 2018). Thus rather than being turned out into the streets or slaughtered as they would be in any other country in the world, cows in India can live out their natural lives in a protected environment (Fig. 2). Most die of old age, and many are well into their teens (Sharma et al., 2020). In a recent study by Sharma et al. (2019a–e) dairy cow welfare was assessed in India shelters.

Some welfare problems for the cows were similar to those in intensive dairies in Western countries, such as the lack of space for the cows and limited access to pasture, giving little freedom of movement (Fig. 3). However, lameness was at much lower levels in the Indian shelters, with a prevalence of 4% (Sharma et al., 2019c), compared with an average incidence rate of 22 cases per 100 cows per year in the UK (Macrae and Esslemont, 2015). The other major scourge of Western dairies, mastitis, was largely absent because the cows were not lactating. Even in peri-urban dairy farms in India, mastitis is only present in less than 10% of the cows (Joshi and Gokhale, 2006) because their milk yields are much lower than in Western dairy farms, which have, in the UK, for example, a median incidence of 33 cases per 100 cows per year (Macrae and Esslemont, 2015). Floor quality tended to be more of an issue than in Western dairies,

**Figure 2** Cows in an Indian shelter (photo courtesy of Dr Arvind Sharma).

**Figure 3** Cows in Indian shelters are sometimes overstocked (photo courtesy of Dr Arvind Sharma).

causing carpal joint hair loss and swellings (Sharma et al., 2019b). Human interactions with cows are almost universally benign, with most cows allowing people to touch them (Sharma and Phillips, 2019), and from scenes I have witnessed in India, some even allowing shelter managers to lie on top of them. This contrasts with Western dairies, where cows are a commodity that must be quickly managed for expeditious milk extraction on a daily basis, and most show considerable fear of people because of the pain inflicted on recalcitrant cows when they are being handled, mostly by being hit on the hind quarters. To an Indian, inflicting pain on a cow to make it move is an anathema. Thus, most cows allowed themselves to be touched in our Indian survey, whereas in a survey of 50 Australian dairy farms median avoidance distance of cows was 1.7 m, with only a very few cows (5% of observations) allowing themselves to be touched (Beggs et al., 2019). The average age of cows in the shelters in our study was 11, which again contrasts with intensive Western dairy production systems, in which most cows last only 2 or 3 lactations, making them 4 or 5 years old when they are slaughtered for meat (De Vries, 2013).

Demonstrating the better welfare of cows in Indian shelters could lead to people and governments in other countries considering their responsibilities towards cows, which are usually killed at an early age in Western dairies because they are infertile, lame, have mastitis or are considered 'too old' (Macrae and Esslemont, 2015). The Indian shelters are often run by charities, have considerable public support, including financial provision and bringing food for the cows (Sharma et al., 2020). In a recent survey conducted in the vicinity of shelters in India, 92% of members of the public said it was very important to shelter cows after their productive life, 88% said that they had no issues with their nearest shelter and 100% said that they supported the shelter (Sharma et al., 2019d). Part of the Indian public commitment to giving dairy cows a good retirement home comes from India's dominant Hindu religion, but quite a few people (33%) reported that they visited them because they enjoyed seeing them, wanted to feed them or check their welfare. Furthermore, their religion espouses kindness to animals generally, which makes managing cows on farms easier.

Considering whether to introduce a similar long-term responsibility towards cows in Western dairy farms would require research to evaluate the attitudes of members of the public. Milk retailers might well be surprised to find that many would tolerate the necessary increase in the milk price to allow cows a healthy and happy retirement. The recent increased popularity of milk alternatives in Western countries (e.g. Grand View Research, 2019), based on soya, almond extracts and others, which are usually considerably more expensive than cows' milk, suggests that the public are prepared to pay for a more ethical source of milk. If people in the most populated country in the world, with one of the lowest per capita incomes, can look after their cows' welfare in this way, it is time that those in Western countries consider their responsibilities in this area, particularly in the light of the negative impact of recent intensification on cow welfare.

## 7 Summary

Dairy farming has, over the last half century, changed from being family based in most parts of the world into an industrial process. There has been an increase in farm size from tens of cows to thousands on each farm, and each animal producing approximately twice as much milk as if did half a century ago. The most obvious upshot of this transition is a relative reduction in milk price, even when compared with other foodstuffs, demonstrating that it has been one of the most successful industrializations of any animal production system. However, the other major outcome has been an increase in the health challenges to cows, arising from the changes, metabolic and biophysical, associated with the industrialized dairy farming system blueprint that has been adopted in many parts of the world. Cows succumb to poor health at much younger ages, and are culled from the herd much earlier than they were in the era of family farms. In the United States cows on average last only two or three lactations in the herd and are slaughtered on average at 4.8 years of age (De Vries, 2013). Although this is the case in much of the world, in the country with the world's largest dairy herd, India, small-scale family farms are still in the majority, and cows are retired in shelters, demonstrating a major commitment to cow welfare that is supported by the public. The concerns of the public about cow welfare in industrialized systems are leading many to buy alternative vegetable-based milk, and cow milk may become a niche product unless there are major changes in cow welfare. An absence of access to pastureland (Fig. 4), big groups of cows standing on concrete for prolonged periods with many associated foot problems (Oehm et al., 2019), breeding cows for high milk yields that predisposes them to mammary gland infections (Bobbo et al., 2019), a diet which is not suited to rumen fermentation but is necessary to provide for the requirements for very high milk yields, are some of the key welfare problems. Unless these are addressed rapidly, it is likely that the public will further transition to cow milk alternatives, reducing demand for cow products and putting further pressure on cow welfare. This long-term forecast contrasts with the current global increase in demand for cow milk that arises principally from growing affluence in Eastern and SE Asian countries whose citizens were formerly unable to afford cow milk in their diet.

## 8 Future trends in research

The future survival of the world's dairy industries will depend on supporting researchers who can address the welfare problems that have been identified.

**Figure 4** Access to pasture is important for good cow health and welfare.

Funding for the research should be at least in part from government and the public since the issue is one of public interest. Industry funding runs the risk of vested interests influencing whether the research is impartial or not (Van der Schott and Phillips, 2013).

Research should attempt to make progress in the welfare impact of key diseases and may simply prevent the current situation becoming even worse. Over the last few years, for example, tremendous strides have been taken in controlling mastitis in dairy cows, which is a serious welfare concern (Ruegg, 2017). However, little progress has been made in reducing the overall welfare impact because, even though better control measures have been developed, the risk factors in modern, intensive dairy farms are greater than before.

The routine use of antibiotics, one of the successful measures of welfare management developed in the last century (Ruegg, 2017), cannot continue without jeopardizing the effectiveness of antibiotics in both humans and other animals. In China, for example, resistance to erythromycin and clindamycin has been found in about one-third of bacterial strains (Zhang et al., 2019), and at a molecular level, antimicrobial resistance genes have been found in nearly all Chinese isolates from dairy cow mastitis samples (Qu et al., 2019). Of particular concern is vancomycin resistance in *S. aureus* from mastitic milk, as vancomycin is used to treat methicillin-resistant staphylococcal infections in humans. Many alternatives are being tested, including bacteriocins (proteinaceous toxins produced by bacteria), teat sealants, immune system stimulants such as ginseng, medicinal plants, as well as cytokines and bacterial cultures (Kitching et al., 2019). Another possibility for mastitis control is the integrated cow-calf system that allows calves to suckle their dams for a period each day, as advocated above for the welfare of both cow and calf (Johnsen et al., 2016; Noudeke et al., 2017). Regular stripping out of the mammary gland by the calf reduces the bacterial contamination of a quarter much more effectively than machine milking. In the immediate future, selective rather than blanket dry cow therapy offers hope of reducing antibiotic use without loss of efficacy (Bucher and Bleul, 2019). It took at least half a century of research to develop antibiotic therapy for mastitis control in dairy cows (Ruegg, 2017), the development of the next stage treatment methods will have to be much faster if we are to avert either an inability to treat some human bacterial infections or a major dairy cow welfare problem.

In relation to lameness, there has already been a considerable research effort to find solutions for digital dermatitis and, more recently, laminitis (Silva et al., 2020). However, the intensification of dairy farming worldwide, with more indoor housing on concrete, larger herds and feeding of preserved feeds, suggests that the problems will only increase.

Another upcoming area of research in dairy cow welfare is the need to address issues of heat stress in many parts of the world. Demand for milk and milk products is increasing most rapidly in east and SE Asia, where hot temperatures prevail. Already there are significant risks to cow welfare posed by heat stress in many parts of the world, particularly the United States, Australia and central and Latin America. Dairy protein demand is also surging in the Middle East, where heat stress management is already needed in many dairies. We know how to avert heat stress in these developed economies (Figs 5 to 7). Yet many of the world's dairy cows are kept by subsistence farmers in India and Africa in the tropics and subtropics, where rising temperatures will jeopardize their livelihoods, food security, and sources of regenerating capital and cooking fuel, in the form of dried faeces. Globally 60% of dairy cow production is in the tropics and subtropics, where it is very vulnerable to global climate change, in particular increased ambient temperatures and in some places higher humidity. Risks to cow health and welfare are likely to be very significant. Dairy cows are the animals that are most prone to heat stress because of their large body size and internal heat production, which results from the fermentation of large amounts of fibrous feed needed to support the high levels of milk production and their lack of protection from the ambient conditions. The Holstein-Friesian cow breed that has come to dominate production systems worldwide, even in the hottest regions, has particularly low resilience to hot environmental conditions. Its relative value in heat-stress-prone areas depends on the potential for mitigation of the risk, through the provision of shade, cool drinking water, showers, and so on, and the importance of having cattle that survive for several years. Some benefits might be possible by incorporating genes associated with heat stress resistance into Holstein cows, but the benefits are yet to be

**Figure 5** Provision of shade for dairy cows in an Australian dairy farm.

**Figure 6** Provision of fans for dairy cows in an Australian dairy farm.

accurately determined (Henry et al., 2018). Natural shade in the form of trees on farms is increasingly advocated (England et al., 2020) (Fig. 8), but will take considerable time to establish. It is also likely that climate change-induced thermal and water stress of forage crops will create potential feed shortages.

**Figure 7** Provision of sprinklers for dairy cows as they enter and leave the milking parlour of an Australian dairy farm.

The search for genetic solutions to health problems will be difficult. The heritabilities of the major diseases are all low (mastitis 0.01–0.03; lameness 0–0.02; metabolic disorders 0–0.006; reproductive disorders 0.005–0.02), and sometimes they are correlated with milk production traits, making it difficult to achieve disease control without penalizing milk production (Abdelsayed et al., 2017; Bobbo et al., 2019). However, it is still important to include health traits in selection indices if there is a continued selection for increased milk production; otherwise cows' health will deteriorate still further.

Research is needed into the acceptability to the public of novel production systems that obviate some of the risks to welfare posed by current systems – integrated cow-calf systems and sheltering cows that have had health problems in dairy farms, for example. Systems of dairy production that have allowed calves access to their dams for part of each day have long been used

**Figure 8** Provision of natural shade from trees is important to help cattle cope with heat stress.

in some tropical and subtropical countries that need both meat and milk from cows (Johnson et al., 2016). Such integrated systems need to be explored for developed countries, where there is particular concern by the public for the welfare of cow and calf following separation. Acceptability is likely to be location specific and may change over time, but determining key drivers of attitudes would help farmers and retailers to plan changes in the systems used for dairy production.

## 9 Where to look for further information

### *9.1 Books*

- Blowey, R. 2016. *The Veterinary Book for Dairy Farmers*, 4th Edition. 5M Publishing. Cockroft, P. D. (Ed.). 2015. *Bovine Medicine*, 3rd Edition. Wiley Blackwell, Chichester, UK.
- Fuller, M., Benevenga, N., Fisher, C., Lall, S., Omed, H. M., Axford, R. F. E. and Phillips, C. J. C. (Eds). 2004. *Encyclopaedia of Farm Animal Nutrition*. CAB International, Wallingford.
- Phillips, C. J. C. 2018. *Principles of Cattle Production*, 3rd Edition. CABI, Wallingford, UK.

- Rushen, J., de Passillé, A. M., von Keyserlingk, M. A. G. and Weary, D. M. 2008. *The Welfare of Cattle*. Springer, Dordrecht, The Netherlands.
- Tucker, C. (Ed.). 2018. *Advances in Cattle Welfare*. Woodhead Publishing, Duxford, UK.
- Webster, J. (Ed.) 2017. *Achieving Sustainable Production of Milk, Volume 3: Dairy Herd Management and Welfare*. Burleigh Dodds Science Publishing, Cambridge, UK (ISBN: 978 1 78676 052 4; www.bdspublishing.com).
- Webster, J. 2013. *Animal Husbandry Regained*. Routledge and Earthscan, London.

## 9.2 Key societies and organizations

- World Association for Animal Production (https://www.waap.it/).
- World Association for Buiatrics (https://buiatrics.com/).
- World Veterinary Association (http://www.worldvet.org/).
- World Organisation for Animal Health (https://www.oie.int).
- Food and Agriculture Organization of the United Nations, Gateway to Farm Animal Welfare, Spotlight on the Welfare of Dairy Cattle (http://www.fao.org/ag/againfo/themes/animal-welfare/spotlight/en/).

## 9.3 Key journals

- *Animals* (https://www.mdpi.com/journal/animals).
- *Animal, the International Journal of Animal Biosciences* (https://www.cambridge.org/core/journals/animal).
- *Applied Animal Behaviour Science* (https://www.journals.elsevier.com/applied-animal-behaviour-science).
- *Animal Welfare* (http://www.UFAWorg.uk).
- *BMC Veterinary Research* (https://bmcvetres.biomedcentral.com/).
- *Journal of Dairy Science* (https://www.journalofdairyscience.org).
- *The Veterinary Journal* (https://www.journals.elsevier.com/the-veterinary-journal/).

## 9.4 Key conferences

- World Buiatrics Congress (http://buiatrics.com/congresses/).
- World Conference for Animal Production (https://www.waap.it/world-conference/).
- World Conference for Farm Animal Welfare (http://wcfaw.com/).
- The World Veterinary Association Congress (https://www.wvac2020.com/).

## 9.5 Some major international research projects

- Pro-Q: On-Farm Research Network for Dairy Cattle (https://www.fibl.org/en/themes/projectdatabase/projectitem/project/884.html).
- Vet Compass (https://www.vetcompass.org/).
- East African Dairy Development Project (https://cgspace.cgiar.org/handle/10568/79437).

## 9.6 Top five research centres

1 Animal Welfare Program, University of British Columbia, Canada.
2 Centre for Animal Welfare and Anthrozoology, University of Cambridge, UK.
3 Department of Animal Science, Aarhus University, Denmark.
4 John Oldacre Centre for Sustainability and Welfare in Dairy Production, Bristol University Veterinary School, UK.
5 OIE Collaborating Centre for Animal Welfare Science and Bioethical Analysis: The David Bayvel consortium.

# 10 References


Abdelsayed, M., Haile-Mariam, M. and Pryce, J. E. 2017. Genetic parameters for health traits using data collected from genomic information nucleus herds. *J. Dairy Sci.* 100(12), 9643-9655.

Battlnl, M., Agostini, A. and Mattiello, S. 2019. Understanding cows' emotions on farm: are eye white and ear posture reliable indicators? *Animals* 9(8), 477. doi:10.3390/ani9080477.

Becker, J., Reist, M. and Steiner, A. 2014. Factors influencing the attitudes of cattle veterinarians, farmers, and claw trimmers towards the pain associated with the treatment of sole ulcers and the sensitivity to pain of dairy cows. *Vet. J.* 200(1), 38-43. doi:10.1016/j.tvjl.2014.01.016.

Beggs, D. S., Jongman, E. C., Hemsworth, P. H. and Fisher, A. D. 2019. The effects of herd size on the welfare of dairy cows in a pasture-based system using animal- and resource-based indicators. *J. Dairy Sci.* 102(4), 3406-3420.

Bertoni, G., Trevisi, E., Houdijk, J., Calamari, L. and Athanasiadou, S. 2016. Welfare is affected by nutrition through health, especially immune function and inflammation. In: Phillips, C. J. C. (Ed.), *Nutrition and the Welfare of Farm Animals*. Springer, Dordrecht, The Netherlands, Chap. 5, pp. 85-113.

Bobbo, T., Roveglia, C., Penasa, M., Visentin, G., Finocchiaro, R. and Cassandro, M. 2019. Genetic relationships of alternative somatic cell count traits with milk yield, composition and udder type traits in Italian Jersey cows. *Anim. Sci. J.* 00, 1-10.

Bucher, B. and Bleul, U. 2019. The effect of selective dry cow treatment on the udder health in Swiss dairy farms. *Schweiz. Arch. Tierheilkd* 161(9), 533-544.

Burton, J. L. and Erskine, R. J. 2003. Immunity and mastitis, some new ideas for an old disease [Review]. *Vet. Clin. North Am. Food Anim. Pract.* 19(1), 1-45, v.

Cockroft, P. D. 2015. Pain management in cattle practice. In: Cockroft, P. D. (Ed.), *Bovine Medicine* (3rd edn). Wiley-Blackwell, Chichester, UK.

Daros, R. R., Costa, J. H. C., von Keyserlingk, M. A. G., Hötzel, M. J. and Weary, D. M. 2014. Separation from the dam causes negative judgement bias in dairy calves. *PLoS ONE* 9(5), e98429.

De Vries, A. 2013. Cow longevity economics: the cost benefit of keeping the cow in the herd. DeLaval, Milkproduction.com. Available at: http://www.milkproduction.com/Library/Scientific-articles/Management/Cow-longevity-economics-The-cost-benefit-of-keeping-the-cow-in-the-herd/. Accessed 4 February 2020.

des Roches, A. D., Lussert, A., Faure, M., Herry, V., Rainard, P., Durand, D., Wemelsfelder, F. and Foucras, G. 2018. Dairy cows under experimentally-induced *Escherichia coli* mastitis show negative emotional states assessed through Qualitative Behaviour Assessment. *Appl. Anim. Behav. Sci.* 206, 1–11. doi:10.1016/j.applanim.2018.06.004.

Désiré, L., Boissy, A. and Veissier, I. 2002. Emotions in farm animals: a new approach to animal welfare in applied ethology. *Behav. Processes* 60(2), 165–180.

DeVries, T. J., Aarnoudse, M. G., Barkema, H. W., Leslie, K. E. and von Keyserlingk, M. A. G. 2012. Associations of dairy cow behavior, barn hygiene, cow hygiene, and risk of elevated somatic cell count. *J. Dairy Sci.* 95(10), 5730–5739. doi:10.3168/jds.2012-5375.

England, J. R., O'Grady, A. P., Fleming, A., Marais, Z. and Mendham, D. 2020. Trees on farms to support natural capital: an evidence-based review for grazed dairy systems. *Sci. Total Environ.* 704, 135345. doi:10.1016/j.scitotenv.2019.135345.

Eriksson, H. K., Daros, R. R., von Keyserlingk, M. A. G. and Weary, D. M. 2020. Effects of case definition and assessment frequency on lameness incidence estimates. *J. Dairy Sci.* 103(1), 638–648. doi:10.3168/jds.2019-16426.

Fitzpatrick, C. E., Chapinal, N., Petersson-Wolfe, C. S., DeVries, T. J., Kelton, D. F., Duffield, T. F. and Leslie, K. E. 2013. The effect of meloxicam on pain sensitivity, rumination time, and clinical signs in dairy cows with endotoxin-induced clinical mastitis. *J. Dairy Sci.* 96(5), 2847–2856.

Frossling, J., Ohlson, A. and Hallen-Sandgren, C. 2017. A systematic review and meta-analyses of risk factors associated with lameness in dairy cows. *J. Dairy Sci.* 100(9), 7368–7378. doi:10.3168/jds.2016-12333.

Gellrich, K., Sigl, T., Meyer, H. H. D. and Wiedemann, S. 2015. Cortisol levels in skimmed milk during the first 22 weeks of lactation and response to short-term metabolic stress and lameness in dairy cows. *J. Anim. Sci. Biotech.* 6(1), 31. doi:10.1186/s40104-015-0035-y.

Gleerup, K. B., Anderson, P. H., Munksgaard, L. and Fjorkman, B. 2015. Pain evaluation in cattle. *Appl. Anim. Behav. Sci.* 171, 25–32.

Goma, A. A., Pearce, G. P., Uddin, J., Rimon, E., Davies, H. and Phillips, C. J. C. 2018. A forced lateralisation test for dairy cows and its relation to their behaviour. *Appl. Anim. Behav. Sci.* 207, 8–19. doi:10.1016/j.applanim.2018.06.008.

Gozho, G. N., Plaizier, J. C., Krause, D. O., Kennedy, A. D. and Wittenberg, K. M. 2005. Subacute ruminal acidosis induces ruminal lipopolysaccharide endotoxin release and triggers an inflammatory response. *J. Dairy Sci.* 88(4), 1399–1403.

Grand View Research. 2019. Soy milk market size, share & trends analysis report by product type (flavored, unflavored), by application (desserts, cheese & snacks, beverages), by region, and segment forecasts, 2019–2025. Available at: https://ww

w.grandviewresearch.com/industry-analysis/soy-milk-market. Accessed 31 January 2020.

Grandin, T. 2013. Genetics and other technologies for increasing the productivity of cattle, sheep and pigs: welfare implications. In: Grandin, T. and Whiting, M. (Eds), *Are We Pushing Animals to Their Biological Limits? Welfare and Ethical Implications*. CABI Publishing, Wallingford, UK, pp. 49–62.

Gross, J. J., Wellnitz, O. and Bruckmaier, R. M. 2015. Cortisol secretion in response to metabolic and inflammatory challenges in dairy cows. *J. Anim. Sci.* 93(7), 3395–3401.

Hemsworth, P. H., Rice, M., Karlen, M. G., Calleja, L., Barnett, J. L., Nash, J. and Coleman, G. J. 2011. Human-animal interactions at abattoirs: relationships between handling and animal stress in sheep and cattle. *Appl. Anim. Behav. Sci.* 135(1–2), 24–33. doi:10.1016/j.applanim.2011.09.007.

Henry, B. K., Eckard, R. J. and Beauchemin, K. A. 2018. Review: Adaptation of ruminant livestock production systems to climate changes. In: *Proceedings of the 10th International Symposium on the Nutrition of Herbivores – Herbivore Nutrition Supporting Sustainable Intensification and Agro-Ecological Approaches*. Clermont Ferrand, France, 2–6 September 2018. *Animal* 12(suppl. 2), S445–S456.

Idrus, M. 2020. Behavioural and physiological responses of beef cattle to hot environmental conditions. PhD thesis. Faculty of Science, University of Queensland, Australia.

Johnsen, J. F., Zipp, K. A., Kalber, T., de Passille, A. M., Knierim, U., Barth, K. and Mejdell, C. M. 2016. Is rearing calves with the dam a feasible option for dairy farms? Current and future research. *Appl. Anim. Behav. Sci.* 181, 1–11. doi:10.1016/j.applanim.2015.11.011.

Joshi, S. and Gokhale, S. 2006. Status of mastitis as an emerging disease in improved and periurban dairy farms in India. *Ann. N. Y. Acad. Sci.* 1081, 74–83.

Kennedy, U., Sharma, A. and Phillips, C. J. C. 2018. The sheltering of unwanted cattle, experiences in India and implications for cattle industries elsewhere. *Animals* 8(5), 64.

Kielland, C., Skjerve, E., Østerås, O. and Zanella, A. J. 2010. Dairy farmer attitudes and empathy toward animals are associated with animal welfare indicators. *J. Dairy Sci.* 93(7), 2998–3006.

Kitching, M., Mathur, H., Flynn, J., Byrne, N., Dillon, P., Sayers, R., Rea, M. C., Hill, C. and Ross, R. P. 2019. A live bio-therapeutic for mastitis, containing *Lactococcus lactis* DPC3147 with comparable efficacy to antibiotic treatment. *Front. Microbiol.* 10, 2220. doi:10.3389/fmicb.2019.02220.

Lecorps, B., Ludwig, B. R., von Keyserlingk, M. A. G. and Weary, D. M. 2019. Pain-induced pessimism and anhedonia: evidence from a novel probability-based judgment bias test. *Front. Behav. Neurosci.* 13, 54. doi:10.3389/fnbeh.2019.00054.

Macrae, A. and Esslemont, R. 2015. The prevalence and cost of important endemic diseases and fertility in dairy herds in the UK. In: Cockroft, P. D. (Ed.), *Bovine Medicine* (3rd edn). Wiley-Blackwell, Chichester, UK, pp. 325–337.

Muller, B. R., Soriano, V. S., Bellio, J. C. B. and Molento, C. F. M. 2019. Facial expression of pain in Nellore and crossbred beef cattle. *J. Vet. Behav.: Clin. Appl. Res.* 34, 60–65. doi:10.1016/j.jveb.2019.07.007.

Neave, H. W., Daros, R. R., Costa, J. H. C., von Keyserlingk, M. A. G. and Weary, D. M. 2013. Pain and pessimism: dairy calves exhibit negative judgement bias following hot-iron disbudding. *PLoS ONE* 8(12), e80556. doi:10.1371/journal.pone.0080556.

Noudeke, N. D., Dotche, I., Ahounou, G. S., Karim, I. Y. A. and Farougou, S. 2017. Inventory of medicinal plants used in the treatment of diseases that limit milk production of cow in Benin. *J. Adv. Vet. Anim. Res.* 4(1), 1–14. doi:10.5455/javar.2017.d183.

Oehm, A. W., Knubben-Schweizer, G., Rieger, A., Stoll, A. and Hartnack, S. 2019. A systematic review and meta-analyses of risk factors associated with lameness in dairy cows. *BMC Vet. Res.* 15(1), 346. doi:10.1186/s12917-019-2095-2.

Peters, M. D. P., Silveira, I. D. B. and Fischer, V. 2015. Impact of subclinical and clinical mastitis on sensitivity to pain of dairy cows. *Animal* 9(12), 2024–2028.

Phillips, C. J. C. 1990. Adverse effects on reproductive performance and lameness of feeding grazing dairy cows partially on silage indoors. *Journal of Agricultural Science, Cambridge* 115, 253–258.

Phillips, C. J. C., Llewellyn, S. and Claudia, A. 2003. Laterality in bovine behavior in an extensive partially suckled herd and an intensive dairy herd. *J. Dairy Sci.* 86(10), 3167–3173. doi:10.3168/jds.S0022-0302(03)73919-8.

Phillips, C. J. C. 2009. *The Welfare of Animals: The Silent Majority*. Springer, Dordrecht, The Netherlands.

Phillips, C. J. C., Oevermans, H., Syrett, K. L., Jespersen, A. Y. and Pearce, G. P. 2015. Lateralization of behavior in dairy cows in response to conspecifics and novel persons. *J. Dairy Sci.* 98(4), 2389–2400. doi:10.3168/jds.2014-8648.

Polat, B., Colak, A., Cengiz, M., Yanmaz, L. E., Oral, H., Bastan, A., Kaya, S. and Hayirli, A. 2010. Sensitivity and specificity of infrared thermography in detection of subclinical mastitis in dairy cows. *J. Dairy Sci.* 93(8), 3525–3532. doi:10.3168/jds.2009-2807.

Qu, Y., Zhao, H. N., Nobrega, D. B., Cobo, E. R., Han, B., Zhao, Z. H., Li, S. M., Li, M. Y., Barkema, H. W. and Gao, J. 2019. Molecular epidemiology and distribution of antimicrobial resistance genes of *Staphylococcus* species isolated from Chinese dairy cows with clinical mastitis. *J. Dairy Sci.* 102(2), 1571–1583. doi:10.3168/jds.2018-15136.

Rizhova, L. Y. and Kokorina, E. P. 2005. Behavioural asymmetry is involved in regulation of autonomic processes: left side presentation of food improves reproduction and lactation in cows. *Behav. Brain Res.* 161(1), 75–81. doi:10.1016/j.bbr.2005.01.007.

Robins, A. and Phillips, C. J. C. 2010. Lateralised visual processing in domestic cattle herds responding to novel and familiar stimuli. *Laterality* 15(5), 514–534. doi:10.1080/13576500903049324.

Robins, A., Goma, A. A., Ouine, L. and Phillips, C. J. C. 2018. The eyes have it: lateralized coping strategies in cattle herds responding to human approach. *Anim. Cogn.* 21(5), 685–702. doi:10.1007/s10071-018-1203-1.

RSPCA (undated). RSPCA welfare standards. Available at: https://www.rspcaassured.org.uk/farm-animal-welfare/meat-chickens/. Accessed 4 February 2020.

Ruegg, P. L. 2017. A 100-year review: mastitis detection, management and prevention. *J. Dairy Sci.* 100(12), 10381–10397.

Sahar, M. W., Beaver, A., Weary, D. M. and von Keyserlingk, M. A. G. 2020. Feeding behavior and agonistic interactions at the feed bunk are associated with hyperketonemia and metritis diagnosis in dairy cattle. *J. Dairy Sci.* 103(1), 783–790. doi:10.3168/jds.2019-16278.

Sharma, A. and Phillips, C. J. C. 2019. Avoidance distance in sheltered cows and its association with other welfare parameters. *Animals* 9(7), 396.

Sharma, A., Umapathy, G., Kumar, V. and Phillips, C. J. C. 2019a. Hair cortisol in sheltered cows and its association with other welfare indicators. *Animals* 9(5), 248.

Sharma, A., Kennedy, U., Schuetze, C. and Phillips, C. J. C. 2019b. The welfare of cows in Indian shelters. *Animals* 9(4), 172.
Sharma, A., Kennedy, U., Schuetze, C. and Phillips, C. J. C. 2019c. The welfare of cows in Indian shelters. *Animals* 9(4), 172. doi:10.3390/ani9040172.
Sharma, A., Schuetze, C. and Phillips, C. J. C. 2019d. Public attitudes towards cow welfare and cow shelters (gaushalas) in India. *Animals* 9(11), 972. doi:10.3390/ani9040172.
Sharma, A., Schuetze, C. and Phillips, C. J. C. 2020. The management of cow shelters (gaushalas) in India, including the attitudes of shelter managers to cow welfare. *Animals* 10(2), 211. doi:10.3390/ani10020211.
Silva, D. C., Queiroz, P. J. B., Borges, P. A. C., da Rosa Pedroso, A. C. B., Arnhold, E., da Cruz, A. S., da Cruz, A. D. and da Silva, L. A. F. 2020. Half a century of research on cattle foot and claw diseases: a scientometric analysis. *Semina: Ciênc. Agrár.* 41(1), 223-235.
Stojkov, J., von Keyserlingk, M. A. G., Marchant-Forde, J. N. and Weary, D. M. 2015. Assessment of visceral pain associated with metritis in dairy cows. *J. Dairy Sci.* 98(8), 5352-5361.
Tadich, N., Tejeda, C., Bastias, S., Rosenfeld, C. and Green, L. E. 2013. Nociceptive threshold, blood constituents and physiological values in 213 cows with locomotion scores ranging from normal to severely lame. *Vet. J.* 197(2), 401-405.
Thomsen, P. T., Anneberg, I. and Herskin, M. S. 2012. Differences in attitudes of farmers and veterinarians towards pain in dairy cows. *Vet. J.* 194(1), 94-97.
Uddin, J., Phillips, C. J. C., Goma, A. A. and McNeill, D. M. 2019. Relationships between infrared temperature and laterality. *Appl. Anim. Behav. Sci.* 220, 104855.
Van der Schott, A. A. and Phillips, C. J. C. 2013. Publication bias in animal welfare scientific literature. *J. Agric. Env. Ethics.* 26(5), 945-958.
Whay, H. R., Waterman, A. E., Webster, A. J. F. and O'Brien, J. K. 1998. The influence of lesion type on the duration of hyperalgesia associated with hindlimb lameness in dairy cattle. *Vet. J.* 156(1), 23-29. doi:10.1016/s1090-0233(98)80058-0.
World Health Organisation 2020. Constitution. Available at: https://www.who.int/about/who-we-are/constitution. Accessed 28 January 2020.
World Organisation for Animal Health (OIE) 2019. Chapter 7.1. Introduction to the recommendations for animal welfare. Available at: https://www.oie.int/index.php?id=169&L=0&htmfile=chapitre_aw:introduction.htm. Accessed 28 January, 2020.
Zhang, Z., Yang, F., Li, X. P., Luo, J. Y., Liu, L. H., Wang, D., Zhang, Y. R. and Li, H. S. 2019. Distribution of serotypes, antimicrobial resistance and virulence genes among *Streptococcus agalactiae* isolated from bovine in China. *Acta Sci. Vet.* 47(1), 1699. doi:10.22456/1679-9216.97254.
Zhao, C., Liu, G., Li, X., Guan, Y., Wang, Y., Yuan, X., Sun, G., Wang, Z. and Li, X. 2018. Inflammatory mechanism of rumenitis in dairy cows with subacute ruminal acidosis. *BMC Vet. Res.* 14(1), 135. doi:10.1186/s12917-018-1463-7.

# Part 3

## Health at different stages in the life cycle

# Chapter 7

# Optimising reproductive management to maximise dairy herd health and production

*Norman B. Williamson, Massey University, New Zealand*



## 1 Introduction

The domestication of cattle has been linked with the beginnings of civilization over 10000 years ago. Keeping cattle for both milk and meat production occurs worldwide. Herds vary in size from one or two cows to many thousands under single management. Two main types of management systems prevail in commercial dairy farms. In one system cattle are housed in buildings or feedlots and are fed rations that are brought to them. Such systems are

http://dx.doi.org/10.19103/AS.2020.0086.09

prevalent in North America and South America, Europe, the Middle East and Asia. The other system has herds living on pasture which they graze for most of their sustenance. Grazing systems occur commonly in Australia, New Zealand, Ireland, parts of Argentina and Chile. These two systems require different management strategies for reproduction that lead to different approaches to breeding and different limits to reproductive health and performance. However, there are similarities in the fundamental principles that apply under both management systems.

Efficient dairy farms produce high volumes of product for the resources employed. Products include milk, male calves for sale, female calves or heifers not required for production, cull cows and sometimes production cows. Farms may also sometimes sell excess feed and, in developing countries, manure can also be a product that is traded. Profitable farms have a large positive difference between income, made up of the amounts of product multiplied by the prices received for them, and costs, which comprise the types of resources used multiplied by their costs.

Reproduction has a major influence on dairy farm efficiency, economics and profitability. Cows calve as a result of mating, insemination or transfer of an embryo. Most modern dairy farms in advanced economies now employ artificial insemination to improve herd genetics and selection of cows for production gain. Calving produces calves that can either be sold or enter the milking herd as replacement cows that produce milk, given that calving initiates lactation (milk production) by the cow.

The interval between consecutive calving events influences milk production efficiency. Shorter calving intervals mean that cows spend relatively more of their milking time in the earlier stages of lactation where production is higher and more efficient. They also produce more calves in a period. Efficient and adequate reproduction provides more offspring that enable discretionary culling of low-producing animals. Increased numbers of replacement heifers enable the retention in the herd of replacements with higher genetic selection differentials which leads to more rapid genetic gain. Efficient reproduction by heifers enables them to join the herd sooner and this limits the overhead costs of rearing. In pasture-based dairy production systems, seasonal calving is often managed so that calving occurs in a restricted period of the year just before peak pasture growth occurs because that allows the requirements of cows for feed to be matched to the available pasture growth.

Inadequate reproduction increases costs and may require animals to be culled. Such enforced culling requires cows to be removed from the herd before they reach their maximum milk production potential, normally at the sixth or seventh lactation, which limits the efficiency of production. Enforced culling also demands more replacement animals, thus reducing the ability of farmers to select only the most genetically superior animals.

Reproductive limits apply at an individual cow level due to variations in reproductive health and physiological performance of individuals. There are also herd reproductive limits associated with approaches to herd breeding, feeding, health and general management.

Reproductive diseases and limitations can occur at all stages of the reproductive cycle, from the onset of puberty until an animal is culled or dies. A failure of or delay in the display of oestrus can occur at expected puberty or after a previous calving. This limits the opportunities for heifers and cows to be mated and delays breeding until oestrus occurs under natural conditions. If oestrus does occur, cows and heifers may not be bred and so the opportunity for reproductive success is delayed. If breeding does occur, conception may or may not occur. If not, a successful reproductive outcome is again delayed. After conception, a pregnancy may terminate in early embryonic death and may not establish. If a pregnancy is established, it can then be aborted. If gestation is completed pregnancy may end in a dystocia or stillbirth and may be followed by a retained foetal membrane.

In the reproductive cycle of cows, these critical events enable reproductive success and are the events about which data must be collected in herd records to allow performance to be monitored and analysed. Data collection and analysis are now normally kept as a part of a computer-based recording and analysis system. Such systems greatly aid the collection and analysis of records that can be used to monitor the performance of individual cows and whole herds of many thousands of cows. Reproductive management using records to detect and identify reproductive limits to production is applied to minimise reproductive disease and reduce the impact of disease and inadequate management on reproductive performance (Williamson, 1987). This can inform interventions that optimise reproductive efficiency, health, performance and productivity (Fetrow et al., 1990).

## 2 Grouping animals to measure individual animal reproduction limits

Reproductive efficiency in a herd can be managed by defining the parameters of expected reproductive performance and selecting cows that fail to meet these parameters. This leads to cows being examined in defined groups during planned reproductive health visits by veterinarians. These groups include:

- Anoestrous cows;
- Nymphomaniacs;
- Post-partum checks;
- Abnormal discharges;

- Pregnancy checks; and
- Pregnancy rechecks.

These are discussed in the following sections.

### 2.1 Anoestrous cows

Failure to display or detect oestrus is a major limit to reproductive efficiency in most dairy regions. Cows that have not shown oestrus within 49 days of calving or those that have not shown a return to a previous unbred oestrus within 35 days (this time can be set at the lower limit of pregnancy diagnosis accuracy) should be selected for examination and treatment in year-round calving herds. These are called NVO for no visible oestrus cows or ONO for oestrus not observed cows. When selected, these cows can be examined and treated appropriately to ensure a rapid return to normal oestrus cycling. If they have a corpus luteum (CL) present (indicating prior-ovulation) they may be injected with prostaglandin to induce a rapid onset of oestrus. Otherwise they may be started on an oestrus induction programme such as one of the progesterone, prostaglandin, gonadotrophin programmes to induce a return to normal oestrus cycling. The presence of a CL also indicates that a previous oestrus is likely to have occurred and to not have been observed or recorded. This indicates that aids to oestrus detection such as dye-containing pressure-sensitive patch detectors, tail painting, activity monitoring or pressure-activated electronic oestrus detection devices should be considered to improve the sensitivity of oestrus detection.

Cows and heifers that are not pregnant at the normal period for pregnancy diagnosis de facto join this group and can be treated in similar ways. Cows in this situation are sometimes called 'phantom' cows. An analogous approach can be taken with heifers that have not demonstrated oestrus by the desired time of first breeding. All heifers not detected in oestrus by 13 or 14 months of age can be examined and treated in a similar way to the NVO cows.

### 2.2 Nymphomaniac cows

If oestrus is observed frequently, that is, three recorded occurrences of oestrus within 36 days, in the absence of treatment with an oestrus cycle shortening drug such as prostaglandin or an oestrogen, cows displaying these signs should be examined so that they can be checked for cystic ovarian disease and appropriately treated.

### 2.3 Post-partum check cows

Post-partum checks are required on cows that have a previous abnormality associated with calving. Abnormalities include induced calving, dystocia, prolapsed uterus, retained foetal membranes and post-partum metritis. These cows are routinely examined from about 20 days after calving even if no external abnormal signs are evident. Some veterinarians, particularly in the northern hemisphere, examine all cows routinely during post-partum or pre-breeding examinations in order to detect clinical or sub-clinical endometritis (Dubuc et al., 2010). This selection for examination became more popular after the introduction of the 'Metricheck' instrument that enables mucus and any pus it contains to be extracted from the cranioventral vagina near the cervix and scored as to the likelihood of endometritis. Cows deemed to have endometritis can then be treated with an antibiotic, commonly cephapirin which is not absorbed parenterally or excreted in the milk. Any cow that has clinical signs of metritis or that has an abnormal vaginal discharge should also be selected to be examined and treated if required.

### 2.4 Cows with abnormal discharges

If abnormal, usually purulent, discharges from the vulva are seen, cows should be selected for examination and treated if required.

### 2.5 Pregnancy check cows

All cows that have been inseminated or bred, and which have not been detected in oestrus for a time that exceeds the lower limit for accurate pregnancy diagnosis (often set at 35 days), should be selected as a pregnancy diagnosis (PD) category. It is a useful practice to confirm and estimate the stage of the pregnancy without knowing the anticipated stage of pregnancy. This is so any discrepancy between the determined and expected stages of pregnancy can be considered and the most appropriate stage of pregnancy allocated.

### 2.6 Rechecks for pregnancy

Rechecks for pregnancy are required if cows that previously were diagnosed pregnant have oestrus or insemination recorded after the positive PD. Rechecks can also be requested by a veterinarian if a pregnancy appears to be abnormal or if it is too difficult to diagnose with confidence at an initial attempt. Cows that exceed their expected calving date by 3 weeks or longer should also be re-examined for pregnancy. Where abortion has occurred, tests to determine possible causes should be conducted.

Pregnancy diagnosis is essential to determine the status of individual cows so they may be managed appropriately. Accumulated pregnancy status information is required to allow key herd reproductive indices to be calculated and to facilitate monitoring, including conception rates and calving to conception intervals.

## 3 Measuring reproductive performance

Despite easy recognition of reproductive inadequacy in herds, achievement of improvement requires a thorough understanding of reproductive processes and relationships in order to specify biological or management reasons limiting reproduction. Farmers and veterinarians frequently recognise that improvements in herd reproductive performance are needed when performance achieved in herds does not match targets.

Reproduction is managed using measures or indices that allow performance to be recorded and then analysed. These performance measures allow targets to be set so that herd reproductive performance can be monitored by comparing the performance achieved against these targets. They can thus help identify inadequate reproductive performance. Where this occurs, some reproductive indices also help to identify likely causes of inadequate reproductive performance.

Indices may be used to measure performance of individual animals or herds. Some indices of reproduction are directly related to production, provide easy measures of farm efficiency and a way to detect that there are problems occurring on a farm, for example, long calving intervals or high culling rates. However, such indices provide no useful information about the causes of problems. Other indices are more focused and reflect performance in specific areas of farm reproductive performance that offer information on what is limiting reproductive performance in a way that allows diagnosis. These can be characterised as diagnostic indices.

To document reproductive performance, complete and accurate herd records are required to be maintained. These records may be found in a variety of forms such as wall sheets, cow cards or books. However, they are more commonly now kept as computer-based records either on the farm, for example, in systems provided by milking equipment companies, or kept centrally by herd test or artificial breeding organisations.

Records consist of details of cows including identification, birth date, the dates and details of all calving events, dates of all observed oestruses and breeding dates and details. Breeding events should record dates, sires or semen used and, when artificial insemination is used, the inseminator's identification. If natural breeding is used and a bull or bulls are run with the herd, the date when bulls are introduced to the herd and date removed should also form a part of the records. If a herd is under veterinary supervision through an ongoing herd reproductive health programme, the dates and findings at all reproductive examinations should form a part of the record. Data recorded will include the pregnancy status and stage, the date and nature of any reproductive abnormality, the dates of all treatments and details of the products used, dry-off dates and the date and reasons for removal from the herd.

Measures of reproductive performance include intervals measured in days which can highlight a potential issue where a discrepancy from target occurs. When reproductive intervals are being recorded and evaluated, it is simple to calculate and consider mean intervals. However, this is not always appropriate as some reproductive intervals have multimodal distributions and are better represented as frequency distributions.

## 4 Production-related reproductive indices for pasture-based seasonally calving herds

In seasonally calving herds, the time and distribution of calving are related to the production of offspring, sale animals and milk and are thus production-related indices. Production-related reproductive indices include:

- Four- and 8-week calving rates;
- Mean calving date and planned start of calving date;
- Proportion of cows calving that were induced (where it is allowed);
- Proportion of calved cows that are culled for inadequate reproduction; and
- Proportion of cows carried over for a season that are non-pregnant.

If these indices fail to be on target, they indicate a potential problem but provide little information on why reproductive performance is not meeting targets. These measures are production related as they influence the amount and efficiency of output of milk and livestock through early compact calving. They do not, however, provide much information on causes or specific factors that lead to inadequate performance. It is known, however, that calving pattern in one season is highly predictive of the calving pattern in the subsequent season.

The season is defined by the planned start of calving date. This sets the desired date for calving to start. The spread of calving around that date is usefully displayed graphically as a frequency distribution of the cows calving around that date. This can be presented as a graph (e.g. Fig. 1).

The InCalf programme as used in New Zealand reports the percentage of cows calved by 3, 6 and 9 weeks after the nominated planned start of calving date and this helps to describe performance in the calving season. Top farmers achieve 60%, 87% and 98%, respectively, for these indices (Fig. 2).

This means of evaluating reproductive performance can be distorted by management interventions such as by the pharmacological induction of late calving cows where this is permitted. If it is permitted, then the proportion of the herd induced to calve early should be recorded and monitored to evaluate the influence of this intervention on the calving pattern. The spread of calving also can be influenced by culling of cows that do not conceive within the planned

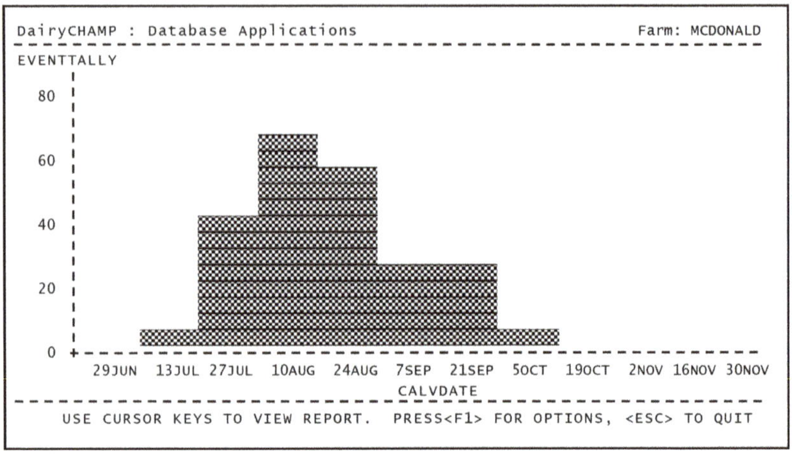


**Figure 1** Histogram of number of cows calving by time.

mating period and that are likely to calve outside the planned season. This can be measured as the empty (non-pregnant) rate at the end of the mating period. Cows and heifers that do not conceive within the mating period and that are then culled influence the subsequent calving spread because they do not figure in it. This makes it important to monitor and report the proportion of cows mated with the intention of having them calve in the next calving season that are culled for reproductive failure because they did not become pregnant in time. This measure includes cows that do not conceive and those that conceive too late to calve in the following defined calving season.

## 5 Diagnostic reproductive indices for pasture-based seasonally calving herds

For seasonally calving herds in Australia and New Zealand, a key index of performance relating to the breeding season that is used by farm advisors

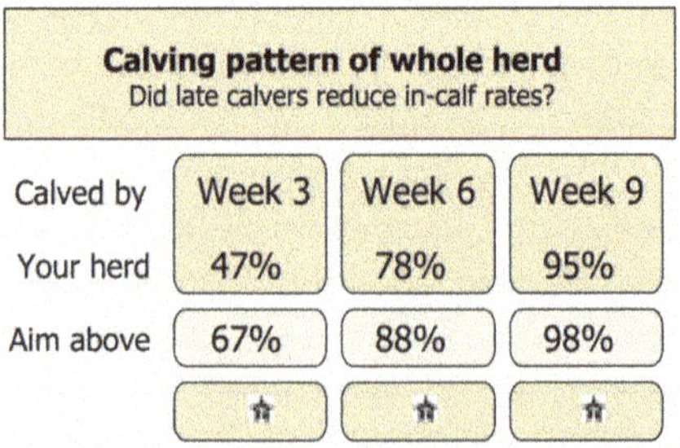
**Calving pattern of whole herd**
Did late calvers reduce in-calf rates?

| Calved by | Week 3 | Week 6 | Week 9 |
|---|---|---|---|
| Your herd | 47% | 78% | 95% |
| Aim above | 67% | 88% | 98% |

**Figure 2** Calving distribution table from the InCalf Fertility Focus Users Guide (2020).

(including veterinarians) is the 6-week in-calf rate. That is the proportion of cows present in the herd at the planned start of mating that are bred and/or inseminated in the first 6 weeks of mating and are diagnosed to be pregnant.

Reproductive diagnostic indices pertinent to seasonally calving herds include the submission rate and the conception (pregnancy) rate or risk. In seasonal dairy herds where conception occurs in a defined breeding period, the conception rate is appropriate. However, more recently, in seasonal and in year-round calving herds, the concept of conception risk has gained favour since it does not require a defined time period (Brownlie et al., 2014; Stevenson et al., 2014; Denis-Robichaud et al., 2016).

The submission rate (or risk) is the proportion of cows present in the herd at the start of mating that are bred (served or inseminated) in a defined time after breeding starts.

The InCalf programme reports this index as a 3-, 6- and 9-week rate after the planned start of mating date, which is the date occurring 282 days before the planned start of calving date (Anon, 2020). This index is influenced by the occurrence of oestrus, the detection of oestrus by farm workers in artificially inseminated cows and herds or bulls in naturally bred herds and then the insemination or breeding of the cows that are detected in oestrus. On high-performing farms, 3-week submission rates of 90% or more are achieved. High performance in this index is required for high in-calf rates to be achieved. The imperative for a high 3-week submission rate can lead to overzealous oestrus detection and the presentation of cows for insemination that are not in oestrus. Evidence for the occurrence of this practise is supported when conception rates are low in the first week of mating then noticeably improve.

The not-in-calf or empty rate is the percentage of the cows within a mating group that fail to become pregnant by the end of a mating period. This represents the proportion of the breeding group that fails to meet the desired outcome by the end of breeding.

Conception rates are the proportion of inseminations occurring in a defined period that result in conception. Since conception can only be measured after pregnancy has been diagnosed, the term pregnancy rate is sometimes used. However, there can be confusion in the use of the term pregnancy rate since, in North America, pregnancy rate can refer to the number of cows that are pregnant in a population of eligible cows that are beyond the voluntary waiting period for a defined time such as 21 days. In New Zealand, some herds can achieve conception rates to first service of 60% or greater, but rates above 53% are currently considered adequate. Conception rates are influenced by female (cow and heifer), male (bull and semen), management and environmental factors.

There are further indices of performance that assist in monitoring reproductive performance and help to pinpoint aspects of breeding that can be improved. Unlike the indices above, these performance measures are common to seasonal and year-round herds and are described below.

## 6 Production-related reproductive indices in year-round calving herds

In year-round calving herds, constraints on production differ from those in seasonal herds since a close tying of pasture availability and milk production does not exist. In these herds the main production-related indices are the calving interval, that is, the time between successive calving events, the culling rate and the proportion of a herd that calves in a year.

Calving intervals of 12 months in cows and up to 13 months in first lactation heifers have been shown to be optimal for milk production efficiency. Shorter calving intervals concentrate periods where cows produce milk at higher levels and with greater efficiency in the earlier parts of the lactation. The reason that heifers have longer recommended intervals is because their lactations persist at higher levels for longer. Shorter calving intervals are associated with a higher output of calves per cow.

The culling rate (for infertility) is another index related to production of animals available for sale – a farm output. The proportion of the herd that calves in a year is another production-related index in year-round calving herds that is not often monitored. A target for this index is greater than 110% due to some cows with intervals less than 12 months and to heifers or other new cows entering the herd only when they calve.

Reproductive intervals used as indices for heifers require analysis separate from those for cows because heifers are not regarded as a part of the milking herd until after their first calving. Heifer indices include several that are measured from birth. The time to puberty is the time from birth until the first oestrus, which should occur at 12–14 months. The time to first service should occur in the 13 to 15-month age range. The main production-related reproductive index for heifers is the time from birth to first calving. A suitable target for this index is from 22 to 24 months (Table 1).

There are analogous intervals for cows that are measured from the last previous calving date. The inter-calving (or calving) interval is ideally 12 months in year-round and seasonally calving herds. This interval index is calculated for cows only and excludes heifers since they do not have a previous calving record. It also excludes culled cows, and those that die or are otherwise disposed of, since no subsequent calving record is available for them. This is a production-related index but is limited because it is insensitive to recent change since it can only be calculated for those cows that do calve again.

The exclusion of cows that do not conceive or that are culled means that the inter-calving index is subject to biases due to omission of information from

**Table 1** Reproductive intervals from birth for maiden heifers and calving for cows that are used to monitor performance

| Interval name | Target | Apply to |
|---|---|---|
| Birth to first oestrus (Time to puberty) | 12-14 months | Heifers |
| Birth to first service | 13-15 months | Heifers |
| Birth to first calving | 22-24 months | Heifers |
| Calving interval | 12 months | Cows |
| Proportion of herd calving in a year | 115% | Cows |
| Proportion of herd culled for reproduction | Less than 10%/year | Cows and heifers separately |
| Proportion of carry-over cows | Nil | Cows |

poorly performing animals. For this reason, the inter-calving interval must be assessed in association with other indices to be validly interpreted. One is the culling rate while another is the number and proportion of cows that remain in the herd but are still not pregnant despite being well beyond the anticipated time of breeding, known as carry-over cows. Both increase the proportion of cows in the herd that is not pregnant and not milking and reduce the proportion that is diagnosed pregnant and milking.

The calving to conception interval is the mean time from calving to conception for a defined cohort of cows. The target for this interval is from 83 to 85 days in order to maintain a 12-month calving interval. Some call this the 'open interval' while others add on days to the last service or days to the current date to calculate the open interval for cows not yet diagnosed pregnant. The calving to conception and open intervals are highly correlated to the calving interval but considerably more current. How current the index is depends on the frequency of pregnancy diagnosis. The calving and open intervals are useful monitoring indices but are not useful in the causal diagnosis of problems because they are the result of male, female and management factors.

## 7 Indices used to diagnose causes of inadequate herd reproduction

The range of reproductive indices used to monitor performance is shown in Table 2. The calving to first service interval is the sum of the intervals to first service for all cows receiving first services, divided by the number of cows that receive a first service. This index helps to separate the causes of long calving intervals into late breeding due to either failure to breed or failure to conceive once bred. If the calving to first service interval is long, reasons for delayed breeding must be investigated. If calving to first service interval is short then long calving intervals can be due to a failure of cows to conceive to the early

**Table 2** Reproductive indices used to monitor performance

| Name of index | Target |
|---|---|
| Calving to first oestrus | 45 days |
| Percentage in oestrus by 60 days after calving | >90% |
| Calving to first service | 65 days |
| Percentage pregnant at pregnancy diagnosis | >85% |
| Oestrus detection rate (risk) | >85% |
| Ratio of single to double inter-oestrus intervals | >7:1 |
| First service conception risk | 45% |
| Services per conception | <1.7 |

breeding or a failure to detect and breed those cows that have not conceived and that return to oestrus after the first breeding. A target for the ideal calving to first service interval will depend on a herd's conception risk but will normally be in the range of 55–70 days. Underlying reasons for a prolonged interval from calving to first service can be anoestrus among the cows to be bred, a failure to detect oestrus by farm workers or a failure to breed cows at detected oestrus.

In order to sort out these potential causes, measures of the occurrence and detection of oestrus must be evaluated. The average calving to first oestrus interval should ideally be within 45 days of calving, but for this to be possible, all oestruses after calving must be recorded. The percentage of cows detected in oestrus by 60 days after calving should exceed 90% to indicate excellent return to oestrus by cows and oestrus detection by farm workers. Another index of the occurrence and detection of oestrus is the percentage of cows that are pregnant at PD in herds where all oestrus events are recorded. This measure reveals the adequacy of oestrus detection because if cows are bred and do not conceive then they should return to oestrus at a normal inter-oestrus interval and be detected. They should only be presented for PD if no oestrus has been observed after the last breeding. If they are not pregnant at PD but are otherwise normal this means that at least one and possibly two returns to oestrus have not been observed or noted. The target proportion pregnant at PD in herds where all observed oestrus events are recorded is greater than 85% if PD occurs at less than 45 days of gestation or over 95% if it occurs at over 45 days. The cycling status of cows can be checked at PD by confirming the presence of a CL on the ovaries of the non-pregnant cows examined.

The oestrus detection rate has also been described as a useful measure of the detection of oestrus. It is calculated as 21 divided by the average days between recorded oestruses times 100. The target for this index is over 85% but it can exceed 100% due to the false detection of oestruses, cows with short inter-oestrus intervals such as nymphomaniacs with cystic ovaries and, more commonly, due to the use of prostaglandins to induce oestrus which shortens ovarian cycles.

Another analytical technique that can be used to evaluate the efficiency of oestrus detection is the ratio of single to double inter-oestrus intervals. Since cows and heifers return to oestrus in a cyclical pattern, oestrus will occur in a peak at periods averaging 21 days after a previous oestrus. If an oestrus is not detected at around 21 days after a previous oestrus, the next occurrence is likely to be at around 42 days or two cycles later. This can be used to examine patterns of oestrus returns to evaluate if farm workers are missing too many occurrences of oestrus. The ratio of single to double inter-oestrus intervals should be from 7 to 1 or greater. This measure is very useful for indicating the efficiency of oestrus detection in large herds but, to be effective, all detected oestruses must be recorded and the use of agents to manipulate the length of oestrus cycles should be minimal.

The major influences on conception efficiency are sire, cow, management and environmental factors. The conception efficiency of cows is monitored as the conception rate now commonly named conception risk since in many cases no time element is included to make it a valid rate. Calculating conception risks to specific sires or inseminators may reveal individuals whose performance is noticeably below target or below that of peers. Enough first services within any groups examined are needed for results to be meaningful. Conception (pregnancy) risks also can be calculated for all services but these can be biased by cows that have many repeat services. Conception risks are often regarded as representing breeding efficiency because they are used to monitor the performance of artificial breeding centres, individual inseminators and individual sires. However, they are less economically important than calving intervals.

Since conception itself is difficult to determine, many proxies exist such as non-return rates and per service pregnancy risks. In view of this, it is critical to determine exactly how the conception index being assessed or monitored has been calculated. The first service conception (by diagnosed pregnancy) risk is the number of first services in a period resulting in a pregnancy divided by the number of first services conducted in the period, expressed as a percentage. It is a measure of the outcome of breeding in a group or herd and not of the performance of individual cows, since the result for individuals can only be 0% or 100%. The first service conception risk is used because it is relatively unbiased by cows that require repeated services due to problems. Frequently it is called the first service conception rate. This index can be focused to examine the outcome of breeding performed on specific age groups, groups mated by specific sires, groups housed together or bred by individual inseminators as the denominator in order to allow associations to be evaluated between these factors and conceptions achieved.

The 24-day non-return rate is the proportion of bred cows that do not return to oestrus by 24 days after breeding. It is sometimes used by farmers and insemination companies as an index of conception but is not an accurate measure of this outcome. A low value of this index does provide an early warning that upcoming conception rates are likely to be low. It overestimates conception risks by 10–25% due to factors such as cows not being re-bred, becoming anoestrus after breeding (phantom cows), subsequent oestrus not being detected or recorded, cows being sold or having died, cows being bred by herd bulls without this being recorded, cows inseminated by a different artificial breeding company and the occurrence of foetal death.

A further measure of breeding efficiency is the number of services per conception. This is the total number of services provided to a herd in a season or period, divided by the number of pregnancies achieved. A target for this index is less than 1.7. This index is useful in estimating the amount of semen required to enable a herd to be bred.

As mentioned above, 'pregnancy rate' in a North American context refers to the number of cows pregnant in a population of eligible cows where an eligible cow is beyond the 'voluntary waiting period' (bred and unbred cows) over a defined period that is usually 21 days, that is, 21-day pregnancy rate (Stevenson and Britt, 2017). This index provides a contemporary measure of breeding efficiency that accommodates high intervention breeding strategies used today. A measure of the occurrence of foetal loss should also be monitored and evaluated when abortions occur in dairy herds. This is discussed later in Section 14.

## 8 Monitoring bull breeding

Comparisons of results based on whether breeding has been by bulls in natural service or by artificial insemination (when both are being used in a herd) may highlight limits such as insemination technique or the ability of cows to conceive when their chances are not limited by the oestrus detection abilities on the farm. In seasonal pasture-based herds, bulls may be grazed together with the cows after the end of the period when artificial insemination is used. It is common that conception rates then improve. This may be due to the bulls' better performance due to their superior oestrus detection accuracy than that of farm workers, better timing of insemination and more frequent and higher sperm dosing. Since the timing of introduction of bulls to the herd commonly coincides with when cows achieve a positive energy balance after calving as production begins to decline, the improved fertility may also be partly due to this. Factors such as breed, age, weight, nutritional adequacy, time since calving, production level, health status and reproductive disease can all influence conception efficiency.

Where specific sires produce low conception rates, investigations or enquiries of performance in other herds using the same sire can be undertaken to further define a problem. The same can be done to examine the results obtained by inseminators. Checks can be conducted into semen quality, such as having the post-thaw motility of semen assessed by semen companies. The technique of inseminators can also be checked including their hygiene, thawing practice and semen placement, especially if they are farm employees, along with checks of their equipment.

The time of mating after calving and the time of mating within an oestrus both have major influences on the conception rates achieved. Early breeding that is used to align planned calving dates in late calving cows results in low conception rates. Breeding early or late within an oestrus also is associated with lower conception rates than when breeding occurs at optimal times, 12 to 18 hours after the commencement of oestrus. Conception rates early in the first 2 weeks of the breeding season in seasonally calving herds can be low due to a lack of specificity in oestrus detection when pre-breeding occurrences of oestrus are not recorded.

Inadequate conception rates by bulls when used in natural service require further investigation. A common management error is for farmers to have an inadequate number of bulls enter the herd to adequately cover the number of still non-pregnant cows after a period of artificial breeding because they over-estimate the conception efficiency that is achieved during the artificial breeding season. This leads to an underestimate of the number of bulls required to adequately breed the remaining non-pregnant cows. Although the number of cows that can be serviced by individual bulls varies, a ratio of 1 bull per 25 non-pregnant cows should be allowed to ensure breeding coverage.

Bull unsoundness can reduce reproductive efficiency and is frequently associated with feet and leg problems but also can be due to low libido, poor semen quality, injury or venereal disease. Bull breeding soundness examination conducted before the bull breeding period can help minimise breeding failure. In addition to locomotor system problems poor bull performance can also be due to low libido, low semen quality or venereal disease. Problems in cows generally do not cause infertility in the bull breeding cleanup period unless a venereal disease exists. This is now rare with the widespread use of artificial breeding because most cow problems have been resolved by the end of the artificial breeding period when bulls are placed with the milking herd.

## 9 Management of herd limits to reproduction: anoestrus

A major limit to herd reproductive efficiency is anoestrus because its occurrence delays breeding and extends calving intervals. This has been observed in studies conducted throughout the world in confined and pastured cattle with high and low production (Zemjanis et al., 1969; Fielden et al., 1973; McDougall et al., 1993; Mwaanga and Janowski, 2000; Macmillan, 2002; Ambrose, 2015). Anoestrus can delay conception with consequent losses in production and increased maintenance costs. To maintain a 365-day calving interval in cattle the average cow must conceive by the 83rd to 85th day after calving. Every oestrus missed after that means that her calving interval extends by about 3 weeks.

Anoestrus is not a disease. It is the manifestation of many conditions which affect the oestrous cycle. Clinically, it is presented as either an individual animal problem or revealed as a herd problem. Herd problems are most often identified after veterinary intervention in the form of an investigation of inadequate herd reproductive performance or by an analysis of performance as a part of an ongoing reproductive health programme.

In either case, the information presented to the herd veterinarian is dependent on the farm managers' abilities, observations and recording skills. When a farmer calls to have a cow examined because she has not been coming into oestrus, the farmer is the one making this assessment. Each case of anoestrus may be due to either a failure of signs of oestrus to occur or a failure of oestrus to be observed or recorded when it does occur.

Differentiation between these two possibilities is not always easy but knowledge of which problem exists is essential for appropriate intervention to manage the situation. Evaluation of records, thorough clinical evaluation and a thorough and detailed examination of oestrus detection and breeding practices are required to determine the likely causes of anoestrus and to allow appropriate intervention.

A common cause of anoestrus in cows at the planned start of mating in seasonally calving pasture-based herds is that insufficient time has elapsed for them to develop normal cycling patterns. Late calving cows are more prone to be anoestrus during the mating period and herds with spread calving patterns thus have more anoestrous cows.

Although ovarian follicles develop by 10 days after calving, ovulation may occur much later. In the United States, cows are reported as ovulating by 15 days after calving. In two Australian herds the author documented that ovulation occurred at approximately 24 days after calving. In New Zealand first ovulation occurred at 36 days in adult cows and at 52 days in 2- and 3-year-old cows. A more recent study documented ovulation at 44 days. The current herd calving pattern is the major determinant of the amount of anoestrus to be dealt with in seasonally calving dairy herds.

Factors that cause true anoestrus or a failure to cycle accounted for only 1/3 of the anoestrus observed in New Zealand dairy herds. About 2/3 of anoestrus (revealed when cows not seen in oestrus were examined a week before the planned start of mating) occurs in cows which are cycling normally (McDougall,

1994). Thus it is highly likely that herd management is at fault and that oestrus detection is the problem, rather than a failure of cows to cycle. Confirmation that the animal is cycling is based on a careful examination of the reproductive tract per rectum. In cycling cattle a CL will be palpable on an ovary for about 80% of the time. During stages when the CL is not palpable, follicular development and changes in tone of the tubular reproductive tract should indicate whether ovarian activity is present. If no evidence of uterine pathology exists cows are unlikely to have a retained CL and repeated examinations should confirm this. Although a cow is cycling, it still may have 'silent' oestrus. In practical terms checking the efficiency of oestrus detection procedures and 'predicting' when oestrus will occur will help clients to breed their animals. Further techniques to determine whether an animal is cycling use serial milk or serum progesterone assays but cost and availability are problems.

Cows presented for examination as being anoestrus may be cycling but not being observed. The many causes of real and apparent anoestrus in herds is called the 'anoestrus syndrome'. The prognosis, treatment and prevention of this problem relate to the cause, and so every effort must be made to determine what this is.

## 10 Clinical examination and treatment of anoestrous cows

Examination for anoestrus should be like any clinical examination of cows and include a thorough exploration of the cervix, uterus, ovaries and other reproductive organs. The examination is to determine if ovarian cycling is occurring and, if not, why not. Anoestrous cows can be divided into three groups based on the structures present on their ovaries. There may be *corpora luteum*, no structures or follicular structures.

### *10.1 Anoestrus with a corpus luteum present*

Many cows presented because oestrus has not been observed are found to be pregnant, even though this finding is not consistent with the cows' records. All cows should be very carefully examined for pregnancy when they are presented as being anoestrous, because some treatments for anoestrus will terminate pregnancy. If there is doubt about the cow's status, re-examination later may clarify her status. Since many apparently anoestrous cows are found to be pregnant, routine examination of so-called anoestrous cows in herd health programmes should be delayed until the interval after a previous oestrus or service when a veterinarian can competently diagnose pregnancy. This helps prevent the inadvertent aborting of pregnant cows.

An apparently retained CL may be due to early embryonic death since, if the embryo dies after the 15th day of the oestrous cycle, the next oestrus is delayed. The small embryo may be aborted unseen or resorbed. This phenomenon can explain the higher incidence of anoestrus after service than anoestrus after oestrus. Progesterone studies indicate that this occurrence is relatively common, since many cows have elevated progesterone levels for 30–35 days after service, which then drop. The CL may also be retained when there is uterine distension or pathology, usually due to pyometra but also sometimes due to mucometra or foetal maceration or mummification.

Other causes of anoestrus with a retained CL are usually seen in barren heifers. Heifers with absence of the endometrial glands do not cycle but oestrus can be induced by removal of the CL manually. Presumably oestrus could be induced by causing luteolysis with prostaglandin, but these animals would probably remain sterile because they would be unlikely to be able to support pregnancy. Zemjanis et al. (1969) found that 2–3% of cows with diagnosed anoestrus had white heifer disease (segmental aplasia of the Mullerian ducts) whilst in New Zealand, the prevalence among cows culled after PD was only 0.09% (Lawton et al., 2000). The disease causes anoestrus if both horns, the uterine body, cervix or vagina are affected. If there is uterus unicornus, ovarian cycling and even pregnancy may occur. When a CL forms on the ovary adjacent to the aplastic horn, the CL may persist indefinitely causing anoestrus. In cases of congenital defects, the rational treatment is to cull affected heifers.

Treatment of anoestrus cows with a CL present will vary depending on other findings. Pyometra, foetal maceration or mummification and mucometra are best treated by prostaglandin-induced luteolysis, which should help to evacuate the uterus and start normal cycling. Other possibilities for treatment include ergonovine, oestradiol and oxytocin. Supportive treatment with antibiotic infusion to the uterus may be given for pyometra. The treatments commonly used are 1–5 g of oxytetracycline-active ingredient or 2 million units of penicillin with or without 2 g of streptomycin. These are generally infused in 20–100 mL of water.

Cows with anoestrus due to suboestrus, weak or silent oestrus also may have a CL palpable on their ovaries. Oestrus may not be observed in these cows for up to 120 days or more after calving when competent herdsmen observe them. On repeated rectal examinations, cows show evidence of cycling with corpora lutea changing sides. Metoestral bleeding or mucus discharge associated with uterine tone may also be found, and some cows may have ovulation depressions (OVDs) on their ovaries. About 85% of normal cycling cows should have readily palpable CLs on their ovaries.

Suboestrus or 'Silent heat' indicates that an animal goes through an ovarian cycle and ovulates but does not show oestrus. Ovulation, without overt signs of oestrus, can be recognised by examination per rectum. These ovulations may be 'fertile cycles' if artificial insemination is used to breed the animal at an appropriate time. Suboestrus occurs, but weak oestrus and inadequate detection are almost indistinguishable. The reported incidence of suboestrus ranges from 18% to 30%. Silent oestrus is common in early ovulations after calving. Data from many studies indicate that 60% of first ovulations after calving are silent. The incidence declines with subsequent cycles until after 60 days post-partum, the incidence is only 6–10%.

Silent oestrus is a convenient diagnosis to make to a farmer when the alternative is to diagnose inadequate oestrus detection, but that does not help the farmer if inadequate detection is indeed the cause of the problem. In herds where oestrus detection is excellent, 21-day submission rates may exceed 95% and mean intervals from calving to first oestrus may be 23–25 days, even in large herds.

The author has an adage called the 'law of silent oestrus' that states 'The incidence of silent oestrus is inversely proportional to the farmer's effort and skill at oestrus detection.'

Several factors influence the expression of oestrus including breed, line, presence of metritis, high production, milking frequency, suckling, high concentrate feeding and lameness. Some of these factors can be eliminated or dealt with while others must be endured. The major influence on the contribution of weak and silent oestrus to an anoestrus problem is the amount of effort and expertise applied in oestrus detection.

## 10.2 Anoestrus with inactive ovaries

Anoestrous cows or heifers with inactive ovaries include those with congenital abnormalities of the genitalia, for example, ovarian hypoplasia and freemartinism and more commonly, those undergoing a nutritional deficiency (stress). Ovarian hypoplasia is reported in the Shorthorn and Swedish Highland breeds of cattle. In them it is genetically transmitted as an autosomal recessive gene with incomplete penetrance. It can also occur sporadically in other breeds. It can be partial or complete, unilateral or bilateral. Anoestrus only occurs when the disease is bilateral and complete. Unilateral and incomplete hypoplasia does not cause sterility so there is a possibility of the disease being spread in the cattle population if carriers are not recognised. The ovaries in affected heifers and cows are thin, firm, narrow structures in the broad ligament. If bilateral, the genitalia are infantile, and the conformation of the heifers is steer-like. Partial hypoplasia involves the medial part of the ovary and the lateral pole may contain normal structures in externally normal-looking cows. Affected animals should be slaughtered.

Freemartinism is the most common congenital defect causing anoestrus. Freemartins comprised 9.5% of clinically diagnosed causes of anoestrus in Minnesota in the study by Zemjanis et al. (1969). The disease is restricted to females born co-twins to bulls and 90% of these are affected. About 8–9% are fertile. The Holstein breed has the highest incidence due to the highest rate of multiple di-zygotic births. The condition is due to a humoral or cellular factor passing from the male to the female twin through anastomosing placental vessels before the stage of sexual differentiation. Freemartins have small undifferentiated ovaries and small and undifferentiated vagina and uterus. A normal vulva and vestibule are present. The condition is diagnosed when there is a history of being a co-twin to a bull and a test tube cannot be inserted more than about 7.5 cm into the genital tract of a freemartin whereas in a normal female the tube will pass from 12 cm to 15 cm into the vagina. There is XX and XY chimerism of leucocyte cultures observable on cytogenetic examination. Affected animals are sterile and therefore culling is the recommended outcome for these animals.

True and pseudo-hermaphrodite conditions that are associated with anoestrus occur in cattle. True hermaphrodites and male pseudo-hermaphrodites may look like females. They are frequently first noticed because of abnormalities in genitalia or male-like body conformation and behaviour. Various types of chimerism may be present in these animals.

The most common cause of anoestrus is due to underfeeding and stress which occur in heifers and cows. In heifers there is an inverse relationship between energy intake during rearing and age at puberty. Many studies in cows show that the level of feeding before calving has a marked influence on the interval until resumption of oestrus after calving. The level of feeding after calving also has an influence on this interval, but not as great as the feeding level before calving which influences the body condition at calving. High nutritional demands for production compete for resources with reproductive requirements and cause changes in body composition associated with reproductive outcomes (Chagas et al., 2007).

Studies of dairy cattle in Australia and New Zealand have found a negative correlation between body condition at calving and the interval to oestrus. Experimental work shows that cows with high body condition at calving are better able to cope with nutritional stresses after calving. Hypoglycaemia has been related to a delay in the onset of oestrus after calving, both in field situations and when hypoglycaemia has been experimentally induced with insulin. Low planes of nutrition, and even short-term fasting, decrease the ovarian response to gonadotrophins which are 'injected' and a similar mechanism may apply in undernourished cows with regard to endogenous gonadotrophins.

The diagnosis of anoestrus due to energy deficiency can frequently be made by assessing body condition and measuring body weight. If cows have low weights and low condition, then low energy intake is the likely cause of their problems. A low blood glucose level in the group of animals affected and high beta-hydroxybuterate level confirm this. Where pasture availability is monitored as a part of nutritional services, a lack of enough feed and an inadequate intake to meet requirements can be documented. An assessment of the diet or pasture intake will reveal undernutrition, when compared to accepted feeding standards.

Treatment is by providing adequate nutrition which should also prevent the problem. In some cases in New Zealand, the problem has been addressed by reducing the nutritional demands on cows. Applying once per day milking until cows are mated can do this by reducing energy requirements. When done in association with supplementary feeding of concentrates, anecdotal reports are that oestrus cycling and conception improved in herds that adopted this strategy in the face of severe feed shortages. Protein deficiency in feed may cause anoestrus but this is due to its effect in depressing appetite and limiting energy utilisation.

Phosphorus deficiency can cause anoestrus through primary or secondary effects. There is a strong association of phosphorus deficiency with poor conception rates and anoestrus that may be due to low protein levels in phosphorus-deficient feeds, low vitamin A levels often associated with phosphorus-deficient feeds, and almost certainly the depressed appetite and wasting which accompanies phosphorus deficiency. Diagnosis can be made in the presence of mean herd blood phosphorus levels of less than 4 mg per 100 mL or when signs of clinical phosphorus deficiency are present. Phosphorus intakes can also be compared with standards. Controlled response trials can confirm a diagnosis of any nutritional deficiency.

Calcium/phosphorus ratio may also be important, and a study showed that anoestrus occurred if there was imbalance in calcium/phosphorus ratios with inadequate vitamin D. It appears that high calcium/low phosphorus ratios inhibit the expression of oestrus and the occurrence of ovulation, particularly if vitamin D is inadequate. Manganese deficiency has also been associated with anoestrus and suboestrus, but always accompanied by the birth of weak calves and calves with arthrogryposis. High dietary calcium may predispose to manganese deficiency.

Copper deficiency has been associated with anoestrus and infertility but the evidence for a direct effect is poor. Anoestrus due to copper deficiency seems only to occur when the signs of copper deficiency are obvious and when anaemia is present. Anaemia with <8 g haemoglobin/100 mL blood is likely to cause anoestrus. The anaemia, weight loss and poor growth, typical of copper deficiency, may also be intermediate causes in the anoestrus seen in copper deficiency. Diagnosis can be made based on signs of copper deficiency and blood levels of <0.07 mg/100 mL and feed levels of <20 ppm dry matter. Supplementation can be used for prevention and control with 1-2 g of $CuSo_4$ daily. Copper glycinate injections of 150 mg for 2-3 times per year were very popular in Australia. Copper wire boluses are popular in New Zealand but are expensive. Inclusion of $CuSO_4$ in topdressing applications is cheap and relatively effective.

Any disease which causes debility or anorexia and loss of body weight can result in a failure of oestrus. The pathogenesis may be due to the loss of body weight, low blood glucose levels, weight loss or the anaemia commonly associated with debilitating diseases.

## 10.3 Anoestrus with follicular or cystic structures

At examination, anoestrous cows may also have follicular structures or cysts present on their ovaries. If cows are found to have follicles on their ovaries at examination for anoestrus, they may be cycling normally, and may not have been observed by the farmer. Ovarian cysts were found to be the most common cause of true or 'organic' anoestrus in the Minnesota study of Zemjanis et al. (1969) that is, non-cycling anoestrus. At least 50% of cows with follicular cystic ovaries are anoestrous. Luteal cysts are usually associated with anoestrus (i.e. luteinised anovulating follicles). Cows with cystic CL which develop after ovulation cycle normally and are not anoestrous for prolonged periods.

There may be one or more cystic follicles on one or both ovaries. They are generally soft, spherical, fluctuating structures which feel like they project from the ovarian surface. They generally exceed 2 cm in diameter. Cysts feel thicker walled than normal follicles. Some are easily ruptured, but luteal cysts may be very difficult to rupture. The mucus from cystic cows is scantier and less clear than from oestrous cows. Some cows show elevation of the tail head and relaxation of the sacro-sciatic ligaments. The uterus may become thick-walled, flaccid and the horns feel shortened. Several treatments are available including gonadotrophin-releasing hormone being the favoured treatment, pituitary gonadotrophins having a luteotrophic effect, human chorionic gonadotrophin for follicular cysts and prostaglandins for cows with luteal cysts. Spontaneous recovery may occur but is less likely with increasing time from calving. Brito and Palmer (2004) extensively reviewed incidence, pathogenesis, diagnosis and treatment of cystic ovarian disease.

## 10.4 Treatment of the anoestrus syndrome

Pregnancy and other physiological conditions must be ruled out before any form of treatment is instituted. Where pathology of the gonads or of the tubular genital tract or intercurrent disease exists, each case must be dealt with on its merits. In many cases improved husbandry will be required. Fortunately, most cases of anoestrus in cattle are temporary and, given time, most affected cows cycle and become pregnant. The problem is in the delays that occur especially under conditions of seasonal breeding.

In managing pre-service anoestrus, the first step is to define the problem in the herd in question. This means careful examination of herd calving and breeding records and clinical examination of problem animals. Pregnancy must be protected if present and obvious 'disease' recognised. Excluding diseased and pregnant cases, the animals being examined fall into two groups:

- those with active ovaries; and
- those with inactive ovaries.

Where ovarian activity exists, this generally means the presence of a CL with no pathology of the genital tract, or cows that are approaching, in, or just going out of, oestrus. The prognosis is thus favourable and nothing need to be done except make the client aware that these cows are cycling; one should endeavour to predict when each will be in oestrus. Once alerted, the client can take more care with detection procedures. If many of these animals are noted in a herd, oestrus detection processes need checking.

Injectable prostaglandins or analogues, progesterone treatments, or Ovsynch programmes are used where circumstances warrant. With these techniques, oestrus occurrence is synchronised and prediction of oestrus and ovulation is controlled such that appointment breeding can be used. This eliminates the need for oestrus detection and fertility may be enhanced in some cases. Cost/benefit must be considered before embarking on such programmes.

Where ovarian inactivity exists, preferential feeding of potential or actual problem animals is required. Factors other than nutrition do influence return to post partum cyclical activity (e.g. the frequency of suckling or milking, sequelae to difficult birth, etc.), but feed intake and treatment of current disease are two factors which may be managerially controlled. Short-term alteration of feed input has remarkably little effect on returns to cyclical activity if the animals concerned are in a 'reproductively unfit' state to recommence the reproductive process again. Cows and heifers calving in poor condition generally divert nutrients to meet the needs of milk production and growth rather than reproduction. Even if ad libitum feed is made available, there is likely to be a considerable lag before a return to cyclical activity occurs. Heifer and cows calving in adequate body condition have the potential to 'survive' this initial lactational stress period and will respond by cycling much sooner unless very severe nutritional deprivation occurs. The establishment of minimum target weights (and/or condition scores) for animals at breeding and calving is a sound practice. Management practices should be examined to ensure that targets can be achieved including the grazing or feeding system, fertiliser practices, 'drying off' dates and animal health interventions.

Mating of heifers 10–14 days before breeding commences in the milking herd is recommended to increase submission rates for primiparous animals. Pre-service anoestrus with inactive ovaries has proven to be significant problem with first calf heifers. By adopting this early mating procedure, the mean calving date for the heifers is usually about 16 days ahead of the mean calving date for the main breeding herd. Their requirement for a relatively long post-partum period before resuming ovarian activity is thus compensated for and they can be expected to have a similar second calving pattern to the rest of the herd. Where satisfactory submission rates with these 2-year old animals are being achieved early breeding of the heifers is unnecessary.

## 11 Improving oestrus detection

Failure to observe oestrus is due to many factors. There is a variation in the length and intensity of oestrus. Studies show that there are periods of intense oestrus activity interspersed by quiet periods. A failure to observe cows at appropriate times may influence oestrus detection and some cows with short duration of oestrus will be missed unless observation coincides with their display of oestrus. Studies show that cows in barns have oestrus which lasts only for 8 hours in contrast to the commonly cited oestrus length of 16–18 hours for cows at pasture. There can be considerable variation between individuals in the duration and intensity of the display of oestrus. Thus while 18 hours is an average for cows to show standing oestrus the range may vary from 4 hours to 30 hours. Since the intensity of the signs shown also varies, it is understandable, particularly in large herds, why the signs may be missed.

Inadequate observation and ignorance of the behavioural and physical signs of oestrus are the major causes of apparent anoestrus. Many farms do not have set periods for oestrus detection, while others only observe once per day, when three times for daily observation is required for optimal oestrus detection. When cows are at pasture, observation for oestrus activity needs to be scheduled as a farm task rather than something which occurs as other farm tasks (such as collecting the cows for milking) are done.

Increased herd sizes make detection more difficult. The facilities needed for larger herds make observation more difficult and the observation time per cow is reduced, since labour availability generally is not increased in proportion to herd size. The importance of heat detection, evaluation of heat detection efficiency, causes of inadequate detection and methods to improve detection, including heat detection aids, were reviewed by Williamson (2005).

Where inadequate oestrus detection is limiting herd reproductive performance the major step in improving the situation is to convince farmers that oestrus detection is an important farm task. To help do this, the oestrus detection efficiency of the farm can be demonstrated by analysing records. The use of aids, like pressure detectors, rubbing detectors or chin-ball markers on teaser bulls for a short period, may show the farmer that many oestruses are being missed. The economic importance of oestrus detection should be stressed to motivate farmers to improve it. False positive detection can be documented by taking serum or milk samples at observed oestrus and analysing for progesterone levels. The presence of a high frequency of progesterone above baseline levels indicates false positive detection.

To achieve improvements the farmer should be encouraged to set aside time to observe cows for oestrus. The time of observation is important. It should be when cows are not being milked and when they are not eating or being moved. A period when the majority of the herd is recumbent and ruminating is the best time. Three periods of 20–30 minutes at times when the herd settles for rumination should be enough to detect most oestrus events. Where cows are restrained in barns the most important step is to let them out for observation. If they are to be fed, then observation should be long enough after they are fed for most of the herd to have settled down and be ruminating. This also applies to situations where bunk feeding occurs in otherwise pastured or lot-housed cattle.

In order to achieve adequate oestrus detection, farmers and farm workers must be educated in the primary or definitive signs of oestrus and the secondary or suggestive signs. Primary signs of oestrus are standing to be mounted and mounting the head of another cow. Recognition of secondary signs greatly assists in detecting oestrus. These include restlessness, mooing and bellowing, increased activity, and standing when most of the herd is lying down. Cows in and near oestrus display seeking activity that includes nudging, sniffing and rubbing their chins on the rear end of other cows. Cows in or near oestrus have wet, pink and swollen vulvae, with long clear mucus discharges. They show a decrease in appetite and milk production. The most visible sign is that they mount other cows. The recording of all observed oestruses allows the prediction of subsequent ones and ensures that non-cycling cows with pathology or those needing therapy to initiate cycling can be identified. Oestrus prediction charts and computer-generated lists are available to aid in predicting upcoming oestrus events. Metoestral bleeding can be used to alert the client to watch for oestrus in 16–22 days.

Oestrus detection aids are invaluable in assisting accurate oestrus detection to occur. Several types of detection patches are available including pressure-sensitive detectors that are glued to the rumps of cows and undergo a colour change when cows are mounted. In a study conducted in Australia, herdsmen detected 56% of oestrus in pastured dairy cattle which did not have detectors fitted and in 90% of cows which did (Williamson et al., 1972). The use of detectors can be highly profitable where oestrus detection is improved.

The use of 'tail paint' or chalk applied over the sacral and first few coccygeal spines of cows acts as a cheap and effective alternative to oestrus detection patches. The paint becomes rubbed and broken when cows are mounted. The paint needs to be 'touched up' or freshened about once per week when latex paints are used and each 2–3 weeks when enamels are used.

Teaser animals can be used to aid oestrus detection and they are especially useful if fitted with Chin-ball harnesses. Teaser animals may be bulls with Chin-ball harnesses and with vasectomies, epididymectomies or deviated penises. They may also be testosterone-primed steers or cows or nymphomaniac cows. Injecting cows or steers with intra-muscular testosterone propionate, 950 mg every 2 days for 20 days, then 500 mg every 2–3 weeks can create teaser animals. These teaser animals mark oestrous cows and help identify them. Cows with chronically cystic ovaries can make effective teaser animals.

Veterinarians may predict oestrus when operating herd health programmes by assessing the state of the genital tract at palpation or scanning. The accuracy of prediction is increased greatly if the animal has a mature mid-cycle CL present on its ovary. This can be made to undergo luteolysis by injecting prostaglandin. Oestrus then usually occurs in 2–4 days and animals can be bred by appointment A.I. at 72 and 96 hours or 80 hours after treatment.

Many further aids to oestrus detection are available including electronic means of reading sensor strips affixed to cows, motion sensors mounted as pedometers or necklets, accelerometers that measure jumping activity, electronic pressure detectors that can link to computers in systems that may alert farmers to the occurrence of oestrus. Some further aids to oestrus detection rely on physiological aids to oestrus detection such as vaginal mucus conductivity probes, progesterone analysis kits to measure low progesterone levels in plasma or serum associated with oestrus. Temperature probes and systems to measure decreases in feed intake and milk production have also been coupled with other means to help detect the occurrence of oestrus in cows.

## 12 Controlled breeding programmes for oestrus synchronisation

There are three main types of oestrus synchronisation programmes used in the management of cattle breeding.

- Those based on extension of the luteal phase of oestrous cycles using exogenous progesterone or progestogens.
- Those based on luteolysis using prostaglandins to shorten the luteal phase of the cycle.
- Those based on follicular wave control by luteinisation and synchronisation of follicular wave development and LH release with gonadotrophin releasing hormone (GnRH) to ensure follicles are young at ovulation.

Many variations on these programmes exist and they may be supplemented using additional treatments, and aspects of the programmes may be mixed together.

When synchronisation programmes are being introduced to farms, adequate management is required for programmes to succeed. Adequate management includes:

- Clarity on the purpose of synchronisation – that is, is it to allow AB, minimise oestrus detection, help treat anoestrus or reduce calving spread?
- Careful planning and documentation of procedures, since programmed steps are required at the appropriate times when people and facilities must be available.
- Complete records to select animals for treatment and conduct the programme.
- High semen quality and serving capacity for natural sires with sufficient inseminators and/or bulls available.
- Experienced and capable inseminators if many animals are to be bred.
- Adequate and efficient facilities.

Animals selected for controlled breeding programmes should be free of disease and abnormality, adequately grown, well nourished, recovered from parturition and preferably cycling. This can be determined by detecting presence of a CL or by oestrus detection prior to implementing the programme. They also should be confirmed as not pregnant.

## 12.1 Progesterone-based programmes to induce and synchronise oestrus

Progestogen or progesterone-based devices are available for controlled breeding in cattle, including intravaginal devices that contain progesterone in silastic coatings or pods and silastic subcutaneous implants. These devices provide a prolonged, measured release of progesterone or a progestogen to simulate the influence of a CL and thereby prolong the luteal phase of the oestrus cycle so that the time of onset of oestrus can be controlled by acute withdrawal of the progesterone or progestogen.

Orally active progestogens were researched to provide a mechanism for controlled breeding, but the programmes of administration proved difficult for farmers to follow, and their use was associated with a reduction of conception rates.

These devices may be used for varying times in cows and result in relatively high synchrony of oestrus and ovulation if the treatment period with progesterone is sufficiently long. However prolonged progesterone exposure lowers the fertility of cows at the controlled oestrus. This is because progesterone or progestogens suppress LH pulse frequency and this causes suppression of dominant follicle growth. Progesterone does not, however, suppress FSH secretion. Progesterone treatment slows the growth of the dominant follicle and will prevent ovulation but does not stop or synchronise regular periodic FSH surges and emergence of follicular waves. Lowered fertility after prolonged progestogen treatment (>14 days) observed in early studies has been attributed to aging of the oocyte within the long-lived oversized follicles. Oestrus tends to occur on the 2nd and 3rd days after removal of the source of progesterone. Combined progesterone and prostaglandin treatments produce synchronisation of oestrus with a high degree of synchrony and high fertility. A progesterone device is used for 7–9 days and a prostaglandin treatment is administered at the time of removal of the progesterone source. Half-doses of prostaglandin are effective in this programme.

Anoestrus, low submission rates and poor breeding outcomes in North America have led to a system that requires hormonally controlled breeding to enable conceptions to occur. Also in seasonal pasture-based herds, controlled breeding is sometimes required to achieve compact calving and schemes exist to enhance the efficiency of breeding through hormonal intervention. Developments have occurred in the use of progestins and prostaglandins as synchronising agents, sometimes supplemented by other hormones such as oestrogens (where legally allowed) and hormones to stimulate follicular development as well as the use of GnRH plus prostaglandin and steroids as a means of controlling follicular wave development and inducing ovulation.

Induction of oestrus cycling as well as synchronisation of oestrus can be achieved through progesterone treatment, often combined with other hormones. If heifers are under-grown and undernourished, such progesterone treatment is ineffective. In such cases correction and prevention of further recurrences should be directed at removing underlying causes.

If farmers wish to treat non-cycling anoestrous cows and heifers to induce ovarian cycling, they may treat them with progesterone devices for 7 days and then remove the devices and inject with 400 IU of PMSG. Cows should then be watched closely for return to oestrus and inseminated on detection of oestrus. If no oestrus is observed within 3 weeks of device insertion, cows can be re-examined and a second treatment regime started if no ovarian activity is detected, or they can be treated with prostaglandin if a CL is palpated on an ovary. A recommended programme for inducing oestrus in non-cycling cows and heifers in New Zealand involves inserting a CIDR-B on day 0, removing the CIDR on day 5 and then injecting with PMSG at the time of CIDR removal. The injection of GnRH at or after CIDR removal causes an LH surge and more synchronous ovulation. Conception rates and pregnancy rates vary but are acceptable and similar to those in control groups. Double insemination at 48–60 hours after removal has produced pregnancy rates up to 75%.

In evaluating the efficacy of induction of oestrus with progesterone devices, results are measured in terms of conceptions and not just submission rates since there is evidence that false positive detections increase. The advantage of induced oestrus is partly its earlier occurrence. In evaluating the economic advantage of induction of oestrus, controls are essential since it has been observed in one study that 41% of control cows showed oestrus within 7 days of treatment of the treatment group cows. In another study, 60% of control cows showed oestrus within 21 days of treatment while an average of about 65% of treated cows showed oestrus within 7 days of treatment.

## 12.2 Prostaglandin-based synchronisation programmes

Several approaches to synchronisation of oestrus with prostaglandin are used. The product used is largely immaterial since available prostaglandin products show similar efficacy in causing luteolysis at recommended dose rates. Prostaglandin causes luteolysis of a mature CL3 on an ovary. Luteolysis only occurs if a mature CL is present. Corpora haemorrhagicum are refractory to prostaglandins until about the 6th day of the cycle. Studies show that maximal response to treatment in lactating dairy cows does not occur until day 8 of the cycle when over 90% undergo luteolysis. Cows respond well to prostaglandin-induced luteolysis between days 8 and 18 of the cycle. Following day 18 the regressing CL is not responsive to injected prostaglandin, but is regressing anyway due to the endogenous hormone.

Prostaglandin $F_{2\alpha}$ (25 mg) or cloprostenol (500 µg) given intramuscularly causes luteolysis and subsequent oestrus. Schemes have been devised to treat groups of animals to synchronise oestrus so that they may then be bred by appointment insemination, thus eliminating the need for oestrus detection. Schemes involving palpation or oestrus detection followed by single injections of these products or double injection programmes 11 days apart and then followed by insemination at 80 hours after the second treatment or twice at 72 and 96 hours have achieved fertility comparable to or better than that achieved in control animals. Controlled oestrus has the potential to reduce the spread of breeding and to achieve early compact calving with reduced labour required for oestrus detection, which results in longer lactations for the treated animals.

A double injection prostaglandin programme allows a high degree of synchrony of oestrus and ovulation to be achieved in randomly cycling animals. Suitable cycling cattle are injected on the first day of the programme. Those in dioestrus will undergo luteolysis and show oestrus, ovulate and form a new CL. Those in metoestrus will continue to develop a CL. Those in proestrus will continue to progress towards oestrus, in synchrony with the dioestrous cows that were injected with prostaglandin. The cows or heifers are then treated again 11 days after the first treatment when all would be expected to have a mature CL3 on their ovaries. Animals can be inseminated at detected oestrus, at 72 and 96 hours after the second injection or once at 72 hours (60 hours for heifers and dry cows). Fertility is highest when oestrus is detected.

In a single/double injection programme, suitable cycling cattle are injected with prostaglandin on day 1 of the programme and observed for oestrus over the next 5–7 days. Those detected in oestrus are inseminated and the remainder receive a second injection 11 days after the first. Animals receiving the second injection may be inseminated by set time breeding as for the double injection programme or after detection of oestrus. In a single injection programme, cows and heifers are palpated and if a mature CL is detected, they are treated with a dose of prostaglandin or analogue. Breeding is as for the previous two treatment programmes.

In the detection plus injection or 'Why Wait program', oestrus is observed carefully (using detectors or tail paint as aids) from 11 days before the planned start of mating (PSM). Cows in oestrus from 11 days to 6 days before the planned start of mating are injected with prostaglandin at the planned start of mating. They are then bred at detected oestrus, which is likely to occur 3–4 days later. Those in oestrus from 5 days to 1 day before the planned start of mating are injected with prostaglandin 6 days into mating and are inseminated at detected oestrus, from 9 days to 10 days into the mating season. Use of this and similar systems has resulted in mean intervals from planned start of mating to first insemination of 7.2 days with a 91% 21-day submission rate in New Zealand herds.

The detection, insemination, injection programme starts with oestrus detection for 5–6 days and cows are inseminated on detection. If over 25% of the group is inseminated, it indicates a high level of cyclicity in the group and the remainder are injected on day 6 with prostaglandin. Insemination then continues for up to a week after injection. This method is most suited to groups of heifers for AI since it allows their cycling status to be evaluated.

## 12.3 Gonadotrophin-prostaglandin programmes

The OvSynch programme uses GnRH followed by a prostaglandin $F_{2\alpha}$ injection 7 days later and a second GnRH treatment 24 or 48 hours after the prostaglandin treatment. This regime of treatment has resulted in synchronised oestrus in both cycling and non-cycling cows. After the second injection of GnRH the cows are bred without regard to oestrus behaviour with the optimum time after GnRH being reported as 16 hours (Pursley et al., 1995). There is considerable variability in response to this treatment programme. Results appear to be influenced by the cycling status of cows, their age (heifers respond poorly) and the precise timing of treatments and insemination. A meta-analysis found that the fertility of cows bred is around 40% conception rates where this regime has been implemented. If this is normal for an area this may be acceptable, but it may represent a reduction in fertility in some environments. The cost and multiple treatments involved in the Ovsynch programme may limit its use in some countries (Rabiee et al., 2005).

### 12.4 Combination programmes

Programmes have become very mixed so that there are various combinations and joint delivery of aspects of the programmes mentioned above. Detailed programmes are too numerous to review here, but include combinations of progesterone, oestradiol or GnRH or HCG and cloprostenol (Gyawu et al., 1991) and combinations of GnRH-prostaglandin programmes with progesterone intravaginal inserts (McDougall and Compton, 2008; Stevenson and Britt, 2017).

### 12.5 Pre-synchronisation and resynchronisation programmes

The various programmes for induction and synchronisation of oestrus followed by timed artificial insemination have been extensively reviewed recently (Stevenson and Britt, 2017; Fricke, 2019). They also describe resynchronisation programmes that may start 7 days before or at the time of a negative PD of cows that have previously been bred by appointment insemination, have not returned to oestrus but are not pregnant (elsewhere called phantom cows). These programmes start with an injection of GnRH or HCG 5 days or 7 days before or on the day of negative PD as the start of the next synchronisation cycle.

## 13 The role of nutrition in limiting and optimising reproduction

Nutritional limits to reproduction are universal in dairy farming. They may involve deficiencies or excesses of nutrients, usually the former, and other ingested matter such as toxins present within feeds. The impact of nutrients on fertility is manifest in many reproductive indicators such as delayed onset of puberty, anoestrus, failure to conceive and early embryonic death. Inadequate levels of specific nutrients in the diet and imbalances of some nutrients may inhibit the occurrence and display of oestrus, cause low conception rates, abortions, dystocia and the birth of weak offspring.

### 13.1 Plant and fungal toxins

Phyto-oestrogens and fungal oestrogens have been associated with the development of cystic ovaries and uterine pathologies that limit conception rates. Phyto-oestrogens depress fertility, possibly due to the cystic endometrial disease that they cause and may cause cystic ovaries. Sources of phyto-estrogens are some species of clovers, alfalfa and moulds that produce zearalenone.

Plant contaminants of pasture or trees consumed, specifically the needles of *Cupressus macrocarpa* and *Pinus ponderosa,* contain isocupressic acid which is an abortifacient in cattle if enough toxin is ingested.

## 13.2 Energy intake and balance

A primary aspect of nutrition with impacts on reproduction is the total feeding amount. Studies have shown that heifers fed inadequate rations and those fed plentiful rations achieve puberty late and early, respectively, but at about the same body size. More recently, it has been shown that increasing the feed level does increase the weight at which puberty occurs. Once oestrous cycling occurs, generally it continues if the rate of feeding is maintained. Oestrous cycling can be stopped by severe nutritional stress and cycling resumes when nutritional levels improve and body condition is restored.

Nutritional anoestrus of young heifers is exhibited as small, flattened, smooth, even-shaped ovaries and an underdeveloped tubular tract if the heifer has not reached puberty. In older heifers the ovaries are more likely to be 'knobbly' as follicular activity starts but with no readily discernible structures. Silent oestrus commonly occurs in the first one or two cycles of heifers reaching puberty. Heifers also weigh less than the weight expected at puberty which in New Zealand is:

- Holstein/Friesians 240–290 kg.
- Jerseys 163–204 kg.
- Shorthorns 240 kg.

Lactating dairy cows have high feed requirements because of the high milk outputs that they achieve and the need to provide nutrition to support this milk production in addition to growth and maintenance requirements. Closely allied with this is the energy required by dairy cows. Energy deficiency can be caused by a lack of feed, inadequate feed intake, poor-quality feed or a deficiency of intake relative to production. In pasture-based systems, cows have poor fertility when feeding is inadequate. Cows also have low body condition scores. Adequate feed intakes and balance underlie adequate reproduction.

## 13.3 Protein

Protein intakes can also be inadequate under some circumstances such as when cows graze dry-standing feed in dry summer and autumn conditions. Protein can also be excessive in some springs when pasture growth is high and pastures are lush. Under conditions of fully feeding cows or zero grazing, protein intakes are frequently controlled because feeds used as protein sources are generally expensive and so are limited to levels unlikely to cause infertility due to excesses in formulated rations.

## 13.4 Macrominerals

Some mineral deficiencies are known to be causal factors for retained placenta and metritis as well as reduced bull fertility. Calcium and phosphorus macrominerals impact many aspects of reproduction. Older studies showed that significant effects of phosphorus deficiency are depression of appetite leading to underfeeding and energy deficiency, anoestrus, suboestrus, irregular oestrus cycles and low conception rates. Calcium to phosphorus ratios have also been implicated in causing infertility when they are well away from 1:1. These effects are exacerbated by low vitamin D. Interactions of calcium and phosphorus and manganese in causing infertility have also been reported in England. These may reflect a previously unrecognised dietary cation-anion difference effect. Calcium deficiency and hypocalcaemia may also reduce fertility due to decreases in muscle tone of the uterus with subsequent dystocias, retained foetal membranes and subsequent metritis or endometritis. Several studies also suggest that an excess dietary cation-anion difference reduces conception rates.

## 13.5 Trace elements

Trace elements also have an impact on fertility. Iodine deficiency has been reported to cause anoestrus, suboestrus, low conception rates, abortion, retained placenta and the birth of dead or weak calves that have arthrogryposis. In addition to its interaction with the calcium to phosphorus ratio, manganese deficiency can cause anoestrus, suboestrus and reduced conception rates. When manganese deficiency is associated with these reproductive signs, it is also associated with the birth of dead or weak calves that have arthrogryposis. This has been reported in New Zealand associated with apple pomice feeding. Many reports associate copper deficiency (direct or secondary) with infertility exhibited as anoestrus and poor conception rates. These effects may be secondary to reduced feed intake, anaemia and reduced body condition that are associated with copper deficiency. It is not likely that copper deficiency will be the cause of poor reproductive performance in the absence of other signs of deficiency. Cobalt deficiency can cause infertility secondary to inappetence, cachexia and debility associated with concurrent anaemia. The reproductive signs are anoestrus, suboestrus and reduced conception rates. Selenium deficiency is a cause of retained placenta, poor uterine involution and metritis. In New Zealand studies, supplementation with selenium failed to improve fertility despite improvements in growth and milk production occurring. Zinc deficiency is a cause of testicular degeneration and resulting decline in conception efficiency of bulls.

Vitamin A deficiency can occur in animals fed stored feeds or dried out pastures. Vitamin A deficiency can cause delayed puberty of both sexes, abortion, stillbirth, the birth of weak blind calves and retained placenta. Intake of fresh green feed prevents deficiency.

Lean et al. (2003) have reviewed the impact of nutritional deficiencies on reproductive performance in temperate dairying conditions.

## 14 Managing abortion

Abortion is one manifestation of the premature death of a conceptus and expulsion of a foetus. Other forms of interrupted pregnancy are early embryonic death that may go unnoticed and be attributed to a failure to conceive, foetal death followed by maceration and absorption, foetal death and mummification and premature birth or stillbirth.

Observed abortions and cases where it is recognised that cows are not pregnant after they have previously been diagnosed pregnant should both be regarded as abortions. In seasonally calving herds, determining the abortion risk as a proportion of confirmed pregnant cows that abort is a satisfactory measure of abortions. The proportion should not exceed 5%, although it can be as low as 0%. Abortion rates are sometimes reported and are variously calculated depending on their intended use. Calculations include the number of abortions occurring in a period as the numerator and the average number of cows pregnant in the period as denominator. The specification of cows in the denominator varies, depending on their pregnancy status and stage, under different definitions of abortion rate.

Abortion is noted when a diagnosed pregnancy is followed by oestrus and then a negative pregnancy diagnosis, when an aborted foetus or foetal membrane is observed or when an abnormal discharge from the vulva is observed after a positive pregnancy diagnosis and on repeat examination, the cow is not pregnant. The observed discharge may be serosanguineous, purulent or contain foetal membranes or their remnants. In the author's experience, 6–7% of pregnancies end in abortion in temperate regions and around 12% in tropical regions.

If abortion is suspected as a herd problem, records should be examined to determine the number and incidence rate of the abortions to confirm that a problem exists. Obtaining a full herd history is required in abortion investigations to aid in diagnosis and to help determine causation, so that remedial and preventive actions can be taken. Possible exposures to infectious causes can be revealed through analysis of animal introductions, events on neighbouring farms, travel of animals, changes in feed and water source. The general health of the herd can be examined and assessed through the herd history as well as the immunisation history and medications used. Corticosteroid and prostaglandin medications are known to cause abortions so their use in the herd should be examined.

Linking epidemiological factors, such as the dates; number of abortions; stage of gestation at abortion; groups affected based on age, location, feeds and water source, can assist in determining causes, as can the identification of the sires of aborted foetuses and an assessment of the fertility of the herd. If feasible, a clinical examination of the whole herd should be undertaken but, in large herds, this may need to be restricted to aborting animals, followed (if possible) by all cows that have been diagnosed as pregnant, in order to identify any cows that have aborted, detect mummies, pyometra and any cows that show signs of illness.

Laboratory examination is highly important in abortion investigations. The diagnostic approaches to infectious causes of reproductive failure, particularly abortion in Australia and New Zealand have been reviewed recently (Reichel et al., 2018). Methods can include diagnostic tests such as PCR and serology for known infectious abortifacient agents, histology and culture of aborted foetal and foetal membrane tissues and examination of feeds and water for infectious agents such as fungi or bacteria. Analyses also can be conducted for toxins. The success of obtaining a diagnosis can be low on submitted samples for several reasons. The necessary cause of the abortion may be gone by the time of abortion and this is even more likely by the time an investigation is undertaken. As an example, feeds containing causal organisms or toxins may have been fully utilised by the time an investigation is undertaken so that there is none of the offending feed left when feed is sampled. Abortion is frequently delayed after foetal death and this can lead to autolysis that masks changes that are characteristic of the cause of abortion. Desired tissues, such as the aborted foetus or foetal membranes, are not always available which limits diagnostic accuracy.

Ideally an entire, fresh foetus can be taken to a diagnostic laboratory for a full examination. If this is not feasible, fresh foetal heart blood, stomach contents and lung samples should be forwarded on ice. Fixed tissue samples of foetal heart, brain, lung, liver and kidney are preferred by pathologists; tissues from cotyledons and caruncles are also preferred. Further useful samples are acute and convalescent maternal serum, samples of feed and water if either is suspected as an associated factor in the abortions, including plant specimens when these are possible causes.

In situations where abortion outbreaks occur, clients are keen to acquire a prognosis for their herd. The prognosis depends on the cause of the abortions, the number of susceptible animals exposed, the stage of pregnancy of animals in the herd and the time between exposure to the abortifacient and diagnosis. In many cases, an outbreak may be almost over before there is veterinary involvement or before a diagnosis is made.

## 15 Conclusion and future trends

Reproductive health and management programmes have changed over the past 45 years by supplementing the previously used intense monitoring and problem identification strategies with ongoing intervention and control. Hormonal manipulation of the reproductive cycles of cows is becoming far more widespread and is used as the standard method of breeding cows in some situations. This change has been necessitated by ongoing increasing herd sizes, increased milk production per cow and reliance on management by employed labour for vital farm functions such as oestrus detection and selection of cows for artificial insemination. Intensive management programmes are assisted by on-farm computers that are often now linked to cloud-based management systems. Systems now allow the rapid daily review and selection of cows requiring management actions and selection of cows for examination and programmed interventions on veterinary visits.

There is little doubt that as computer artificial intelligence is advanced, machine-based management will increase and cause reduction in the need for inputs by people. Small examples of this exist as in automated oestrus detection with automated diverting of cows to be held for artificial insemination and computer recognition of mastitis and lameness with similar diversion for further examination and treatment.

It is highly likely that treatment regimens delivered by a single device will be developed for oestrus induction and synchronisation. Previous attempts to develop such devices include an intelligent breeding device that delivers hormonal treatments intravaginally to synchronise and induce oestrus through a computer-controlled insert. Micronisation and other developing technologies make such single device treatment programs increasingly feasible and likely. They would offer benefits of precise timing, precise dosing and a decreased chance of environmental contamination or accidental exposure to the veterinary medicines used in hormonal breeding control programs.

## 16 Where to look for further information

Reproductive health and management of cattle in Australia and New Zealand in seasonal and year-round calving herds are described well in Chapter 11 'Reproduction and disorders of the reproductive system' by Parkinson, T. J., Vermunt, J. J., Malmo, J. and Weston, J. F. in *Diseases of Cattle in Australasia*, Editors Parkinson, T. J., Vermunt, J. J. and Malmo, J. Published by New Zealand Veterinary Association Foundation for Continuing Education (VetLearn), Wellington, New Zealand (2010).

Postpartum anoestrus and its management are extensively reviewed in Chapter 52 'Postpartum anestrus and its management in dairy cattle' by

Ambrose, D. J. in *Bovine Reproduction*, Editor Hopper, R. M. Published by John Wiley and Sons (2015). https://onlinelibrary.wiley.com/doi/book/10.1002/9781118833971.

Reproductive management and changes occurring in herd management of pastured seasonally calving herds are described in 2 papers by Macmillan, K. L. (2002). Advances in bovine theriogenology in New Zealand. 1. Pregnancy, parturition and the postpartum period. New Zealand Veterinary Journal, 50:sup3, 67–73 and 2. Breeding management and technologies for improved reproduction. *New Zealand Veterinary Journal*, 50:sup3, 74–80.

Reproductive management and veterinary intervention have been reviewed using a Hazards Analysis Critical Control Point (HACCP) approach by Lean, I. J., Rabiee, A. R. and Moss, N. (2003). A hazards analysis critical control point approach to improving reproductive performance in lactating dairy cows. Proceedings of the 20th Annual Seminar, Society of Dairy Cattle Veterinarians of the NZVA. Pp. 419–444.

Protocols for synchronisation and re-synchronisation programs for dairy cows are available on the web site of the Dairy cattle Reproduction Council at https://www.dcrcouncil.org/wp-content/uploads/2019/04/Dairy-Cow-Protocol-Sheet-Updated-2018.pdf.

Key journals addressing reproductive management of dairy herds include:

- Animal Reproduction Science. https://www.sciencedirect.com/journal/animal-reproduction-science.
- Journal of Dairy Science. https://www.journalofdairyscience.org/.
- Theriogenology. https://www.journals.elsevier.com/theriogenology.

Congresses addressing dairy cattle reproduction include the two listed below. There are also many national veterinary associations for buiatrics or cattle health and management associated with the World Buiatrics Association that hold national congresses where reproductive health and management are popular topics of presentations:

- International Congress on Animal Reproduction. Normally held every 4 years. http://animalreproduction.org/icar/
- World Association for Buiatrics Congress and its proceedings. Normally held biennially. http://Buiatrics.com

## 17 References

Ambrose, D. J. (2015) Chapter 52. Postpartum anestrus and its management in dairy cattle. In: *Bovine Reproduction* Hopper, R. M. (Ed.). John Wiley and Sons, Hoboken, NJ.

Anon. (2020). *The InCalf Fertility Focus Users Guide*. DairyNZ, Hamilton, New Zealand. Available at: https://www.dairynz.co.nz/publications/animal/incalf-fertility-focus-users-guide/. Accessed on 28 Sept 2020.

Brito, L. F. C. and Palmer, C. W. (2004). Cystic ovarian disease in cattle. *Large Animal Vet. Rounds* 4(10), 7 pp. Department of Large Animal Clinical Sciences, Western College of Veterinary Medicine, Saskatoon.

Brownlie, T. S., Morton, J. M., Heuer, C., Hunnam, J. and McDougall, S. (2014). Reproductive performance of seasonal-calving, pasture-based dairy herds in four regions of New Zealand. *N. Z. Vet. J.* 62(2),77–86.

Chagas, L. M., Bass, J. J., Blache, D., Burke, C. R., Kay, J. K., Lindsay, D. R., Lucy, M. C., Martin, G. B., Meier, S., Rhodes, F. M., Roche, J. R., Thatcher, W. W. and Webb, R. (2007). Invited review: new perspectives on the roles of nutrition and metabolic priorities in the subfertility of high producing dairy cows. *J. Dairy Sci.* 90(9), 4022–4032.

Denis-Robichaud, J., Cerri, R. L. A., Jones-Bitton, A. and LeBlanc, S. J. (2016). Survey of reproductive management on Canadian dairy farms. *J. Dairy Sci.* 99(11), 9339–9351.

Dubuc, J., Duffield, T. F., Leslie, K. E., Walton, J. S. and LeBlanc, S. J. (2010). Risk factors for postpartum uterine diseases in dairy cows. *J. Dairy Sci.* 93(12), 5764–5771.

Fetrow, J., McClary, D., Harman, R., Butcher, K., Weaver, L., Studer, E., Erlich, J., Etrherington, W., Guterbock, W., Klingborg, D., Reneau, J. and Williamson, N. (1990). Calculating selected reproductive indices: recommendations of the American Association of Bovine Practitioners. *J. Dairy Sci.* 73(1), 78–90.

Fielden, E. D., Macmillan, K. L. and Watson, J. D. (1973). The anoestrous syndrome in New Zealand dairy cattle. 1. A preliminary investigation. *N. Z. Vet. J.* 21(5), 77–81.

Fricke, P. (2019). Evolution of fertility programs for lactating cows. *Clin. Theriogenol.* 11, 317–328.

Gyawu, P., Ducker, M. J., Pope, G. S., Saunders, R. W. and Wilson, G. D. (1991). The value of progesterone, oestradiol benzoate and cloprostenol in controlling the timing of oestrus and ovulation in dairy cows and allowing fixed time insemination. *Br. Vet. J.* 147(2), 171–182.

Lawton, D. E. B., Mead, F. M. and Baldwin, R. R. (2000). Farmer record of pregnancy status pre-slaughter compared with actual pregnancy status post-slaughter and prevalence of gross genital tract abnormalities in New Zealand dairy cows. *N. Z. Vet. J.* 48(6), 160–165.

Macmillan, K. L. (2002). Advances in bovine theriogenology in New Zealand. 1. Pregnancy, parturition and the postpartum period. *N. Z. Vet. J.* 50(3) (Suppl.),67–73.

McDougall, S. (1994). Postpartum anoestrum in the pasture grazed New Zealand dairy cow. Ph.D. Thesis, Massey University, Palmerston North.

McDougall, S. and Compton, C. (2008). 'Effects of treatment of 'not detected in oestrus' cows with gonadotrophin releasing hormone, prostaglandin and progesterone. Proceedings of the Society of Dairy Cattle Veterinarians of the NZVA, Pp 35–50.

McDougall, S., Leijnse, P., Day, A. M., Macmillan, K. L. and Williamson, N. B. (1993). A case control study of anoestrum in New Zealand dairy cows. Proceedings of the New Zealand of Society of Animal Production 53, 101–103.

Mwaanga, E. S. and Janowski, T. (2000). Anoestrus in dairy cows: causes, prevalence and clinical forms. *Reprod. Domest. Anim.* 35(5),193–200.

Pursley, J. R., Mee, M. O. and Wiltbank, M. C. (1995). Synchronisation of ovulation in dairy cows using PGF2α and GnRH. *Theriogenology* 44(7), 915–923.

Rabiee, A. R., Lean, I. J. and Stevenson, M. A. (2005). Efficacy of Ovsynch program on reproductive performance in dairy cattle: a meta-analysis. *J. Dairy Sci.* 88(8), 2754-2770.

Reichel, M. P., Wahl, L. C. and Hill, F. I. (2018). Review of diagnostic procedures and approaches to infectious causes of reproductive failures of cattle in Australia and New Zealand. *Front. Vet. Sci.* 5(222), 15 pp. Available at: https://www.ncbi.nlm.nih.gov/pmc/articles/PMC6176146/. Accessed on 28 Sept 2020.

Stevenson, J. S. and Britt, J. H. (2017). A 100 year review: practical female reproductive management. *J. Dairy Sci.* 100(12), 10292-10313.

Stevenson, J. S., Hill, S. L., Nebel, R. L. and DeJarnette, J. M. (2014). Ovulation timing and conception risk after automated activity monitoring in lactating dairy cows. *J. Dairy Sci.* 97(7), 4296-4308.

Williamson, N. B. (1987). The interpretation of herd records and clinical findings for identifying and solving problems of infertility. *Compend. Contin. Educ. Pract. Vet.* 9(1), F15, 14-23.

Williamson, N. B. (2005). Heat detection. Proceedings of the Society of Dairy Cattle Veterinarians of the NZVA, Pp 141-151.

Williamson, N. B., Morris, R. S., Blood, D. C. and Cannon, C. M. (1972). A study of oestrous behaviour and oestrus detection methods in a large commercial dairy herd. *Vet. Rec.* 91(3),50-58.

Zemjanis, R., Fahning, M. L. and Schultz, R. H. (1969). Anestrus. The practitioners dilemma. *Vet. Scope* XIV, 14-21.

# Chapter 8

# Managing dry cow udder health

*Päivi J. Rajala-Schultz, University of Helsinki, Finland; and Tariq Halasa, University of Copenhagen, Denmark*



## 1 Introduction

Providing cows a 6- to 8-week dry period between lactations is a common management practice in dairy herds. The dry period lays a foundation for the next lactation, and research has clearly demonstrated its importance for the future udder health and productivity of dairy cows (Eberhart, 1986, Dingwell et al., 2003, Bradley and Green, 2004). The dry period is crucial for the renewal of senescent mammary epithelial cells (Capuco and Akers, 1999) and consequently for milk production in the following lactation. It also provides an excellent opportunity to treat existing subclinical intramammary infections (IMIs), as treatment of such infections is often not cost-effective during lactation (Gussmann et al., 2019).

At the same time, the dry period is a high-risk time for cows to acquire new IMIs (Neave et al., 1950, Cousins et al., 1980, Smith et al., 1985, Bradley and Green, 2000). The bovine udder is highly susceptible to new infections during the dry period, when the udder transitions first from a lactating to a non-lactating state and then at the end of the dry period, back to milk synthesis as the next calving approaches. During this time period, cows typically also encounter several management changes, including potentially several feeding, location and social group changes. These adjustments can be stressful for the cow and impair her immune function, further increasing her susceptibility to

http://dx.doi.org/10.19103/AS.2020.0086.10

infections. This makes management of cows around dry-off and during the dry period crucially important but simultaneously challenging.

Management practices associated with drying cows off at the end of lactation include various methods used to stop the cow's milk production (changes in milking frequency and in feeding) and the use of antibiotic dry cow therapy (DCT) and internal teat sealants (TSs) to treat existing IMIs and to prevent new ones during the dry period. Often location changes in the barn, regrouping of cows and subsequent social group changes are also part of the drying-off procedures. Milk production of cows has steadily increased over the past decades due to improvements in genetics and management (Blayney, 2002, Oltenacu and Broom, 2010), and cows can still produce considerable amounts of milk at the end of lactation, making cessation of milking a challenge. The way we stop milking cows at the end of lactation can have a substantial impact on udder health and cow comfort, but this topic has received surprisingly little attention among researchers over the years, considering that the drying-off of cows is almost a daily routine in dairy herds.

From the udder health perspective, antibiotic DCT has been a crucial part of dry cow management, and it has long played an important role in mastitis control. However, due to increasing global concerns about antibiotic resistance, the situation is currently evolving.

This chapter will first briefly describe the involution process the mammary gland undergoes at the end of lactation and will then describe the most common drying-off procedures and their impact on involution, udder health and welfare of dairy cows. Then the practice of treating cows with DCT at the end of lactation will be reviewed and discussed. Finally, potential future directions in managing dry cow udder health (especially related to methods of milk cessation and DCT) are discussed.

## 2 Mammary gland involution

Female mammals often give birth multiple times and complete several lactation cycles during their reproductive life. In the bovine mammary gland, milk synthesis increases after calving until the peak production and then declines gradually until milk removal is ceased, either at weaning of the offspring or as in modern dairy herds, as a management decision before the next calving (Hurley, 1989). For dairy cows, the non-lactating (dry) period between two successive lactations is important for the renewal and regeneration of the secretory cells and a dry period of 45–60 days has been recommended for maximising the future milk production (Hurley, 1989, Kuhn et al., 2006).

Involution is a complex process during which the mammary gland transitions from a lactating to a non-lactating state. Despite the fact that the involution process of dairy animals differs distinctly from that of rodents, rodent models have considerably increased our overall understanding of the mammary gland involution (Capuco and Akers, 1999). Three types of involutions have been described. Gradual involution is the regression of lactation during a normal progression of lactation, senile involution occurs at the end of an animal's reproductive life and initiated involution is the process that occurs in the mammary gland when milk removal stops, at weaning or in the beginning of the dry period (Hurley, 1989). The dry period of dairy cows is divided into three stages in relation to the involution process: active involution (the first 2-3 weeks of the dry period, a transition to a non-lactating state), steady state (the mid-dry period, a fully involuted gland) and colostrogenesis (1-2 weeks before next calving, a transition back to milk synthesis and formation of colostrum) (Hurley, 1989).

Prolactin is an important hormone involved in milk secretion and synthesis, and it is released in conjunction with manipulation of the udder and teats. Prolactin secretion and milk synthesis diminish if milk is not removed within 16 h (Davidson and Stabenfeldt, 2007), and the involution process is initiated within 12-24 h after the cessation of milking (Hurley, 1989). During active involution, after milk removal from the udder stops, changes in morphology and function of the bovine mammary tissue occur. In contrast to involution occurring in the rodent mammary gland, a very evident morphological feature during involution in the bovine udder is the maintenance of alveolar structure throughout the period. This is possibly because of the concurrent pregnancy of dairy cows (Capuco and Akers, 1999). While no detectable loss of mammary cells is evident in dairy cows, the dry period is important for cell turnover and regeneration of senescent mammary epithelial cells (Capuco and Akers, 1999, Capuco et al., 1997). After the final milking at the end of lactation, the synthesis of milk continues for a period before the loss of secretory capacity of mammary cells occurs. The majority of epithelial cells reach a non-secretory state in about 21 days (Hurley, 1989).

Milk accumulation, distension of the secretory tissue, systemic and local regulatory hormones, pregnancy status as well as the methods used for milk cessation, all influence the mammary gland involution process (Capuco and Akers, 1999, Stefanon et al., 2002). The changes that occur during early involution in the composition of mammary secretion reflect changes in the mammary epithelial cell function and increased tight junction permeability between adjacent cells (Stelwagen et al., 1994b). The concentration of mammary-derived components, casein, lactose and fat decrease and blood-derived factors, including leucocytes and immunoglobulins, increase (Oliver and Sordillo, 1989). During involution, non-specific antibacterial substances, such as lactoferrin, complement and lysozyme, increase (Sordillo et al., 1997). Lactoferrin is a protein in milk that binds iron, making it unavailable to many bacterial species that need it for their growth, such as coliforms and many staphylococci. *Streptococcus* spp., on the other hand, do not need iron for their growth, which may be a reason why they are common causal agents in new IMIs over the dry period (Smith et al., 1985, Rainard, 1986).

### 2.1 Susceptibility to intramammary infections in the dry period

The mammary gland contains several defense mechanisms, including the formation of the keratin plug and the immune cells that reside in the mammary gland (Nickerson, 1989). These immune cells are mainly lymphocytes, polymorphoneuclear cells (PMN) and macrophages (Damm et al., 2017). The task of lymphocytes is to regulate the initiation and suppression of the immune response, whereas macrophages ingest bacteria and cellular debris (Sordillo et al., 1997). In addition, macrophages recognise invading pathogens and trigger an immune reaction by the rapid recruitment of PMN (Paape et al., 2002, Oviedo-Boyso et al., 2007). The PMN attack the invading pathogen and defend the mammary gland at the start of an acute inflammatory reaction (Oviedo-Boyso et al., 2007). Once the mammary gland is infected, a surge in PMN occurs following a trigger from the resident cells (lymphocytes and macrophages) to fight invading bacteria (Schukken et al., 2003). If the cells move fast from the blood to the mammary gland and clear the bacteria, cell recruitment ceases. Nevertheless, if the bacteria are capable of surviving the immune reactions, the inflammation continues and may become severe, leading to damage to the mammary gland tissue (Sordillo et al., 1997).

The early dry period is a high-risk time for new IMIs for several reasons: (1) the flushing effect of milking on bacteria from the teat canal and cistern ceases; (2) disinfection of the teat ends (i.e. teat dipping in conjunction with milking) stops; (3) milk continues to be synthesized and intramammary pressure increases, potentially leading to milk leakage, keeping the teat canals open and facilitating bacterial entry into the mammary gland; and (4) mammary secretions in early dry period contain low levels of natural protective factors and low numbers of immune system cells, such as macrophages, lymphocytes and polymorphonuclear neutrophils (Oliver and Sordillo, 1989). Additionally, the accumulation of milk and milk components in the mammary gland interferes with the phagocytic activity of the immune cells and further increases the susceptibility of the cow to IMIs (Pezeshki et al., 2010).

## 3 Milk cessation methods: impacts of gradual versus abrupt milk cessation

Protocols to dry cows off at the end of lactation vary among countries, herds and sometimes even among cows within herds. No uniform procedure or recommendation to dry cows off to optimise udder health, milk production and cow welfare exist, likely because such practices are highly dependent on herd and cow characteristics. Two main methods, abrupt and gradual dry-off, are commonly in use in dairy herds. With abrupt dry-off, all drying-off procedures are implemented on one day; milking of the cow stops suddenly on a predetermined day, based on the cow's expected calving date and a desired dry period length in a herd without any particular preparation for the dry-off.

With gradual dry-off, on the other hand, milking frequency is gradually reduced, and the procedure can last from a few days to several weeks. Protocols for intermittent milking and reduced milking frequency vary, but many researchers have used a once-daily (1×/day) milking schedule for a week before dry-off (Gott et al., 2017, Kelly et al., 1998, Natzke et al., 1975). Also, a 5-day protocol using 1×/day milking on days 1, 2, 3 and 5 has been described (Dancy et al., 2019, Zobel et al., 2013). Gradual dry-off is accomplished by reducing milking frequency either with or without changes in feeding. A reduced-energy diet has often served as an intervention in studies evaluating drying-off methods: feeding cows only ad libitum hay or straw (Bushe and Oliver, 1987, Odensten et al., 2005, Ollier et al., 2014), restricting dry matter intake (Tucker et al., 2009) and feeding higher versus lower nutrient-density diets at the end of lactation (Dancy et al., 2019).

Internationally, drying-off procedures vary greatly among countries. Abrupt cessation of milking is common in numerous major dairy-producing countries (Bertulat et al., 2015, Fujiwara et al., 2018, USDA, 2016) and it is often coupled with blanket DCT. Abrupt dry-off is relatively easy to implement, as everything can be accomplished in a day. However, if a large number of cows needs to be dried off on the same day or within a very short time period (as in large seasonal herds), labour requirements may impose some constrains. Gradual dry-off, where milking frequency is reduced over a period of time, can also be challenging to implement in smaller or medium-sized free-stall barns, where facilities may not allow the creation of additional management groups of cows before dry-off. On the other hand, as automatic milking systems are becoming more common in many regions (Barkema et al., 2015), the potential to automatically reduce milking frequency and energy-content in the diet through the milking robot can make gradual dry-off easy to implement. Gradual dry-off is common, for example, in the Nordic countries in Europe, even though no standard protocol for it exists (Anon., 2020, Vilar et al., 2018). These two drying-off methods may have distinctly different effects on cow comfort and udder health.

Besides the traditional methods of reducing milking frequency and changing ration and reducing nutrient/energy intake of cows, pharmaceutical products as potential aids for drying off cows have recently been studied. Such products, however, are not widely commercially available. These include systemically injected prolactin inhibitors, such as cabergoline or quinagolide, to decrease milk synthesis (Boutinaud et al., 2016, Ollier et al., 2013) and intramammary infusions of inflammatory mediators, such as casein hydrolyzate to accelerate involution process (Shamay et al., 2003).

### 3.1 Milk yield at dry-off

With milk production per cow continuously increasing over the decades (Blayney, 2002, Oltenacu and Broom, 2010), dairy cows can still produce large quantities of milk at the end of lactation - transition to the non-lactating state can therefore be challenging. Because milk accumulates in the udder for several days and increases intramammary pressure after the final milking, the transition to the dry period can be painful for the cow, especially if the dry-off is abrupt. Drying-off practices should transition cows into the dry period as smoothly as possible, reduce any discomfort caused by discontinued removal of milk from the udder, accelerate mammary involution and enhance the cow's ability to eliminate existing IMIs and prevent new ones. Reducing milk yield prior to the last milking serves as a key to reaching these goals.

Several studies have described significantly lower milk yield at the last milking after intermittent milking and reduced milking frequency compared to performing abrupt dry-off. Once-daily milking during the final week of lactation can reduce milk yield by 20% up to 47% (Bushe and Oliver, 1987, Gott et al., 2016, Oliver et al., 1990, Tucker et al., 2009). During the last week of lactation, milk yield decreased from 23.9 kg/day to 14.3 kg/day among cows milked once daily compared to a decrease from 24.7 kg to 22.7 kg in abruptly dried-off cows (Rajala-Schultz et al., 2018). With a 5-day intermittent milking protocol (once-daily milking on days 1, 2, 3 and 5), milk yield dropped from 23.8 kg/day before the intervention to 10.9 kg at dry-off in one study (Zobel et al., 2013). An average decrease in milk yield over the drying-off period was 10 kg/day in another study using the same protocol (Dancy et al., 2019). Similarly, Newman et al. (2009) found that cumulative milk yield for the last week of lactation was significantly lower in gradually dried-off cows than in abruptly dried-off cows (78.7 kg vs. 125.8 kg). When intermittent milking was combined with a ration change, milk production declined over 60% (Oliver et al., 1990).

### 3.2 Accelerated involution

Gradual cessation of milking during drying-off has been associated with increased concentrations of natural protective factors in the mammary gland (Bushe and Oliver, 1987). Comparing groups of cows with similar milk production prior to drying-off, the concentration of lactoferrin increased significantly among intermittently milked cows during a 1-week drying-off procedure compared to lactoferrin concentration in abruptly dried-off cows (Newman et al., 2009). Similarly, cows milked once daily during late lactation had higher concentrations of immunoglobulin G in milk at dry-off than cows with similar milk yield levels but milked twice daily (Lacy-Hulbert et al., 1999). Moreover, somatic cell count (SCC) levels were higher at dry-off in quarters of intermittently milked cows than in quarters of cows dried off abruptly (Stelwagen et al., 1994a, Zobel et al., 2013), independent of quarter infection status. Silanikove et al. (2013) reported that involution in cows that naturally produced low amounts of milk (14 L/day or less) before dry-off was characterised by metabolic and immunological features that allowed the cows to effectively clear apoptotic cells and eliminate or prevent new infections. Among cows with high milk yield (25 L/day or more), on the other hand, involution was associated with distress responses and a less effective ability to fight bacterial invasion into the udder. These studies thus indicate that reduced milking frequency and low milk production at the end of lactation accelerate the involution process and lead to increases in natural protective factors in milk, enhancing the innate defense system during the high-risk period for new IMIs.

Feeding and ration changes can also hasten the involution process. Reduced-energy diets during drying-off have resulted in increased SCC at the final milking: During a drying-off procedure with restricted feeding and energy intake that lasted 5 days, milk SCC was significantly higher at the final milking than in the beginning of the drying-off period (Odensten et al., 2007b). Also, Lacy-Hulbert et al. (1999) reported that cows under feed restriction in pasture conditions at the end of lactation showed increased milk SCC by the time of dry-off compared to cows that stayed on their regular ration. Bushe and Oliver (1987) reported that combining intermittent milking and feeding of a reduced-energy ration resulted in higher concentrations of immunoglobulin G at dry-off compared to concentrations in cows dried off abruptly or milked intermittently but without feeding changes. Bushe and Oliver (1987) and Kelly et al. (1998) also observed that, in contrast to cows dried off without any ration changes, cows dried off with intermittent milking combined with restricted energy intake showed a more rapid lactoferrin increase in mammary secretions.

### 3.3 Improved udder health

The number of studies evaluating the effects of milk cessation methods on udder health is surprisingly small; however, results from the studies on the topic conducted during different decades are quite consistent. One of the first studies reporting on different drying-off procedures and their effects on udder health was conducted in 1930 (Wayne and Macy, 1933). The milk yields were much lower and the number of cows enrolled in the study was limited, but the study demonstrated interest in the topic already at that time. In the 1950s, Oliver et al. (1956a,b) reported that healthy quarters of abruptly dried-off cows were more likely to contract new IMIs during the dry period than quarters of cows milked intermittently before the final milking. New infection rate over the dry period also increased with higher milk yield at drying-off. Similarly, roughly 20 years later, Natzke et al. (1975) reported that quarters of abruptly dried-off cows developed more new infections over the dry period than did the quarters of gradually dried-off cows, especially if quarters were not treated with antibiotic dry cow products. These early studies suggested that cows that dried off abruptly were more susceptible to new IMIs than cows dried off by intermittent milking. It is noteworthy that milk yields at the time were significantly lower than they are now.

### *3.3.1 Keratin-plug formation and risk of milk leakage*

Most mastitis cases are caused by bacteria entering the udder through the teat canal, which is the cow's first line of defense against IMIs. During the dry period, one of the most important natural defense mechanisms against new IMIs is the formation of a keratin plug in each teat canal. Keratin is a waxy, antibacterial proteinaceous substance produced by the epithelial cells lining the teat canal. After milking is ceased, keratin accumulates in the teat canal, physically closes the opening and prevents the entry of bacteria into the mammary gland. During the early dry period, when the milk synthesis still continues, the accumulating milk in the cistern raises internal pressure in the mammary gland (Bertulat et al., 2013) and this may lead to milk leakage and delays in keratin-plug formation.

Information about the length of time needed for keratin-plug formation after dry-off is limited. Comalli et al. (1984) reported that teat-canal closure occurs approximately 16 days after cessation of milking. According to Dingwell et al. (2004), the median time for teat-canal closure ranged from 1 week to 2 weeks; however, 6 weeks after dry-off, 23% of teat canals were still open. They also reported that cows producing more than 21 kg/day of milk at the time of dry-off had over twice the odds of open teat canals compared to lower-producing cows. Furthermore, at 2–3 weeks after dry-off, Odensten et al. (2007b) observed more open teat canals in high-yielding than in low-yielding cows (milk production of 17.8–29.5 kg/day versus 5.0–11.4 kg/day during the last week prior to dry-off, respectively). Similarly, teat-canal closure at 7 and 14 days after cessation of milking was more common among cows under restricted dry matter intake after the final milking compared to cows without feed restriction (Summers et al., 2004).

The risk of milk leakage is greater among cows with high milk production at dry-off than among cows with low production (Bertulat et al., 2013, Gott et al., 2016, Hop et al., 2019, Silanikove et al., 2013). In some studies, milk leakage has been associated with the milk cessation method. Zobel et al. (2013) reported that the probability of leaking milk was higher in cows dried off abruptly than gradually, potentially because cows dried off abruptly produced significantly more at dry-off. While Gott et al. (2016) did not observe any significant difference in milk leakage between abrupt and gradual dry-off groups, they reported that every 4.5-kg increase in milk yield at dry-off above 18 kg raised the odds of leaking milk 2.1-fold. Quarters with open teat canals within the first 3 weeks of the dry period have higher odds of acquiring a new IMI during the dry period (Dingwell et al., 2004). In a New Zealand study, the majority (97%) of clinical infections during the dry period occurred in quarters identified as having open teat canals (Williamson et al., 1995). Likewise, cows that leaked milk after dry-off had a four-fold greater risk of clinical mastitis in the dry period (Schukken et al., 1993).

### 3.3.2 Milk yield at dry-off and risk of intramammary infections

High milk yield at dry-off elevates the risk of IMIs during the dry period and after calving. Higher cumulative milk yield during the last week of lactation was associated with a significantly greater probability of IMIs at calving for quarters that were uninfected prior to dry-off. Uninfected quarters of cows producing more than 115 kg during the last week of lactation (on average 16 kg/d) had over 7 times higher odds to be infected at calving than uninfected quarters of cows producing less than 75 kg (on average 10 kg/day) (Newman et al., 2010). This occurred despite the fact that all quarters of all cows were treated with antibiotic DCT in that study. Similarly, every 5-kg increase in milk yield at dry-off above 12.5 kg raised the odds of having IMIs at calving by 77% (Rajala-Schultz et al., 2005). Correspondingly, Odensten et al. (2007a) observed a smaller proportion of IMI cases in cows with low milk yield than in those with medium or high milk yield at dry-off.

In the period from dry-off to calving, Gott et al. (2017) observed a higher reduction in the proportion of cows with infected quarters in the gradual (16.8%) than in the abrupt cessation group (12.2%). This was despite the fact that all cows in both groups received DCT and produced similar amounts of milk on the final test day before the intervention started (22.2 kg in the gradual vs. 21.8 kg in the abrupt group). At the final milking, however, cows in the gradual dry-off group yielded significantly less than did cows dried off abruptly. This supports the concept that methods that reduce milk production also hasten involution and elevate natural protective factors in the udder, and this can play an important role in the prevention and control of new IMIs during the dry period (Oliver and Sordillo, 1989). Table 1 summarises the results from studies conducted after 2000, looking at effects of milk yield at dry-off and its impact on udder health and cow welfare.

In summary, cows with a higher milk yield at dry-off are at a greater risk of having open teat canals and leaking milk than are their lower-producing herd mates. This, on the other hand, increases the risk of new IMIs and clinical mastitis. Drying-off practices that reduce milk yield prior to the final milking

**Table 1** Milk production levels at dry-off and their effect on udder health and cow welfare

| Reference | Milk yield threshold for grouping cows | Effect of increasing milk yield at dry-off or effect of high compared to low milk yield at dry-off |
|---|---|---|
| Dingwell et al., 2004 | Low: <21 kg/day<br>High: ≥21 kg/day | Odds of open teat canals 1.8× higher |
| Rajala-Schultz et al., 2005 | 12.5 kg, with 5 kg increase | Odds of IMI at calving 1.8× higher |
| Odensten et al., 2007a | Low: 5–11.4 kg/day<br>High: 17.8–29.5 kg/day | Higher plasma cortisol levels<br>Higher percentage of open teat canal<br>Higher percentage of IMI at calving |
| Newman et al., 2010 | Low: <10.7 kg/day<br>High: ≥16.4 kg/day | Odds of IMI at calving 7× higher |
| Bertulat et al., 2013 | Low: <15 kg/day<br>High: >20 kg/d | Higher udder pressure<br>Odds of milk leakage 5.1× higher<br>Higher increase of faecal glucocorticoid |
| Silanikove et al., 2013 | Low: ≤14 L/day<br>High: 25–35 L/day | Lower lactoferrin concentration<br>Lower proportion of milk leukocyte population |
| Gott et al., 2016 | 18.1 kg, with 4.5 kg increase | Odds of milk leakage 2× higher |
| Hop et al., 2019 | Low: 13–18 kg/day<br>High: >18 kg/day | Odds of milk leakage 3.2× higher |

*Source*: Vilar and Rajala-Schultz (2020).

at the end of lactation promote faster teat-canal closure and keratin-plug formation. Gradual drying-off reduces milk yield before the final milking and accelerates the involution process, reducing the risk of IMIs.

## 3.4 Improved cow comfort

During the past decades, animal welfare has become a great consumer interest (Boogaard et al., 2011, Vanhonacker et al., 2012). This is something that livestock producers increasingly need to consider in their daily practices. Besides having an impact on udder health, milk cessation practices and especially high milk yield at dry-off can also affect cow welfare and cow comfort.

Researchers have measured cortisol and its metabolites and observed cow behaviour around dry-off to assess potential stress or discomfort in animals (Bach et al., 2015, Bertulat et al., 2013, Bertulat et al., 2017, Odensten et al., 2007b, Rajala-Schultz et al., 2018, Silanikove et al., 2013, Tucker et al., 2009, Valizaheh et al., 2008, Zobel et al., 2013). Bertulat et al. (2013) measured greater intramammary pressure and higher faecal glucocorticoid metabolite levels after dry-off in cows with higher milk yield at the time of dry-off than in lower-yielding cows. They concluded that high-producing cows experienced some level of discomfort, and potentially pain, for several days, due to the increasing udder pressure after abrupt cessation of milking. Similarly, during a 5-day drying-off procedure and after the final milking, Odensten et al. (2007a) observed elevated plasma cortisol concentrations in cows with medium (11.5–17.7 kg/day) and high (17.8–29.5 kg/day) milk yield, whereas cortisol levels did not rise in low-yielding cows (5.0–11.4 kg/day). In concordance with Bertulat et al. (2013), they also suggested that the high plasma cortisol levels were associated with high intramammary pressure in cows with high milk production at dry-off. In addition, the concentration of daily plasma cortisol during drying-off was higher in cows fed only straw than in cows fed straw and silage (Odensten et al., 2007b).

Changes in lying behaviour have also been used as potential indicators of discomfort around dry-off, due to udder engorgement and increased intramammary pressure (Chapinal et al., 2014). Some studies have reported shorter daily lying times and shorter lying bouts after dry-off in cows with higher milk yield compared to lower-producing cows (Zobel et al., 2013, Chapinal et al., 2014, Rajala-Schultz et al., 2018). While lying behaviour was not directly associated with a milk cessation method, gradually dried-off cows with lower milk yield at dry-off tended to have longer lying bouts than had abruptly dried-off cows during the first days of the dry period (Rajala-Schultz et al., 2018). Such an observation, however, did not apply to grazing cows under pasture condition where milking cows once instead of twice daily during the last week of lactation had little effect on behaviour before or after the final milking at dry-off (Tucker et al., 2009). Lower milk yields in general in a pasture system could partly explain this.

Vocalisations provide information about cows' physiological and psychological functioning, and they can be behavioural indicators for such issues as hunger, negative feelings and pain (Watts and Stookey, 2000). Silanikove et al. (2013) suggested that increased vocalisation in cows with high milk yield at dry-off after abrupt milk cessation was suggestive of pain from udder engorgement. While ration changes and restriction of energy and nutrient intake of cows are effective ways of reducing milk yield before dry-off, they may negatively impact cow welfare. Cows on more restrictive diets during drying-off vocalised more, which scientists interpreted to indicate that the cows were experiencing hunger (Tucker et al., 2009, Valizaheh et al., 2008).

### *3.5 What is an optimal milk production level at dry-of?*

As reviewed above, several studies have reported associations between high milk yield at dry-off and udder health or cow comfort after dry-off. These studies have used different cut-offs for milk yield to show an increased risk of milk leakage and IMIs with higher milk yield (Table 1). None of these studies, however, was designed or even attempted to determine the optimal milk production level at the time of the final milking with respect to the best udder health or cow comfort. However, based on the published literature, it can be concluded that to minimise the risk of milk leakage, new IMIs and discomfort and pain due to udder engorgement milk yield on the day of dry-off should be below 15 kg/day, or even closer to 10 kg/day (Vilar and Rajala-Schultz, 2020).

## 4 Dry cow therapy

The use of antibiotics at the end of the lactation to cure existing IMIs and to prevent new ones was encouraged following studies conducted in the late 1960s and early 1970s (Smith et al., 1967, Pearson and Wright, 1969, Langley et al., 1971, Eberhart and Buckalew, 1972). These studies pointed out the beneficial effects of DCT to cure existing IMIs at dry-off and to prevent new IMIs during the dry period. The so-called 5-point plan for mastitis control, developed in the 1960s, included a recommendation for treating all quarters of all cows with antibiotics at dry-off (blanket DCT), and it has contributed to successful mastitis control (Neave et al., 1969, Hillerton and Booth, 2018). The use of DCT was specifically encouraged to reduce the prevalence of contagious mastitis pathogens, especially *Streptococcus agalactiae* and *Staphylococcus aureus* (Natzke, 1981, Smith, 1983). As the relative importance of mastitis pathogens originating from environmental sources increased following the control of contagious pathogens, environmental mastitis has become important to manage, however, using other means besides antibiotics (Smith and Hogan, 1993, Klaas and Zadoks, 2018).

### *4.1 Application and efficacy of dry cow therapy to treat and prevent intramammary infections*

Following overwhelming evidence about the effectiveness of DCT to prevent new IMIs and cure existing IMIs caused by contagious Gram-positive pathogens (Natzke, 1981), the broad application of DCT (referred to as blanket DCT) within dairy cattle herds was recommended and became a custom in several countries, such as USA, UK and Canada (Ruegg, 2017). The adoption and implementation of this practice have been very high, reaching 80% to over 90% of cows and herds in many countries (Bertulat et al., 2015, USDA, 2016, Fujiwara et al., 2018). Nevertheless, already in the 1970s, the application of selective DCT was suggested. Bratlie (1973) suggested the use of selective DCT on a quarter level. Rindsig et al. (1978) suggested selective DCT as a response to authorities who believed that blanket DCT must be applied only when a cow has an IMI problem. Poutrel and Rainard (1981) proposed the use of the California Mastitis Test to guide the application of selective DCT, in which treatment is applied on a quarter level. The argumentation at that time was that DCT would not only eliminate the major pathogens (*Streptococcus agalactiae* and *Staphylococcus aureus*) but also minor mastitis pathogens such as coagulase negative staphylococci. The presence of the minor pathogens was believed to prevent IMIs with the major pathogens, and hence these pathogens should not be eliminated. Interestingly, argumentation against blanket DCT due to potential antibiotic resistance was not pronounced until the 1990s, when concerns related to increasing antibiotic resistance became a prominent argumentation for the use of selective DCT (Østerås et al., 1991, Østerås et al., 1999).

## 4.2 Blanket dry cow therapy

Blanket DCT, also referred to as complete DCT (Rindsig et al., 1978) or total DCT (Smith, 1983), includes treating all dairy cows at dry-off with long-acting antibiotics in all lactating quarters of the udder, regardless of their infection status. Blanket DCT is one of the main recommended control measures in the NMC mastitis control plan (NMC, 2006).

The efficacy of blanket DCT has been assessed in several studies. Halasa et al. (2009b) conducted a meta-analysis on the protective capability of blanket DCT against IMIs. Based on 19 studies, they calculated that cows treated with DCT had 0.61 (95% CI 0.53–0.71) times lower risk to have new IMIs in the beginning of the subsequent lactation than untreated cows, but this effect differed between different aetiologic agents of the IMIs. The treated cows had a 0.76 (95% CI 0.54–1.07) times lower risk to develop new IMIs caused by *Staphylococcus* spp. compared to the untreated cows. Given the inclusion of one (1) within the confidence interval, the difference is not statistically significant. However, cows that had received DCT had 0.39 (0.30–0.51) times lower risk to develop new IMIs caused by *Streptococcus* spp. compared to untreated cows. This very low value confirms the usefulness of DCT as a measure to control IMIs in herds with streptococcal IMI problem. For coliform pathogens, DCT had clearly no protective effect as the RR was 0.95 (0.81–1.10). McDougall (2010) studied the use of two different DCT products compared to no treatment and had comparable findings to those of Halasa et al. (2009a,b), as shown in Table 2. The author indicated that treated quarters that had an IMI caused by a minor pathogen had a significantly higher cure rate than quarters that had an IMI caused by a major pathogen, regardless of the antibiotic formulation used.

Several studies have compared different antibiotic preparations for DCT, and while some have reported differences between the products, others have found none (Table 2). Petzer et al. (2009), for example, attributed differences in cure rates to the concentration of antibiotics in the products. Arruda et al. (2013) compared three different DCT products for the risk of new IMIs following calving and cure from existing IMIs at dry-off. The authors found no significant difference among the preparations: the risk of new infection was 14.7% (the

**Table 2** Risk of new IMI and cure of existing IMI under different dry cow treatment protocols and products

| Study | Level[a] | Risk of new IMI | Cure risk of IMI | Treatment protocol |
|---|---|---|---|---|
| Petzer et al., 2009 | Quarter | 0.20 | 0.79 | BDCT: Procaine benzylpenicillin 297.92 mg (300 000 IU), nafcillin 100 mg and hydrostreptomycin 100 mg |
| Petzer et al., 2009 | Quarter | 0.14 | 0.86 | BDCT: Cephalexin 250 mg and neomycin sulphate 250 mg |
| Petzer et al., 2009 | Quarter | 0.16 | 0.72 | BDCT: Procaine benzylpenicillin 4.9% m/m and dihydrostreptomycin SO4 6.5% m/m |
| Petzer et al., 2009 | Quarter | 0.24 | 0.92 | BDCT: cloxacillin 600 mg in a long-acting base with 3% aluminium mono-stearate |
| Petzer et al., 2009 | Quarter | 0.13 | 0.75 | BDCT: Cephalonium, 250 mg in a long-acting base |
| Petzer et al., 2009 | Quarter | 0.17 | 0.84 | BDCT: Cloxacillin 500 mg, ampicillin 250 mg in a long-acting base with 3% aluminium stearate |
| McDougall, 2010 | Cow | 0.16 (0.12–0.21) | 0.92 (0.89–0.96) | BDCT: 250 mg of novel cephalonium |
| McDougall, 2010 | Cow | 0.13 (0.1–0.17) | 0.88 (0.83–0.93) | BDCT: 250 mg of cephalonium DCT |
| McDougall, 2010 | Cow | 0.56 (0.49–0.62) | - | No treatment |
| McDougall, 2010 | Quarter | 0.04 (0.03–0.06) | 0.90 (0.87–0.93) | BDCT: 250 mg of novel cephalonium |
| McDougall, 2010 | Quarter | 0.04 (0.03–0.06) | 0.93 (0.90–0.96) | BDCT: 250 mg of cephalonium |
| McDougall, 2010 | Quarter | 0.18 (0.15–0.23) | - | No treatment |
| Bradley et al., 2010 | Quarter | 286/830 | 0.81 | High SCC infected cows, DCT: 250 mg of cephalonium |
| Bradley et al., 2010 | Quarter | 243/831 | 0.86 | High SCC infected cows, DCT+ITS: 250 mg of cephalonium, 65% bismuth subnitrate |
| Bradley et al., 2010 | Quarter | 226/777 | 0.72 | Low SCC uninfected cows, ITS: 65% bismuth subnitrate |
| Bradley et al., 2010 | Quarter | 220/779 | 0.78 | Low SCC uninfected cows, DCT+ITS: 250 mg of cephalonium 65% bismuth subnitrate |
| Bradley et al., 2011 | Quarter | 0–0.21[b] | 0.73–1[b] | DCT: 150 mg of cefquinome |

| | | | | |
|---|---|---|---|---|
| Bradley et al., 2011 | Quarter | 0–0.19[b] | 0.34–1[b] | DCT+ITS: 600 mg of cloxacillin and 65% of bismuth subnitrate |
| Bradley et al., 2011 | Quarter | 0–0.21[b] | 0.67–1[b] | DCT: 600 mg of cloxacillin |
| Arruda et al., 2013 | Quarter | 0.16 | 0.89 | BDCT: 1 000 000 IU of procaine penicillin G and 1 g of dihydrostreptomycin |
| Arruda et al., 2013 | Quarter | 0.14 | 0.88 | BDCT: 500 mg of ceftiofur hydrochloride |
| Arruda et al., 2013 | Quarter | 0.16 | 0.90 | BDCT: 300 mg of cephapirin benzathine |
| Vasquez et al., 2018 | Quarter | 0.06 | 0.93 | Low-risk cows, DCT+TS: 300 mg of cephapirin benzathine and external TS |
| Vasquez et al., 2018 | Quarter | 0.07 | 0.84 | Low-risk cows, TS: application of external TS |
| Kabera et al., 2020 | Quarter | 0.16 | – | BDCT: 200 000 IU of penicillin G procaine and 400 mg of novobiocin |
| Kabera et al., 2020 | Quarter | 0.13 | – | BDCT + ITS: 200 000 IU of penicillin G procaine and 400 mg of novobiocin and 65% bismuth subnitrate |
| Kabera et al., 2020 | Quarter | 0.16 | – | SDCT: 200 000 IU of penicillin G procaine and 400 mg of novobiocin to on-farm-culture positive quarters and ITS to healthy quarters |
| Kabera et al., 2020 | Quarter | 0.15 | – | SDCT+ITS: 200 000 IU of penicillin G procaine and 400 mg of novobiocin to on-farm-culture positive quarters and ITS to all quarters. |
| Rowe et al., 2020a | Quarter | 0.20 | 0.87 | BDCT+ITS: 500 mg of ceftiofur hydrochloride and ITS in all quarters |
| Rowe et al., 2020a | Quarter | 0.21 | 0.88 | Culture-based SDCT+ITS: 500 mg of ceftiofur hydrochloride to culture positive quarters and ITS |
| Rowe et al., 2020a | Quarter | 0.20 | 0.88 | Algorithm-based SDCT+ITS, cows with SCC >200 000 or with ≥2 clinical mastitis cases during the current lactation: 500 mg of ceftiofur hydrochloride and ITS in all 4 quarters |

The table shows results from studies that were not reviewed by Halasa et al. (2009a,b) and Winder et al. (2019).
[a] Intramammary infection measurement level.
[b] The cure rate varied depending on the IMI causative pathogen.
BDCT, blanket dry cow therapy; SDCT, selective dry cow therapy; ITS, internal teat sealant; TS, teat sealant.

risk varied between 14% and 16% among the three products), while the cure risk varied between 88% and 90%, with no significant differences among the treatments (Table 2). The highest cure rates (100%) were achieved for minor pathogens, while the lowest was observed for *S. aureus* (varied between 42.9% and 88.9% depending on the DCT product). Similarly, Johnson et al. (2016) found no differences in quarter-level cure risk (84.5% vs. 85.7%) or risk of new IMIs over the dry period (16.6% vs. 14.1%) or in risk of a cow experiencing clinical mastitis within 100 DIM (17% vs. 15.3%) between cows treated either with DCT containing cloxacillin or ceftiofur, respectively. Also, Bradley et al. (2011) compared the use of narrow-spectrum antibiotics (cloxacillin) at dry-off, with or without internal teat sealants (ITS), to broad-spectrum antibiotics (cefquinome, a fourth-generation cephalosporin). The authors found no significant differences in the cure rates or risk of new IMIs caused by major pathogens after calving between the treatment groups. However, they found a higher risk of clinical mastitis after calving in the group that was treated with narrow-spectrum antibiotics without ITS, but no difference between the other two groups. These studies suggest that narrow-spectrum antibiotics provide equal efficacy in the treatment and prevention of IMIs over the dry period as do broad-spectrum antibiotics, especially if combined with the use of ITS.

### 4.3 Blanket dry cow therapy and internal teat sealants

The application of DCT at dry-off does not guarantee the complete prevention of new IMIs. This is partly because the effect of antibiotics may not last until calving, even following the use of long-acting antibiotic products (Bates and Dohoo, 2016). In addition, formation of the keratin plug may be delayed or not achieved during the dry period (as discussed above), leaving the teat canal open for bacterial introduction into the udder. Internal TS has therefore been suggested as an extra measure together with DCT to prevent IMIs, especially in the late dry period, where the effect of DCT is limited.

Based on a meta-analysis, Halasa et al. (2009b) showed that cows that had DCT and ITS had 0.39 (95% CI 0.18–0.82) times lower risk of contracting new IMIs compared to cows that had only DCT, but no ITS. Similarly, a meta-analysis of Rabiee and Lean (2013) indicated that the application of ITS with or without DCT at dry-off significantly reduced the incidence of IMIs and clinical mastitis in lactating dairy cows after calving. Bradley et al. (2010), however, found significant benefits of using DCT with ITS compared to using only DCT among high-SCC cows, but the improved effect of the combination treatment was less clear among low-SCC cows. In a subsequent study, Bradley et al. (2011) showed that the combination of DCT and ITS was actually significantly better in preventing new IMIs after calving than using only DCT, but this depended on the IMIs causing pathogens. A recent study by Kabera et al. (2020) found no significant difference in the risk of new IMIs post-calving (Table 2), persistency of IMIs throughout the dry period, clinical mastitis, milk production and SCC post-calving between using DCT alone or DCT combined with ITS, implemented on a quarter level.

Runciman et al. (2010) showed that cows that had received DCT and ITS had significantly lower odds (OR = 0.52 and 95% CI 0.32–0.85) to develop clinical mastitis within 21 days post-calving compared to cows that had received only DCT. Furthermore, the risk of subclinical mastitis (defined as an SCC > 250 000 cells/mL) was also significantly lower in cows that had received DCT and ITS compared to cows that had received only DCT. Similarly, Bates et al. (2016) found that cows that had DCT together with ITS had 0.53 (95% CI 0.32–0.89) times lower odds to have elevated SCC (>150 000) within 60–80 days after calving. The cows also had significantly lower odds of developing clinical mastitis in the first 100 days. Golder et al. (2016), however, found no significant difference in the number of clinical mastitis cases post-calving for cows that had received DCT in combination with ITS compared to cows that had received only DCT. Nevertheless, the group of cows that had received only DCT had significantly more cases of subclinical mastitis (defined as SCC >250 000 cells/mL) compared to the cows that had received DCT and ITS.

In summary, evidence of the efficacy of DCT combined with ITS compared to the application of blanket DCT alone points out to both directions, but with sufficiently more studies demonstrating a beneficial effect of using ITS combined with blanket DCT. Nevertheless, from an economic stand point, an earlier study found that the use of blanket DCT resulted in lower net costs due to IMIs compared to the combined use of blanket DCT and ITS (Halasa et al., 2010).

## 4.4 Selective dry cow therapy with or without internal teat sealants

Selective DCT involves the administration of DCT to either the entire udder of selected individual animals within the herd or only to selected quarters of the udder. Both approaches have been suggested earlier (Bratlie, 1973, Serieys and Roguinsky, 1975), but the former is perhaps more frequently applied than the latter.

In an earlier meta-analysis of four published studies, Halasa et al. (2009b) showed a significant favourable effect of using blanket DCT compared to selective DCT in preventing new IMIs post-calving (RR = 1.83; 95% CI 1.23–2.71). A recent meta-analysis that pooled data from nine different studies (Winder et al., 2019) showed a comparable effect (RR = 1.34; 95% CI 1.13–1.59). Nevertheless, when ITS was used together with DCT in the selective DCT group, selective and blanket DCT did not differ significantly (RR = 1.09; 95% CI 0.92–1.28). This pooled effect was based on three studies (Cameron et al., 2014, Patel et al., 2017, Seeth et al., 2017) that all separately showed the same insignificant difference between selective DCT and blanket DCT, when ITS was applied in combination with DCT in case of selective DCT.

Rajala-Schultz et al. (2011) investigated the effect of selective DCT on SCC and milk production. The authors found that low SCC cows that had received DCT had significantly lower SCC (16%) than the untreated low SCC cows, but milk production did not differ significantly between the groups. Scherpenzeel et al. (2014) found that using selective DCT compared to blanket DCT in low SCC cows significantly increased the incidence of clinical mastitis and SCC in these cows in the subsequent lactation. Vasquez et al. (2018), on the other hand, found no significant difference between selective DCT combined with external TS and external TS alone for low-risk cows (cows with an average SCC over the last three tests before dry-off of ≤200 000 cells/mL, an SCC ≤200 000 cells/mL on the last test, and no more than 1 case of clinical mastitis in the current lactation). In addition, the risk of clinical mastitis (in the first 30 days) and hazard of culling in the subsequent lactation did not differ significantly between the two treatment groups.

In a recent study by Kabera et al. (2020), selective DCT was compared to blanket DCT in different scenarios, as detailed in Table 2. The authors found no significant differences between selective and blanket DCT in the risk of new IMIs and clinical mastitis post-calving, persistency of IMIs throughout the dry period, milk production and SCC post-calving. In another recent study, by Rowe et al. (2020a), blanket DCT was compared to selective DCT based on two selection methods. The authors found no significant differences between selective and blanket DCT in the risk of new IMIs in the subsequent lactation or in the cure probability of IMI cases after calving (Table 2). In a subsequent study and using the same data, Rowe et al. (2020b) found no significant differences between the selective and blanket DCT groups in the risk of clinical mastitis and the hazard of cow removal from the herd during the subsequent lactation. In addition, the authors found a negligible difference in log SCC of cows that had received blanket DCT compared to those that had received selective DCT. This was also the case for milk production. These findings are also supported by recent findings by Niemi et al. (2020), who found that it is possible to maintain low herd-average SCC and good milk production when using selective DCT.

In summary, earlier studies reported that selective DCT was suboptimal compared to the use of blanket DCT. This is perhaps because teat sealants were not used in many of the earlier studies that compared selective to blanket DCT. However, a substantial reduction in antibiotic usage in dairy herds occurred when switching from blanket to selective DCT. The use of teat sealants with a proper selection of cows for DCT can substitute the blanket use of DCT, enhancing prudent use of antibiotics. Furthermore, a majority of IMIs at dry-off and at calving are caused by Gram-positive pathogens (up to 96%), and most studies have not found differences in udder health outcomes when comparing narrow- and broad-spectrum antibiotics, especially when ITS were used (Arruda et al., 2013, Cameron et al., 2014, Johnson et al., 2016, Kabera et al., 2020). Thus, the application of selective DCT and choosing dry cow products containing narrow-spectrum antibiotics are opportunities for promoting prudent antibiotic use.

## 4.5 Selection of cows for dry cow therapy

One of the challenges with selective DCT is the need to accurately distinguish cows (and quarters) that are infected and need antibiotic treatment from cows that are healthy and can and should be left without DCT at the end of lactation. For selective DCT to be widely adopted, information to support these treatment decisions should be readily available and cost-effective to farmers. Thus, the selection of cows for DCT has primarily been based on their SCC and history of clinical mastitis. In addition, in some countries (the Nordic countries), bacteriologic testing of milk samples at dry-off is commonly used to select cows for DCT (Vilar et al., 2018).

Although SCC is widely used as an indicator to select cows for DCT, the application of this criterium varies. The International Dairy Federation suggests the use of 200 000 cells/mL as a cut-off to define subclinical mastitis (IDF, 2013). This threshold was used in many studies (Bradley and Green, 2005, Green et al., 2007, Pantoja et al., 2009). Several studies have evaluated the use of the California Mastitis Test (CMT), monthly testday SCC values and clinical mastitis (CM) history of the cow, alone and in different combinations, as criteria to select cows for DCT (Rindsig et al., 1979, Poutrel and Rainard, 1981, Torres et al., 2008, Scherpenzeel et al., 2014, Vasquez et al., 2018, Rowe et al., 2020a) (Table 3).

Scherpenzeel et al. (2016) investigated the effects of using different cut-off values of SCC before dry-off on the risk of CM, antibiotic treatment and economic effects on the herd level and found an increased risk of CM and economic damage with increasing the cut-off value of SCC. Nevertheless, antibiotic usage at the herd level decreased. The authors recommended a selection of a cut-off value that balances between economics, CM occurrence and antibiotics usage. Torres et al. (2008) evaluated combinations of different SCC cut-offs, the different number of testday SCC records and CM history as selection criteria. They concluded that infected and uninfected cows at dry-off were most efficiently identified using the last 3 months' testday SCC records with a threshold of 200 000 cells/mL for cows without CM during the lactation and a threshold of 100 000 cells/mL during the rest of lactation for cows with CM during the first 90 days in milk. Vasquez et al. (2018) used the 200 000 cells/mL as a cut-off value for the selection of cows for DCT and combined it with a history of CM. In a study by Zecconi et al. (2019), the use of a cut-off value of

100 000 cells/mL for primiparous cows and 200 000 cells/mL for multiparous cows was recommended.

Some studies have suggested the use of on-farm culturing of milk samples to guide the application of selective DCT (Cameron et al., 2013, Kabera et al., 2020, Rowe et al., 2020a) (Table 3). Cameron et al. (2013, 2014) selected cows based on SCC >200 000 cells/mL and history of CM for a randomised study on the administration of blanket DCT or selective DCT guided by the Petrifilm based on-farm culture. Similarly, based on the SCC level >200 000 cells/mL and history of CM, Rowe et al. (2020b) selected cows for bacteriological testing using the Minnesota Easy4Cast plate.

Diagnostic tests are not perfect, and when using them for selective dry cow therapy (SDCT) decision support, some infected cows will likely be missed (false-negative result) and some healthy cows are treated when they would

**Table 3** Criteria used for identifying cows for treatment (or no treatment) when implementing selective dry cow therapy (SDCT). Bacterial culture was used in most studies as the reference test to detect intramammary infection (IMI), but definitions for IMI varied among studies

| Reference | Criteria for SDCT | Comments |
|---|---|---|
| Rindsig et al., 1979 | CMT, test day SCC (different cut-offs) and clinical mastitis (CM), singly and in combinations<br>*CMT, SCC >500 000 cells/mL for 2 months | |
| Poutrel and Rainard, 1981 | CMT, 4 and 8 weeks prior to dry-off | |
| Torres et al., 2008 | SCC from 1 to 6 last test days, different SCC cut-offs, CM history<br>*SCC <200 000/mL in last 3 test days, CM history during lactation | Cow-level DCT |
| Cameron et al., 2013 | SCC <200 000/mL during the last 3 test days, no CM, on-farm culture of composite cow milk samples (Petrifilm) | Cow-level DCT, ITS |
| Scherpenzeel et al., 2014 | SCC <150 000 cells/mL (primipar. cows)<br>SCC <250 000 cells/mL (multipar. cows) | Split-udder design |
| Vasquez et al., 2018 | average SCC <200 000 cells/mL in last 3 test days, no CM, SCC <200 000 cells/mL in last test day | Cow-level DCT, external TS |
| Kabera et al., 2020 | BTM SCC < 250 000 cells/mL for last 12 months, on-farm culture of quarter milk samples (3M Petrifilm) | Quarter-level DCT, ITS |
| Rowe et al., 2020a,b | Algorithm-based: any test day SCC >200 000 cells/mL and/or ≥2 CM cases<br>Culture-based: on-farm rapid culture (Minnesota Easy 4Cast plate) | Cow-level DCT, ITS quarter-level DCT, ITS |

*The criterium that was determined as the most efficient to identify the infected cows in need of DCT. SCC, somatic cell count; CMT, California Mastitis Test; CM, clinical mastitis; BTM, bulk tank milk.

actually not need DCT (false-positive result). If the emphasis of a selective DCT programme is to ensure that all truly infected cows get treatment, the criteria or the diagnostic test should have high sensitivity (ability to correctly identify infected cows) (Sanford et al., 2006). In that case, the number of false negatives (infected cows that are missed by the test) would be minimal. If, on the other hand, the goal is to reduce the use of antibiotics and not to treat healthy cows, then high specificity (ability to correctly identify uninfected cows/quarters) would be desirable, with a minimal number of false positives. Most often, however, it is necessary to find a balance between these. The choice of criteria to select cows for DCT may depend on the farmers' goals and the udder health situation in a herd, such as the prevalence of IMIs and predominant mastitis-causing bacteria in the herd.

## 5 Other dry cow management practices

### *5.1 Dry period length*

Providing dairy cows a 45–60-day dry period between lactations is a common management practice in dairy herds as it is considered necessary for mammary secretory cell renewal and for maximising future milk production. As discussed above, high milk yield at the time of dry-off creates a potential risk for the udder health and well-being of the cow. With increasing milk production/cow, drying-off of modern high-producing dairy cows has become a challenge, and thus a shortened or even completely omitted dry period has been discussed and studied. For cows that still produce large amounts of milk 60 days before calving, shortening the dry period to 35 days could be a reasonable solution for easier drying-off (Santschi and Lefebvre, 2014). Some potential advantages that may follow from a short or omitted dry period are an extended lactation with additional milk and financial benefits; fewer dietary and management changes and less stress to the cow; less severe negative energy balance and improved health and metabolic status of cows in early lactation (Bachman and Schairer, 2003). So far, despite improved energy balance in early lactation, current evidence for improvements in health and fertility of dairy cows as a consequence of shortened or omitted dry period is marginal. Milk yield in the following lactation, however, has declined in many studies, especially for primiparous cows (Watters et al., 2008, Santschi et al., 2011, van Knegsel et al., 2013, O'Hara et al., 2020).

### *5.2 Housing and social group changes*

Dry cows are often considered the non-productive part of a dairy herd, and consequently, they typically do not receive the same attention as lactating cows. They may receive poor-quality, leftover ration, and they may be moved to older, less comfortable, poorly ventilated and potentially crowded facilities. In free-stall barns, cows may experience several group changes during the dry and transition period, and this can be stressful, especially for socially lower-ranking cows (von Keyserlingk et al., 2008, Chapinal et al., 2014). Understanding and considering the behavioural needs of dairy cows in facility design, housing arrangements and overall management, especially during the dry period when many physiological changes occur, is crucial for maintaining good health and welfare (Cook and Nordlund, 2004, von Keyserlingk et al., 2008). All the managerial and environmental stressors during the dry period, coupled with metabolic and hormonal changes, are likely to negatively affect the immune function of cows. Preparation for the next lactation starts at dry-off and the dry period lays the foundation for successful lactation. Therefore, feeding and housing of dry cows should receive the same attention as those of lactating cows - cows cannot reach their potential after calving if a solid foundation for it is not laid down already during the dry period. A clean, dry, comfortable environment, with appropriate stocking density, is crucial for dry cows.

## 5.3 Feeding

Feeding changes (reduced energy and/or nutrient diets) prior to the final milking at the end of lactation help the actual cessation of milk production by the cow and make the drying-off process easier for the cow. Drastic feeding changes or too restrictive diets, however, can cause metabolic disturbances (Odensten et al., 2005, 2007b) and impair the welfare of cows (Valizaheh et al., 2008). Good balanced ration that meets the cow's daily demands during the dry period is a prerequisite for optimal immune function and a successful next lactation. Monitoring body condition score and avoiding big changes in it during the dry period is important to prevent increased disease risk postpartum (Mulligan and Doherty, 2008). Decreased dry matter intake and negative energy balance are common phenomena postpartum, and different feeding strategies during the dry period can be used to avoid or minimise these, as reviewed by Friggens et al. (2004). Energy and protein nutrition management of dry and transition dairy cows is crucial for the maintenance of health and has been reviewed (Mulligan and Doherty, 2008, Leblanc, 2010).

## 5.4 Motivating a change

A major challenge when seeking to change a management strategy on a farm is motivating the farmer to adopt the new strategy. People may have strong beliefs in their existing knowledge even if this knowledge may be based on weak evidence. Higgins et al. (2017) studied how veterinarians rationalise their prescribing decisions for intramammary treatments at dry-off and identified barriers and motivators that veterinarians perceive for the application of selective DCT. The authors indicated that trust between farmers and veterinarians is a major element to enhance a change in farmers' behaviour from using blanket DCT to selective DCT. In addition, the authors indicated that veterinarians might keep prescribing blanket DCT as a way to manage risk and uncertainty, to protect themselves, their relationship with the farmer and animals against possible negative outcomes. Finally, the authors recommended that veterinarians must work together to enhance a prompt shift to selective DCT. In a similar study, several years after the prophylactic use of antibiotics in animal production was banned in the Netherlands, Scherpenzeel et al. (2018b) investigated the attitude of veterinarians towards the application of selective DCT instead of blanket DCT. The authors generally found a positive attitude of Dutch veterinarians towards switching to selective DCT and towards prudent use of antibiotics, despite a proportion of veterinarians with a negative attitude. This indicates that veterinarians can adapt relatively quickly to prompt changes, when necessary.

Normally economic incentives can be used to encourage farmers to change management. In the case of changing from blanket to selective DCT, economic assessments are quite limited. An earlier study showed that the net costs of contagious IMIs were lowest when blanket DCT was used compared to selective DCT (Halasa et al., 2010). Huijps and Hogeveen (2007), however, indicated that the optimal strategy for DCT must be herd-specific, as it depends on herd-level factors and goals. Scherpenzeel et al. (2016, 2018a) indicated that selective DCT could economically outperform blanket DCT, but a balance should be found between economic profit, antibiotic usage and udder health to fulfil the objectives of the farm, considering potential restrictions.

## 6 Conclusion and future trends

Drying cows off at the end of lactation is a routine management practice in dairy herds. Cows are typically given a 45-60-day dry period between two lactations. This is an opportunity for the renewal of senescent secretory mammary cells to optimise future production. Modern dairy cows, however, may still produce considerable amounts of milk at the end of lactation, at the time when milking should be halted. This makes cessation of milking and drying off cows challenging, especially when dry-off occurs abruptly. Several studies have shown an association between high milk yield at dry-off and increased risk of milk leakage, open teat canals and IMIs during the dry period and at calving. Milk production can be effectively lowered with gradual dry-off by reducing milking frequency and/or limiting energy and nutrient intake of cows prior to the final milking at the end of lactation.

The benefits from gradual dry-off and reduced milk yield at the time of the last milking are considered multifold: involution is accelerated, resistance to IMIs is enhanced, and cow comfort also improves. Gradual dry-off accelerates mammary involution, making the udder more resistant to infections faster. It is also associated with less udder engorgement and a lower risk for milk leakage and open teat canals. All this results in fewer IMIs during the dry period and at calving, improving udder health. Lower milk production and less udder engorgement after gradual dry-off also improve cow welfare. No studies, however, have specifically been designed to determine the optimal milk production at dry-off, but published results suggest that milk production should be below 15 kg/day at the time of the final milking for the best outcome. However, further research on determining the optimal milk production level at dry-off is warranted. Similarly, as automatic milking systems have become more common, research on gradual dry-off implemented through the milking robot and its settings on milking frequency and ration changes is warranted.

Because it is well known that the dry period is a high-risk time for cows to acquire new IMIs, antibiotic DCT has played a crucial part of udder health management of dry cows. Antibiotic DCT has been widely used and routinely given to all quarters of all cows (blanket DCT) to eliminate existing infections and to prevent new infections during the dry period. The use of antibiotics in the treatment of bacterial infections will continue, as long as effective antibiotics are available. Nevertheless, antibiotic resistance is a major global concern that demands the prudent use of antibiotics. Selective DCT must therefore be used to ensure that cows that need treatment are treated, while applying antibiotics prudently. However, antibiotics are only a part of managing udder health during the dry period. We need to ensure that cows' own defense mechanisms are maximised and exposure of teat ends to mastitis-causing pathogens is minimised. This means providing the cow a clean, comfortable, stress-free environment and balanced nutrition to meet her needs to maintain steady and appropriate body condition across the dry period.

In conclusion, future work should focus on optimal drying-off procedures coupled with the identification of nutritional regimens that optimise health, comfort and production of the cow. For a long time, dairy advisors have tended to create 'one-size-fits-all' recommendations, as they are often easy to implement, especially in increasingly larger herds. However, all herds are different, including their cows, facilities and managers, and therefore farmers together with their veterinarians and other advisors should focus on finding herd-specific solutions. As cows differ from each other, herds also differ greatly in the prevalence and incidence of IMIs and the distribution of the mastitis-causing pathogens. Knowledge about the pathogens causing IMIs in a herd is crucial for creating effective treatment protocols and for dry cow and udder health management in general. Research should focus on finding optimal criteria for the selection of cows for DCT, considering both cow and herd factors, including mastitis-causing pathogens in a herd. The DCT should be administered to cows that need the treatment and would most likely benefit from it. Future research should also evaluate the combined effect of different milk cessation methods and DCT strategies together. A cow- and herd-specific DCT strategy is expected to result in improved udder health and productivity, while minimal antibiotics are used for DCT. Finally, researchers should continue investigating the potential of selective DCT on quarter level, rather than the whole udder, perhaps coupled with investigating the value of using on-farm testing for selection.

## 7 Where to look for further information

The following organisations have a long history on working for improved animal and udder health and milk quality and one can find a lot of useful information on their websites:

- International Dairy Federation (IDF) (https://www.fil-idf.org).
- National Mastitis Council (NMC) (https://www.nmconline.org).

Textbooks that provide solid information on dairy herd health management and food animal medicine, including udder health:

- Green, M. and M. J. Green. 2012. *Dairy Herd Health*. CABI, Oxfordshire, UK.
- Radostits, O. M. 2001. *Herd Health: Food Animal Production Medicine*. Philadelphia: Saunders.

Practical guides to udder health and dry cow management have been published in the 'Cow signals' series:

- Hulsen, J. 2012. Dry period, special needs cows and treatments.
- Hulsen, J. and Lam T. 2012. Udder Health. A practical guide to first-rate udder health.

## 8 References

Anon. 2020. Animal health care report - Redogörelse för husdjursorganisationernas Djurhälsovård 2018/2019. Available at: https://www.vxa.se/globalassets/dokument/statistik/redogorelse-for-husdjursorganisationernas-djurhalsovard-2018-2019.pdf (accessed 19 September 2020).

Arruda, A. G., Godden, S., Rapnicki, P., Gorden, P., Timms, L., Aly, S. S., Lehenbauer, T. W. and Champagne, J. 2013. Randomized noninferiority clinical trial evaluating 3 commercial dry cow mastitis preparations: I. Quarter-level outcomes. *J. Dairy Sci.* 96(7), 4419–4435.

Bach, A., De-Prado, A. and Aris, A. 2015. Short communication: the effects of cabergoline administration at dry-off of lactating cows on udder engorgement, milk leakages, and lying behavior. *J. Dairy Sci.* 98(10), 7097-7101.

Bachman, K. C. and Schairer, M. L. 2003. Invited review: bovine studies on optimal lengths of dry periods. *J. Dairy Sci.* 86(10), 3027-3037.

Barkema, H. W., von Keyserlingk, M. A., Kastelic, J. P., Lam, T. J., Luby, C., Roy, J. P., Leblanc, S. J., Keefe, G. P. and Kelton, D. F. 2015. Invited review: changes in the dairy industry affecting dairy cattle health and welfare. *J. Dairy Sci.* 98(11), 7426-7445.

Bates, A. J., Chambers, G. and Laven, R. A. 2016. Comparison of cephalonium alone and in combination with an internal teat sealant for dry cow therapy in seasonally calving dairy cows. *N. Z. Vet. J.* 64(2), 95-100.

Bates, A. J. and Dohoo, I. 2016. Risk factors for peri-parturient farmer diagnosed mastitis in New Zealand dairy herds: findings from a retrospective cohort study. *Prev. Vet. Med.* 127, 70-76.

Bertulat, S., Fischer-Tenhagen, C. and Heuwieser, W. 2015. A survey of drying-off practices on commercial dairy farms in northern Germany and a comparison to science-based recommendations. *Vet. Rec. Open* 2(1), e000068.

Bertulat, S., Fischer-Tenhagen, C., Suthar, V., Mostl, E., Isaka, N. and Heuwieser, W. 2013. Measurement of fecal glucocorticoid metabolites and evaluation of udder characteristics to estimate stress after sudden dry-off in dairy cows with different milk yields. *J. Dairy Sci.* 96(6), 3774-3787.

Bertulat, S., Isaka, N., De Prado, A., Lopez, A., Hetreau, T. and Heuwieser, W. 2017. Effect of a single injection of cabergoline at dry off on udder characteristics in high-yielding dairy cows. *J. Dairy Sci.* 100(4), 3220-3232.

Blayney, D. P. 2002. The changing landscape of U. S. Milk production - electronic report from the Economic Research Service. Available at: www.ers.usda.gov.

Boogaard, B. K., Oosting, S. J., Bock, B. B. and Wiskerke, J. S. 2011. The sociocultural sustainability of livestock farming: an inquiry into social perceptions of dairy farming. *Animal* 5(9), 1458-1466.

Boutinaud, M., Isaka, N., Lollivier, V., Dessauge, F., Gandemer, E., Lamberton, P., Taranilla, A. I., Deflandre, A. and Sordillo, L. M. 2016. Cabergoline inhibits prolactin secretion and accelerates involution in dairy cows after dry-off. *J. Dairy Sci.* 99(7), 5707-5718.

Bradley, A. and Green, M. 2005. Use and interpretation of somatic cell count data in dairy cows. *In Practice* 27(6), 310-315.

Bradley, A. J., Breen, J. E., Payne, B. and Green, M. J. 2011. A comparison of broad-spectrum and narrow-spectrum dry cow therapy used alone and in combination with a teat sealant. *J. Dairy Sci.* 94(2), 692-704.

Bradley, A. J., Breen, J. E., Payne, B., Williams, P. and Green, M. J. 2010. The use of cephalonium containing dry cow therapy and an internal teat sealant, both alone and in combination. *J. Dairy Sci.* 93(4), 1566-1577.

Bradley, A. J. and Green, M. J. 2000. A study of the incidence and significance of intramammary enterobacterial infections acquired during the dry period. *J. Dairy Sci.* 83(9), 1957-1965.

Bradley, A. J. and Green, M. J. 2004. The importance of the non-lactating period in the epidemiology of intramammary infection and strategies for prevention. *Vet Clin North Am Food Anim. Pract.*, 20(3), 547-568.

Bratlie, O. 1973. Letter: Dry cow therapy. *Vet. Rec.* 93(15), 430-431.

Bushe, T. and Oliver, S. P. 1987. Natural protective factors in bovine mammary secretions following different methods of milk cessation. *J. Dairy Sci.* 70(3), 696–704.

Cameron, M., Keefe, G. P., Roy, J. P., Dohoo, I. R., MacDonald, K. A. and McKenna, S. L. 2013. Evaluation of a 3M Petrifilm on-farm culture system for the detection of intramammary infection at the end of lactation. *Prev. Vet. Med.* 111(1–2), 1–9.

Cameron, M., McKenna, S. L., MacDonald, K. A., Dohoo, I. R., Roy, J. P. and Keefe, G. P. 2014. Evaluation of selective dry cow treatment following on-farm culture: risk of postcalving intramammary infection and clinical mastitis in the subsequent lactation. *J. Dairy Sci.* 97(1), 270–284.

Capuco, A. V. and Akers, R. M. 1999. Mammary involution in dairy animals. *J. Mammary Gland Biol. Neoplasia* 4(2), 137–144.

Capuco, A. V., Akers, R. M. and Smith, J. J. 1997. Mammary growth in Holstein cows during the dry period: quantification of nucleic acids and histology. *J. Dairy Sci.* 80(3), 477–487.

Chapinal, N., Zobel, G., Painter, K. and Leslie, K. E. 2014. Changes in lying behavior after abrupt cessation of milking and regrouping at dry-off in freestall-housed cows: A case study. *J. Vet. Behav. Clin. Appl. Res.* 9(6), 364–369.

Comalli, M. P., Eberhart, R. J., Griel, L. C., Jr. and Rothenbacher, H. 1984. Changes in the microscopic anatomy of the bovine teat canal during mammary involution. *Am. J. Vet. Res.* 45(11), 2236–2242.

Cook, N. B. and Nordlund, K. V. 2004. Behavioral needs of the transition cow and considerations for special needs facility design. *Vet. Clin. North Am. Food Anim. Pract.* 20(3), 495–520.

Cousins, C. L., Higgs, T. M., Jackson, E. R., Neave, F. K. and Dodd, F. H. 1980. Susceptibility of the bovine udder to bacterial infection in the dry period. *J. Dairy Res.* 47(1), 11–18.

Damm, M., Holm, C., Blaabjerg, M., Bro, M. N. and Schwarz, D. 2017. Differential somatic cell count-A novel method for routine mastitis screening in the frame of Dairy Herd Improvement testing programs. *J. Dairy Sci.* 100(6), 4926–4940.

Dancy, K. M., Ribeiro, E. S. and Devries, T. J. 2019. Effect of dietary transition at dry off on the behavior and physiology of dairy cows. *J. Dairy Sci.* 102(5), 4387–4402.

Davidson, A. P. and Stabenfeldt, G. H. 2007. Reproduction and lactation. In: Cunningham, J. G. and Klein, B. G. (Eds) *Textbook of Veterinary Physiology* (4th edn.). St. Louis, MO: Saunders Elsevier.

Dingwell, R. T., Kelton, D. F. and Leslie, K. E. 2003. Management of the dry cow in control of peripartum disease and mastitis. *Vet. Clin. North Am. Food Anim. Pract.* 19(1), 235–265.

Dingwell, R. T., Leslie, K. E., Schukken, Y. H., Sargeant, J. M., Timms, L. L., Duffield, T. F., Keefe, G. P., Kelton, D. F., Lissemore, K. D. and Conklin, J. 2004. Association of cow and quarter-level factors at drying-off with new intramammary infection during the dry period. *Prev. Vet. Med.* 63(1–2), 75–89.

Eberhart, R. J. 1986. Management of dry cows to reduce mastitis. *J. Dairy Sci.* 69(6), 1721–1732.

Eberhart, R. J. and Buckalew, J. M. 1972. Evaluation of a hygiene and dry period therapy program for mastitis control. *J. Dairy Sci.* 55(12), 1683–1691.

Friggens, N. C., Andersen, J. B., Larsen, T., Aaes, O. and Dewhurst, R. J. 2004. Priming the dairy cow for lactation: a review of dry cow feeding strategies. *Anim. Res.* 53(6), 453–473.

Fujiwara, M., Haskell, M. J., Macrae, A. I. and Rutherford, K. M. D. 2018. Survey of dry cow management on UK commercial dairy farms. *Vet. Rec.* 183(9), 297.

Golder, H. M., Hodge, A. and Lean, I. J. 2016. Effects of antibiotic dry-cow therapy and internal teat sealant on milk somatic cell counts and clinical and subclinical mastitis in early lactation. *J. Dairy Sci.* 99(9), 7370–7380.

Gott, P. N., Rajala-Schultz, P. J., Schuenemann, G. M., Proudfoot, K. L. and Hogan, J. S. 2016. Intramammary infections and milk leakage following gradual or abrupt cessation of milking. *J. Dairy Sci.* 99(5), 4005–4017.

Gott, P. N., Rajala-Schultz, P. J., Schuenemann, G. M., Proudfoot, K. L. and Hogan, J. S. 2017. Effect of gradual or abrupt cessation of milking at dry off on milk yield and somatic cell score in the subsequent lactation. *J. Dairy Sci.* 100(3), 2080–2089.

Green, M. J., Bradley, A. J., Medley, G. F. and Browne, W. J. 2007. Cow, farm, and management factors During the dry period that determine the rate of clinical mastitis After calving. *J. Dairy Sci.* 90(8), 3764–3776.

Gussmann, M., Steeneveld, W., Kirkeby, C., Hogeveen, H., Farre, M. and Halasa, T. 2019. Economic and epidemiological impact of different intervention strategies for subclinical and clinical mastitis. *Prev. Vet. Med.* 166, 78–85.

Halasa, T., Nielen, M., van Werven, T. and Hogeveen, H. 2010. A simulation model to calculate costs and benefits of dry period interventions in dairy cattle. *Livest. Sci.* 129(1–3), 80–87.

Halasa, T., Nielen, M., Whist, A. C. and Osteras, O. 2009a. Meta-analysis of dry cow management for dairy cattle. Part 2. Cure of existing intramammary infections. *J. Dairy Sci.* 92(7), 3150–3157.

Halasa, T., Osteras, O., Hogeveen, H., van Werven, T. and Nielen, M. 2009b. Meta-analysis of dry cow management for dairy cattle. Part 1. Protection against new intramammary infections. *J. Dairy Sci.* 92(7), 3134–3149.

Higgins, H. M., Golding, S. E., Mouncey, J., Nanjiani, I. and Cook, A. J. C. 2017. Understanding veterinarians' prescribing decisions on antibiotic dry cow therapy. *J. Dairy Sci.* 100(4), 2909–2916.

Hillerton, J. E. and Booth, J. M. 2018. The five-point mastitis control plan–A revisory tutorial. 57th Annual Meeting of National Mastitis Council, Tucson AZ, USA. National Mastitis Council, 3–19.

Hop, G. E., De Prado-Taranilla, A. I., Isaka, N., Ocak, M., Bertet, J., Supre, K., Velthuis, A., Schukken, Y. H. and Deflandre, A. 2019. Efficacy of cabergoline in a double-blind randomized clinical trial on milk leakage reduction at drying-off and new intramammary infections across the dry period and postcalving. *J. Dairy Sci.* 102(12), 11670–11680.

Huijps, K. and Hogeveen, H. 2007. Stochastic modeling to determine the economic effects of blanket, selective, and no dry cow therapy. *J. Dairy Sci.* 90(3), 1225–1234.

Hurley, W. L. 1989. Mammary gland function during involution. *J. Dairy Sci.* 72(6), 1637–1646.

IDF. 2013. *Guidelines for the Use and Interpretation of Bovine Milk Somatic Cell Counts (SCC) in the Dairy Industry*. Brussels, Belgium: International Dairy Federation.

Johnson, A. P., Godden, S. M., Royster, E., Zuidhof, S., Miller, B. and Sorg, J. 2016. Randomized noninferiority study evaluating the efficacy of 2 commercial dry cow mastitis formulations. *J. Dairy Sci.* 99(1), 593–607.

Kabera, F., Dufour, S., Keefe, G., Cameron, M. and Roy, J. P. 2020. Evaluation of quarter-based selective dry cow therapy using Petrifilm on-farm milk culture: a randomized controlled trial. *J. Dairy Sci.* 103(8), 7276-7287.

Kelly, A. L., Reid, S., Joyce, P., Meaney, W. J. and Foley, J. 1998. Effect of decreased milking frequency of cows in late lactation on milk somatic cell count, polymorphonuclear leucocyte numbers, composition and proteolytic activity. *J. Dairy Res.* 65(3), 365-373.

Klaas, I. C. and Zadoks, R. N. 2018. An update on environmental mastitis: challenging perceptions. *Transbound. Emerg. Dis.* 65 (Suppl. 1), 166-185.

Kuhn, M. T., Hutchison, J. L. and Norman, H. D. 2006. Dry period length to maximize production across adjacent lactations and lifetime production. *J. Dairy Sci.* 89(5), 1713-1722.

Lacy-Hulbert, S. J., Woolford, M. W., Nicholas, G. D., Prosser, C. G. and Stelwagen, K. 1999. Effect of milking frequency and pasture intake on milk yield and composition of late lactation cows. *J. Dairy Sci.* 82(6), 1232-1239.

Langley, O. H., Meaney, W. J., Cullen, N. P. and Cunningham, J. F. 1971. The control of mastitis. *Vet. Rec.* 89(11), 315-316.

Leblanc, S. 2010. Monitoring metabolic health of dairy cattle in the transition period. *J. Reprod. Dev.* 56 (Suppl.), S29-S35.

McDougall, S. 2010. A randomised, non-inferiority trial of a new cephalonium dry-cow therapy. *N. Z. Vet. J.* 58(1), 45-58.

Mulligan, F. J. and Doherty, M. L. 2008. Production diseases of the transition cow. *Vet. J.* 176(1), 3-9.

Natzke, R. P. 1981. Elements of mastitis control. *J. Dairy Sci.* 64(6), 1431-1442.

Natzke, R. P., Everett, R. W. and Bray, D. R. 1975. Effect of drying off practices on mastitis infection. *J. Dairy Sci.* 58(12), 1828-1835.

Neave, F. K., Dodd, F. H. and Henrichs, E. 1950. Udder infections in the dry period. *J. Dairy Res.* 17, 37-49.

Neave, F. K., Dodd, F. H., Kingwill, R. G. and Westgarth, D. R. 1969. Control of mastitis in the dairy herd by hygiene and management. *J. Dairy Sci.* 52(5), 696-707.

Newman, K. A., Rajala-Schultz, P. J., Degraves, F. J. and Lakritz, J. 2010. Association between milk yield and infection status at dry-off and intramammary infections at subsequent calving. *J. Dairy Res.* 77(1), 99-106.

Newman, K. A., Rajala-Schultz, P. J., Lakritz, J. and Degraves, F. J. 2009. Lactoferrin concentrations in bovine milk prior to dry-off. *J. Dairy Res.* 76(4), 426-432.

Nickerson, S. C. 1989. Immunological aspects of mammary involution. *J. Dairy Sci.* 72(6), 1665-1678.

Niemi, R. E., Vilar, M. J., Dohoo, I. R., Hovinen, M., Simojoki, H. and Rajala-Schultz, P. J. 2020. Antibiotic dry cow therapy, somatic cell count, and milk production: retrospective analysis of the associations in dairy herd recording data using multilevel growth models. *Prev. Vet. Med.* 180, 105028.

NMC. 2006. NMC. Recommended mastitis control program. Available at: https://www.nmconline.org/wp-content/uploads/2020/04/RECOMMENDED-MASTITIS-CONTROL-PROGRAM-International.pdf.

O'Hara, E. A., Holtenius, K., Bage, R., Von Bromssen, C. and Emanuelson, U. 2020. An observational study of the dry period length and its relation to milk yield, health, and fertility in two dairy cow breeds. *Prev. Vet. Med.* 175, 104876.

Odensten, M. O., Berglund, B., Waller, K. P. and Holtenius, K. 2007a. Metabolism and udder health at dry-off in cows of different breeds and production levels. *J. Dairy Sci.* 90(3), 1417–1428.

Odensten, M. O., Holtenius, K. and Waller, K. P. 2007b. Effects of two different feeding strategies during dry-off on certain health aspects of dairy cows. *J. Dairy Sci.* 90(2), 898–907.

Odensten, M. O., Chilliard, Y. and Holtenius, K. 2005. Effects of two different feeding strategies during dry-off on metabolism in high-yielding dairy cows. *J. Dairy Sci.* 88(6), 2072–2082.

Oliver, J., Dodd, F. H. and Neave, F. K. 1956a. Udder infections in the dry period. III. The method of drying-off cows in the dry period. *J. Dairy Res.* 23(2), 197–203.

Oliver, J., Dodd, F. H. and Neave, F. K. 1956b. Udder infections in the dry period. IV. The relationship between the new infection rate in early dry period and the daily milk yield at drying-off when lactation was ended by either intermittent or abrupt cessation of milking. *J. Dairy Res.* 23(2), 204–211.

Oliver, S. P., Shull, E. P. and Dowlen, H. H. 1990. Influence of different methods of milk cessation on intramammary infections during the periparturient period. International Symposium on Bovine Mastitis, September 13–16, 1990. Indianapolis, IN: National Mastitis Council, American Association of Bovine Practitioners, 92–97.

Oliver, S. P. and Sordillo, L. M. 1989. Approaches to the manipulation of mammary involution. *J. Dairy Sci.* 72(6), 1647–1664.

Ollier, S., Zhao, X. and Lacasse, P. 2013. Effect of prolactin-release inhibition on milk production and mammary gland involution at drying-off in cows. *J. Dairy Sci.* 96(1), 335–343.

Ollier, S., Zhao, X. and Lacasse, P. 2014. Effects of feed restriction and prolactin-release inhibition at drying off on metabolism and mammary gland involution in cows. *J. Dairy Sci.* 97(8), 4942–4954.

Oltenacu, P. A. and Broom, D. M. 2010. The impact of genetic selection for increased milk yield on the welfare of dairy cows. *Anim. Welf.* 19, 39–49.

Østerås, O., Martin, S. W. and Edge, V. L. 1999. Possible risk factors associated with penicillin-resistant strains of Staphylococcus aureus from bovine subclinical mastitis in early lactation. *J. Dairy Sci.* 82(5), 927–938.

Østerås, O., Sandvik, L., Aursjo, J., Gjul, G. G. and Jorstad, A. 1991. Assessment of strategy in selective dry cow therapy for mastitis control. *J. Vet. Med.* 38(7), 513–522.

Oviedo-Boyso, J., Valdez-Alarcon, J. J., Cajero-Juarez, M., Ochoa-Zarzosa, A., Lopez-Meza, J. E., Bravo-Patino, A. and Baizabal-Aguirre, V. M. 2007. Innate immune response of bovine mammary gland to pathogenic bacteria responsible for mastitis. *J. Infect.* 54(4), 399–409.

Paape, M., Mehrzad, J., Zhao, X., Detilleux, J. and Burvenich, C. 2002. Defense of the bovine mammary gland by polymorphonuclear neutrophil leukocytes. *J. Mammary Gland Biol. Neoplasia* 7(2), 109–121.

Pantoja, J. C., Hulland, C. and Ruegg, P. L. 2009. Somatic cell count status across the dry period as a risk factor for the development of clinical mastitis in the subsequent lactation. *J. Dairy Sci.* 92(1), 139–148.

Patel, K., Godden, S. M., Royster, E. E., Timmerman, J. A., Crooker, B. A. and McDonald, N. 2017. Impact of using a culture-guided selective dry cow therapy program targeting quarter-level treatment on udder health and antibiotic use. *Bov. Pract.* 51, 48–57.

Pearson, J. K. L. and Wright, C. L. 1969. Dry cow therapy as a means of controlling bovine mastitis. *Vet. Rec.* 84(12), 294–298.

Petzer, I. M., Lourens, D. C., van der Schans, T. J., Watermeyer, J. C., van Reenen, R., Rautenbach, G. H. and Thompson, P. 2009. Intramammary infection rate during the dry period in cows that received blanket dry cow therapy: efficacy of 6 different dry-cow intra-mammary antimicrobial products. *J. S. Afr. Vet. Assoc.* 80(1), 23–30.

Pezeshki, A., Capuco, A. V., De Spiegeleer, B., Peelman, L., Stevens, M., Collier, R. J. and Burvenich, C. 2010. An integrated view on how the management of the dry period length of lactating cows could affect mammary biology and defence. *J. Anim. Physiol. Anim. Nutr. (Berl)* 94(5), e7–e30.

Poutrel, B. and Rainard, P. 1981. California Mastitis Test guide of selective dry cow therapy. *J. Dairy Sci.* 64(2), 241–248.

Rabiee, A. R. and Lean, I. J. 2013. The effect of internal teat sealant products (Teatseal and Orbeseal) on intramammary infection, clinical mastitis, and somatic cell counts in lactating dairy cows: a meta-analysis. *J. Dairy Sci.* 96(11), 6915–6931.

Rainard, P. 1986. Bacteriostatic activity of bovine milk lactoferrin against mastitic bacteria. *Vet. Microbiol.* 11(4), 387–392.

Rajala-Schultz, P. J., Gott, P. N., Proudfoot, K. L. and Schuenemann, G. M. 2018. Effect of milk cessation method at dry-off on behavioral activity of dairy cows. *J. Dairy Sci.* 101(4), 3261–3270.

Rajala-Schultz, P. J., Hogan, J. S. and Smith, K. L. 2005. Association between milk yield at dry-off and probability of intramammary infections at calving. *J. Dairy Sci.* 88(2), 577–579.

Rajala-Schultz, P. J., Torres, A. H. and Degraves, F. J. 2011. Milk yield and somatic cell count during the following lactation after selective treatment of cows at dry-off. *J. Dairy Res.* 78(4), 489–499.

Rindsig, R. B., Rodewald, R. G., Smith, A. R. and Spahr, S. L. 1978. Complete versus selective dry cow therapy for mastitis control. *J. Dairy Sci.* 61(10), 1483–1497.

Rindsig, R. B., Rodewald, R. G. Smith, A. R., Thomsen, N. K. and Spahr, S. L. 1979. Mastitis history, California Mastitis Test, and somatic cell counts for identifying cows for treatment in a selective dry cow therapy program. *J. Dairy Sci.* 62(8), 1335–1339.

Rowe, S. M., Godden, S. M., Nydam, D. V., Gorden, P. J., Lago, A., Vasquez, A. K., Royster, E., Timmerman, J. and Thomas, M. J. 2020a. Randomized controlled non-inferiority trial investigating the effect of 2 selective dry-cow therapy protocols on antibiotic use at dry-off and dry period intramammary infection dynamics. *J. Dairy Sci.* 103(7), 6473–6492.

Rowe, S. M., Godden, S. M., Nydam, D. V., Gorden, P. J., Lago, A., Vasquez, A. K., Royster, E., Timmerman, J. and Thomas, M. J. 2020b. Randomized controlled trial investigating the effect of 2 selective dry-cow therapy protocols on udder health and performance in the subsequent lactation. *J. Dairy Sci.* 103(7), 6493–6503.

Ruegg, P. L. 2017. A 100-year Review: mastitis detection, management, and prevention. *J. Dairy Sci.* 100(12), 10381–10397.

Runciman, D. J., Malmo, J. and Deighton, M. 2010. The use of an internal teat sealant in combination with cloxacillin dry cow therapy for the prevention of clinical and subclinical mastitis in seasonal calving dairy cows. *J. Dairy Sci.* 93(10), 4582–4591.

Santschi, D. E., Lefebvre, D. M., Cue, R. I., Girard, C. L. and Pellerin, D. 2011. Complete-lactation milk and component yields following a short (35-d) or a conventional

(60-d) dry period management strategy in commercial Holstein herds. *J. Dairy Sci.* 94(5), 2302-2311.

Santschi, D. E. and Lefebvre, D. M. 2014. Review: practical concepts on short dry period management. *Can. J. Anim. Sci.* 94(3), 381-390.

Sanford, C. J., Keefe, G. P., Sanchez, J., Dingwell, R. T., Barkema, H. W., Leslie, K. E. and Dohoo, I. R. 2006. Test characteristics from latent-class models of th California Mastitis Test. *Prev. Vet. Med.* 77(1-2), 96-108.

Scherpenzeel, C. G., Den Uijl, I. E., van Schaik, G., Olde Riekerink, R. G., Keurentjes, J. M. and Lam, T. J. 2014. Evaluation of the use of dry cow antibiotics in low somatic cell count cows. *J. Dairy Sci.* 97(6), 3606-3614.

Scherpenzeel, C. G. M., Den Uijl, I. E. M., van Schaik, G., Riekerink, R. G. M. O., Hogeveen, H. and Lam, T. J. G. M. 2016. Effect of different scenarios for selective dry-cow therapy on udder health, antimicrobial usage, and economics. *J. Dairy Sci.* 99(5), 3753-3764.

Scherpenzeel, C. G. M., Hogeveen, H., Maas, L. and Lam, T. J. G. M. 2018a. Economic optimization of selective dry cow treatment. *J. Dairy Sci.* 101(2), 1530-1539.

Scherpenzeel, C. G. M., Santman-Berends, I. M. G. A. and Lam, T. J. G. M. 2018b. Veterinarians' attitudes toward antimicrobial use and selective dry cow treatment in the Netherlands. *J. Dairy Sci.* 101(7), 6336-6345.

Schukken, Y. H., Vanliet, J., Vandegeer, D. and Grommers, F. J. 1993. A randomized blind study on dry cow antibiotic infusion in a low somatic cell count herd. *J. Dairy Sci.* 76(10), 2925-2930.

Schukken, Y. H., Wilson, D. J., Welcome, F., Garrison-Tikofsky, L. and Gonzalez, R. N. 2003. Monitoring udder health and milk quality using somatic cell counts. *Vet. Res.* 34(5), 579-596.

Seeth, Mt, Wente, N., Paduch, J. H., Klocke, D., Vries, E. M. D., Hoedemaker, M. and Kromker, V. 2017. Different selective dry cow therapy concepts compared to blanket antibiotic dry cow treatment. *Tierarztl Prax. Ausg. G* 45(6), 343-349.

Serieys, F. and Roguinsky, M. 1975. Comparative tests with dry cow therapy for some or all cows in a herd. IDF Seminar of Mastitis Control. Bruxelles: IDF, 349-351.

Shamay, A., Shapiro, F., Leitner, G. and Silanikove, N. 2003. Infusions of casein hydrolyzates into the mammary gland disrupt tight junction integrity and induce involution in cows. *J. Dairy Sci.* 86(4), 1250-1258.

Silanikove, N., Merin, U., Shapiro, F. and Leitner, G. 2013. Early mammary gland metabolic and immune responses during natural-like and forceful drying-off in high-yielding dairy cows. *J. Dairy Sci.* 96(10), 6400-6411.

Smith, A., Westgarth, D. R., Jones, M. R., Neave, F. K., Dodd, F. H. and Branner, G. C. 1967. Methods of reducing the incidence of udder infection in dry cows. *Vet. Rec.* 81(20), 504-510.

Smith, K. L. 1983. Mastitis control: a discussion. *J. Dairy Sci.* 66(8), 1790-1794.

Smith, K. L. and Hogan, J. S. 1993. Environmental mastitis. *Vet. Clin. North Am. Food Anim. Pract.* 9(3), 489-498.

Smith, K. L., Todhunter, D. A. and Schoenberger, P. S. 1985. Environmental pathogens and intramammary infection during the dry cow period. *J. Dairy Sci.* 68(2), 402-417.

Sordillo, L. M., Shafer-Weaver, K. and Derosa, D. 1997. Immunobiology of the mammary gland. *J. Dairy Sci.* 80(8), 1851-1865.

Stefanon, B., Colitti, M., Gabai, G., Knight, C. H. and Wilde, C. J. 2002. Mammary apoptosis and lactation persistency in dairy animals. *J. Dairy Res.* 69(1), 37-52.

Stelwagen, K., Davis, S. R., Farr, V. C., Eichler, S. J. and Politis, I. 1994a. Effect of once daily milking and concurrent somatotropin on mammary tight junction permeability and yield of cows. *J. Dairy Sci.* 77(10), 2994-3001.

Stelwagen, K., Politis, I., White, J. H., Zavizion, B., Prosser, C. G., Davis, S. R. and Farr, V. C. 1994b. Effect of milking frequency and somatotropin on the activity of plasminogen activator, plasminogen, and plasmin in bovine milk. *J. Dairy Sci.* 77(12), 3577-3583.

Summers, E. L., Lacy-Hulbert, S. J., Williamson, J. H. and Sugar, B. P. 2004. Influence of feeding level after drying off on incidence of mastitis and keratin plug formation in dairy cows. *N. Z. Soc. Anim. Prod.* 64, 48-52.

Torres, A. H., Rajala-Schultz, P. J., Degraves, F. J. and Hoblet, K. H. 2008. Using dairy herd improvement records and clinical mastitis history to identify subclinical mastitis infections at dry-off. *J. Dairy Res.* 75(2), 240-247.

Tucker, C. B., Lacy-Hulbert, S. J. and Webster, J. R. 2009. Effect of milking frequency and feeding level before and after dry off on dairy cattle behavior and udder characteristics. *J. Dairy Sci.* 92(7), 3194-3203.

USDA. 2016. *Dairy 2014: Milk Quality, Milking Procedures, and Mastitis on U.S. Dairies, 2014*. Fort Collins, CO: USDA-APHIS-VS-CEAH-NAHMS, #704.0916.

Valizaheh, R., Veira, D. M. and Von Keyserlingk, M. A. G. 2008. Behavioural responses by dairy cows provided two hays of contrasting quality at dry-off. *Appl. Anim. Behav.* 109(2-4), 190-200.

Vanhonacker, F., Verbeke, W., van Poucke, E., Pieniak, Z., Nijs, G. and Tuyttens, F. 2012. The concept of farm animal welfare: citizen perceptions and stakeholder opinion in Flanders, Belgium. *J. Agric. Environ. Ethics* 25(1), 79-101.

van Knegsel, A. T., van der Drift, S. G., Cermakova, J. and Kemp, B. 2013. Effects of shortening the dry period of dairy cows on milk production, energy balance, health, and fertility: a systematic review. *Vet. J.* 198(3), 707-713.

Vasquez, A. K., Nydam, D. V., Foditsch, C., Wieland, M., Lynch, R., Eicker, S. and Virkler, P. D. 2018. Use of a culture-independent on-farm algorithm to guide the use of selective dry-cow antibiotic therapy. *J. Dairy Sci.* 101(6), 5345-5361.

Vilar, M. J., Hovinen, M., Simojoki, H. and Rajala-Schultz, P. J. 2018. Short communication: drying-off practices and use of dry cow therapy in Finnish dairy herds. *J. Dairy Sci.* 101(8), 7487-7493.

Vilar, M. J. and Rajala-Schultz, P. J. 2020. Dry-off and dairy cow udder health and welfare: effects of different milk cessation methods. *Vet. J.* 262, 105503.

von Keyserlingk, M. A. G., Olenick, D. and Weary, D. M. 2008. Acute behavioral effects of regrouping dairy cows. *J. Dairy Sci.* 91(3), 1011-1016.

Watters, R. D., Guenther, J. N., Brickner, A. E., Rastani, R. R., Crump, P. M., Clark, P. W. and Grummer, R. R. 2008. Effects of dry period length on milk production and health of dairy cattle. *J. Dairy Sci.* 91(7), 2595-2603.

Watts, J. M. and Stookey, J. M. 2000. Vocal behaviour in cattle: the animal's commentary on its biological processes and welfare. *Appl. Anim. Behav. Sci.* 67(1-2), 15-33.

Wayne, R. and Macy, H. 1933. The effect of various methods for drying up cows on the bacterial and cell content of milk. *J. Dairy Sci.* 16(1), 79-91.

Williamson, J. H., Woolford, M. W. and Day, A. M. 1995. The prophylactic effect of a dry-cow antibiotic against *Streptococcus uberis*. *N. Z. Vet. J.* 43(6), 228-234.

Winder, C. B., Sargeant, J. M., Kelton, D. F., Leblanc, S. J., Duffield, T. F., Glanville, J., Wood, H., Churchill, K. J., Dunn, J., Bergevin, M. D., Dawkins, K., Meadows, S. and O'Connor,

A. M. 2019. Comparative efficacy of blanket versus selective dry-cow therapy: a systematic review and pairwise meta-analysis. *Anim. Health Res. Rev.* 20(2), 217–228.

Zecconi, A., Sesana, G., Vairani, D., Cipolla, M., Rizzi, N. and Zanini, L. 2019. Somatic cell count as a decision tool for selective dry cow therapy in Italy. *Ital. J. Anim. Sci.* 18(1), 435–440.

Zobel, G., Leslie, K., Weary, D. M. and von Keyserlingk, M. A. 2013. Gradual cessation of milking reduces milk leakage and motivation to be milked in dairy cows at dry-off. *J. Dairy Sci.* 96(8), 5064–5071.

# Chapter 9

# Managing calves/youngstock to optimise dairy herd health

*John F. Mee, Teagasc, Ireland*



## 1 Introduction

The focus of this chapter is on how better female calf/youngstock management can improve dairy herd productivity and health; 'the calves of today are the cows of tomorrow'. Youngstock diseases incur both direct costs of prevention and therapy and indirect costs associated with subsequent reduced dairy herd health and performance, a theme that is emphasised here. In addition, calf morbidity and mortality is a welfare issue assuming greater importance now due to increased consumer scrutiny of management practice on modern dairy farms (Mee, 2013).

Here, calves are defined as cattle from birth to weaning (fed colostrum, milk or milk replacer) and replacement heifers as cattle from weaning to first calving; youngstock encompasses both (from birth to calving). Unless specified, the material refers to female Holstein-Friesian cattle managed in temperate dairy industries internationally.

The emphasis is on modifiable risk factors and management practices with a view to improving youngstock immunity and reducing dependence on antimicrobial use (AMU). Although not specifically addressed here, the role of

http://dx.doi.org/10.19103/AS.2020.0086.11

good animal welfare as preventative medicine has recently attracted attention as a drug-free way of improving immune function (Dawkins, 2019). Successful management of youngstock is the start of the lifecycle to optimal dairy herd health (Lorenz et al., 2011). However, farmers are often blind to this axiom.

## 2 Costs of heifer rearing

One way to convince farmers of the value of good youngstock management is to demonstrate the high, but variable costs of rearing replacement heifers and the effects of management on these costs. It is accepted, however, that farmers are also motivated by the principles of good calf welfare and pride in producing healthy, well-grown heifers as well as by economics. Heifer rearing is the second largest expense on a dairy operation (15–20% of total costs). The average costs of rearing a replacement dairy heifer in recent international calculations, both in confinement- and pasture-based systems vary between €1545 (Shalloo et al., 2014) and €2450 (CAFRE, 2020). However, there can be more than 95% difference in costs between the top and the bottom 25% of farmers (CAFRE, 2020). Most of this extra cost is due to delayed first calving and it takes, on average, two lactations to recoup these costs (Boulton et al., 2017). The rearing costs of a heifer that experienced disease at least once were, on average, €95 higher than those of healthy heifers (Modh Nor et al., 2012). For example, the most recent estimate of the costs of respiratory disease in calves put the short-term costs alone at nearly €40/case (Dubrovsky et al., 2020). With recent international improvements in dairy cow fertility and increasing use of sexed semen, it may be prudent in future to shift away from producing excess heifers for herds that are not expanding and concentrate on producing fewer, higher quality, healthier heifers (Overton and Dhuyvetter, 2020). The number of heifers required and the number generated can be calculated from Equations 1 and 2 (Dechow, 2020).

**Equation 1.** Number of replacement dairy heifers required:

$$\text{Heifers required} = \text{herd size} \times \text{AFC}/24 \times \text{herd turnover rate} \times (1 - \text{heifer cull rate})$$

where herd size = lactating plus dry cows, AFC = age-at-first calving, herd turnover rate = culling and mortality rate.

**Equation 2.** Number of replacement dairy heifers generated:

$$\text{Heifers generated} = \text{herd size} \times 12/\text{CI} \times \%\ \text{heifers} \times (1 - \text{calf mortality}) \times 24/\text{AFC}$$

where CI = calving interval.

Thus youngstock rearing is an expensive investment, so to reduce costs and optimise genetic potential, good early life management is essential. One of the first steps in cost reduction and improved herd health is having achievable, meaningful performance goals.

## 3 Targets for heifer rearing

An essential component of youngstock management is data management, data recording, target-setting and benchmarking - if you don't measure, you can't monitor. While ideally the goals of youngstock rearing should be farm-specific, there are generic, within the management system, targets which should be achievable based on the top 25% of farmers' benchmarked performance. Examples of some critical youngstock targets are listed in Table 1. Because of breed variation in some of these standards, relative, rather than absolute, targets can be more useful, for example, achieving double birth weight pre-weaning (8 weeks), 55-60% of mature body weight (MBW) pre-breeding and 85-90% of MBW pre-calving. Targets will vary by country and in some cases can be even more stringent than those listed here, for example, a target of <2% pre-weaning (<3 month) calf mortality per year in the UK (Hyde et al., 2020). Absolute values will also vary by the management system, reflecting their different goals, for example, higher production targets in hyperintensive confinement systems and lower in pasture-based systems (Mee and Boyle, 2020).

While the most appropriate target values are those generated from within the national dairy industry cohort (using data from the top quartile of farmers; norm-referenced thresholds), as opposed to those generated empirically

**Table 1** Targets for Holstein-Friesian youngstock rearing

| Metric | Target |
|---|---|
| Perinatal (0-48 h) mortality (%/year) | <5 |
| Calf umbilical infection therapy (%) | <0.5 |
| Calf diarrhoea therapy (%) | <5 |
| Calf respiratory disease therapy (%) | <1 |
| Pre-weaning mortality (%/year) | <3 |
| Heifer mortality (%/year) | <3 |
| Pre-pubertal ADG (kg) | 0.6-0.7 |
| Post-pubertal ADG (kg) | 0.8-0.9 |
| First breeding age (mo) | 13-14 |
| First breeding weight (kg) | 325-350 |
| Calving age (mo) | 22-24 |
| Calving weight (kg) | 550-600 |
| Calving BCS (1-5 scale) | 3.25-3.50 |

(criteria-referenced thresholds), the latter are often more accessible than the former.

In the context of this chapter, setting and achieving targets for youngstock rearing have direct impacts on subsequent dairy herd health as outlined hereunder.

## 4 Start of the dairy herd health lifecycle

Management of the perinate (and the foetus) is the start of the dairy herd health lifecycle. This management begins prior to calving but has significant effects on young calf health both directly through foetal nutrition and indirectly though effects on colostrogenesis (Abuelo, 2020).

### *4.1 Effects of pre-calving management of calf health*

The data on the effects of pre-calving feeding on neonatal calf health come from both dairy and beef cow studies. Restriction of energy (Carstens et al., 1987) or protein (Funston et al., 2010) during the last third of pregnancy in beef heifers resulted in the birth of calves with reduced thermogenesis and increased susceptibility to hypothermia. Studies in dairy (Nakao et al., 2000) and in beef cows (Bull et al., 1978) attributed reduced perinate vigour ('weak calf syndrome') to pre-calving protein and energy restriction. However, the calves of beef cows in which body condition score (BCS) increased during the dry period and which were in overcondition at calving, took longer to stand and to initiate suckling (Hamilton and Giesen, 1996). Diarrhoea and neonatal calf mortality rates were higher in beef calves born to dams subjected to pre-calving protein and energy restriction (Corah et al., 1975). The diarrhoea may have been due to affected calves being slower to suckle or having reduced immunoglobulin (Ig) absorption. In addition, it has been suggested that diarrhoea in calves from dams with disturbed acid-base balance may be due to altered colostral osmolarity resulting in higher abomasal pH which prolongs casein precipitation time (Gotowiecka et al., 2012). A study in beef cows showed a detrimental effect of maternal protein undernutrition in late pregnancy on calf health after weaning; more treatments for respiratory disease attributed to foetal programming (Larson et al., 2009). Additionally, in recent research, feeding grass silage alone to pregnant dairy cows has been associated with increased incidence of enteritis in dairy calves when compared with feeding grass silage and concentrates (Dunn et al., 2017).

Maternal micronutrient status pre-calving is the major determinant of the micronutrient status and risk of micronutrient disorders in neonatal calves through transplacental (trace elements) and colostral (trace elements and vitamins) transfer of nutrients. Imbalances (deficiency and excess) of micronutrients in calves are associated with micronutrient-specific disorders (e.g. anaemia, CJLD, goitre, nutritional muscular dystrophy). Micronutrient imbalances are also associated with reduced calf immunity and resultant increased incidences of diarrhoea, respiratory diseases and ill thrift, as reviewed by Enjalbert (2009) and Mee (2004). Both the concentration and the chemical form of the micronutrient in the pre-calving diet can influence its effects on the neonate. For example, organic selenium fed to beef cows pre-calving reduced the incidence of diarrhoea in their calves compared to cows fed inorganic selenium (Guyot et al., 2007). While a study in dairy cows showed no detrimental effect of different iodine concentrations in the pre-calving diet on calf health (Connelly et al., 2014), previous studies in sheep had demonstrated negative effects (Boland et al., 2005). Micronutrient disorders are probably of greater significance in suckler beef cows than in dairy cows (Gilmore and Mee, 2012). In addition, a recent study in dairy cows showed that peripartum maternal hypocalcaemia was associated with increased incidence of neonate enteritis compared to calves from eu-calcaemic dams (Hunter, 2015).

## 4.2 Perinatal management

The critical risks to the calf during the perinatal period (0–48 h) are dystocia, perinatal mortality (from pre-, intra- and post-parturient causes) and infectious diseases (both clinical infections and subclinical infections) (Mee et al., 2013a). Perhaps more importantly, perinatal management has a major impact on subsequent calf health, for example, through the role of colostrum in mitigating the risk of occurrence and the severity of neonatal morbidities (e.g. infectious enteritis and respiratory disease).

Management of each of these critical control points, dystocia, perinatal mortality and infectious diseases is discussed below.

### 4.2.1 Management of dystocia

Dystocia (defined here as abnormal calving) has both a direct and indirect impact on calf health. It is the primary cause of perinatal mortality. In addition, dystocia indirectly influences the immunity of the newborn calf. This effect is due to the impact of dystocia-induced foetal asphyxiation on reduced colostral Ig absorption and on abnormal perinate behaviour (delayed time from birth to sternal recumbence, standing and sucking). The cumulative effect of these indirect effects of dystocia on the calf is increased risk of failure of passive transfer (FPT) (Furman-Fratczak et al., 2011), increased youngstock morbidity, health therapy, age-at-first calving and mortality (Barrier et al., 2013, Heinrichs et al., 2005). Additionally, dystocia has significant impacts on the dam. Hence, management of dystocia is critical to dairy herd health.

In order to manage dystocia the associated (modifiable) risk factors and causes need to be recognised. The primary causes of dystocia are relative foetal oversize in primiparae and maldispositions (malpostures, malpresentations and malpositions in order of frequency) in pluriparae (Mee, 2008). Given that many of the significant risk factors for dystocia are not under management control (e.g. primiparity, pluriparity, previous dystocia, etc.), farmers need to focus on the modifiable risk factors (Mee et al., 2011).

Management of relative foetal oversize involves ensuring that both foetal body weight and maternal pelvic area are congruent. Thus the modifiable factors associated with calf birth weight such as sire genetic merit for dystocia, breed, nutrition and gestation length need to be managed through genetic selection, feeding management and, where necessary, pharmacological induction of parturition. For Holstein-Friesian cows the optimal calf birth weight range to reduce the risk of dystocia is from 42 kg to 45 kg due to relative foetal oversize (Mee, 2008). Similarly, the modifiable factors associated with the maternal pelvic area such as bodyweight at service and age and body condition at calving, need to be managed through feeding management (Mee, 2014). Neither internal nor external pelvimetry around service is sufficiently predictive of dystocia to select optimal pelvic area-heifers. Of the two approaches, optimising calf birth weight (primarily by genetic selection) is currently considered the more important and feasible long-term management strategy to reduce the occurrence of dystocia in dairy herds. In future, advanced neural network models using recorded cow data (e.g. parity, body condition score, breed, etc.) may be used to predict the likelihood of calving assistance and difficulty; current models have an accuracy of 75% (Fenlon et al., 2017).

Unlike relative foetal oversize, management of dystocia due to maldispositions is more challenging, particularly because the aetiologies of malpostures, malpresentations and malpositions differ. Factors associated with the occurrence of malpostures include pluriparity, prematurity, uterine inertia and foetal ankyloses and mortality. Of these, only uterine inertia secondary to hypocalcaemia can be managed nutritionally to reduce risk of malpostures. Factors associated with the occurrence of malpresentations include older dam age, sire, sire breed, males, lighter foetuses, pluriparity and foetal ankyloses (Holland et al., 1993). Of these, only sire/breed selection can be managed to reduce future risk of occurrence.

Irrespective of the cause of dystocia, good periparturient management can reduce the associated sequelae. Key management practices are prediction of calving time, good calving supervision and good obstetrical technique. The impacts of night-time feeding of pregnant cows to reduce dystocia associated with night calvings are equivocal (Gleeson et al., 2007). Ongoing formal training of farm staff periparturient management can improve outcomes.

### 4.2.2 Management of perinatal mortality

The prevalence of perinatal mortality can vary internationally between 2 and almost 10%. Given the newborn calf has the best genetics in the herd, the loss of such valuable genetic material for future productivity needs to be minimised. In addressing the issue of perinatal mortality one needs to disaggregate the losses into those which occur prepartum, peripartum and postpartum, as different risk factors, aetiologies and management strategies apply in each category of loss. The primary causes of perinatal mortality are non-infectious and vary by calving assistance score with anoxia and maldispositions the most common causes of death. An exemplar dataset from Irish commercial dairy herds is shown in Table 2 where the predominant cow breed is Holstein-Friesian but there are also, in decreasing order, Jersey and Norwegian Reds, and their cross breeds, and the predominant sire breed is Holstein-Friesian but there are also, in decreasing order, Jersey, Aberdeen Angus, Norwegian Red, Hereford and other beef breed sires (Belgian Blue, Simmental, Limousin, etc.).

Thus management to reduce perinatal mortality should focus on the dystocia-associated modifiable risk factors for stillbirth/perinatal mortality (Mee et al., 2014) and asphyxia during calving. However, given the variability between farms, and years in the causes of mortality, investigation of losses (Mock et al., 2020) should be a standard protocol on all dairy farms.

Foetal asphyxiation is a major cause of perinatal mortality; hence, management of this problem is recommended. Approaches include shortening the duration of stage two of calving (by avoiding disturbance or movement of animals during late stage one of calving, judiciously timed intervention during stage two or promptly performing a Caesarean section) and calf resuscitation (Mee, 2018).

Other causes of perinatal mortality can be managed as appropriate to the cause, for example, prepartum macronutrient and micronutrient supplementation of the dam (Mee, 2014). However, in herds with high perinatal mortality, farmers and their veterinarians need to focus on periparturient management (Mee et al., 2013b).

### 4.2.3 Management of perinatal infections

Perinatal infections need to be categorised into those which occur *in utero* and after birth as their management differs.

#### 4.2.3.1 In utero infections

*In utero* infections occur in approximately 20% of perinatal mortalities and contribute to death in approximately 10% of cases (Jawor et al., 2017). Common foetopathogens include *Neospora caninum*, bovine viral diarrhoea

**Table 2** Causes of perinatal mortality (0–48 h) (%) categorised by calving assistance score ($n$=259) (Mee and Kenneally, 2017)

| Cause of calf death (%) | Unobserved calving ($n$=70) | Observed, no assistance ($n$=46) | Easy assistance ($n$=57) | Moderate difficulty ($n$=47) | Severe difficulty ($n$=39) | All calvings ($n$=259) |
|---|---|---|---|---|---|---|
| Anoxia | 30 | 17.4 | 14 | 12.8 | 20.5 | 19.7 |
| Mal-disposition | 0 | 2.2 | 17.5 | 34 | 43.6 | 17 |
| Congenital defect(s) | 10 | 21.7 | 12.3 | 10.6 | 12.8 | 13.1 |
| Haemorrhage/Anaemia | 14.3 | 15.2 | 10.5 | 2.1 | 2.6 | 9.7 |
| Premature placental separation | 1.4 | 2.2 | 19.3 | 23.4 | 2.6 | 9.7 |
| Other | 22.9 | 21.7 | 14.1 | 12.8 | 17.9 | 18.1 |
| No significant findings | 21.4 | 19.6 | 12.3 | 4.3 | 0 | 12.7 |

(BVD) virus, *Leptospira, Listeria, Escherichia, Trueperella* and *Salmonellae* spp. and pathogenic fungi. While for some of these infections management is possible, for others (e.g. *Escherichia, Trueperalla*, mycoses) the organisms are ubiquitous or so sporadic that effective management is not possible. Control of neosporosis involves necropsy examination/sampling (PCR) of all foetal and perinatal calf mortalities, herd screening (e.g. bulk tank milk serology) to establish likely prevalence, whole herd serology to detect positive animals and culling or breeding to beef sires of detected cows. In addition, implementation of both effective bioexclusion (e.g. blood testing purchased animal), and biocontainment measures (e.g. preventing canid faeces from getting into the cows' feed) are recommended (Mee et al., 2012). If a national BVD eradication programme is not in place, sampling of all newborn calves (e.g. ear tag biopsy), herd screening, bioexclusion and vaccination are recommended to manage this disease. For pathogenic *Leptospira* and *Salmonella* spp., vaccination is recommended both to control disease in calves but also to reduce associated zoonotic risks.

### 4.2.3.2 Neonatal infections

Infections contracted by newborn calves include both those likely to result in clinical signs of enteritis, septicaemia (e.g. *Escherichia coli, Salmonella* spp.), omphalophlebitis, arthritis, meningitis and subclinical infections (e.g. *Mycobacterium bovis* subspecies *paratuberculosis* (MAP) and *Mycoplasma bovis*). These neonatal infections can have long-term consequences for youngstock growth, productivity and health (vide infra).

Central to control of all neonate infections is reducing infectious challenge, for example, environmental hygiene in the calving and calf housing and calf feeding equipment and improving host immunity, for example, passive transfer of immunity via best practice colostrum management.

## 4.2.4 Colostrum management

Providing an adequate volume of hygienic, good quality, first milking colostrum as soon as possible after calving to the neonatal calf is the bedrock of good colostrum management. Numerous mnemonics have been devised to embed this axiom in farmer's management standard operating procedures (SOPs), for example, *Colostrum 1-2-3*; colostrum is only the first (1) milking which should be given within two (2) hours of birth in a feed of three (3) or more litres (body weight-dependent) (Animal Health Ireland). In the absence of any/ sufficient natural colostrum, colostrum replacers or supplements have a role to play especially if the cow's colostrum quality or quantity is inadequate, or in Johne's test-positive herds, and, critically, the product, and number of doses, provides the necessary amount of total Ig (Williams et al., 2014, Mee et al.,

1996). While the roles of colostrum in preventing calf diseases and improving calf growth are well established, recently the role of colostrum in transferring signalling molecules (microRNAs) which are important for neonatal intestinal development and the complete immune system has been identified (Van Hese et al., 2020). Harvesting of good-quality colostrum begins with the management of the pregnant cow.

### 4.2.4.1 Effects of pre-calving management on colostrogenesis

The studies on the effects of pre-calving feeding of cows on colostrum quantity and quality are derived from dairy and beef cow studies. Altering dietary energy content or restricting DMI during the far-off (Nowak et al., 2012a) or close-up dry period (Nowak et al., 2012b) in dairy cows had no effect on colostrum or calf serum Ig concentrations. Feeding diets high in undegradable protein during the dry period to dairy cows did not alter colostrum or calf serum Ig concentrations (Bland et al., 2007). However, a negative correlation was found between the amount of concentrates fed to dairy cows pre-calving and colostrum IgG content by Gulliksen et al. (2008) attributed to a possible dilution effect with increased colostrum yield. Earlier studies by Stockdale and Smith (2004) in dairy and Shell et al. (1995) in beef cows found a similar effect with increasing energy intake/plane of nutrition during gestation. Stockdale and Smith (2004) also found that increasing dietary crude protein content increased colostrum Ig content in dairy cows. Both Santos et al. (2001) and Hook et al. (1989) found no effect of diet crude protein content and colostrum volume or IgG content in dairy cows. Similar results on diet crude protein content were found in beef cattle (Quigley and Drewry, 1998, Funston et al., 2010) though the latter authors did find lower serum IgG content in calves from dietary protein-restricted beef cows. While beef cows fed straw only in the last fortnight pre-calving had lower colostrum and calf serum IgG content (McGee et al., 2006), in general dietary nutrient restriction or supplementation did not alter colostrum quantity and quality (Quigley and Drewry, 1998, Hough et al., 1990, Burton et al., 1984). However, one recent study in dairy cattle found increased colostrum yield (though not Ig concentration) in cows fed grass silage and concentrates pre-calving compared to those fed grass silage alone (Dunn et al., 2017). Ig absorption was reduced in calves fed colostrum from restricted dairy (Burton et al., 1984) and beef cows (Rytkonen et al., 2004, Hough et al., 1990). This may be associated with lower colostral tri-iodothyronine, which is involved in Ig absorption, or altered foetal intestinal vascularity and foetal enterocyte proliferation, in restricted cows. A low-energy diet fed pre-calving, though reducing colostrum volume, increased colostral Ig concentration (Mann et al., 2013). Interestingly, 'focus feeding' in sheep has been shown to increase colostrum volume (Banchero et al., 2004), though effects on Ig content were not reported; similar results are not available in dairy or beef cows. Selenium supplementation of the pre-calving diet has been shown to increase colostrum volume and Ig content (Mee, 2004) and calf serum Ig content (Illek, 2010). For example, selenium supplementation of pregnant cows increased calf Ig absorption efficiency and hence calf serum Ig content (Hall et al., 2014). However, parenteral selenium administration to dairy cows (Leyan et al., 2004) and high dietary iodine (Boland et al., 2005) and selenium content (Hammer et al., 2007) in pregnant ewes has been shown to reduce offspring serum Ig content. Supplementing the diet of pre-calving cows with mannan oligosaccharides tended to improve calf colostrum-derived rotavirus antibody titres in response to maternal vaccination (Franklin et al., 2005). There are conflicting results on the effects of feeding anionic diets to dairy cows pre-calving on calf acidosis, colostrum Ig absorption and serum Ig concentration (Morrill et al., 2010, Quigley and Drewry, 1998).

In addition to these nutritional management factors, farmers can increase the quality of the colostrum they feed to their calves in other ways. Vaccination of the dam pre-calving against common calfhood enteropathogens (e.g. rotavirus) can produce pathogen-specific, hyperimmune colostrum (Civra et al., 2019). Other factors such as dam parity and temperature-humidity-index (THI) can also affect colostral Ig content (Shivley et al., 2018). But perhaps of more practical import is the fact that short dry periods (less than 8 weeks) reduce colostrum quality and colostrum Ig content decreases by approximately 4% for every hour after calving that the cow is not milked (Morin et al., 2010). These factors emphasise the importance of colostrum-focussed dry cow and periparturient management in enhancing colostrum quality for the newborn calf. Unfortunately the measurement of colostrum quality is not common on dairies. While the 'gold standard' colostrum immunity quality test is the measurement of the colostrum immunoglobulin G concentration by single radial immunodiffusion (SRID), this is an expensive laboratory assay only used in research studies. However, surrogate tests have been validated against this assay which can be used by farmers (Quigley et al., 2013). One of the most commonly used tests is refractometry. This can be carried out with a Brix optical refractometer. This inexpensive device has the advantage over a colostrometer (which measures colostrum specific gravity as a proxy for total protein content) that colostrum temperature or foaming (when a foam top develops on the colostrum sample after pouring) does not affect the reading. Good-quality colostrum has a Brix value of greater than or equal to 21%, equivalent to an SRID value of 50 g/L of IgG (Quigley et al., 2013). A recent review indicated that only 10–20% of farmers internationally implement this simple practice of measuring colostrum quality (Mee, 2020).

#### 4.2.4.2 Colostrum hygiene

While producing an adequate volume of high Ig content colostrum is a basic management goal, this good work can be undone by bad colostrum hygiene. Microbial contamination of colostrum can impair Ig absorption, accelerate gut closure and cause diarrhoea or sepsis and reduced weight gain in neonatal calves. Such contamination is surprisingly common. For example, recent surveys of the colostrum microbiome showed that only a quarter of samples met the total plate count target while over 75% showed evidence of faecal contamination (Nejedla et al., 2019, McAloon et al., 2016). Thus, management of neonatal calf immunity begins with a collection of high microbial quality colostrum and either refrigeration for short term (a day) or freezing for long-term (up to a year) storage or heat treatment (pasteurisation) to ensure 'clean' colostrum is fed to calves.

#### 4.2.4.3 Failure of passive transfer (FPT)

Newborn calves do not have an active immune system and so, colostral antibodies are the only source of immunoglobulins to protect them from infectious diseases. Method of colostrum feeding (e.g. suckling vs. bucket feeding vs. stomach tubing) can affect the risk of FPT (suckling>stomach tubing>nipple bucket feeding; but, volume-dependent) (Godden et al., 2009). However, one study found no effect on subsequent clinical health scores or average daily gain (ADG), (Ebert et al., 2007).

Measurement of calf blood Ig concentrations (or surrogate measures, for example, total protein TP, ZST, Brix readings) (Hogan et al., 2015) can be used as a management tool to screen (on a group, not an individual basis) for adequacy of passive transfer (in healthy calves 2–7 days old) and to predict future calf health and growth (Cuttance et al., 2018). For example, calves with a TP <50 g/L had a 2.4-times greater chance of dying than calves with a concentration between 50 and 60 g/L (McCorquodale et al., 2013). Recommended targets include 70% of samples above 60 g/L for TP and above 20 units for ZST. A recent study concluded that to completely avoid FPT, calves should receive at least 2.5 L of high-quality colostrum (Ig >88 g/L) within one hour of birth (Lora et al., 2018). Though not the focus of this chapter, male calves, whose health and welfare should be treated similarly to that of female calves, are at greater risk of FPT through receiving a lower volume of colostrum and of receiving pooled colostrum (Renaud et al., 2020).

#### 4.2.4.4 Snatch calving

In addition to avoiding pooled colostrum (colostrum from more than one cow), implementation of 'snatch calving' (immediate removal of the calf from the cow after calving) is currently recommended to reduce the risk of transmission of MAP between the cow and her calf. However, this practice is coming under scrutiny due to the associated welfare implications for the cow-calf dyad (Beaver et al., 2019).

It is clear from this overview that ensuring that calves survive the perinatal period and do not suffer detrimental sequelae due to suboptimal neonatal management, especially FPT, can best prepare the calf for subsequent development, growth and health. However, both calfhood nutritional management and diseases can undermine the good work carried out in the perinatal period.

## 5 Impacts of calfhood nutritional management on subsequent dairy herd health

There is now an extensive corpus of literature showing that management of the neonatal and later developmental stages of the calf's life impacts subsequent dairy herd health, age-at-first calving, lifetime productivity and survival. For the latest review, see Hammon et al. (2020). The latest adjacent literature on the effects of management of the pregnant cow (e.g. milk production level, heat stress, ration energy density) on foetal developmental programming ('The Barker Hypothesis'; Barker et al., 1989), specifically, immune function (Cooke, 2019) and youngstock health (Perry et al., 2019), support this hypothesis. Nutritional programming of calves alters their long-term metabolic and productive performance as adults. This is a permanent effect on life-long gene expression of the release of hypothalamic neuropeptides which control feed intake and weight gain (Taylor and Poston, 2007). These related streams of research, foetal and calfhood programming, have altered the paradigm of management as a factor just affecting current performance to one potentially affecting future performance also. Specifically in relation to milk feeding, this research has highlighted the limitations of restricted milk supply and demonstrated the short- and long-term benefits of intensive/accelerated milk feeding protocols (Hammon et al., 2020, Hu et al., 2020). However, despite most dairy farmers rearing their own replacement heifers, the impacts of calfhood management and disease on subsequent productivity and longevity are often unrecognised (or unknown, for example, the lactocrine hypothesis; Bagnella and Bartol, 2019) by both farmers and their veterinarians. This has implications for farmer perception and management of calfhood nutrition and health hazards and veterinary communication on calfhood nutritional and disease sequelae with lag dynamics.

### *5.1 Age at first calving (AFC)*

Age at first calving (AFC) is important because it affects farm profitability through heifer rearing costs and effects on subsequent reproductive and productive performance. AFC is correlated with milk production (Eastham et al., 2018), calving (Mee, 2008) and reproductive performance (Ettema and Santos, 2004) and longevity (Berry and Cromie, 2009). There is a linear association between AFC and first lactation peak and 305-day milk yield (Berry and Cromie, 2009). Target/optimal values for AFC vary widely internationally between 21 (Cole et al., 2013) and 30 months (Haworth et al., 2008), however, a recent knowledge summary literature review concluded that the optimum range should be narrower; from 22 months to 25 months (Steele, 2020). Actual median values vary between 23 months (Stanton et al., 2012) and 29 months (Haworth et al., 2008). Lower AFC is associated with better udder health, better reproductive performance and increased lifetime daily milk production (Eastham et al., 2018). Given the time lag between calfhood nutritional management and first calving, farmers or their veterinary practitioners are rarely aware of the potential impacts of former on the latter.

### 5.1.1 Calf nutritional management factors affecting age at first calving (AFC)

Feeding a higher volume of colostrum has been shown to be associated with a lower AFC in some (Mohd Nor et al., 2013) but not in other studies (Faber et al., 2005; DeNise et al., 1989). However, feeding a higher volume of colostrum (Brickell et al., 2009) and higher serum Ig concentrations after birth (Robison et al., 1988) have been shown to significantly increase the ADG and girth in the first six months of life, factors which reduce AFC. For example, a recent study showed an inverse linear relationship between ADG of heifers at the start of the breeding season and subsequent days open (Hayes et al., 2019). Additionally, calves with higher serum Ig concentrations at 1–2 days of age had a lower age at insemination (Furman-Fratczak et al., 2011), which affects AFC. Not feeding waste milk to calves has been associated with increasing the odds (OR 3.656; CI95 0.953–14.034, $P$=0.059) of heifers calving at ≤24 months compared to >24–<27 months (Mohd Nor et al., 2013).

Higher nutrient intake pre-weaning has been shown to affect AFC. Increasing the volume of milk fed increased the odds (OR 1.776; CI95 0.932–3.384, $P$=0.081) of heifers calving at ≤24 months compared to >24–<27 months (Mohd Nor et al., 2013). This is in agreement with previous data from the United States where maximum milk intake positively influenced AFC (Heinrichs et al., 2005) and from the UK where *ad libitum* milk feeding significantly increased ADG and heifer height from 1 to 6 months of age compared to restricted feeding (Brickell et al., 2009). However, a high ADG (1.15 kg/d) prior to breeding (8–10 months) delayed the onset of puberty (Silper et al., 2013) which could increase the AFC. In addition, heifers fed poorer quality forage (higher ADF) tended to have increased AFC ($P$=0.081)

(Heinrichs et al., 2005) and heifers in Sweden which spent more time grazing had increased AFC (Hultgren et al., 2008).

Only a limited number of studies compared the effect of whole milk (WM) versus milk replacer (MR) on AFC. In a Dutch study the median AFC was lower for calves fed MR compared to WM ($P$=0.012) (Mohd Nor et al., 2013). This is in agreement with a study from the UK where calves fed MR had significantly increased ADG, girth, crown-rump length and height from 1 to 6 months of age compared to calves fed WM (Brickell et al., 2009). In contrast, in Israeli studies age at puberty was lower ($P$<0.01) (Shamay et al., 2005) and AFC tended to be lower where calves were fed WM compared to MR ($P$<0.07) (Moallem et al., 2010). Differences in the nutrient composition and feeding regimes of MR between studies probably account for these differences.

The effects of milk replacer feeding regimes on AFC are also conflicting. While increasing MR feeding level (g/day and crude protein (CP) content) tended to reduce both age at first observed oestrus and age at first service, AFC was not altered in a Northern Ireland study (Morrison et al., 2012). Similar results were found by Terre et al. (2009) in a Spanish study where feeding a higher volume of a high dry matter (DM) MR increased BW at weaning but not thereafter or age at first breeding, compared to a lower feeding rate of a lower DM MR. Drackley et al. (2007) did not change AFC by utilising a conventional limit-fed or an intensified MR programme. In contrast, a high-CP, high-feeding volume MR regime reduced AFC numerically ($P$=0.11) compared to a conventional MR regime (Davis Rincker et al., 2011) and an intensive high-solid (16.7%), high-feeding MR regime reduced AFC compared to a conventional MR feeding regime ($P$=0.05) (Raeth-Knight et al., 2009). It may be concluded that reduced AFC is associated with increased nutrient intake and ADG pre-weaning.

## 5.2 Lifetime productivity

The lifetime productivity (defined as from first calving to death) of a cow is short (3–4 years) relative to natural life expectancy (~20 years); this is a welfare concern (De Vries and Marcondes, 2020). It is influenced by many interrelated factors such as age at first calving, calving intervals, milk yield/day, availability of replacement heifers, and age at removal but also by nutritional management of the pre-weaned calf. Heifers that calve at a younger age have more days of productive life (higher longevity index) and higher lifetime milk yields (Lin et al., 1988). A heifer that first calves at three years of age and is culled in her third lactation spends only 40% of her life being productive, whereas one that calves at two years and survives to her fifth lactation produces milk for more than 60% of her life thus increasing her lifetime productivity (Garnsworthy, 2012). Given the high (30%, Mohd Nor et al., 2012) and increasing (Bach, 2011) culling, and hence, replacement rates on modern dairy farms it is becoming increasingly difficult for farmers to increase cow lifetime productivity profitably. Hence any impacts of calfhood management on lifetime productivity are increasingly important.

### 5.2.1 Calf nutritional management factors affecting lifetime productivity

Given the multitude of nutritional, genetic and environmental factors which impact lifetime productivity it can be difficult to detect early life effects such as dairy calf nutritional management. However, such influences have been reported. The beneficial effects of feeding a larger volume of colostrum (4 vs. 2 L) (Faber et al., 2005) and of higher serum Ig status (DeNise et al., 1989) have been shown for first and second lactation milk production. Higher nutrient intake pre-weaning has been reported to affect lifetime productivity though results are conflicting. Drackley et al. (2007) found a response in first lactation milk production to an intensive MR programme ($P<0.01$) and an Israeli study found that calves offered ad libitum WM compared to MR produced 1.3 kg/d more 3.5% fat-corrected milk in first lactation ($P<0.005$) (Shamay et al., 2005). However, Morrison et al. (2012), Raeth-Knight et al. (2009) and Terre et al. (2009) failed to find significant effects of enhanced milk feeding regimes. Gleeson and O'Brien (2012) found no effect of milk feeding frequency (once vs. twice daily) on ADG or first lactation milk production. A review stated that greater nutrient supply through increased volume of milk appears to increase milk yield, but more research is needed to confirm this effect (Khan et al., 2011). This opinion was supported by a meta-analysis which concluded that calves fed high nutrient liquid diets were twice more likely to have greater milk yield in first lactation (Soberon and Van Amburgh, 2013).

The ADG between birth and breeding can affect lifetime productivity. Soberon et al. (2012) showed that pre-weaning ADG, weaning weight ($P<0.03$) and ADG from birth to breeding ($P<0.01$) was positively correlated with first lactation milk yield. For every kg of pre-weaning ADG, heifers produced 850 kg more milk during their first lactation. Significant positive correlations were also found for ADG and milk production in lactations one through three indicating impacts on lifetime productivity. The authors concluded that metabolic programming during early life has lifelong implications for milk production possibly through epigenetic changes. The nutritional management of the calf pre-weaning has the greatest impact on the development of the mammary gland which is associated with higher subsequent milk production (Geiger et al., 2016). It is also possible that a lactocrine effect of bioactive factors in WM affects milk yield as Moallem et al. (2010) reported that calves fed WM had a 10% higher first lactation milk yield than those fed MR ($P<0.001$). These findings are supported by those of Heinrichs and Heinrichs (2011) who found that for every 1 kg increase in DMI at weaning, there was a corresponding increase of 287 kg in first lactation milk yield. While decreased mammary development has been observed as prepubertal ADG is increased, a meta-analysis by Zanton and Heinrichs (2005) concluded that first lactation milk production is increased as prepubertal ADG (150–320 kg) increased up to 799 g/d whereas further increases in ADG were associated with lower milk yield indicating a curvilinear relationship. A meta-analysis concluded that for every kg of pre-weaning ADG, first lactation milk yield increased by 1550 kg (Soberon and Van Amburgh, 2013). The beneficial effects of larger, well-grown heifers on milk production were also shown by Archbold et al. (2012) who reported that heavier heifers and those in better BCS at mating start date in a seasonal calving system had significantly higher milk production in lactation one through three indicating effects on lifetime productivity.

## 6 Impacts of calfhood diseases on subsequent dairy herd health

Similar to the effects of youngstock nutritional management, calfhood disease management has both short- and long-term impacts on subsequent dairy herd health and performance. Calfhood disease can reduce the benefits of nutritional programming, with calves which develop respiratory diseases having poor subsequent average daily gain despite earlier intensive feeding management (Maccari et al., 2015).

### *6.1 Age at first calving (AFC)*

Three aspects of youngstock health management have been shown to affect AFC, calving management, calf diarrhoea management and respiratory disease management.

Dystocia has short-term detrimental effects on calf health such as increasing the risk of FPT or incidence of umbilical cord pathologies (Moscuzza et al., 2014). However, often not recognised are the long-term sequelae. For example, a North American study showed that as calving assistance score of a calf increased by one unit, her subsequent AFC increased by 17 days ($P$=0.048) (Heinrichs et al., 2005).

The effects of calf diarrhoea on AFC are conflicting. Some studies (e.g. Morrison et al., 2010; Hultgren et al., 2008) found no influence of diarrhoea pre-weaning on AFC. In contrast, there is a tendency ($P$=0.064) for calves treated for diarrhoea or pneumonia to have increased AFC in a study by Heinrichs et al. (2005), and Waltner-Toews et al. (1986) found that calves treated for diarrhoea pre-weaning had a higher AFC.

The effects of bovine respiratory disease (BRD) on AFC are generally reported as detrimental. For example, Stanton et al. (2012) in Canada showed that calves with BRD were 0.8-times as likely to calve as heifers without BRD (*P*=0.02) and the odds of calving by 25 months of age were 0.6-times lower in calves with BRD (*P*=0.01). Donovan et al. (1998) had shown that days treated for BRD before 6 months of age decreased heifer ADG between birth and 6 months of age and between 6 and 14 months of age (*P*<0.025). This is in agreement with earlier studies which showed that calves with BRD were less likely to calve and were calved later (Warnick et al., 1994; Correa et al., 1988; Waltner-Toews et al., 1986). In contrast, Morrison et al. (2010) found no effect of BRD pre- or post-weaning on AFC in one study in Northern Ireland. Interestingly, high-housing maximum relative humidity (*P*=0.032) and mean ambient temperature (*P*=0.028) were both associated with increased AFC in one North American study (Heinrichs et al., 2005).

## 6.2 Lifetime productivity

Similar to effects on AFC, calving management and calfhood morbidity have been shown to detrimentally impact dairy herd productivity.

The more difficult a heifer calf's birth, the lower her subsequent milk production. As calving assistance score of a calf increased by one unit, her first lactation milk production decreased by 285 kg (*P*=0.03) and lifetime production tended to decrease (*P*=0.06) in one North American study (Heinrichs and Heinrichs, 2011). Previous studies had shown that the risk of FPT was significantly higher (OR = 2.6) for calves on operations where farmstaff would not seek veterinary assistance when unable to correct a calf during dystocia (Beam et al., 2009). In addition, a recent study has shown that calves which survive dystocia experience greater physiological stress, lower passive immunity transfer and higher mortality (Barrier et al., 2013). Dystocia-associated FPT may increase the risk of digestive and respiratory diseases in calves (Lombard et al., 2007), which may in turn contribute to lower subsequent productivity.

Most studies did not find a significant influence of calf diarrhoea on milk production (Soberon et al., 2012; Morrison et al., 2010; Warnick et al., 1995; Britney et al., 1984). However, one Swedish study found that calves that had diarrhoea during their first 3 months of life had 344 kg lower milk production in first lactation (Svensson and Hultgren, 2008).

The evidence for effects of calfhood BRD on milk yield is more convincing. Morrison et al. (2010) found that BRD pre-weaning tended (*P*=0.10) to and BRD post-weaning did (*P*<0.05) decrease first lactation milk production (by 6%). The same group reported that while calves treated for a single episode of BRD did not have lower milk yield, calves treated for multiple cases of BRD had lower first (*P*<0.05) and second lactation (*P*<0.01) milk production (Morrison et al., 2013). A Canadian study showed that first test milk production tended to be 1.1 kg lower for heifers with BRD (*P*=0.06) but 305-day milk production was not affected (Stanton et al., 2012). A Swedish study reported an association between severe BRD before 90 days of age and lower first lactation milk production (*P*=0.02) (Svensson and Hultgren, 2006). Perhaps, not surprisingly, vaccinating against bovine respiratory syncytial virus (BRSV) was shown in one study to be associated with an increase in first lactation milk production of 493 kg (P=0.063) (Mohd Nor et al., 2013). Earlier studies by Warnick et al. (1995) and Britney et al. (1984) did not show an association between BRD and milk production.

However, a number of studies have shown that calf morbidity is associated with lower milk production. Soberon et al. (2012) showed that calves receiving antibiotics (primarily for BRD) had 493 kg lower milk yield in first lactation (*P*<0.01). Similarly, Heinrichs and Heinrichs (2011) reported that the number of days that calves had diarrhoea or BRD in the first 4 months of life had a negative effect on first lactation milk production (*P*<0.01). 'Other' calf health events were associated with lower milk yield in first lactation in an earlier study by Miller and Faust (2000).

## 6.3 Survival

Calving management and calf-hood morbidity, especially BRD, have been shown to significantly reduce heifer survival in the dairy herd.

Calves which had a difficult calving had a three-fold greater risk of dying before weaning (*P*<0.05) and before first service (*P*<0.001) compared with heifers born unassisted. Survival to first calving was not affected by the difficulty experienced at birth (Barrier et al., 2012). Calves with poor neonatal immunity are less likely to survive. McCorquodale et al. (2013) found that calves with serum total protein (TP) values of <50 g/L were 2.4-times more likely to die within 4 months than calves with higher TP concentrations. These results are in agreement with earlier findings on survival up to six months of age (Robison et al., 1988).

Heifer calves which developed omphalo-arthritis had significantly poorer survival distribution up to the second lactation than unaffected calves (Britney et al., 1984). Disease treatment during the first week of life was associated with a 1.3-times greater risk of mortality and calves that received three or more treatments were 3.5-times more likely to die before 4 months of age (McCorquodale et al., 2013). However, the number of days of illness did not affect survival or culling age (Heinrichs and Heinrichs, 2011) and calfhood disease was not associated with length of herd life in an earlier American study (Warnick et al., 1997).

The most consistent relationships between calf health and survival were reported for BRD (Van Der Fels-Klerx et al., 2002). In a Canadian study only 66% of calves with BRD survived to first calving compared to 84% without BRD.

However, BRD did not affect survival to 120 DIM for calved heifers (Stanton et al., 2012). Similar detrimental effects of BRD on survival to first calving were reported by Warnick et al. (1995) and by Correa et al. (1988). These findings are consistent with those of Waltner-Toews et al. (1986) who found that heifers treated for BRD in the first 3 months of life were 2.5-times more likely to die between 3 months of age and calving. While Bach (2011) found no association between incidence of BRD and survival they did find that heifers that experienced four or more cases of BRD had greater odds (OR 1.87) of not surviving the first lactation. In addition, the accumulated DIM throughout productive life and the proportion of productive days with respect to recorded days of life decreased linearly ($P<0.05$) as the number of BRD episodes increased. Previously Moriarty et al. (2007) had shown the association between calfhood BRD and necropsy lesions of unresolved pneumonia and stunting due to lack of compensatory growth in later life.

Given these relationships between calfhood health traits and future dairy herd health it would appear logical to incorporate calfhood traits in future genetic selection indices, as has been suggested (Mousa et al., 2015). However, a recent study indicated that different SNP markers contributed to disease in calves and in cows; hence, such selection may not be appropriate with current knowledge (Mahmoud et al., 2017).

## 7 Role of vet in communicating best practice in youngstock management

Veterinary practitioners play a central role in better youngstock management ideally as part of a herd health and productivity management programme (Boersma et al., 2010). Farmers expect their consultants to identify calf care issues, inform farmers about the issue and provide practical steps to remediate these issues (Croyle et al., 2019). One of the issues veterinary practitioners and consultants need to be cognisant of is farm-blindness. Farm-blindness may be defined as 'a misperception by farmers that what they see every day on their own farm is normal, particularly when it is not; a new normal' (Mee, 2020). This paradigm applies to all aspects of farm management including other important animal health issues, for example, lameness. Evidence of farm-blindness in youngstock management may be found in failure to implement recommended practices, for example, prompt colostrum feeding, or failure to recognise high calf morbidity or mortality rates. Farmers' youngstock management hazard ranking perceptions differ significantly from those of their veterinarians; hence, the latter need to recognise this asymmetry when addressing the core issue of farm blindness. De-normalising poor youngstock management can be achieved by first creating awareness of the existence of the problem (e.g. through external audit of farm performance and national calf health campaigns. e.g. CalfCare[R] - Animal Health Ireland), secondly, by providing farmers with locally relevant peer benchmarks against which to compare their own youngstock performance (e.g. re-norming using regional/ national calf mortality rates), thirdly, by using agri-technology to provide farmer-independent alerts to deviations in animal performance and finally, effectively communicating current recommended best practices to farmers (e.g. good colostrum management) through peers, practitioners or domain experts.

A critical concept veterinarians need to grasp is the principle that with individual animal cases the treatment protocol is under the control of the vet. However, with herd problems, the resolution of the issue resides with the farmer (More et al., 2017). Hence, effective communication of prioritised, best-practice recommendations (in particular what good farmers are doing - use of social contagion herd effect nudges) is critical to enable most farmers to realign their management with current norms. Veterinary trouble-shooting of problems in animals and herds will still be required but the central role of the farmer needs to be recognised both by the vet and by the farmer. Realignment needs to address both activity (processes) and performance (output). Veterinarians and specialised advisory staff have a critical role in extending calf-rearing messages to farmers (Svensson and Hultgren, 2008) but must be cognisant of the limitations of one-step thinking which may miss the more subtle web of causality. Thinking in knowledge transfer is changing significantly from the former linear, top-down approach to a more sharing, bottom-up, design solutions approach (Hennessy and Heanue, 2012). Peer-to-peer learning is particularly effective in this regard via farmer action or discussion groups (Morgans et al., 2018; Hennessy and Heanue, 2012). As farms get larger and evolve from an owner-operator model to a distributed model, staff numbers/job titles increase. Inter-staff communication becomes more complex as the goals of care staff may be process-orientated (e.g. calf hutch hygiene) while those of managers are outcome-orientated (e.g. calf mortality rate) (Pereira et al., 2014). Also, educational, generational and language barriers may need to be overcome. Effective communication results in better stockmanship and good calf management stockmanship knowledge, attitudes and perceptions (KAPs) are positively associated with better calf health outcomes (Adler et al., 2019; Vaarst and Sorensen, 2009). Novel models of farming may also contribute to better youngstock management where calves are sent to specialised/custom contract rearers until the point of calving (Mee et al., 2018). In addition to addressing issues of poor youngstock management, education and public engagement (EPE) on best practice in youngstock management will be a pillar of sustainable agricultural systems ('social license to farm') in the future to avoid reputational risk to the dairy industry.

## 8 Conclusion and future trends

The material reviewed in this chapter clearly demonstrates that youngstock management can play a critical role in optimising dairy herd health. While the immediate impact of better calf management is visible to farmers in better youngstock health and growth, the long-term benefits in dairy herd productivity and survival need to be emphasised more by veterinarians and agricultural advisers.

The most consistently reported management factors significantly associated with reduced age at first calving (AFC), higher milk production and longer survival were feeding a larger volume of colostrum, offering a higher liquid feed volume pre-weaning, offering a higher solids milk replacer, offering whole milk and a higher average daily gain. The three health disorders most frequently associated with increased AFC, lower milk production and shorter survival were dystocia, calf diarrhoea and respiratory disease. It is concluded from this review that dairy calf health and heifer management can significantly affect age at first calving, lifetime productivity and survival. These results have implications for producer perception and management of calfhood nutrition and health hazards and veterinary communication on calfhood disease sequelae with lag dynamics.

Based on recent changes in the dairy industry worldwide, it is predicted that some of the important future trends in youngstock management will include more emphasis on reducing antimicrobial use in calf rearing (and perhaps increased emphasis on direct-fed microbials and prebiotics and probiotics, for example, as colostrum, supplements), improving calf welfare by external certification ('happy calves'; calf comfort will be the new cow comfort), incorporation of precision livestock farming concepts in calf rearing for pre-clinical diagnostics (e.g. wearable wellness sensors), increased use of SOPs for calf/heifer management on 'traditional' smaller dairy farms, the growth of specialised vertically integrated contract calf/heifers rearing and increased use of social science to 'nudge' stakeholders towards best practice of youngstock management. National infectious disease eradication schemes (e.g. bovine viral diarrhoea virus and possibly *Mycoplasma bovis*) will tangentially improve calf immunity and health, while if gene editing is approved in food-producing animals the resultant chimeras will dramatically change calf health.

## 9 Where to look for further information

### 9.1 Further reading

- *Farm health and productivity management of dairy young stock* by S.-J. Boersema, J. Cannas da Silva, J. F. Mee and J. Noordhuizen.

- *Dairy heifer management*, Veterinary Clinics of North America, Edited by S. Godden and S. M. McGuirk.

## 9.2 Key journals/conferences

- *Journal of Dairy Science* - international peer-reviewed research results/ reviews; researcher-focused.
- *World Buiatrics Congress* - international cattle disease congress; veterinary practitioner-focused.
- *Smart Calf Rearing Conference* - international calf rearing/health conference; farmer/industry-focused.
- *Healthy Calf Conference* - Canadian veal calf rearing and health conference; farmer/industry-focused.
- *Dairy Calf & Heifer Association Conference* - North American calf rearing and health conference; farmer/industry-focused.
- *European Calf Conference (ECC)* - European calf rearing and health conference; farmer/industry-focused.

## 9.3 Youngstock associations internationally

- *Calfcare* - Irish national calf rearing and health technical working group (www.animalhealthireland.ie) - stakeholder-focused.
- *KalfOK* - Dutch national calf management programme (www.gddiergezondheid.nl) - farmer-focused.
- *DCHA* - American Dairy Calf & Heifer Association - national calf rearing and health organisation (www.calfandheifer.org) - farmer-focused.
- *Healthy Calves* - New Zealand commercial calf health advice (www.healthycalves.co.nz) - farmer-focused.
- *CalfNotes* - American calf nutrition and rearing advice (www.calfnotes.com) - farmer-focused.
- *Calf Care* - Canadian calf rearing organisation (www.calfcare.ca) - farmer-focused.
- *CalfMatters* - UK commercial calf health plan advice (www.calfmatters.co.uk) - farmer-focused.

## 9.4 Major international research projects

- *The BRD 10K study* - a major North American research study of respiratory disease in dairy youngstock (www.sciencedirect.com/journal/journal-of-dairy-science).
- *The ProYoungStock project* - a European research study examining the effects of natural rearing and feeding strategies on health, welfare and economic performance (https://www.proyoungstock.net/about.html).

## 10 References

Abuelo, A. (2020). Symposium review: late-gestation maternal factors affecting the health and development of dairy calves. *Journal of Dairy Science* 103(4): 3882–3893.

Adler, F., Christley, R. and Campe, A. (2019). Examining farmers' personalities and attitudes as possible risk factors for dairy cattle health, welfare, productivity and farm management: a systematic scoping review. *Journal of Dairy Science* 102: 3805–3824.

Archbold, H., Shalloo, L., Kennedy, E., Pierce, K. M. and Buckley, F. (2012). Influence of age, body weight and body condition score before mating start date on the pubertal rate of maiden Holstein-Friesian heifers and implications for subsequent cow performance and profitability. *Animal* 6(7): 1143–1151.

Bach, A. (2011). Associations between several aspects of heifer development and dairy cow survivability to second lactation. *Journal of Dairy Science* 94(2): 1052–1057.

Bagnella, C. A. and Bartol, F. F. (2019). Relaxin and the'Milky Way': the lactocrine hypothesis and maternal programming of development. *Molecular and Cellular Endocrinology* 487: 18–23.

Banchero, G. E., Quintas, G., Martin, G. B., Milton, J. T. and Lindsay, D. R. (2004). Nutrition and colostrum production in sheep. 2. Metabolic and hormonal responses to different energy sources in the final stages of pregnancy. *Reproduction, Fertility, and Development* 16(6): 645-653.

Barker, D. J., Winter, P. D., Osmond, C., Margetts, B. and Simmonds, S. J. (1989). Weight in infancy and death from ischaemic heart disease. *Lancet* 2(8663): 577–580.

Barrier, A. C., Dwyer, C. M., Macrae, A. I. and Haskell, M. J. (2012). Short communication: survival, growth to weaning, and subsequent fertility of live-born dairy heifers after a difficult birth. *Journal of Dairy Science* 95(11): 6750–6754.

Barrier, A. C., Haskell, M. J., Birch, S., Bagnall, A., Bell, D. J., Dickinson, J., Macrae, A. I.. and Dwyer, C. M. (2013). The impact of dystocia on dairy calf health, welfare, performance and survival. *The Veterinary Journal* 195(1): 86–90.

Beam, A. L., Lombard, J. E., Kopral, C. A., Garber, L. P., Winter, A. L., Hicks, J. A. and Schlater, J. L. (2009). Prevalence of failure of passive transfer of immunity in newborn heifer calves and associated management practices on US dairy operations. *Journal of Dairy Science* 92(8): 3973–3980.

Beaver, A., Meagher, R. K., von Keyserlingk, M. A. G. and Weary, D. M. (2019). Invited review: a systematic review of the effects of early separation on dairy cow and calf health. *Journal of Dairy Science* 102(7): 5784–5810.

Berry, D. P. and Cromie, A. R. (2009). Associations between age at first calving and subsequent performance in Irish spring-calving Holstein-Friesian dairy cows. *Livestock Science* 123(1): 44–54.

Bland, I., Lang, P. and Hill, J. (2007). The consequences of offering diets high in rumen undegradable protein during the dry period on immunoglobulin G concentrations in serum of neonates, mature dairy cows and colostrum. In: *Proceedings of the 3rd Australasian Dairy Science Symposium*, Melbourne, pp. 378–386.

Boersma, S.-J., Cannas da Silva, J., Mee, J. and Noordhuizen, J. P. T. M. (2010). Diseases of young stock. In: *Farm Health and Productivity Management of Dairy Young Stock*, Boersma, S.-J., Cannas da Silva, J., Mee, J. and Noordhuizen, J. P. T. M. (Eds), Wageningen Academic Publishers, Wageningen, pp. 131–192.

Boland, T. M., Keane, N., Nowakowski, P., Brophy, P. O. and Crosby, T. F. (2005). High mineral and vitamin E intake by pregnant ewes lowers colostral immunoglobulin G absorption by the lamb. *Journal of Animal Science* 83(4): 871-878.

Boulton, A. C., Rushton, J. and Wathes, D. C. (2017). An empirical analysis of cost of rearing dairy heifers from birth to first calving and the time taken to repay these costs. *Animal* 11(8): 1372-1380.

Brickell, J. S., McGowan, M. M. and Wathes, D. C. (2009). Effect of management factors and blood metabolites during the rearing period on growth in dairy heifers on UK farms. *Domestic Animal Endocrinology* 36(2): 67-81.

Britney, J. B., Martin, S. W., Stone, J. B. and Curtis, R. A. (1984). Analysis of early calf-hood health status and subsequent dairy herd survivorship and productivity. *Preventive Veterinary Medicine* 3(1): 45-52.

Bull, R., Ross, R., Olson, D., Blecha, F. and Curtis, S. (1978). Effects of maternal protein restriction on blood composition and pathologic lesions in neonatal calves. *Journal of Animal Science* 47(Suppl. 1): 173.

Burton, J. H., Hosein, A. A., Mcmillan, I., Grieve, D. G. and Wilkie, B. N. (1984). Immunoglobulin absorption in calves as influenced by dietary protein intakes of their dams. *Canadian Journal of Animal Science* 64(5): 185-186.

CAFRE (2020). *Heifer Summary Report*. College of Agriculture, Food & Rural Enterprise, Antrim, Northern Ireland.

Carstens, G. E., Johnson, D. E., Holland, M. D. and Odde, K. G. (1987). Effects of prepartum protein nutrition and birth weight on basal metabolism in bovine neonates. *Journal of Animal Science* 65(3): 745-751.

Civra, A., Altomare, A., Francese, R., Donalisio, M., Aldini, G. and Lembo, D. (2019). Colostrum from cows immunised with a veterinary vaccine against bovine rotavirus displays enhanced in vitro anti-human rotavirus activity. *Journal of Dairy Science* 102(6): 4857-4869.

Cole, J., Hutchinson, J., Bickhart, D. and Null, D. (2013). Optimal age at first calving for US dairy cattle. *Journal of Dairy Science* 91(E-Suppl. 2): 289.

Connelly, M., Berry, D., Sayrs, R., Murphy, J., Lornz, I., O'Doherty, E. and Kennedy, E. (2014). Does iodine supplementation of the pre-partum dairy cow diet affect the serum IgG concentrations, iodine and health status of her calf? *Journal of Dairy Science* 8: 5120-5139.

Cooke, R. F. (2019). Effects on animal health and immune function. *Veterinary Clinics of North America: Food Animal Practice* 35(2): 331-341.

Corah, L. R., Dunn, T. G. and Kaltenbach, C. C. (1975). Influence of prepartum nutrition on the reproductive performance of beef females and the performance of their progeny. *Journal of Animal Science* 41(3): 819-824.

Correa, M. T., Curtis, C. R., Erb, H. N. and White, M. E. (1988). Effect of calf-hood morbidity on age at first calving in New York Holstein herds. *Preventive Veterinary Medicine* 6(4): 253-262.

Croyle, S. L., Belage, E., Khosa, D. K., LeBlanc, S. J., Haley, D. B. and Kelton, D. F. (2019). Dairy farmers' expectations and receptivity regarding animal welfare advice: a focus group study. *Journal of Dairy Science* 102(8): 7385-7397.

Cuttance, E. L., Mason, W. A., Laven, R. A. and Phyn, C. V. C. (2018). The relationship between failure of passive transfer and mortality, farmer-recorded animal health events and body weights of calves from birth until 12 months of age on pasture-based, seasonal calving dairy farms in New Zealand. *The Veterinary Journal* 236: 4-11.

Davis Rincker, L. E., VanderHaar, M. J., Wolf, C. A., Liesman, J. S., Chapin, L. T. and Weber Nielsen, M. S. (2011). Effects of intensified feeding of heifer calves on growth, pubertal age, calving age, milk yield, and economics. *Journal of Dairy Science* 94(7): 3554–3567.

Dawkins, M. (2019). Animal welfare as preventative medicine. *Animal Welfare* 28(2): 137–141.

Dechow, C. (2020). "Rightsize" replacement inventories. *Hoard's Dairyman* 165: 140.

DeNise, S. K., Robison, J. D., Stott, G. H. and Armstrong, D. V. (1989). Effects of passive immunity on subsequent production in dairy heifers. *Journal of Dairy Science* 72(2): 552–554.

De Vries, A. and Marcondes, M. I. (2020). Review: overview of factors affecting productive lifespan of dairy cows. *Animal* 14(S1): 155–164.

Donovan, G. A., Dohoo, I. R., Montgomery, D. M. and Bennett, F. L. (1998). Calf and disease factors affecting growth in female Holstein calves in Florida, USA. *Preventive Veterinary Medicine* 33(1–4): 1–10.

Drackley, J., Pollard, B., Dann, H. and Stamey, J. (2007). First-lactation milk production for cows fed control or intensified milk replacer programs as calves. *Journal of Dairy Science* 1: 614.

Dubrovsky, S., Van Eenennaam, A., Aly, S., Karle, B., Rossitto, P., Overton, M. and Lehenbauer, T. and Fadel, J. (2020). Preweaning cost of bovine respiratory disease (BRD) and cost-benefit of implementation of preventative measures in calves on California dairies: The BRD 10K study. *Journal of Dairy Science* 103: 1583–1597.

Dunn, A., Ashfield, A., Earley, B., Welsh, M., Gordon, A., McGee, M. and Morrison, S. J. (2017). Effect of concentrate supplementation during the dry period on colostrum quality and effect of colostrum feeding regimen on passive transfer of immunity, calf health and performance. *Journal of Dairy Science* 100(1): 357–370.

Eastham, N. T., Coates, A., Cripps, P., Richardson, H., Smith, R. and Oikonomou, G. (2018). Associations between age at first calving and subsequent lactation performance in UK Holstein and Holstein-Friesian dairy cows. *PLoS ONE* 13(6): e0197764.

Ebert, A., Heyn, E., Erhard, M. and Klee, W. (2007). Influence of method of colostrum application on IgG status, incidence of diseases, and weight gain in neonatal calves. In: *Proceedings of the 40th AABP*, p. 221.

Enjalbert, F. (2009). The relationship between trace elements stouts and health of calves. *Revue de Medecine Veterinaire* 160: 429–435.

Ettema, J. F. and Santos, J. E. P. (2004). Impact of age at calving on lactation, reproduction, health, and income in first-parity Holsteins on commercial farms. *Journal of Dairy Science* 87: 2730–2742.

Faber, S. N., Faber, N. E., McCauley, T. C. and Ax, R. L. (2005). Effects of colostrum ingestion on lactation performance. *Professional Animal Scientist* 21(5): 420–425.

Fenlon, C., O'Grady, L., Mee, J. F., Butler, S. T., Doherty, M. L. and Dunnion, J. (2017). A comparison of 4 predictive models of calving assistance and difficulty in dairy heifers and cows. *Journal of Dairy Science* 100(12): 9746–9758.

Franklin, S. T., Newman, M. C., Newman, K. E. and Meek, K. I. (2005). Immune parameters of dry cows fed mannan oligosaccharide and subsequent transfer of immunity to calves. *Journal of Dairy Science* 88(2): 766–775.

Funston, R. N., Larson, D. M. and Vonnahme, K. A. (2010). Effects of maternal nutrition on conceptus growth and offspring performance: implications for beef cattle production. *Journal of Animal Science* 88(13 Suppl.): E205–E215.

Furman-Fratczak, K., Rzasa, A.. and Stefaniak, T. (2011). The influence of colostral immunoglobulin concentration in heifer calves' serum on their health and growth. *Journal of Dairy Science* 94(11): 5536–5543.

Garnsworthy, P. (2012). The contribution of youngstock to production efficiency and environmental impact of dairy systems. In: *Proceedings of the XXVII World Buiatrics Congress*, Lisbon, Portugal, pp. 161–164.

Geiger, A. J., Parsons, C. L. M., James, R. E. and Akers, R. M. (2016). Growth, intake, and health of Holstein heifer calves fed an enhanced preweaning diet with or without postweaning exogenous estrogen. *Journal of Dairy Science* 99(5): 3995–4004.

Gilmore, J. and Mee, J. F. (2012). Transition period management of the suckler cow and calf. *Journal of the Irish Grassland Association* 43: 143–156.

Gleeson, D. and O'Brien, B. (2012). Effect of milk feed source, frequency of feeding and age at turnout on calf performance, live-weigh at mating and 1st lactation milk production. *Irish Veterinary Journal* 65(1): 18.

Gleeson, D. E., O'Brien, B. and Mee, J. F. (2007). Effect of restricting silage feeding prepartum on time of calving, dystocia and stillbirth in Holstein-Friesian cows. *Irish Veterinary Journal* 60(11): 667–671.

Godden, S. M., Haines, D. M., Konkol, K. and Peterson, J. (2009). Improving passive transfer of immunoglobulins in calves. II: Interaction between feeding method and volume of colostrum fed. *Journal of Dairy Science* 92(4): 1758–1764.

Gotowiecka, M., Nicpon, J., Twardon, J., Ochota, M. and Gotowiecka, M. (2012). The influence of alkalosis and acidosis in cows on the content of colostrum and the osmolarity of the gastric juice and blood in calves. In: *Proceedings of the Abstract Book of the XXVII World Buiatrics Congress*, Lisbon, Portugal, p. 208.

Gulliksen, S. M., Lie, K. I., Solverod, L. and Osteras, O. (2008). Risk factors associated with colostrum quality in Norwegian dairy cows. *Journal of Dairy Science* 91(2): 704–712.

Guyot, H., Spring, P., Andrieu, S. and Rollin, F. (2007). Comparative responses to sodium selenite and organic selenium supplements in Belgian Blue cows and calves. *Livestock Science* 111(3): 259–263.

Hall, J. A., Bobe, G., Vorachek, W. R., Estill, C. T., Mosher, W. D., Pirelli, G. J. and Gamroth, M. (2014). Effect of supranutritional maternal and colostral selenium supplementation on passive absorption of immunoglobulin G in selenium-replete dairy calves. *Journal of Dairy Science* 97(7): 4379–4391.

Hamilton, T. and Giesen, L. (1996). Effect of cow condition on calving, cow and calf behaviour, colostrum quality and rebreeding. Agriculture and Agri-Food Canada. *Ontario Beef Research Update*, pp. 30–31.

Hammer, C., Vonnahme, K., Taylor, J., Redmer, D., Luthar, J., Neville, L., Reed, J., Caton, J. and Reynolds, L. (2007). Effects of maternal nutrition and selenium supplementation on absorption of IgG and survival of lambs. *Journal of Dairy Science* 90(Suppl.1): 391.

Hammon, H. M., Liermann, W., Frieten, D. and Koch, C. (2020). Review: importance of colostrum supply and milk feeding intensity on gastrointestinal and systemic development in calves. *Animal* 14(S1): s133–s143.

Haworth, G. M., Tranter, W. P., Chuck, J. N., Cheng, Z. and Wathes, D. C. (2008). Relationships between age at first calving and first lactation milk yield, and lifetime productivity and longevity in dairy cows. *Veterinary Record* 162(20): 643–647.

Hayes, C. J., McAloon, C. G., Carty, C. I., Ryan, E. G., Mee, J. F. and O'Grady, L. (2019). The effect of growth rate on reproductive outcomes in replacement dairy heifers in seasonally calving, pasture-based systems. *Journal of Dairy Science* 102(6): 5599-5611.

Heinrichs, A. J. and Heinrichs, B. S. (2011). A prospective study of calf factors affecting first-lactation and lifetime milk production and age of cows when removed from the herd. *Journal of Dairy Science* 94(1): 336-341.

Heinrichs, A. J., Heinrichs, B. S., Harel, O., Rogers, G. W. and Place, N. T. (2005). A prospective study of calf factors affecting age, body size, and body condition score at first calving of Holstein dairy heifers. *Journal of Dairy Science* 88(8): 2828 -2835.

Hennessy, T. and Heanue, K. (2012). Quantifying the effect of discussion group membership on technology adoption and farm profit on dairy farms. *Journal of Agricultural Education and Extension* 18(1): 41-54.

Hogan, I., Doherty, M., Fagan, J., Kennedy, E., Connelly, M., Brady, P., Ryan, C. and Lorenz, I. (2015). Comparison of rapid laboratory tests for failure of passive transfer in the bovine. *Irish Veterinary Journal* 68(1): 18.

Holland, M., Speer, N., Le Fever, D., Taylor, R., Field, T. and Odde, K. (1993). Factors contributing to dystocia due to fetal malpresentation in beef cattle. *Theriogenology* 39: 898-908.

Hook, T. E., Odde, K. G., Aguilar, A. A. and Olson, J. D. (1989). Protein effects on fetal growth, colostrum and calf immunoglobulins and lactation in dairy heifers. *Journal of Animal Science* 67(S1): 539.

Hough, R. L., McCarthy, F. D., Kent, H. D., Eversole, D. E. and Wahlberg, M. L. (1990). Influence of nutritional restriction during late gestation on production measures and passive immunity in beef cattle. *Journal of Animal Science* 68(9): 2622-2627.

Hu, W., Hill, T. M., Dennis, T. S., Suarez-Mena, F. X., Aragona, K. M., Quigley, J. D. and Schlotterbeck, R. L. (2020). Effects of milk replacer feeding rates on growth, performance of Holstein dairy calves to 4 months of age, evaluated via a meta-analytical approach. *Journal of Dairy Science* 103(3): 2217-2232.

Hultgren, J., Svensson, C., Maizon, D. O. and Oltenacu, P. A. (2008). Rearing conditions, morbidity and breeding performance in dairy heifers in southwest Sweden. *Preventive Veterinary Medicine* 87(3-4): 244-260.

Hunter, A. (2015). *Association of Serum Calcium Status at Calving on Survival, Health and Performance of Post-Partum Holstein Cows and Calves. MSc*, Ohio State University.

Hyde, R. M., Green, M. J., Sherwin, V. E., Hudson, C., Gibbons, J., Forshaw, T., Vickers, M. and Down, P. M. (2020). Quantitative analysis of calf mortality in Great Britain. *Journal of Dairy Science* 103(3): 2615-2623.

Illek, J. (2010). The effect of organic and inorganic selenium supplementation on selenium concentrations in colostrum and blood of beef heifers and their calves. In: *Proceedings of the XXVI World Buiatrics Congress*, Chile, p. 89.

Jawor, P., Krol, D., Mee, J. F., Soltysiak, Z., Dzimira, S., Larska, M. and Stefaniak, T. (2017). Infection exposure, detection and causes of death in perinatal mortalities in Polish dairy herds. *Theriogenology* 103: 130-136.

Khan, M. A., Weary, D. M. and von Keyserlingk, M. A. (2011). Effects of milk ration on solid food intake, weaning and performance in dairy heifers. *Journal of Dairy Science* 94(3): 1071-1081.

Larson, D. M., Martin, J. L., Adams, D. C. and Funston, R. N. (2009). Winter grazing system and supplementation during late gestation influence performance of beef cows and steer progeny. *Journal of Animal Science* 87(3): 1147-1155.

Leyan, V., Wittwer, F., Contreras, P. and Kruze, J. (2004). Serum colostrum immunoglobulin concentrations from selenium deficient cows and in the blood of their calves. *Archivos de Medicina Veterinaria* 36: 155-162.

Lin, C. Y., McAllister, A. J., Batra, T. R., Lee, A. J., Roy, G. L., Vesely, J. A., Wauthy, J. M. and Winter, K. A. (1988). Effects of early and late breeding of heifers on multiple lactation performance of dairy cows. *Journal of Dairy Science* 71(10): 2735-2743.

Lombard, J. E., Garry, F. B., Tomlinson, S. M. and Garber, L. P. (2007). Impacts of dystocia on health and survival of dairy calves. *Journal of Dairy Science* 90(4): 1751-1760.

Lora, I., Barberio, A., Contiero, B., Paparella, P., Bonfanti, L., Brscic, M., Stefani, A. L. and Gottardo, F. (2018). Factors associated with passive immunity transfer in dairy calves: combined effect of delivery time, amount and quality of the first colostrum meal. *Animal* 12(5): 1041-1049.

Lorenz, I., Mee, J. F., Earley, B. and More, S. J. (2011). Calf health from birth to weaning. I. General aspects of disease prevention. *Irish Veterinary Journal* 64(1): 10.

Maccari, P., Wiedemann, S., Kunz, H. J., Piechotta, M., Sanftleban, P. and Kaske, M. (2015). Effects of two different rearing protocols for Holstein bull calves in the first 3 weeks of life on health status, metabolism and subsequent performance. *Journal of Animal Physiology and Animal Nutrition* 99(4): 737-746.

Mahmoud, M., Yin, T., Brugemann, K. and Konig, S. (2017). Phenotypic, genetic and single nucleotide polymorphism marker associations between calf diseases and subsequent performance and disease occurrences of first-lactation German Holstein cows. *Journal of Dairy Science* 100(3): 2017-2031.

Mann, S., Leal Yepes, F., Overton, T. and Nydam, D. (2013). The influence of different nutritional planes in the dry period on immunoglobulin G concentration of bovine colostrum. In: *Proceedings of the 46th AABP Conference*, Milwaukee, Wisconsin, p. 170.

McAloon, C., Doherty, M., Donlon, J., Lorenz, I., Meade, J., O'Grady, L. and Whyte, P. (2016). Microbial contamination of colostrum on Irish dairy farms. *Veterinary Record* 1-3. doi: 10/1136/vr.103641.

McCorquodale, C. E., Sewalem, A., Miglior, F., Kelton, D., Robinson, A., Koeck, A. and Leslie, K. E. (2013). Short communication: Analysis of health and survival in a population of Ontario Holstein heifer calves. *Journal of Dairy Science* 96(3): 1880-1885.

McGee, M., Drennan, M. and Caffrey, P. (2006). Effect of age and nutrient restriction pre partum on beef suckler cow serum immunoglobulin concentrations, colostrum yield, composition and immunoglobulin concentration and immune status of their progeny. *Irish Journal of Agricultural and Food Research* 45: 157-171.

Mee, J. F. (2004). The role of micronutrients in bovine periparturient problems. *Cattle Practice* 12: 95-108.

Mee, J. F. (2008). Prevalence and risk factors for dystocia in dairy cattle: a review. *The Veterinary Journal* 176(1): 93-101.

Mee, J. F. (2013). Why do so many calves die on modern dairy farms and what can we do about calf welfare in the future? *Animals* 3(4): 1036-1057.

Mee, J. F. (2014). Impacts of nutrition pre-calving on periparturient dairy cow health and neonatal calf health. *Recent Advances in Animal Nutrition* 45: 37-59.

Mee, J. F. (2018). Intensive care of the newborn dairy calf - knowledge into practice. In: *Proceedings of the 30th World Buiatrics Congress*, Sapporo, Japan, pp. 85-89.

Mee, J. F. (2020). Invited review: denormalising poor dairy youngstock management - dealing with 'farm-blindness'. *Journal of Animal Science* 98(S1): 140-S149.

Mee, J. F. and Kenneally, J. (2017). Why Do Calves Die at Calving and What Can You Do about It? In: *Irish Dairying Resilient Technologies*, Butler, S., Horan, B., Mee, J. F. and Dillon, P. (Eds), Teagasc, Moorepark, Ireland, pp. 150–151.

Mee, J. F. and Boyle, L. (2020). Invited review: Assessing whether dairy cow welfare is 'better' in pasture- than in confinement-based management systems. *New Zealand Veterinary Journal* 68(3): 168–177.

Mee, J. F., Berry, D. P. and Cromie, A. R. (2011). Risk factors for calving assistance and dystocia in pasture-based Holstein-Friesian heifers and cows in Ireland. *The Veterinary Journal* 187(2): 189–194.

Mee, J. F., Geraghty, T., O'Neill, R. and More, S. J. (2012). Bioexclusion of diseases from dairy and beef farms: risks of introducing infectious agents and risk reduction strategies. *The Veterinary Journal* 194(2): 143–150.

Mee, J. F., Grant, J., Sanchez-Miguel, C. and Doherty, M. (2013b). Pre-calving and calving management practices in dairy herds with a history of high or low bovine perinatal mortality. *Animals* 3(3): 866–881.

Mee, J. F., McCarthy, M. and O'Grady, L. (2018). Contract rearing – a disaster waiting to happen or a fundamental component of herd expansion? *Irish Veterinary Journal* 8: 474–476.

Mee, J. F., O'Farrell, K. J., Reitsma, P. and Mehra, R. (1996). Effect of a whey protein concentrate used as a colostrum substitute or supplement on calf immunity, weight gain and health. *Journal of Dairy Science* 79(5): 886–894.

Mee, J. F., Sanchez-Miguel, C. and Doherty, M. (2013a). An international Delphi study of the causes of death and the criteria used to assign cause of death in bovine perinatal mortality. *Reproduction in Domestic Animals*48(4): 651–659.

Mee, J. F., Sanchez-Miguel, C. and Doherty, M. (2014). Influence of modifiable risk factors on the incidence of stillbirth/perinatal mortality in dairy cattle. *The Veterinary Journal* 199(1): 19–23.

Miller, M. and Faust, M. (2000). Effects of performance and physiological characteristics of dairy heifers on first lactation yield and lifetime performance. *Journal of Dairy Science* 83(Suppl. 1): 230.

Moallem, U., Werner, D., Lehrer, H., Zachut, M., Livshitz, L., Yakoby, S. and Shamay, A. (2010). Long-term effects of ad libitum whole milk feeding prior to weaning and prepubertal protein supplementation on skeletal growth rate and first-lactation milk production. *Journal of Dairy Science* 93(6): 2639–2650.

Mock, T., Mee, J. F., Dettwiler, M., Rodriguez-Campos, S., Husler, J., Michel, B., Hafliger, I. M., Drogemuller, C., Bodmer, M. and Hirsbrunner, G. (2020). Evaluation of an investigative model in dairy herds with high calf perinatal mortality rates in Switzerland. *Theriogenology* 148: 48–59.

Modh Nor, N., Steenveld, W., Mourits, M. C. and Hogeveen, H. (2012). Estimating the costs of rearing young dairy cattle in the Netherlands using a simulation model that accounts for uncertainty related to diseases. *Preventive Veterinary Medicine* 106(3-4): 214–224.

Mohd Nor, N., Steeneveld, W., van Werven, T., Mourits, M. C. and Hogeveen, H. (2013). First-calving age and first lactation milk production on Dutch dairy farms. *Journal of Dairy Science* 96(2): 981–992.

More, S. J., Doherty, M. L. and O'Grady, L. (2017). An investigative framework to facilitate epidemiological thinking during herd problem-solving. *Irish Veterinary Journal* 70: 11.

Morgans, L., Bolt, S., van Dijk, L., Reyher, K. and Main, D. (2018). Farmer action groups - A participatory approach to instigating change on farm. In: *Proceedings of the 30th WBC*, Sapporo, Japan, p. 132.

Moriarty, J., Toolan, D., Casey, M., Barry, J. and McLaughlin, J. (2007). Stunting in cattle on three farms associated with calf-hood diseases. *Irish Veterinary Journal* 60: 742–746.

Morin, D. E., Nelson, S. V., Reid, E. D., Nagy, D. W., Dahl, G. E. and Constable, P. D. (2010). Effect of calving and management factors on colostral IgG concentration in dairy cows. *Journal of the American Veterinary Medical Association* 237(4): 420–428.

Morrill, K. M., Marston, S. P., Whitehouse, N. L., Van Amburgh, M. E., Schwab, C. G., Haines, D. M. and Erickson, P. S. (2010). Anionic salts in the prepartum diet and addition of sodium bicarbonate to colostrum replacer, and their effects on immunoglobulin G absorption in the neonate. *Journal of Dairy Science* 93(5): 2067 2075.

Morrison, S., Carson, A., Matthews, D., Kilpatrick, D., McCluggage, I. and Mulholland, M. (2010). Latest research on calf management for performance. In: *Proceedings of the AgriSearch Seminar on Dairy Genetics and Heifer Management*, Hillsborough, Northern Ireland, pp. 68–98.

Morrison, S., Scoley, G. and Barley, J. (2013). The impact of calf health on future performance. *Irish Veterinary Journal* 3: 264–268.

Morrison, S. J., Wicks, H. C., Carson, A. F., Fallon, R. J., Twigge, J., Kilpatrick, D. J. and Watson, S. (2012). The effects of calf nutrition on the performance of dairy herd replacements. *Animal* 6(6): 909–919.

Moscuzza, C., Milicich, H., Alvarez, G., Gutierrez, B. and Nahum, M. (2014). Calving assistance influences the occurrence of umbilical cord pathologies treated surgically in calves. *Turkish Journal of Veterinary and Animal Sciences* 38: 405–408.

Mousa, M., Seykora, A., Abdellatif, M. and Mousa, E. (2015). Genetic and phenotypic correlations of age at first calving with some calf-hood fitness and health traits in Holstein heifers. *Egyptian J. Animal Production* 52(Suppl. Issue): 37–43.

Nakao, T., Zhang, W., Kida, K., Moriyoshi, M. and Nakada, K. (2000). Effects of nutrition during gestation on birth weight and viability of calves, and weight and expulsion time of placenta in dairy cattle. In: *Proceedings of the 33rd Annual Conference of the American Association of Bovine Practitioners, Rapid City, South Dakota, United States of America*.

Nejedla, E., Stanek, S., Zouharova, M., Fleischer, P., Nedbalcova, K. and Slosarkova, S. (2019). Microbial quality of colostrum from Czech dairy herds. *Cattle Practice* May, p. 48.

Nowak, W., Mikula, R., Kasprowicz-Potocka, M., Ignatowicz, M., Zachwieja, A., Paczynska, K. and Pecka, E. (2012a). Effect of cow nutrition in the far-off period on colostrum quality and immune response of calves. *Bulletin of the Veterinary Institute in Pulawy* 56(2): 241-246.

Nowak, W., Mikula, R., Zachwieja, A., Paczynska, K., Pecka, E., Drzazga, K. and Slosarz, P. (2012b). The impact of cow nutrition in the dry period on colostrum quality and immune status of calves. *Polish Journal of Veterinary Sciences* 15(1): 77–82.

Overton, M. W. and Dhuyvetter, K. C. (2020). An abundance of replacement heifers: what is the economic impact of raising more than are needed? *Journal of Dairy Science* 103(4): 3828–3837.

Pereira, R., Siler, J., Sischo, W., Davis, M. and More, D. (2014). Intra-farm communication about calf health on US dairy farms. In: *Proceedings of the 28th WBC*, Cairns, Australia, p. 185.

Perry, V. E. A., Copping, K. J., Miguel-Pacheco, G. and Hernandez-Medrano, J. (2019). The effects of developmental programming upon neonatal mortality. *Veterinary Clinics of North America: Food Animal Practice* 35(2): 289–302.

Quigley, J. D. and Drewry, J. J. (1998). Nutrient and immunity transfer from cow to calf pre- and postcalving. *Journal of Dairy Science* 81(10): 2779–2790.

Quigley, J. D., Lago, A., Chapman, C., Erickson, P. and Polo, J. (2013). Evaluation of the Brix refractometer to estimate immunoglobulin G concentration in bovine colostrum. *Journal of Dairy Science* 96(2): 1148–1155.

Raeth-Knight, M., Chester-Jones, H., Hayes, S., Linn, J., Larson, R., Ziegler, D., Ziegler, B. and Broadwater, N. (2009). Impact of conventional or intensive milk replacer programs on Holstein heifer performance through six months of age and during first lactation. *Journal of Dairy Science* 92(2): 799–809.

Renaud, D. L., Waalderbos, K. M., Beavers, L., Duffield, T. F., Leslie, K. E. and Windeyer, M. C. (2020). Risk factors associated with failed transfer of passive immunity in male and female dairy calves: a 2008 retrospective cross-sectional study. *Journal of Dairy Science* 103(4): 3521–3528.

Robison, J. D., Stott, G. H. and DeNise, S. K. (1988). Effects of passive immunity on growth and survival in the dairy heifer. *Journal of Dairy Science* 71(5): 1283–1287.

Rytkonen, A., Hanninen, L. and Manninen, M. (2004). Effect of restricted feeding on the liveweight of suckler cows and their calves, quality of colostrum and calf serum IgG concentration. *Suomen Elainlaakarilehti* 110: 427–433.

Santos, J. E., DePeters, E. J., Jardon, P. W. and Hubert, J. T. (2001). Effect of prepartum dietary protein level on performance of primigravid and multiparous Holstein dairy cows. *Journal of Dairy Science* 84(1): 213–224.

Shalloo, L., Cromie, A. and McHugh, N. (2014). Effect of fertility on the economics of pasture-based dairy systems. *Animal* 8(Suppl. 1): 222–231.

Shamay, A., Werner, D., Moallem, U., Barash, H. and Bruckental, I. (2005). Effect of nursing management and skeletal size at weaning on puberty, skeletal growth rate and milk production during first lactation of dairy heifers. *Journal of Dairy Science* 88(4): 1460–1469.

Shell, T. M., Early, R. J., Carpenter, J. R. and Buckley, B. A. (1995). Prepartum nutrition and solar radiation in beef cattle: 2. Residual effects on postpartum milk yield, immunoglobulin, and calf growth. *Journal of Animal Science* 73(5): 1303 –1309.

Shivley, C. B., Lombard, J. E., Urie, N. J., Haines, D. M., Sargent, R., Kopral, C. A., Earlywine, T. J., Olson, J. D. and Garry, F. B. (2018). Preweaned heifer management on US dairy operations: Part II. Factors associated with colostrum quality and passive transfer status of dairy heifer calves. *Journal of Dairy Science* 101(10): 9185–9198.

Silper, B., Madureira, A., Burnett, T., de Passeille, A., Rushen, J. and Cerri, R. (2013). Relationships between pre- and postweaning growth on estrus behavior and reproductive parameters of Holstein heifers. *Journal of Dairy Science* 91(E-Suppl. 2): 671.

Soberon, F., Raffrenato, E., Everett, R. W. and Van Amburgh, M. E. (2012). Preweaning milk replacer intake and effects on long-term productivity of dairy calves. *Journal of Dairy Science* 95(2): 783–793.

Soberon, F. and Van Amburgh, M. E. (2013). The effect of nutrient intake from milk or milk replacer of preweaned dairy calves on lactation milk yield as adults: a meta-analysis of current data. *Journal of Animal Science* 91(2): 706–712.

Stanton, A. L., Kelton, D. F., LeBlanc, S. J., Wormuth, J. and Leslie, K. E. (2012). The effect of respiratory disease and a preventative antibiotic treatment on growth, survival, age at first calving, and milk production of dairy heifers. *Journal of Dairy Science* 95(9): 4950–4960.

Steele, M. (2020). Age at first calving in dairy cows: which months do you aim for to maximise productivity? *Veterinary Evidence* 5(1): 1–22.

Stockdale, C. and Smith, C. (2004). Effect of energy and protein nutrition in late gestation on immunoglobulin G in the colostrum of dairy cows with varying body condition scores. *Animal Production in Australia* 25: 176–179.

Svensson, C. and Hultgren, J. (2006). The effect of calf rearing factors on first-lactation milk production. In: *Proceedings of the World Buiatrics Congress*, Nice, France, CD-ROM, OS37-1.

Svensson, C. and Hultgren, J. (2008). Associations between housing, management, and morbidity during rearing and subsequent first-lactation milk production of dairy cows in Southwest Sweden. *Journal of Dairy Science* 91(4): 1510–1518.

Taylor, P. D. and Poston, L. (2007). Developmental programming of obesity in mammals. *Experimental Physiology* 92(2): 287–298.

Terre, M., Tejero, C. and Bach, A. (2009). Long-term effects on heifer performance of an enhanced-growth feeding programme applied during the preweaning period. *Journal of Dairy Research* 76(3): 331–339.

Vaarst, M. and Sorensen, J. T. (2009). Danish farmers' perceptions and attitudes related to calf management in situations of high versus no calf mortality. *Preventive Veterinary Medicine* 89(1–2): 128–133.

Van der Fels-Klerx, H. J., Martin, S. W., Nielen, M. and Huirne, R. B. M. (2002). Effects on productivity and risk factors of bovine respiratory disease in dairy heifers: a review for the Netherlands. *NJAS – Wageningen Journal of Life Sciences* 50(1): 27–45.

Van Hese, I., Goossens, K., Vandaele, L. and Opsomer, G. (2020). Invited review: microRNAs in bovine colostrum – focus on their origin and potential health benefits for the calf. *Journal of Dairy Science* 103(1): 1–15.

Waltner-Toews, D., Martin, S. W. and Meek, A. H. (1986). The effects of early calf-hood health status on survivorship and age at first calving. *Canadian Journal of Veterinary Research* 50(3): 314–317.

Warnick, L., Erb, H. and White, M. (1994). The association of calf-hood morbidity with first lactation calving age and dystocia in New York Holstein herds. *Kenya Veterinarian* 18: 177–179.

Warnick, L. D., Erb, H. N. and White, M. E. (1995). Lack of association between calf morbidity and subsequent first lactation milk production in 25 New York Holstein herds. *Journal of Dairy Science* 78(12): 2819–2830.

Warnick, L., Erb, H. and White, M. (1997). The relationship of calf-hood morbidity with survival after calving in 25 New York Holstein herds. *Preventive Veterinary Medicine* 31: 263–266, 268–273.

Williams, D. R., Pithau, P., Garcia, A., Champagne, J., Haines, D. M. and Aly, S. S. (2014). Effect of three colostrum diets on passive transfer of immunity and preweaning health in calves on a California dairy following colostrum management training. *Veterinary Medicine International* 2014: 1-9. Article ID 98741.

Zanton, G. I. and Heinrichs, A. J. (2005). Meta-analysis to assess effect of prepubertal average daily gain of Holstein heifers on first-lactation production. *Journal of Dairy Science* 88(11): 3860–3867.

# Chapter 10

## Managing replacement and culling in dairy herds

*Albert De Vries, University of Florida, USA*



### 1 Introduction

As all life ends, so does that of the dairy animal. It is estimated that the natural longevity of dairy cows is approximately 20 years (Nowak and Walker, 1999). The average actual lifespan is approximately only 4.5-6 years in most developed dairy industries (De Vries and Marcondes, 2020). When the approximate age at first calving is about 2 years, the productive lifespan after first calving can be taken to be 2.5- 4 years. Before 1950, the average length of productive life was about 3.5 years in the USA and various other countries (Asdell, 1951).

This chapter focuses on factors that affect replacement and culling in dairy herds with a focus on cows. The act of culling is simple, but the risk factors and economic considerations are complex. The chapter first presents some data on culling risks and reasons, then explores more in depth the effects of poor health on culling, and finally presents aspects of economic decision-making regarding culling and replacement decisions.

http://dx.doi.org/10.19103/AS.2020.0086.12

## 2 Culling definitions, culling risks and culling reasons

Culling (exiting) is the departure of cows from the herd because of sale, slaughter (salvage) or death (Fetrow et al., 2006). Sale implies that the cow was sold alive with the expressed goal of continuing to provide income while alive, such as by producing milk. Slaughter is a term used for cow sales intended for human consumption, while salvage is used for disposal for nonhuman consumption. Salvaged cows may have been in the withholding period after drugs were given, been culled shortly after major surgery, or been culled with diseases that exclude them from human consumption. The distinction between these two categories may not be known on the dairy farm. Finally, some cows die on the farm.

Over 3.9 million milk recording lactation records (slightly less than half of all milk cows) by the Dairy Herd Information Association (DHI) in the USA show that in 2019 on average 5.2% of cows exited dairy farms for sale, 26.1% slaughtered and 5.3% died (Norman et al., 2020) for a total herd turnover rate of 37%. Excluding the cows that exited herds for sale, 31% of cows were left with no productive future. This annual cull rate (defined here as herd turnover rate based on slaughter or death but not sale) varied between 25% for crossbreds and 34% for Guernsey, and was 32% for Holsteins (80% of all records). Focusing on Holsteins, the annual cull rate increased from 22% in parity 1 to 61% in parity 6. As cows become older, their annual risks of slaughter and death both increase. The cull rate for slaughter increased from 19% in parity 1 to 50% in parity 6. The annual cull rate for death increased from 3% in parity 1 to 12% in parity 6. Given the annual cull rate of 32%, the average productive life therefore was 1/32% = 3.1 years until culling for slaughter or death. Other studies based on large datasets report annual cull rates that are several percentage points different from these numbers (De Vries and Marcondes, 2020). The Council on Dairy Cattle Breeding reported productive lifespan of 28.4 months in 2018, whereas this was 34.4 months for cows born in 1975 (Norman, 2019), despite the greater ability of cows to avoid forced culling. The variation in annual cull rates among dairy farms is large. The total lifespan of dairy cattle is, on average, only a quarter of the natural longevity.

Annual culling rates in other countries than the USA are similar or tend to be somewhat lower. For example, in Canada the average reported annual cull rate for 2019 was 33% (slaughter only) and 41% when all reasons for exiting herds were included (Canadian Dairy Information Center, 2020). In the Netherlands, an annual cull rate of 28% (slaughter) was reported for 2010 (Mohd Nor et al., 2014), which appears to have slightly increased to approximately 30% (CRV, 2020). For Estonia, an overall culling rate of 26% was reported (Rilanto et al., 2020). Annual cull rates in New Zealand appear to be somewhat lower, based on a greater percentage of cows surviving each lactation (DairyNZ and LIC,

2019). Annual cull rates for pasture-based beef cows are approximately only 10 –15% (Ramsey et al., 2017). Intense selection for milk yield has been blamed for higher metabolic demand experienced by high-yielding cows, particularly during early lactation, which negatively affects their health and survival (Rauw et al., 1998; Bergeå et al., 2016). This negative relationship is debated, however.

Milk recording programs often allow farmers to choose the most important cull reason from a predefined list. In the 2019 DHI data (Norman et al., 2020), the main reason in the category slaughter was unspecified (36%), followed by low production (26%), reproductive problems (17%), mastitis of high somatic cell count (16%) and locomotion (5%). The primary cull reason was death (21% of all culling), followed by reproduction (18%), injury/other (14%) and low production and mastitis (both 12%) in a dataset of cows calving between 2001 and 2006 in 2054 herds in the eastern USA (Pinedo et al., 2010). Hadley et al. (2006) reported that up to 80% of all culling was due to health issues. In Canada, primarily culling reasons for slaughter are unknown (24%), followed by reproductive reasons (16%), mastitis (10%), low milk yield (8%) and feed and leg problems (6%). Other several poor health-related reasons sum up to an additional 9%. In the Netherlands, primary reasons for culling were reproduction (19%), mastitis or high cell count (18%) and feed and legs (15%) (WUR-LR, 2013). Estonian farmers stated that primary reasons for culling were feet/claw disorders (26%), udder disorders (23%), metabolic and digestive disorders (18%) and fertility problems (13%) (Rilanto et al., 2020). Culling reasons are associated with culling risk depending on parity, stage of lactation and reproductive status (Pinedo et al., 2010). Cows generally avoid culling when they get pregnant on time, are healthy, produce enough quality milk and have functional udders, feet and legs (Fetrow et al., 2006).

These primary culling reasons point to failures in health, reproduction, locomotion or udder functionality as main causes of cows leaving the herd. Reported culling reasons cannot be reliably compared across farms, however. Many farms report no or just a few reasons (Pinedo et al., 2010). Cases are not uniformly defined, and often culling is a decision based on multiple factors and causes (Bascom and Young, 1998). Relying upon culling records and reasons to monitor disease is, therefore, an ineffective management strategy (Fetrow et al., 2006).

## 3 Poor health and conformation as risk factors for culling

Poor health is a risk factor for culling. For example, lactations with a diagnosis of downer cow, clinical mastitis or teat problems were 3.5, 2.0 and 2.7 times, respectively, more likely to end in culling than lactations without the disease in New York dairy herds (Milian-Suoza et al., 1988). Cows that were culled following being diagnosed with milk fever, downer cow syndrome, left displaced abomasum, teat problems, foot and leg program and some cases of clinical mastitis were often culled within 30 days of the diagnosis (Milian-Suoza et al., 1988). Beaudeau et al. (2000) presented a table with the impact of 13 health disorders on the risk of culling. All except one study on ketosis showed increases in or no effects on culling risk. Although these results are somewhat dated now, diseases generally increase the risk of culling. In one herd of cows that had a single case of clinical disease developed within 21 days after calving had a 79% greater risk of culling than cows without a clinical disease in the first 305 days after calving. Cows with multiple clinical diseases had a 206% greater risk of culling in the first 305 days (Carvalho et al., 2019). At the time of culling, the primary culling reason reported may not be linked to the health disorder that predisposed the cow to culling, however. A brief literature survey of some health and conformation factors follows.

Fresh cow diseases like retained placenta, metritis, displaced abomasum and ketosis are associated with the cull reason 'disease'. Fresh cow diseases lead to lower milk production and lower reproduction rate. Affected cows may be culled later in lactation, and the link with the primary disease cause may not be made. For example, several studies reported no significant effect of retained placenta on culling risk (Dubuc et al., 2011; Probo et al., 2018). Retained placenta has been reported as a risk factor for metritis (LeBlanc, 2008), resulting in lower reproductive performance, and was therefore indirectly linked to increased culling.

Metritis has been associated with lower milk production and increased days between first service and conception. Cows with metritis had greater risk to be culled when compared with cows without metritis (Oltenacu et al., 1990). Other studies did not find an association between metritis and culling risk (Probo et al., 2018; Dubuc et al., 2011; Hertl et al., 2011).

Some studies (Seifi et al., 2011; Probo et al., 2018) reported that ketosis was associated with increased risk of culling in multiparous cows; however, a study by Hertl et al. (2011), conducted on primiparous cows, did not find an association between ketosis and culling.

Milk fever is strongly associated with the risk for culling within the first 120 days after calving, as revealed by some studies (Probo et al., 2018; Seifi et al., 2011). Still others like Cha et al. (2013) did not observe any relationship between milk fever and culling. There are positive associations between milk fever and other peripartum diseases such as dystocia and metritis (Rajala-Schultz and Gröhn, 1999).

Cows with displaced abomasum often have higher risks of culling (Seifi et al., 2011; Cha et al., 2013). Displaced abomasum was the second most determinant metabolic disease for culling, followed by mastitis and ketosis in one study (Probo et al., 2018).

The daily risk of culling with primary reason 'mastitis' closely follows the occurrence of clinical mastitis throughout lactation (Pinedo et al., 2010). The occurrence of mastitis was the disease factor that most influenced culling in dairy herds in New York (Gröhn et al., 1998). Culling with primary cause mastitis may not be reported as such. Cows with subclinical mastitis produce less milk, are more difficult to get pregnant and have a greater risk of miscarriage (Dahl et al., 2020). A review of the literature revealed that on average 26% of the total failure costs of mastitis (the cost incurred when mastitis occurs, excluding expenses to prevent mastitis) was due to culling. However, estimates differed widely among studies, ranging from 0% to 60% (Hogeveen et al., 2019).

Pregnancy greatly reduces the risk of culling (Cha et al., 2013; De Vries et al., 2010). The risk of culling for failure to get pregnant increases with the stage of lactation (Pinedo et al., 2010). Failure to get pregnant is linked to several preceding disorders such as twinning, dystocia (Roberts et al., 2012), metritis and retained placenta. Failure to get pregnant can also be related to management factors such as poor estrus detection, low body condition score and insufficient heat abatement. Cows with twins had higher culling rates (Bicalho et al., 2007; Probo et al., 2018) than cows with singletons. The risk of culling increased for cows that had greater calving difficulty, gave birth to males, were in herds with fewer days to the first insemination or had longer days to conception (De Vries et al., 2010). Cows that became pregnant later in lactation had increased risk of culling in the subsequent lactation (Pinedo and De Vries, 2010). Lameness is known to negatively impact reproductive efficiency and milk production and is often a reason for culling in dairy herds (Bicalho et al., 2007).

Functional udders and feet and legs are an important factor for a dairy farmer to keep cows. Cows that are difficult to milk in the parlor are at a greater risk to get culled. Correct conformation especially of udder depth, fore udder attachment, front teat placement and udder support protect against culling of cows (Caraviello et al., 2003).

Forced culling as preceded by poor health or lameness can be considered as an indicator of animal welfare (De Vries et al., 2011; Langford and Stott, 2012; Rilanto et al., 2020). A reduction in forced culling through better health will likely be associated with better animal welfare, which is of growing importance to society (Von Keyserlingk et al., 2009; Barkema et al., 2015).

## 4 Herd effects on the risk of culling

Older studies showed large herd effects on the risk of culling. This implies that culling of cows for health disorders does not only depend on individual factors such as age, stage of lactation, severity of the disease and performance but also on factors lying outside of the individual cow, such as the availability of replacements, milk price and slaughter price (Beaudeau et al., 2000). Herd-level factors also include factors such as herd size, facilities design, organic production and labor. For example, herds with fewer cows per employee and a greater percentage of labor supplied by family members tended to have a lower risk of forced culling of profitable cows (Weigel et al., 2003). These authors also found that high-producing cows in herds with fans, sprinklers, self-locking manger stalls, palpation rails and maternity pens had a lower risk of culling than cows in herds without such facilities.

## 5 Heifers

Approximately 10–20% of dairy female calves born alive and intended to become cows do not reach the first calving. Morbidity and mortality in preweaned dairy heifer calves based on various health, feeding and management practices, as well as environmental factors, were recently summarized by Urie et al. (2018). In this study, calves born with a higher birth weight had a lower risk of mortality. An increase in serum IgG concentration decreased the risk of mortality. Mortality was greater in calves fed ≤0.15 kg of fat/d in the liquid diet compared with calves fed ≥0.22 kg of fat/d. Calves that experienced any disease throughout the preweaning period had 4.7 times higher odds of mortality than in calves with no disease. Calves born to dams having dystocia were at greater risk of culling by 120 days after birth (Lombard et al., 2007).

In addition, factors early in the female's life before first calving affect her risk of culling after first calving (Bach, 2011; Heinrichs et al., 2017). For example, calf growth is associated with lifetime production and affects culling (Heinrichs and Heinrichs, 2011; Soberon and Van Amburgh, 2013). Calves with at least one respiratory disease treatment in the first 120 days of life produced less milk in the first lactation and doubled the likelihood of failing to reach the first lactation (Schaffer et al., 2016). Heat stress during late gestation in cows negatively affects the performance of the cow and her calf (Tao et al., 2019) and leads to an increased risk of culling. The environment at conception affects the female conceptus's performance after she calves and her risk of culling (Pinedo and De Vries, 2017). In one herd, parity and health status of the dam in the previous lactation were associated with morbidity, survival and performance of their daughters from birth through 305 days in milk of the first lactation (Carvalho et al., 2020). Documenting the impact of early life events and performance on lifetime performance is an active area of research currently.

## 6 Cost of culling

Cows that are culled are typically replaced with calving heifers. The cost to raise a heifer may be over $2000 per head in the USA (Tranel, 2019), but varies greatly between farms. Purchase prices of heifers depend on supply and demand and have been as low as $1000 in 2019. In the western USA, the cost of a purchased heifer was approximately $1700 and the price received for a slaughter cow was approximately $800 in 2017 (Frazer, 2018). Cows are well into their second lactation before their purchase or raising costs have been recovered through milk sales.

Cow depreciation is the basis of herd replacement costs. Herd replacement costs, defined as the cost to maintain a fixed herd demographic, are approximately 10% of the total cost of operations on dairy farms in the USA (Frazer, 2018). In 2017, these costs were on average $3.68 per 100 kg of milk sold and $424 per cow per year. Herd replacement costs are calculated as the annual expenses to obtain calving heifers minus the revenues from culled cows, divided by the amount of milk sold. In these data (Frazer, 2018), annual cull rate was 38%, equivalent to 31.4 months of productive life when assuming that culled cows were slaughtered.

The cost of a forced cull due to a health problem is the opportunity cost of not keeping the cow in the herd until the optimal time of replacement (Huirne et al., 1997). To determine this cost, one needs to estimate the cash flow that is generated from keeping the healthy cow until a future (economically optimal) time of replacement and subtract the estimated cash flow following the immediate cull and replacement. In both cases, the cash flow from replacement heifers needs to be included until the cull decision made today has no longer any effect on the cash flow at some point in the future. For cull vs. keep decisions, this is approximately 5 years (De Vries, 2006). Dynamic programming has often been used to estimate future cash flows and determine the future optimal time of replacement (Groenendaal et al., 2004; De Vries, 2006). The difference between both cash flows is termed future value, cow value, keep value or retention pay-off. The higher the keep value, the greater the value to keep the cow in the herd. A negative keep value indicates that culling (and replacement) is economically optimal. The keep value is also the amount of money that can be spent on a sick cow if she would immediately and fully recover. The keep value is generally high after calving and then gradually declines with days after calving if the cow fails to get pregnant. Pregnancy increases the keep value and then generally stays hundreds of dollars above $0. De Vries and Marcondes (2020) contains a figure that is a schematic trajectory of keep values (retention payoffs) by days in milk, level of milk yield and pregnancy status. Higher-producing cows have greater keep values than low-producing cows, which decrease below $0 sooner. Keep values make clear that forced culling early in lactation is very expensive.

Figure 1 shows keep values for a dairy herd of 2531 dairy cows where 510 cows are open, 757 are bred and 1264 cows are pregnant. The average keep value is $899. There is a large variation within a day after calving and reproductive status. The keep values in this dairy herd are used to rank cows

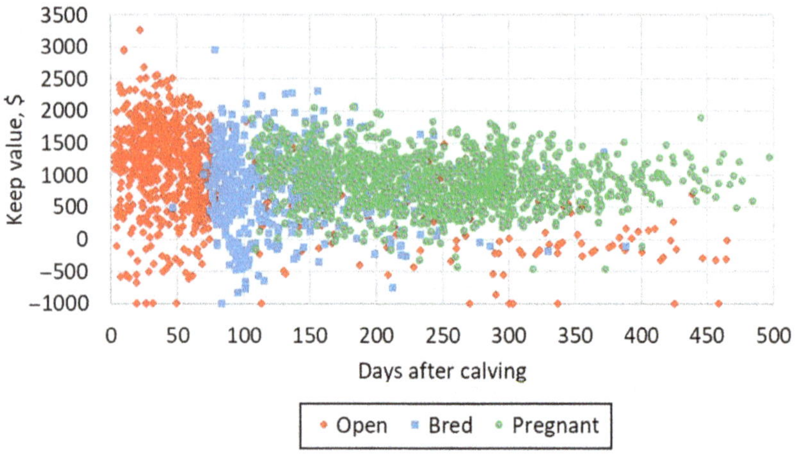


**Figure 1** Keep values for a herd of 2531 dairy cows. The average keep value is $899. A keep value below $0 means it is economically best to replace the cow with a calving heifer.

for culling decisions. Nevertheless, cows with negative keep values are not necessarily immediately culled and cows with positive keep values are not necessarily kept. Culling depends on the availability of replacement heifers and is also based on factors that are not included in the calculation of the keep values. For example, salvage value is only roughly estimated because individual body weights are not available. Body condition score is also not included (because not available) but affects dairy cow production and health (Roche et al., 2009) and is subjectively included by dairy producers when making culling decisions. This limitation also exists with the lack of past and current disease events in the prediction of future cash flows. Although the principle of using keep decisions to inform treatment vs. culling decisions is valid, improvements in the accuracy of cash flow projections are necessary, for example, with the help of precision dairy technologies (De Vries, 2016).

## 7 Economic decision-making

Traditionally, culling for reasons other than low milk production and sale are labeled as involuntary, as if the farmer had no choice. However, only a small part of these involuntary culls are truly forced in the sense that no possible productive future exists for the cow, for example because she is permanently sterile, irreparably injured and positive for tuberculosis (Fetrow et al., 2006). Economic culls are those cows for which a decision has been made that replacing them with another cow increases profitability for the dairy farm. Culling is primarily the outcome of an economic decision-making process. This process does not necessarily lead to the same culling decision by different farmers. Culling criteria and decision-making are based on sociological variables (attitudes, education, degree of involvement in dairy groups, demographic characteristics and so forth) in addition to economic or forced criteria (Beaudeau et al., 1996). Over 70 years ago, it was recognized that most of the variation in culling was due to economic conditions (Asdell, 1951).

As noted earlier, culling is primarily the outcome of an economic decision-making process (Fetrow et al., 2006). This process is driven by cow characteristics (current status and outlook for reproduction, milk yield, locomotion, disease, salvage value, etc.), is associated with sociological variables (Beaudeau et al., 1996) and is dependent on external factors such as the cost and availability of replacement heifers or cows (Overton and Dhuyvetter, 2020).

Improvement in the health, fertility and functional ability of cows will result in fewer culls that are truly forced but may not necessarily lead to different cull rates. One reason is that advancements in transition and reproductive management, coupled with the widespread use of sexed semen in dairy heifers and cows, have led to an oversupply of dairy replacement heifers in the USA (Overton and Dhuyvetter, 2020). This oversupply results in the culling of cows that would otherwise have been kept longer to make room for calving heifers. The oversupply can be reduced by breeding fewer replacement heifers, either by selling surplus heifers earlier in life or by more breeding with beef semen. Beef semen use in dairy herds has doubled from 2017 to 2019 in the USA (NAAB, 2020) but is still a relatively new phenomenon. The dairy industry in the USA is still searching for the optimal supply of heifers. The use of beef semen in dairy herds has been common longer in other countries such as the Netherlands and may have led there earlier to a reduction in the number of dairy heifers and consequently lower cull rates.

Another reason for higher cull rates, related to the issue of an oversupply of heifers, is the belief that improved genetic merit in heifers warrants a high cow replacement rate. A high rate results in a young herd with superior genetics on average compared to an older herd. Genetic merit for individual traits such as milk production and economic selection indices has indeed been increasing for decades in the USA (Cole and VanRaden, 2018; CDCB, 2020), although the trends are not favorable for all traits. The advent of routine use of genomic testing in 2009 has resulted in a reduction in the generation interval for bull selection (García-Ruiz et al., 2016) and consequently a doubling in the rate of increase of genetic merit of an economic selection index such as Lifetime Net Merit (CDCB, 2020). In Canada, the average rate of genetic gain of the selection indices - Lifetime Performance Index and Pro$ - has also more than doubled since 2009 (Lactanet, 2019).

Many genetic evaluation organizations have defined traits of dairy cows that directly protect against culling. In the USA, this trait is Productive Life. In Canada, a similar trait is Herd Life. The average breeding value for a productive life for cows has increased by approximately 18 months since 1960 (CDCB, 2020). This implies that today's cows would stay in herds 18 months longer than cows in 1960, adjusted for changes in other traits related to culling such as milk yield, fertility, etc., if these old and new cows would be in herds at the same time. Assuming that the average annual cull rate in 1960 was similar to that of 1950 (29%; Asdell, 1951), an increase of 18 months in the breeding value of productive life would result in an annual cull rate of 1/(1/29% × 12 months + 18 months) = 20% today. This low annual cull rate has not been realized today, which is more evidence that factors other than cow characteristics affect culling decisions. Alternatively, the cows from 1960 would be culled much quicker today because the competition of improved cows is much stronger.

Economic theory says that assets (cows) should be replaced sooner when the challenger (heifers) is technically superior (Groenendaal et al., 2004). An increase in the rate of change of an economic selection index should therefore lead to a higher cull rate. The effect of a doubling of the rate of change on the economically optimal cull rate was shown to be several percentage points at most (De Vries, 2020). However, this result depends on many assumptions and may not include a full accounting of all consequences of an increased rate of change. Nevertheless, an increase in culling driven by an increase in the rate of genetic gain is in opposition with an emerging perception in society that productive life should be prolonged.

Many cows are prematurely culled for reasons that can be traced back to health disorders including non-pregnancy. Keep values are an indication of the financial loss of a forced cull and the expenses that can be made to return the cow back to health.

Improving health in a dairy herd leads to reduction in forced culling and consequently changes the herd structure. This does not, however, imply that lower annual herd replacement rates are always economically optimal. Following some literature review, economically optimal cull rates have been estimated to be below 30% in the past (De Vries, 2017). Past estimates assumed that forced culling was at least 20% and could not be affected by the dairy farmer making the culling decision. None of the reviewed models includes the increased rate of genetic merit observed in the past 10 years, and other inputs such as the risk of forced culling and prices are dated as well.

An illustration of the key factors that determine the optimal herd replacement rate from an economic perspective was presented in De Vries (2020). This simple model assumes that cows are healthy and calve once per year. The total cost of herd structure in this model is the sum of 1) replacement cost, 2) lack of maturity cost, 3) aged cow cost, 4) genetic opportunity cost and 5) calf value opportunity cost. In this model, the probability of culling is the same in every lactation. This probability is varied such that a target annual herd replacement rate, for example, 30%, is reached. This assumption implies that culling is random. For simplicity, one lactation is set to 1 year. The economically optimal herd replacement rate then is the one where the sum of the five costs is the lowest. The model illustrates the trade-offs between the factors that determined the optimal herd replacement rate. The five cost factors are explained as follows. Reasonable values are assumed, and the final result is shown in Fig. 2.

### *7.1 Replacement cost*

Replacement cost is calculated as the heifer replacement cost minus the average price received when the cow leaves the herd, divided by the time the cow spent in the herd. For example, assuming \$1700 for a replacement heifer and \$800 for exit price, the replacement cost is \$360/year when the herd replacement rate is 40% and \$180/year when the herd replacement rate is 20%. The total replacement cost per year decrease with a longer productive life.

### *7.2 Lack of maturity cost*

'Lack of maturity cost' is associated with younger cows that are in their first through fourth lactation and are less efficient milk producers than mature cows. Dairy cows reach maturity around their fifth lactation (Mißfeldt et al., 2015) and have then the highest annual net revenue calculated as milk sales minus feed cost. Milk production in first lactation cows may be 80% or less of maturity. Feed intake will also be lower than in mature cows. Based on data from several dairy farms, lack of maturity cost in first through fourth lactation

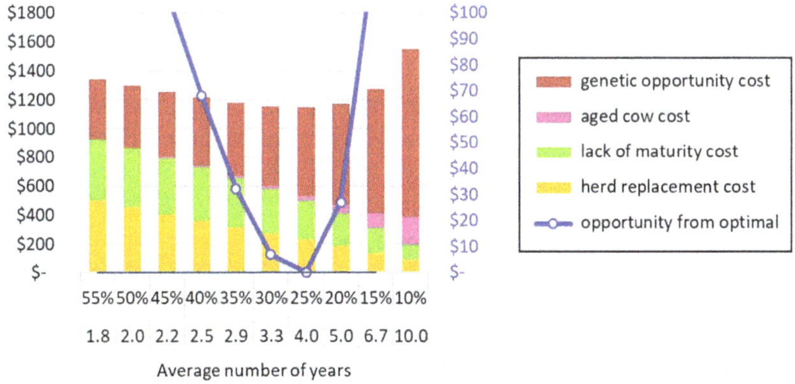


**Figure 2** An illustration of four key drivers of the total cost of maintaining herd structure: genetic opportunity cost, herd replacement cost, lack of maturity cost and aged cow cost. The optimal herd replacement rate is the one where the total cost is the lowest - 25% in this example. The opportunity cost from optimal is the increase in cost per cow per year from the optimal herd replacement rate.

animals were set at $500, $428, $340 and $182, respectively, compared to mature fifth and sixth lactation cows ($0). Lack of maturity costs is assumed to be free from genetic gain over time. Herds with greater herd replacement rates have relatively more cows in their first through fourth lactation compared to older cows and, therefore, there is a greater lack of maturity cost per cow in the herd. In the simple model, lack of maturity cost is $368/year when the herd replacement rate is 40% and $228/year when the herd replacement rate is 20%. Total lack of maturity cost decreases with a lower herd replacement rate.

## 7.3 Aged cow cost

Aged cow cost is the result of older cows in the herd that are in their seventh or later lactations. These cows are assumed to have lower net revenues (milk sales minus feed cost) compared to mature cows due to advanced age. It was assumed that aged cow cost increased by $50/year after the sixth lactation. For example, cows in their ninth lactation were assumed to have $150 aged cow cost compared to mature cows ($0). Aged cow cost is difficult to estimate in practice because few cows remain past their sixth lactation and those that do have survived culling for low production in their earlier lactations. Aged cow cost is also assumed to be free from genetic gain over time. In the simple model, aged cow cost was $6 per year when the herd replacement rate was 40%. Aged cow cows was 57% when the herd replacement rate was 20%. These values are low because few animals reach the aged cow stage. Total aged cow costs increase with lower herd replacement rates.

## 7.4 Genetic opportunity cost

Genetic opportunity cost is the loss from keeping older, genetically less improved cows in the herd instead of younger, genetically more improved cows. Genetic gain was set at $70 per year in predicted transmitting ability for Lifetime Net Merit, based recent trends (CDCB, 2020). This rate is approximately double the rate of gain observed 15 years ago. The $70 predicted transmitting ability equals $140 estimated breeding value per year. Lifetime is approximately 3 years, thereby making the genetic opportunity cost approximately $47 per year. With the genetic opportunity cost of the sire of the cow set to $0, the genetic opportunity cost of the cow is $480 per year when the herd replacement rate was 40% and $706 per year when the herd replacement rate was 20%. Genetic opportunity cost increases with an older herd and is a major argument used by dairy farmers to justify high herd replacement rates (Norman, 2019).

### 7.5 Calf value opportunity cost

Finally, calf value opportunity cost is the cost of not producing higher valued calves that can be sold because the calf is needed to replace the cow to be culled. An example is crossbred (beef sire, dairy dam) calves that receive a premium when sold. Calf value opportunity cost decreases with lower herd replacement rates, but their influence is small (De Vries, 2020). For simplicity, calf value opportunity cost will not be considered further in this chapter.

Using the first four drivers as the total cost of herd structure, Fig. 2 shows an optimal herd replacement rate of 25%. This is reached when the sum of the four cost factors is the lowest. The total cost per cow per year is $7 greater when herd replacement rate is 30% and $27 greater when the herd replacement rate is 20%. Fig. 2 shows that genetic gain itself does not warrant the 35-40% herd replacement rate often observed in the USA and when forced culling is low. These results should be interpreted cautiously because the model is simple and individual farm conditions vary.

A sensitivity analysis of the four key inputs is shown in Table 1. The first row is equal to the results in Fig. 2. Without herd replacement cost, the optimal herd replacement rate is the maximum considered (50%). Excluding genetic opportunity cost, the optimal herd replacement rate is the minimum considered (10%). Setting all inputs at 50% of their default values reduced the opportunity costs to 50% of the default, and the optimal herd replacement rate remains 25%. Six of the eight input combinations evaluated put the optimal herd replacement rate between 20% and 30%.

## 8 Environmental impact

Increased productive life will reduce the environmental impacts of dairy production because fewer heifers need to be reared per cow (De Vries and Marcondes, 2020). Society may demand that forced culling is reduced and the average productive lifetime is extended (Van Knegsel et al., 2014).

Herds with lower herd replacement rates will need fewer replacement heifers and will have more mature cows that are higher producing. A herd with a high proportion of heifers and primiparous cows emits more methane and excretes more phosphorus in the environment per unit of milk when compared with a herd with a greater proportion of multiparous cows (Hristov et al., 2013). In one study, the contribution of replacement heifers to the herd enteric methane emissions ranged from 20% to 33%, depending on age at first calving and herd replacement rate (Knapp et al., 2014). This study also found that a decrease in the herd replacement rate would decrease the amount of beef contributed by the dairy industry and would require an increase in beef animals to maintain the same beef supply (Knapp et al., 2014). Consequently, reducing the herd

**Table 1** Sensitivity analysis of optimal annual herd replacement rate and opportunity cost from the optimal herd replacement rate ($ per cow per year) for combinations of herd replacement cost, lack of maturity cost, aged cow cost and genetic opportunity cost[1]

| Herd replacement cost (%) | Lack of maturity cost (%) | Aged cow cost (%) | Genetic opportunity cost (%) | 50% | 45% | 40% | 35% | 30% | 25% | 20% | 15% | 10% |
|---|---|---|---|---|---|---|---|---|---|---|---|---|
| 100 | 100 | 100 | 100 | 152 | 109 | 68 | 32 | 7 | 0 | 27 | 127 | 404 |
| 100 | 0 | 0 | 100 | 60 | 35 | 15 | 2 | 0 | 16 | 62 | 171 | 431 |
| 0 | 100 | 100 | 100 | 0 | 1 | 5 | 15 | 34 | 72 | 144 | 289 | 612 |
| 100 | 100 | 100 | 0 | 478 | 415 | 350 | 281 | 213 | 146 | 82 | 27 | 0 |
| 100 | 100 | 100 | 50 | 261 | 207 | 154 | 102 | 55 | 18 | 0 | 22 | 148 |
| 100 | 50 | 50 | 100 | 103 | 68 | 38 | 14 | 0 | 4 | 41 | 145 | 414 |
| 50 | 100 | 100 | 100 | 56 | 34 | 16 | 3 | 0 | 16 | 65 | 187 | 487 |
| 50 | 50 | 50 | 50 | 76 | 54 | 34 | 16 | 3 | 0 | 14 | 63 | 202 |

[1] Herd replacement cost: $1700 heifer price and $800 salvage value.
The column with the value 0 indicates the optimal herd replacement rate. Values greater than 0 are opportunity cost. Rate of genetic merit increase: $70 predicted transmitting ability per year. Lack of maturity and aged cows cost: maximum net revenue in the fifth and sixth parity.

replacement rate is less effective at reducing greenhouse gas emissions per kilogram of milk when the amount of beef produced is accounted for (Vellinga and De Vries, 2018).

These assertions, however, need to be verified, because a lower herd replacement rate does not imply that fewer calves are being born. Dairy bull calves, surplus dairy heifer calves and crossbred calves would be raised as beef animals and contribute to the beef supply. However, beef from dairy animals produces less greenhouse gas per kilogram of meat when the greenhouse gas allocated to milk is accounted for, compared to the meat produced in feedlot or extensive production systems by none or low-milk-producing cows (Zehetmeier et al., 2012).

## 9 Future trends in research

Future trends in research will include continued efforts to improve the health, fertility, performance and environmental impact of dairy cattle. This will happen through advances in veterinary medicine, nutrition, reproduction, genetic selection, and housing and care of animals. Much emphasis related to replacement and culling will be directed toward better understanding the factors preceding or early in the life of an animal and her later performance including health and fertility.

Research will also continue to be directed at supporting decision-making. This includes a better ranking of dairy cattle for their future value for culling, insemination and treatment decisions. Algorithms that combine data from more sources and at greater frequencies to predict the future phenotypic performance of dairy cattle will continue to be developed. More effort will also be directed at fine-tuning the optimal number and quality of replacement heifers. These efforts will be influenced by the societal and environmental aspects of the management of replacement and culling in dairy herds. Finally, research findings need to be extended to dairy farmers.

## 10 Where to look for further information

Management of culling and replacement in dairy herds is a topic that is mostly addressed at the cross-sections of agricultural economics, animal sciences and veterinary medicine. Journals in these fields contain most of the literature on culling and replacement. There is not a single journal that is primarily focused on this topic, however. Key research and data on culling can be found in the following journals:

- *Journal of Dairy Science*, https://www.journalofdairyscience.org/.
- *Animal*, https://www.cambridge.org/core/journals/animal.

- *Computers and Electronics in Agriculture*, https://www.journals.elsevier.com/computers-and-electronics-in-agriculture.
- *Preventive Veterinary Medicine*, https://www.journals.elsevier.com/preventive-veterinary-medicine.
- *Livestock Science*, https://www.journals.elsevier.com/livestock-science.
- A variety of dairy herd management factors that affect culling and replacement can be found in the third edition of *Large Dairy Herd Management* (editor Dr David Beede), published as an e-book by ADSA in 2017: https://ldhm.adsa.org/.

## 11 References

Asdell, S. A. 1951. Variations in amount of culling in DHIA herds. *Journal of Dairy Science* 34(6), 529–535.

Bach, A. 2011. Associations between several aspects of heifer development and dairy cow survivability to second lactation. *Journal of Dairy Science* 94(2), 1052–1057.

Barkema, H. W., Von Keyserlingk, M. A. G., Kastelic, J. P., Lam, T. J. G. M., Luby, C., Roy, J. P., LeBlanc, S. J., Keefe, G. P. and Kelton, D. F. 2015. Invited review: changes in the dairy industry affecting dairy cattle health and welfare. *Journal of Dairy Science* 98(11), 7426–7445 https://doi.org/10.3168/jds.2015-9377.

Bascom, S. S. and Young, A. J. 1998. A summary of the reasons why farmer cull cows. *Journal of Dairy Science* 81(8), 2299–2305.

Beaudeau, F., van der Ploeg, J. D., Boileau, B., Seegers, H. and Noordhuizen, J. P. T. M. 1996. Relationships between culling criteria in dairy herds and farmers' management styles. *Preventive Veterinary Medicine* 25(3–4), 327–342.

Beaudeau, F., Seegers, H., Ducrocq, V. P., Fourichon, C. and Bareille, N. S. 2000. Effect of health disorders on culling in dairy cows: a review and a critical discussion. *Annales de Zootechnie* 49(4), 293–311.

Bergeå, H., Roth, A., Emanuelson, U. and Agenäs, S. 2016. Farmer awareness of cow longevity and implications for decision-making at farm level. *Acta Agriculturae Scandinavica, Section A – Animal Science* 66(1), 25–34.

Bicalho, R. C., Vokey, F., Erb, H. N. and Guard, C. L. 2007. Visual locomotion scoring in the first seventy days in milk: Impact on pregnancy and survival. *Journal of Dairy Science* 90, 4586–4591.

Caraviello, D. Z., Weigel, K. A. and Gianola, D. 2003. Analysis of the relationship between type traits, inbreeding, and functional survival in Jersey cattle using a Weibull proportional hazards model. *Journal of Dairy Science* 86, 2984–2989.

Carvalho, M. R., Aboujaoude, C., Penagaricano, F., Santos, J. E. P., De Vries, T. J., McBride, B. W. and Ribeiro, E. S. 2020. Associations between maternal characteristics and health, survival and performance of dairy heifers from birth through first lactation. *Journal of Dairy Science* 103(1), 823–839.

Carvalho, M. R., Peñagaricano, F., Santos, J. E. P., De Vries, T. J., McBride, B. W. and Ribeiro, E. S. 2019. Long-term effects of postpartum clinical disease on milk production, reproduction, and culling of dairy cows. *Journal of Dairy Science* 102(12), 11701–11717. https://doi.org/10.3168/jds.2019-17025.

CDCB (Council on Dairy Cattle Breeding) 2020. Genetic and phenotypic trend. Accessed June 13, 2020. Available at: https://queries.uscdcb.com/eval/summary/trend.cfm.

CDIC (Canadian Dairy Information Centre) 2020. Culling and replacement rates in dairy herds in Canada. Accessed June 20, 2020. Available at: https://www.dairyinfo.gc.ca/eng/dairy-statistics-and-market-information/dairy-animal-genetics/culling-and-replacement-rates-in-dairy-herds-in-canada/?id=1502475693224.

Cha, E., Hertl, J. A., Schukken, Y. H., Tauer, L. W., Welcome, F. L. and Gröhn, Y. T. 2013. The effect of repeated episodes of bacteria-specific clinical mastitis on mortality and culling in Holstein dairy cows. *Journal of Dairy Science* 96(8), 4993–5007.

Cole, J. B. and VanRaden, P. M. 2018. Symposium review: possibilities in an age of genomics: the future of selection indices. *Journal of Dairy Science* 101(4), 3686–3701. https://doi.org/10.3168/jds.2017-13335.

CRV (Coöperatie CRV) 2020. Bedrijven en koeien in cijfers Nederland (Statistics of Farms and Cows in the Netherlands). Accessed June 12, 2020. Available at: https://www.cooperatie-crv.nl/downloads/stamboek/bedrijven-en-koeien-in-cijfers/.

Dahl, M. O., De Vries, A., Galvao, K. N., Maunsell, F. P., Risco, C. A. and Hernandez, J. A. 2020. Combined effect of mastitis and parity on pregnancy loss in lactating Holstein cows. *Theriogenology* 143, 57–63. https://doi.org/10.1016/j.theriogenology.2019.12.002.

DairyNZ and LIC 2019. New Zealand dairy statistics 2017–18. Accessed June 12, 2020. Available at: https://www.lic.co.nz/documents/450/NZ_DAIRY_STATISTICS_2017-18-WEB-10_OCT.pdf.

De Vries, A. 2006. Economic value of pregnancy in dairy cattle. *Journal of Dairy Science* 89, 3876–3885.

De Vries, A. 2016. Enhanced accuracy of cows' retention, breeding, and milking values with precision dairy technologies. pp. 245–250. In: Proceedings of the International Conference on Precision Dairy Farming, Leeuwarden, The Netherlands, June 21–23. Available at: https://doi.org/10.3920/978-90-8686-829-2.

De Vries, A. 2017. Economic trade-offs between genetic improvement and longevity in dairy cattle. *Journal of Dairy Science* 100(5), 4184–4192.

De Vries, A. 2020. Symposium review: why revisit dairy cattle productive lifespan? *Journal of Dairy Science* 103(4), 3838–3845. https://doi.org/10.3168/jds.2019-17361.

De Vries, A. and Marcondes, M. I. 2020. Review: overview of factors affecting productive lifespan of dairy cows. *Animal: An International Journal of Animal Bioscience* 14(Suppl.1), s155–s164. https://doi.org/10.1017/s1751731119003264.

De Vries, A., Olson, J. D. and Pinedo, P. J. 2010. Reproductive risk factors for culling and productive life in large dairy herds in the eastern United States between 2001 and 2006. *Journal of Dairy Science* 93, 613–623.

De Vries, M., Bokkers, E. A., Dijkstra, T., Van Schaik, G. and De Boer, I. J. 2011. Invited review: associations between variables of routine herd data and dairy cattle welfare indicators. *Journal of Dairy Science* 94(7), 3213–3228. https://doi.org/10.3168/jds.2011-4169.

Dubuc, J., Duffield, T. F., Leslie, K. E., Walton, J. S. and LeBlanc, S. J. 2011. Effects of postpartum uterine diseases on milk production and culling in dairy cows. *Journal of Dairy Science* 94(3), 1339–1346.

Fetrow, J., Nordlund, K. V. and Norman, H. D. 2006. Invited review: culling: nomenclature, definitions, and recommendations. *Journal of Dairy Science* 89(6), 1896–1905.

Frazer, L. L. C. 2018. Dairy farm operating trends for the year ended December 31, 2017. Accessed June 14, 2020. Available at: https://frazerllp.com.

García-Ruiz, A., Cole, J. B., VanRaden, P. M., Wiggans, G. R., Ruiz-López, F. J. and Van Tassell, C. P. 2016. Changes in genetic selection differentials and generation intervals in US Holstein dairy cattle as a result of genomic selection. *Proceedings of the National Academy of Sciences of the United States of America* 113(28), E3995–E4004.

Groenendaal, H., Galligan, D. T. and Mulder, H. A. 2004. An economic spreadsheet model to determine optimal breeding and replacement decisions for dairy cattle. *Journal of Dairy Science* 87(7), 2146–2157.

Gröhn, Y. T., Eicker, S. W., Ducrocq, V. and Hertl, J. A. 1998. Effect of diseases on the culling of Holstein dairy cows in New York State. *Journal of Dairy Science* 81, 966–978.

Hadley, G. L., Wolf, C. A. and Harsh, S. B. 2006. Dairy cattle culling patterns, explanations, and implications. *Journal of Dairy Science* 89(6), 2286–2296.

Heinrichs, A. J. and Heinrichs, B. S. 2011. A prospective study of calf factors affecting first-lactation and lifetime milk production and age of cows when removed from the herd. *Journal of Dairy Science* 94(1), 336–341.

Heinrichs, A. J., Zanton, G. I., Lascano, G. J. and Jones, C. M. 2017. A 100-year Review: a century of dairy heifer research. *Journal of Dairy Science* 100(12), 10173–10188. https://doi.org/10.3168/jds.2017-12998.

Hertl, J. A., Schukken, Y. H., Bar, D., Bennett, G. J., González, R. N., Rauch, B. J., Welcome, F. L., Tauer, L. W. and Gröhn, Y. T. 2011. The effect of recurrent episodes of clinical mastitis caused by gram-positive and gram-negative bacteria and other organisms on mortality and culling in Holstein dairy cows. *Journal of Dairy Science* 94(10), 4863–4877.

Hogeveen, H., Steeneveld, W. and Wolf, C. A. 2019. Production diseases reduce the efficiency of dairy production: a review of the results, methods, and approaches regarding the economics of mastitis. *Annual Review of Resource Economics* 11(1), 289–312. https://www.annualreviews.org/doi/10.1146/annurev-resource-100518-093954.

Hristov, A. N., Ott, T., Tricarico, J., Rotz, A., Waghorn, G., Adesogan, A., Dijkstra, J., Montes, F., Oh, J., Kebreab, E., Oosting, S. J., Gerber, P. J., Henderson, B., Makkar, H. P. S. and Firkins, J. L. 2013. SPECIAL TOPICS—mitigation of methane and nitrous oxide emissions from animal operations: III. A review of animal management mitigation options. *Journal of Animal Science* 91(11), 5095–5113.

Huirne, R. B. M., Dijkhuizen, A. A., Van Beek, P. and Renkema, J. A. 1997. Dynamic programming to optimize treatment and replacement decisions In: Dijkhuizen, A. A. and Morris, R. S. (Eds.). *Animal Health Economics; Principles and Applications. Post Graduate Foundation in Veterinary Science*. University of Sydney, Sydney, Australia. 85–97

Knapp, J. R., Laur, G. L., Vadas, P. A., Weiss, W. P. and Tricarico, J. M. 2014. Invited review: enteric methane in dairy cattle production: quantifying the opportunities and impact of reducing emissions. *Journal of Dairy Science* 97(6), 3231–3261.

Lactanet 2019. 10 years of genomic selection: what's next? Accessed June 13, 2020 Available at: https://www.cdn.ca/document.php?id=530.

Langford, F. M. and Stott, A. W. 2012. Culled early or culled late: economic decisions and risks to welfare in dairy cows. *Animal Welfare* 21(1), 41–55.

LeBlanc, S. J. 2008. Postpartum uterine disease and dairy herd reproductive performance: a review. *The Veterinary Journal* 176, 102–114.

Lombard, J. E., Garry, F. B., Tomlinson, S. M. and Garber, L. P. 2007. Impacts of dystocia on health and survival of dairy calves. *Journal of Dairy Science* 90(4), 1751–1760. https://doi:10.3168/jds.2006-295.

Mißfeldt, F., Mißfeldt, R. and Kuwan, K. 2015. Economic lifetime of milk cows (in German). *Züchtungskunde* 87, 120–143.

Milian-Suazo, F., Erb, H. N. and Smith, R. D. 1988. Descriptive epidemiology of culling in dairy cows from 34 herds in New York State. *Preventive Veterinary Medicine* 6(4), 243–251.

Mohd Nor, N. M., Steeneveld, W. and Hogeveen, H. 2014. The average culling rate of Dutch dairy herds over the years 2007 to 2010 and its association with herd reproduction, performance and health. *Journal of Dairy Research* 81(1), 1–8. https://doi.org/10.1017/s0022029913000460.

NAAB (National Association of Animal Breeders) 2020. Annual reports of semen sales and custom freezing. Accessed June 13, 2020. Available at: https://www.naab-css.org/semen-sales.

Norman, H. D. 2019. Why do cows leave the milking herd earlier than in the past? Council on Dairy Cattle Breeding. November 2019. Accessed June 14, 2020. Available at: https://www.uscdcb.com/wp-content/uploads/2019/11/Why-Do-Cows-Leave-Earlier-11_2019.pdf.

Norman, H . D., Walton, L. M. and Dürr, J. W. 2020. Reasons that cows in Dairy Herd Improvement programs exit the milking herd (2019). Council on Dairy Cattle Breeding. Available at: https://queries.uscdcb.com/publish/dhi/cull.html

Nowak, R. M. and Walker, E. P. 1999. *Walker's Mammals of the World* (Vol. 1). JHU Press, Baltimore, MD.

Oltenacu, P. A., Frick, A. and Lindhé, B. 1990. Epidemiological study of several clinical diseases, reproductive performance and culling in primiparous Swedish cattle. *Preventive Veterinary Medicine* 9(1), 59–74.

Overton, M. W. and Dhuyvetter, K. C. 2020. Symposium review: an abundance of replacement heifers: what is the economic impact of raising more than are needed? *Journal of Dairy Science*. 103(4), 3828–3837. https://doi.org/10.3168/jds.2019-17143.

Pinedo, P. J. and De Vries, A. 2010. Effect of days to conception in the previous lactation on the risk of death and live culling around calving. *Journal of Dairy Science* 93, 968–977.

Pinedo, P. J. and De Vries, A. 2017. Season of conception is associated with future survival, fertility, and milk yield of Holstein cows. *Journal of Dairy Science* 100, 6631-6639.

Pinedo, P. J., De Vries, A. and Webb, D. W. 2010. Dynamics of culling risk with disposal codes reported by Dairy Herd Improvement dairy herds. *Journal of Dairy Science* 93(5), 2250–2261.

Probo, M., Pascottini, O. B., LeBlanc, S., Opsomer, G. and Hostens, M. 2018. Association between metabolic diseases and the culling risk of high-yielding dairy cows in a transition management facility using survival and decision tree analysis. *Journal of Dairy Science* 101(10), 9419-9429.

Rajala-Schultz, P. J. and Gröhn, Y. T. 1999. Culling of dairy cows. Part II. Effects of diseases and reproductive performance on culling in Finnish Ayrshire cows. *Preventive Veterinary Medicine* 41, 279–294.

Ramsey, J., Tisor, L., Hulsman Hanna, L. and Ringwall, K. 2017. Cow Herd Appraisal Performance Software (CHAPS): 15 years of beef production benchmarks. 2017 North Dakota Beef Report (AS1862). Accessed June 13, 2020. Available at: https://www.ag.ndsu.edu/publications/livestock/2017-north-dakota-beef-report

Rauw, W. M., Kanis, E., Noordhuizen-Stassen, E. N. and Grommers, F. J. 1998. Undesirable side effects of selection for high production efficiency in farm animals: a review. *Livestock Production Science* 56(1), 15–33.

Rilanto, T., Reimus, K., Orro, T., Emanuelson, U., Viltrop, A. and Mõtus, K. 2020. Culling reasons and risk factors in Estonian dairy cows. *BMC Veterinary Research* 16(1), 173. Published 2020 Jun 1. https://doi.org/10.1186/s12917-020-02384-6.

Roberts, T., Chapinal, N., LeBlanc, S. J., Kelton, D. F., Dubuc, J. and Duffield, T. F. 2012. Metabolic parameters in transition cows as indicators for early-lactation culling risk. *Journal of Dairy Science* 95, 3057–3063.

Roche, J. R., Friggens, N. C., Kay, J. K., Fisher, M. W., Stafford, K. J. and Berry, D. P. 2009. Invited review: body condition score and its association with dairy cow productivity, health, and welfare. *Journal of Dairy Science* 92(12), 5769–5801, https://doi.org/10.3168/jds.2009-2431.

Schaffer, A. P., Larson, R. L., Cernicchiaro, N., Hanzlicek, G. A., Bartle, S. J. and Thomson, D. U. 2016. The association between calfhood bovine respiratory disease complex and subsequent departure from the herd, milk production, and reproduction in dairy cattle. *Journal of the American Veterinary Medical Association* 248(10), 1157–1164. https://doi.org/10.2460/javma.248.10.1157.

Seifi, H. A., LeBlanc, S. J., Leslie, K. E. and Duffield, T. F. 2011. Metabolic predictors of post-partum disease and culling risk in dairy cattle. *The Veterinary Journal* 188(2), 216–220.

Soberon, F. and Van Amburgh, M. E. 2013. Lactation Biology Symposium: the effect of nutrient intake from milk or milk replacer of preweaned dairy calves on lactation milk yield as adults: a meta-analysis of current data. *Journal of Animal Science* 91(2), 706–712.

Tao, S., Dahl, G. E., Laporta, J., Bernard, J. K., Orellana Rivas, R. M. and Marins, T. N. 2019. Physiology symposium: effects of heat stress during late gestation on the dam and its calf. *Journal of Animal Science* 97, 2245–2257.

Tranel, L. 2019. Heifer Raising Costs in 2019. Iowa State University Publication LT-18-01. Available at: https://www.extension.iastate.edu/dairyteam/files/page/files/whats_it_cost_to_raise_heifers_2019_0.pdf.

Urie, N. J., Lombard, J. E., Shivley, C. B., Kopral, C. A., Adams, A. E., Earleywine, T. J., Olson, J. D. and Garry, F. B. 2018. Preweaned heifer management on US dairy operations: Part V. Factors associated with morbidity and mortality in preweaned dairy heifer calves. *Journal of Dairy Science* 101(10), 9229-9244.

Van Knegsel, A. T. M., Hammon, H. M., Bernabucci, U., Bertoni, G., Bruckmaier, R. M., Goselink, R. M. A., Gross, J. J., Kuhla, B., Metges, C. C., Parmentier, H. K., Trevisi, E., Tröscher, A. and van Vuuren, A. M. 2014. Metabolic adaptation during early lactation: key to cow health, longevity and a sustainable dairy production chain. *CAB Reviews: Perspectives in Agriculture, Veterinary Science, Nutrition and Natural Resources* 9(2), 1–15. https://doi.org/10.1079/pavsnnr20149002.

Vellinga, T. V. and de Vries, M. 2018. Effectiveness of climate change mitigation options considering the amount of meat produced in dairy systems. *Agricultural Systems* 162, 136–144.

Von Keyserlingk, M. A. G., Rushen, J., de Passillé, A. M. and Weary, D. M. 2009. Invited review: the welfare of dairy cattle–key concepts and the role of science. *Journal of Dairy Science* 92(9), 4101–4111. https://doi.org/10.3168/jds.2009-2326.

Weigel, K. A., Palmer, R. W. and Caraviello, D. Z. 2003. Investigation of factors affecting voluntary and involuntary culling in expanding dairy herds in Wisconsin using survival analysis. *Journal of Dairy Science* 86(4), 1482–1486.

WUR-LR (Wageningen UR Livestock Research) 2013. Routekaart Levensduur (Road to Longevity). Report 668. Available at: https://edepot.wur.nl/275131.

Zehetmeier, M., Baudracco, J., Hoffmann, H. and Heißenhuber, A. 2012. Does increasing milk yield per cow reduce greenhouse gas emissions? a system approach. *Animal: An International Journal of Animal Bioscience* 6(1), 154–166.

# Part 4

## Particular health issues

# Chapter 11

## Optimizing udder health in dairy cattle

*Theo J. G. M. Lam, Royal GD Animal Health and Utrecht University, The Netherlands; and Sarne De Vliegher, M-team, Ghent University and MEX™, Belgium*



### 1 Introduction

Every dairy farmer would like to have as little mastitis as possible. Mastitis, Inflammation of the mammary gland, does not only lead to milk loss, antimicrobial use and costs for treatment, culling and labor, it also disturbs effective working schedules, which are crucial in cost- and time-efficient modern dairy farming. Not all dairy farmers, however, do everything possible in order to minimize the number of mastitis cases on their farm. Reasons for that range from ignorance to rationally underbuilt management decisions. Obviously, from an economic point of view, the best decision is not to strive for as little mastitis as possible. Every dairy herd will have a case of mastitis once in a while, and if mastitis really is an exception, investing time and money to further optimize udder health may not be sensible (Huijps et al., 2010).

In many farms, however, whether or not to put effort in improved mastitis management is a subjective decision. When mastitis problems are obvious, taking action may be unavoidable, for example, to avoid paying penalties. But what is an obvious problem for one farmer may be daily routine for another. And apart from obvious (clinical) mastitis problems, there are subclinical mastitis problems, and there is management in order to prevent mastitis problems. Part

http://dx.doi.org/10.19103/AS.2020.0086.13

of the latter and an important tool to objectify mastitis problems are mastitis performance indicators. These should be evaluated at regular intervals in order to objectively decide whether or not to put more effort in optimizing mastitis management or to give priority to other aspects of dairy farming.

Although it is sometimes tempting to directly try and solve the obvious problem, it is important to first analyze the herd and the available data, before diagnosing the cause and providing solutions to mastitis problems. You may miss important findings – critical control points – that can be important for long-term solutions and may affect other parts of the management in the herd.

By far the most time in a herd is spent on operational activities such as feeding, milking, cleaning and if necessary solving problems, the so-called day-to-day work. The least time is spent on the strategic level of investment decisions, the real entrepreneurial part of the business. These latter decisions, however, have a huge impact on farm results yet cost little working hours. Once these decisions are taken, it generally is what it is and, unless you go back to the investment level, you have to live and work in the given situation. The linkage between the strategic level and the day-to-day work is called 'management' (Fig. 1). Not different from any other business, farm management means that, given the circumstances you work in, you try to optimize the results of the day-to-day work, not by working harder but by doing things smarter. An effective management is crucial for effective day-to-day work, but is often not taken serious enough in dairy herds (Penry et al., 2011).

Some people take management decisions based on their gut feeling, for most it helps to quantify data and rationalize the process. To do so, one first has to have goals, a part that is often lacking in mastitis management. 'A low somatic cell count (SCC)' or 'as little mastitis as possible' may feel ambitious, but these are no goals because one is unable to realize them because they are

**Figure 1** The time pyramid of a business (source: Lam and De Vliegher, 2017).

just not specific enough. In a rational management cycle (Fig. 2) one has to set specific goals and to decide whether the performance indicators meet these goals or not. If not, one has to decide what to change in the day-to-day work to improve the results in order to meet the goals set (Lam et al., 2011a).

Setting goals is simple and rewarding also if, for whatever reason, it is not possible to fully elaborate a management cycle on a farm. Just ask when a farmer would be satisfied with respect to udder health. It tells you whether or not he is motivated to take action. A more sophisticated approach is through quantifying the current mastitis situation. This is preferably based on a one-year period, because shorter periods may be influenced by seasonal variability. Although available data differ from country to country and from farm to farm, the most important udder health parameters are bulk milk and individual SCC and clinical mastitis data. In many countries nowadays antimicrobial usage, is added (Lam et al., 2017). In specific herd situations it may be advisable to add parameters such as number of cows culled for udder health, or heifers with mastitis. Based on these data, goals can be set which ought to be SMART: specific, measurable, acceptable, realistic and time-bound. If not available through management software, a simple sheet as presented in Table 1 may help to visualize results and goals in order to take rational management decisions. How exactly these parameters are defined and quantified differs for different situations, but the exact definition is not as important as is the fact that they are used.

Reviewing results on a regular basis and comparing them to goals set as presented in Fig. 2 help to pick up herd-level problems at an early stage to prevent serious problems such as chronically infected cows. In addition, evaluating the day-to-day work before deviating parameters force you to do so would be a true preventive approach. This is often described as Hazard Analysis

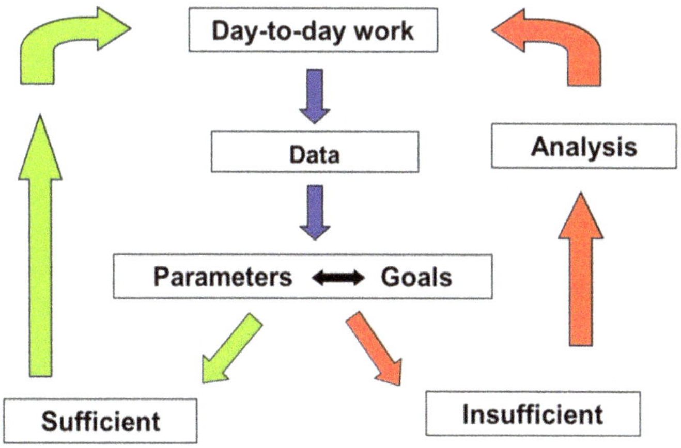


**Figure 2** Management cycle (source: Lam and De Vliegher, 2017).

**Table 1** An example of a simple overview of results and goals to quantify udder health on a dairy farm

| | Results 2021 | Goal 2022 |
|---|---|---|
| Number of cases of clinical mastitis/100 cows/year | ...... | ...... |
| Bulk milk somatic cell count | ...... | ...... |
| Percentage of cows with SCC > 200 000 | ...... | ...... |
| Antimicrobial usage for clinical mastitis | ...... | ...... |
| Antimicrobial usage for drying off | ...... | ...... |
| .................. | ...... | ...... |

*Source*: Lam and De Vliegher (2017).

of Critical Control Points (HACCP). Although in dairy farms HACCP cannot fully replace an approach in which results are evaluated as described above, it is of added value (Lam et al., 2011a). Through the years several mastitis prevention and control plans have been developed and adapted based on changes in herd sizes, importance of pathogens, available knowledge and technology, to mention a few. The current international standard for mastitis prevention is the 10-point plan of the National Mastitis Council (NMC), which is described in more detail on the NMC website (www.nmconline.org, assessed January 2, 2020).

## 2 Mastitis diagnosis

In order to fight and solve mastitis problems effectively, first a proper diagnosis has to be made. This is true both at the level of the individual cow as well as at the herd level. Intramammary infections lead to mastitis as will be described in more detail in the paragraph on immunology. Mastitis diagnostics often starts with clinical inspection. A cow may have mastitis, but as long as there are no clinical symptoms, we define that as subclinical mastitis. Once clinical symptoms like redness, pain or swelling of a quarter are visible, or if it loses its normal function and produces abnormal milk, we talk about clinical mastitis. Although this may seem as a clear difference, practice teaches that in fact the difference is gradual. Not all milkers strip cows before milking and not all milkers have the same judgment of what they think is abnormal milk. Additionally, circumstances may be different, such as in the case of automatic milking systems. Thus, different types of mastitis diagnostics are of importance and will be discussed. Although in some situations mild clinical mastitis cases may be missed, clinical mastitis is a fact of life in virtually all dairy herds. Based on severity of symptoms, clinical mastitis is classified as mild (grade 1) when only the milk is abnormal, moderate (grade 2) when abnormal milk is accompanied by swelling or redness of a quarter, or severe (grade 3) when the cow exhibits systemic signs of illness such as depression, anorexia, dehydration and/or fever (Pinzón-Sánchez and Ruegg, 2011). This classification is helpful in treatment considerations as will be discussed later.

The most important cause of mastitis in dairy cattle is bacteria, with different bacteria having a different ecology and epidemiology and causing different disease patterns. Thus, a bacteriological diagnosis is important. Such a diagnosis is generally based on bacteriologic culturing, although other approaches to come to a diagnosis of mastitis, such as somatic cell counting (ten Napel et al., 2009), measuring of electrical conductivity (Kamphuis et al., 2008) or lactate dehydrogenase (Friggens et al., 2007) are used.

## *2.1 Somatic cell count*

Of all parameters used to diagnose mastitis, somatic cell counting is by far the most used. Although the information from SCC is not very specific the measurement is cheap, easily available and lots of experience interpreting it exists. The SCC reflects the level of inflammation of the mammary gland and the number of cells in the milk, being neutrophils, epithelial cells and mononuclear cells. It is expressed as number of cells per mL of milk and can be measured at quarter, cow and herd level. New approaches in which SCC is further specified by differentiating somatic cells (Damm et al., 2017) are promising to further increase the added value of SCC, especially in early mastitis diagnostics.

In many countries there is a maximum level for bulk milk SCC (EU > 400 000 cells/mL; US > 750 000 cells/mL) above which milk is not considered fit for human consumption. That threshold obviously is the absolute maximum, while there are many reasons to strive for a much lower bulk milk SCC. The SCC of a healthy quarter is approximately 70 000 cells/mL, slowly increasing with age and days in milk and decreasing milk production (Schukken et al., 2003). Of these cells roughly 80% are immune cells, mainly macrophages, with some 20% epithelial cells. After an immune response in a quarter, the number of neutrophils rapidly increases with the percentage of immune cells increasing to 99%. The SCC, often, is not a very stable parameter. If repeatedly measured, SCC may vary in time as is obvious in Fig. 3, which is an example of a quarter infected with *Staphylococcus aureus* (Lam and De Vliegher, 2017).

The level of the SCC is highly associated with loss of milk production. In a Canadian study on 2800 dairy herds it was found that specifically the high-producing cows lose most milk because of subclinical mastitis, both in kilograms as well as percentage-wise. Specifically high-producing older cows may easily lose 10% of their daily milk production if the individual SCC exceeds 1 500 000 cells/mL leading to high production losses if the increased SCC remains for a longer period of time (Hand et al., 2012).

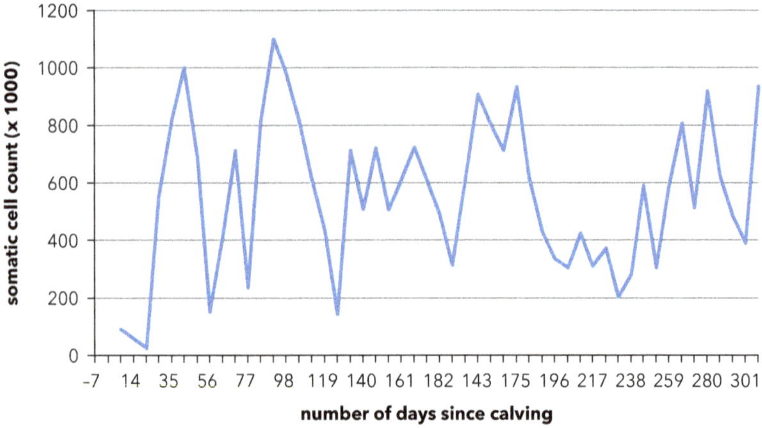


**Figure 3** Example of the SCC of a quarter that became infected with *Staphylococcus aureus* shortly after calving (source: Lam and De Vliegher, 2017).

## *2.2 Bacteriological culture*

To know the cause of mastitis the ultimate approach is bacteriological culturing. Although other methods such as polymerase chain reaction (PCR) have their value, classical culturing is most used and is a valuable approach. Bacteriological culturing can be done with bulk milk, composite cow as well as individual quarter samples, with increasing sensitivity and specificity but also with obviously increasing costs. Bacteriological culturing helps to identify the most important pathogens at the herd level and determines the focus of the herd level approach as will be discussed later. At the individual quarter or cow level, bacteriological results give information on prognosis and thus on possible treatment of mastitis cases. The most important bacteria causing mastitis are presented in Table 2. In addition to bacteria, yeasts, molds and algae, and exceptionally, viruses can also cause mastitis.

Before bacteriological culturing can be executed, milk samples have to be collected. The importance of proper sample collection cannot be exaggerated. Although it seems simple to collect some milk in a tube, it is very easy to contaminate such a sample with bacteria from the environment or the exterior of the udder or cow, leading to worthless results and much frustration. It is important that milk samples are collected before a cow is treated. Antimicrobials in treated cows or quarters will eliminate bacteria from the milk sample, which then obviously will not be found through culturing. One has to realize that antimicrobials that may not be effective in vivo may still suppress bacterial growth in vitro. The simple conclusion that a culture negative sample with antimicrobials means the treatment is effective therefore is incorrect.

**Table 2** The most frequently found causative agents of mastitis, classified based on morphology and Gram-staining

| Gram-positive cocci | Gram-negative rods |
|---|---|
| Staphylococci | • *Escherichia coli* |
| • *Staphylococcus aureus* | • *Klebsiella* species |
| • Non-*aureus* staphylococci | • *Enterobacter* species |
| Streptococci (like) | • *Serratia* species |
| • *Streptococcus agalactiae* | • *Pseudomonas* species |
| • *Streptococcus dysgalactiae* | • *Pasteurella* species |
| • *Streptococcus uberis* | • *Acinetobacter* species |
| • Other *Streptococcus* species | • *Proteus* species |
| • *Enterococcus* species | |
| Gram-positive rods | Other microorganisms |
| • *Corynebacterium bovis* | • *Mycoplasma* species |
| • Other *Corynebacterium* species | • *Prototheca* species |
| • *Trueperella pyogenes* | • Yeasts and molds |
| • *Nocardia* species | |
| • *Bacillus* species | |

*Source*: Lam and De Vliegher (2017).

At the start of the sampling process, the milk tube should be stamped with cow and quarter identification as well as the sampling date. The teats should be thoroughly cleaned and disinfected with alcohol or spirit. Avoid contamination of disinfected teats while sampling by first disinfecting the teats that are furthest away and first sampling the teats that are closest. Two or three streams of milk should be discarded before the tube is filled for approximately three-quarter. Make sure that during this process the cap does not touch the cow and gets contaminated. The likelihood of contamination decreases if the whole process is executed correctly and quickly; practice makes the professional. If milk samples are transported to an external laboratory, then that ought to be done under cooled conditions or frozen. Ideally, milk samples are cultured within 24 h and preferably should not have been frozen for longer than one month before culturing.

Milk samples are typically cultured on solid culture media. Bacteria multiply to colony-forming units that have specific morphological characteristics which help to identify them. For culturing milk samples generally three culture media are used: (1) a non-selective blood agar, often based on sheep blood, (2) a selective agar for streptococci, and (3) a selective agar for gram-negative bacteria (NMC, 2017). Bacteria generally grow best on the blood agar, where their specific hemolytic pattern based on destruction of erythrocytes helps identification. Selective media for streptococci generally are Edwards medium or TKT agar, for gram-negative bacteria often MacConkey agar no. 3 is used.

For more details on the culturing of milk samples we refer to the Laboratory Handbook on Bovine Mastitis of the NMC (2017).

Modern laboratory technology, specifically matrix-assisted laser desorption/ionization time-of-flight mass spectrometry (MALDI-TOF MS), is fast and cheap once the equipment is available and can be very helpful in identifying mastitis pathogens. Using MALDI-TOF MS bacteria generally can be identified to the species level, based on comparison of mass spectra from verified species with an available database. This technique is not only much quicker (hours vs. days) than classical phenotypical and biochemical testing, but is also more reliable and more specific, that is, to identify non-aureus staphylococcal species. As in classical techniques, MALDI-TOF MS requires pure culture specimens, because it is not yet possible to identify polymicrobial samples (Nonnemann et al., 2019).

In recent years bacteriological culturing of milk samples of individual mastitis cases executed at the dairy farm, the so called on-farm culturing, has increased. Farmers appreciate a quick result and indicate it helps them to make treatment decisions (Griffioen, 2020). Basically the same procedure as in the laboratory can be followed on farm, although differentiation to the species level can be a challenge. Various commercial tests specifically developed for on-farm use have become available to tackle this issue. To simplify methods, so-called bi-, tri- and quad-plates were developed with several culture media in one agar plate. Some of these work on the same selective principle as classical bacteriological culturing in the laboratory. Others use chromogenic media that change color if specific types of bacteria, for example, Gram-negative bacteria, grow. A specific method is the use of Petrifilm, which is a ready-to-use compact dry culture medium that can be preserved easily on farm. In this approach several films can be used at the same time, leading to different diagnostic approaches, dependent on the purpose and the expected outcome of the culturing. More details on different types of on-farm mastitis tests can be found in Martins et al. (2019).

## 3 Mastitis immunology

Mastitis is an immunological reaction of the udder tissue against irritation, generally induced by an intramammary infection caused by bacteria. The udder is exposed to many different bacteria, with the teat being the connection between the udder and the outside world. The teat orifice is the opening of the udder through which two or three times a day during approximately 5 min a large amount of milk passes, while it is supposed to keep bacteria out in between milkings. The first defense mechanism to refrain mastitis-causing bacteria from entering the mammary gland is the teat orifice and the flow of milk.

If bacteria manage to pass this first mechanical barrier, they meet the immune system, which consists of an innate and an adaptive part, that both cooperate in order to get rid of invading pathogens. At the proximal side of the teat canal, the Fürstenberg rosette is located, which has a specific role with respect to local immune defense activities. Generally, the immune system is able to successfully capture and remove bacteria that invade the udder. If, however, the immune response is not effective, bacteria will be able to multiply and infect the epithelium of the mammary gland, resulting in potential persistent intramammary infections. In some cases an immune response leads to clinical mastitis, characterized by observable symptoms of the milk or the udder, or even by systemic symptoms or in exceptional cases death of a cow. If an intramammary infection is established in the absence of visible symptoms, the mastitis cases are considered subclinical.

Apart from the mechanical part, the innate immune response has a humoral and a cellular component. The humoral response consists of proteins and chemokines that inhibit bacterial growth and sometimes kill bacteria. Important parts of the innate humoral system are substances such as lactoferrin, complement, transferrin, lysozyme and cytokines.

Immune cells such as lymphocytes, macrophages and neutrophils are present in the milk of healthy quarters to take appropriate measures if needed to establish an early immune response (Alnakip et al., 2014). Macrophages are the first immunological cells to recognize bacterial infections. They do so with the help of receptors, with the most important ones being the Toll-like receptors TLR-2, TLR-4 and TLR-6. After recognition of specific bacterial proteins, the NF-KB system is activated, the macrophages start producing pro-inflammatory cytokines such as interleukins (IL-1, IL-6, IL-8, IL 12), tumor necrosis factor α (TNF α) and interferon γ (IFN γ), and anti-inflammatory cytokines such as IL-4 and IL-10. Cytokine production differs per type of bacteria, where a good balance between pro- and anti-inflammatory cytokines is important to effectively combat the bacterial infection without an immunological overreaction. Macrophages also play a role in phagocytosis (engulfing and digesting) of bacteria, but their main function is to attract neutrophils. This type of immune cells ingests the bacteria and kills them with enzymes and oxidants (Alnakip et al., 2014).

The adaptive immune system consists of a humoral response through specific antibodies (IgM and IgG) formed by B lymphocytes against parts of bacteria after an initial intramammary infection. After a second infection these antibodies may recognize the bacteria and attach to its surface, which is called opsonization and which facilitates phagocytosis through neutrophils. The role of specific immunity in mastitis seems to be limited. Challenge studies showed that after an initial infection, cows are approximately two weeks protected against reinfection with sometimes even opposite effects.

Contrary to the limited protective effect of natural infections (Zadoks et al., 2001), vaccination with antigens of mastitis bacteria does seem efficacious (Pomeroy et al., 2016).

### 3.1 Vaccination

Mastitis vaccines are used to prevent mastitis caused by *E. coli* and *Klebsiella* spp. mastitis, to a lesser extent by *S. aureus* and non-*aureus* staphylococci and in recent years also by *S. uberis*. Because of the many different mastitis pathogens and strains it is not easy to develop an efficacious, multivalent mastitis vaccine. Vaccines aim to stimulate the specific humoral response through memory B lymphocytes that are activated to quickly produce specific antibodies after exposure to a specific mastitis pathogen. Another part of the working mechanism of a vaccine is through stimulation of cellular immunity of cytotoxic T lymphocytes (Riollet et al., 2000; Denis et al., 2006).

Vaccines for different types of bacteria have different working mechanisms. Vaccines for environmental pathogens such as coliforms and *S. uberis*, primarily aim to limit severity of disease in order to decrease costs and culling (Wilson et al., 2009). The most frequently described vaccine against *E. coli* is the J5 vaccine, which is frequently used in the United States (González et al., 1989). Vaccines for contagious mastitis such as *S. aureus* focus more on decreased shedding of bacteria, decreasing transmission and therefore the number of new infections (Schukken et al., 2014).

## 4 Antimicrobial treatment of mastitis

Although prevention of intramammary infections is the ultimate way to improve udder health, mastitis cases will occur on dairy farms. This means action has to be taken in order to treat the diseased animals and to prevent spread of bacteria causing new cases. With respect to the latter, actions such as segregating of infected cows through creating separate groups, adapting the order of milking, or culling of cows can be helpful. Nevertheless, treating cows with mastitis is an important part of mastitis management on many farms. Because mastitis almost always is caused by bacterial infections, antimicrobials play an important role in mastitis treatment.

Although there is no indication of serious problems with antimicrobial resistance in mastitis pathogens, it goes without saying that antimicrobial use related to udder health should also be prudent. That means that antimicrobials should not be used if no added value is to be expected compared to other approaches, including no treatment at all. The success of treatment depends on factors such as the bacteria involved, the duration of infection and immunological characteristics of the cow. Whether or not it is wise to apply an antimicrobial treatment differs from case to case and from herd to herd. In dairy herd health programs it is therefore wise to develop herd-level treatment protocols (standard operating procedures) for clinical mastitis, subclinical mastitis and dry cow treatment, that are updated at least once every year or whenever new relevant information is available (Lam et al., 2009).

Different antimicrobials have different working mechanisms of which the most important, related to mastitis treatment, are inhibiting the cell wall synthesis, inhibiting protein synthesis and interfering with the DNA of the bacteria (Giguère et al., 2013). The most important type of antibiotics that inhibits cell wall synthesis is the β-lactam antibiotics. Representatives of this group that are frequently used in mastitis treatment are presented in Table 3.

The β-lactam antibiotics disturb the formation of peptidoglycan in the cell wall of both Gram-positive and Gram-negative bacteria, killing these bacteria while multiplying. Thus, these antibiotics are considered bactericidal. One of the mechanisms bacteria use to protect themselves against β-lactams is the production of the enzyme β-lactamase, which breaks down β-lactam. The effect of β-lactamase can be neutralized by inhibitors such as clavulanic acid that binds irreversibly to β-lactamase (Kapoor et al., 2017).

Antimicrobials that inhibit protein synthesis disturb bacterial growth through interfering with the ribosomes. Several antimicrobials with this mode of action that are frequently used in mastitis treatment are presented in Table 4. In the ribosomes, bacterial proteins are synthesized through connecting amino acids

**Table 3** The most important antimicrobials inhibiting cell wall synthesis that are relevant for the treatment of bovine mastitis

| | | |
|---|---|---|
| Penicillin | Narrow spectrum | Penicillin G (benzyl penicillin) |
| | | Penethamate hydriodide |
| | Semi-synthetic | Cloxacillin |
| | | Nafcillin |
| | Broad spectrum | Amoxicillin |
| | | Ampicillin |
| Cephalosporins | First generation | Cefapirin |
| | | Cefalexin |
| | | Cefazolin |
| | | Cefalonium |
| | Third generation | Cefoperazone<br>Ceftiofur |
| | Fourth generation | Cefquinome |
| β- lactam inhibitors | | Clavulanic acid |

*Source*: Lam and De Vliegher (2017).

**Table 4** The most important antimicrobials inhibiting cell protein synthesis that are relevant for the treatment of bovine mastitis

| | |
|---|---|
| Lincosamides | Lincomycin |
| Macrolides | Erythromycin |
| | Tylosin |
| Aminoglycosides | Kanamycin |
| | Neomycin |
| | Framycetin |
| | Dihydrostreptomycin |

*Source*: Lam and De Vliegher (2017).

**Table 5** The most important antimicrobials interfering with the DNA of bacteria that are relevant for the treatment of bovine mastitis

| | |
|---|---|
| Fluoroquinolones | Enrofloxacin |
| | Danofloxacin |
| | Marbofloxacin |
| Sulfonamides | Sulfadiazine |
| | Sulfadoxine |
| Diaminopyrimidines | Trimethoprim |
| Rifamycins | Rifamyxin |

to a growing peptide chain, which is disturbed by the invading antimicrobials. When lincosamides and macrolides are used, bacteria are no longer able to grow, which is described as a bacteriostatic affect. Aminoglycosides also work on the ribosomes, but lead to synthesis of divergent proteins that are built in the bacterial cell wall, which kills the bacteria and which is called a bactericidal effect (Giguère et al., 2013).

Antimicrobials that interfere with the DNA of the bacteria such as fluoroquinolones have an effect on replication and transcription of DNA (Table 5). This has an effect on multiplication of sensitive bacteria and, through release of pieces of DNA, to oxidative stress and death of bacteria. Sulfonamides and trimethoprim are bactericidal, disturbing the synthesis of folium acid which is essential for bacterial growth (Kapoor et al., 2017).

Combining different antimicrobials may have a synergistic as well as an antagonistic effect. The combination of β-lactam antibiotics and aminoglycosides is a well-known synergistic combination. After β-lactam antibiotics make the cell wall permeable, the aminoglycosides enter the bacteria more easily. Although it is sometimes disputed, combining bacteriostatic and bactericidal antimicrobials, however, may work antagonistically because bactericidal antimicrobials like β-lactam antibiotics have an effect only on growing bacteria, while growth is inhibited by bacteriostatic antimicrobials (Ocampo et al., 2014). Combining antimicrobials can be very effective but does need veterinary consideration.

### 4.1 Antimicrobial resistance

An important issue related to antimicrobial use is antimicrobial resistance. Bacteria can be intrinsically resistant to antimicrobials because of the working mechanism of that specific antimicrobial. Narrow spectrum penicillins, macrolides and lincosamides, for instance, are active against Gram-positive bacteria, but not against coliforms. Cloxacillin is a penicillin that can be used to treat penicillinase-producing *S. aureus*, because it is penicillinase stable. First-generation cephalosporins are mainly active against Gram-positives but also have a working spectrum against *Enterobacteriaceae* such as *E. coli*. Just like cloxacillin and nafcillin they are, however, resistant to penicillinase. *Streptococcus* spp. are not very sensitive to aminoglycosides, and *Klebsiella* spp., *Enterobacter* spp., *Serratia* spp., and *Pseudomonas* spp. are resistant to (amino) penicillins. More details on antimicrobial resistance in mastitis pathogens can be found in the publication of Owens and Murinda (2012) and in the standard work of Giguère et al. (2013).

Bacteria can acquire antimicrobial resistance through mutations in their DNA or by picking up DNA parts. Subsequent selection of these adapted bacteria may lead to a population of resistant bacteria. *Staphylococcus aureus* can acquire resistance against lincomycin and ampicillin. We refer to cross resistance if bacteria develop resistance against a certain antimicrobial (class), and by the same resistance mechanism became resistant to other antibiotics. A well-known example of cross resistance is resistance against ampicillin and against penicillin, an important resistance mechanism in *S. aureus* and non-*aureus* staphylococci (Barkema et al., 2006).

Streptococci are often resistant to lincomycin, while they are generally sensitive to penicillin and cephalosporins. Streptococci are variable-sensitive to macrolides. Required resistance is limited in *E. coli* and *Klebsiella* spp., with the exception of required ampicillin resistance in *E. coli*. When using β-lactam antibiotics, it has to be taken into account that, specifically in *Enterobacteriaceae*, genes encoding for the production of extended-spectrum β-lactamase (ESBL) and AmpC type β-lactamases can be selected for. Specifically the use of third- and fourth-generation cephalosporins has been described to cause this effect (Snow et al., 2012). Although the occurrence of ESBL/AmpC-producing bacteria in the udder seems of minor importance, prudent use of (β-lactam) antimicrobials is also important in dairy, in order to minimize development and spread of antimicrobial resistance.

### 4.2 Treatment considerations

In most countries veterinarians play an important role in advice on mastitis treatment, whereas farmers often do the actual treatments. Although applying mastitis tubes or injecting drugs seems a relatively simple action, there is quite some knowledge behind it, both with respect to the choices in drugs to be made as well as to the application. It is therefore of importance that farmer and veterinarian decide on the optimal approach of treatment of mastitis and develop a herd level mastitis treatment plan. In such a plan, guidelines are described with respect to the treatment of clinical and subclinical mastitis during lactation and subclinical mastitis at drying off (Lam et al., 2011a). Although there is an obvious difference in the approach of clinical and subclinical mastitis, there is only a gradual difference from the bacteriological perspective. Intramammary infections lead to inflammation that leads to non-visible (subclinical) or visible (clinical) inflammation of the mammary gland.

It is advisable to decide on a default treatment for each type of mastitis, including the classes of clinical mastitis (grade 1–3) as described above. Based on the specific herd situation and locally available drugs a specific herd treatment plan can be tailor made. Such a plan preferably is based on bacteriological culture results from the herd. Additionally the results of on-farm mastitis culturing can be included in such a protocol (Griffioen, 2020). Today good evidence exists that not using antibiotics to treat non-severe cases of clinical mastitis caused by Gram-negative bacteria (or when the test shows no growth) does not come with negative consequences for udder health or milk yield, allowing to implement this approach to reduce antibiotic consumption (Lago et al., 2011a,b). If that type of culture results is not available, it is important to realize that in case of grade 2 or 3 mastitis, it is more likely the mastitis case is caused by Gram-negative bacteria such as *E. coli* (Verbeke et al., 2014).

Antimicrobial treatment of subclinical mastitis has, specifically for *S. aureus*, been intensively studied with respect to the potential effect of treatment on bacteriological cure, as well as on transmission of intramammary infections (van den Borne et al., 2010). Treatment of subclinical mastitis cases ought to be done based on bacteriological culturing results in order to target treatment. This is also true for dry-off treatment, although in many countries antimicrobial use at drying-off is either based on Dairy Herd Improvement SCC results or blanket dry cow treatment is applied, a practice by which all cows are treated with long-acting antibiotics, irrespective of their infectious status (Scherpenzeel et al., 2016). In recent years the implementation of selective dry cow treatment has gained traction in many countries. That is partly based on political and societal pressure (Lam et al., 2017), but also on the fact that the application of selective dry cow treatment does not have a deteriorating effect on udder health while antimicrobial usage decreases significantly which often is also attractive from an economic point of view (Scherpenzeel et al., 2017; Lipkens, 2019).

## 5 Preventive management

What is true for many other diseases is also true for mastitis: the best way to solve mastitis problems is by preventing them. The level of occurrence of new mastitis problems is based either on the infection pressure or on the susceptibility of cows for new infections, the host resistance. Several factors influence both, of which milking will receive specific attention in this chapter. Host resistance may be specifically optimized through vaccination as described above, yet most measures affecting host resistance are non-specific.

A sometimes underestimated way to improve host resistance against mastitis is via breeding. Although selecting bulls based on a low mastitis incidence in their offspring does not solve acute mastitis problems it does have an important effect in the long run. In spite of the low heritability of clinical mastitis, the cumulative effect of breeding for a better udder health was proven effective (Heringstadt et al., 2003). Breeding values for udder health can be measured based on the occurrence of clinical and subclinical mastitis in the first second and third lactation of the offspring. Because doing so would result in six different breeding values, it seems efficient to combine that information in one udder health index, which was found to be effective (de Haas et al., 2008). Based on breeding, udder health can improve (or become worse), specifically if the same approach is used for several generations. At the same time we do have to realize there are many factors influencing the choices made with respect to breeding, of which udder health is only one. An important advice is to - if available - include mastitis breeding values in the decision-making process and to definitely not worsen the breeding value for udder health (Lam et al., 2017).

### *5.1 Housing and hygiene*

An important part of preventive mastitis management is also laid in housing and hygiene. Obviously hygiene is directly related to infection pressure, which can be quantified through hygiene scoring of cows (Schreiner and Ruegg, 2003). Housing has an effect on both the infection pressure as well as on host resistance. Apart from the negative effect on climate, suboptimal housing will lead to effects such as too little lying time, which will affect food intake and therefore host resistance. Cows should be able to lay down for 12–14 h each day, divided in 6–9 laying periods. That means a cow has to lay down and stand up frequently, and the size and comfort of cubicles should allow them to. Sand is generally accepted as an ideal bedding material for dairy cows in a loose housing system and leads to longer laying times than, for instance, matrasses (Cook and Nordlund, 2004).

Apart from cubicles, the quality of the walking paths in free stalls is also of utmost importance. Cows should be able to move freely and be able to, if necessary, evade other cows. Two cows ought to be able to easily pass each other and therefore a width of 3–4 m is needed, without dead ends (Rodenburg, 2003). Floors should be clean and dry and definitely not slippery.

Another point that needs serious attention with respect to housing is heat stress. The optimal temperature for a dairy cow is between −5°C and +18°C, with a substantial effect of humidity on the level of heat stress when temperatures are higher. Heat stress leads to lower food intake, lower milk production and more mastitis and claw problems. This is largely influenced by 2–3 h less laying time that cows in heat stress may have. To prevent heat stress, barns should not be overcrowded (there should be as many cubicles as cows), roofs should be isolated and ventilators can be used. In specific hot days sprinklers may be used, as long as there is enough ventilation in order to keep humidity low.

## 5.2 Nutrition

A central role in the preventive mastitis management is reserved for nutrition because of its effect on host resistance. Specifically in the dry and the transition period, optimal nutrition is very important. In the period around calving the dry matter intake of dairy cows is decreased, which is the start of a period of negative energy balance, which has a negative effect on the competence of the immune system resulting in more clinical disease such as mastitis (Pérez-Báez et al., 2019). The decreased capacity of the immune system during the transition period is the effect of a complex interrelationship between a decreased dry matter intake, a lack of vitamins, trace minerals and/or antioxidants, a negative energy and protein balance, increasing levels of free fatty acids, hypocalcemia and insufficient dietary fiber (Goff, 2006).

Dairy cows have to be fed according to their requirements with respect to energy and protein, and other nutritional aspects such as vitamins and trace minerals (Nutrient Requirements of Dairy Cattle, 2001). The energy requirements during the dry period do not always receive the attention they deserve and differ from the lactational period. It all starts with a daily milk production at drying off, which ought not to be higher than 12 kg per day. Milk production before drying off can be decreased through a timely decrease in concentrate fed at the end of lactation, decreasing energy gift through diluting the ration with hay or straw and by a lower frequency of milking. During the dry period, the body condition of dairy cows ought to be stable, which can be realized through feeding according to the energy requirements of the dry period, which is lower in the far-off (>3 weeks) than in the close-up (≤3 weeks) periods relative to calving (Roche et al., 2009). If the dry matter intake at the end of the dry period is insufficient, this may already lead to a mild negative energy balance, fat mobilization and an increase in circulating free fatty acids before calving, ultimately leading to (subclinical) ketosis (Goff, 2006). A general advice for the day of calving is to give 2 kg of concentrate which is gradually built up in the following period in order to meet the energy requirements of the lactating cow as good as possible. The same is true for protein gift during the dry and the transition period, with sufficient essential amino acids such as lysine and methionine (van der Drift et al., 2012).

## 6 Milking machine and milking

Because of the direct contact with the udder, the milking machine influences udder health to a large extent. In that respect, regular maintenance as well as a correct adjustment of the milking machine, be it a conventional system or an automated system, are of paramount importance. The basic principle of machine milking is to remove milk from the udder quickly, in an acceptable way from the perspective of animal welfare and with minimal risk to introduce or spread contagious mastitis pathogens. The milk-out process involves vacuum at the teat end, allowing the teat canal to open and the milk to flow out. The vacuum generated by the vacuum pump should be constant throughout milking whereas the pulsation system, causing the teat liners to open and collapse, ensures adequate blood circulation around the teat (Blowey and Edmondson, 2010).

Essential parts of the milking machine comprise the vacuum pump (the heart of the milking machine, generating vacuum by extracting air from the milking system), the regulator (leaking air into the system to maintain a stable vacuum), the vacuum gauge (indicating at what vacuum level the machine is operating), a sanitary trap (preventing milk from entering the pipelines that only carry air), the pipelines (either carrying air, milk or both), the milking cluster (combining the clawpiece and the teat cups containing the teat liners), the receiver vessel (containing the milk that was evacuated from the udders) and the automatic cluster removers (removing the cluster when milk flow has dropped below a specific threshold during a certain amount of time). The teat liners should allow fast and gentle milking, avoiding teat-end damage and should fit well with the average teat size of the herd. Because liners lose their elasticity over time, they should be replaced every 2500 milking when from rubber or every 5000–7500 milkings when made of silicone.

To check whether the milking machine is properly functioning, two types of test can be carried out: a static test, carried out in between milkings and a dynamic test, carried out during milking. The former may be easier to perform, but the latter generates much more information, allowing to pinpoint specific problems that occur during milking such as irregular vacuum fluctuations. An additional advantage of the dynamic test is that milking procedures can also be evaluated at the same time. Proper milking procedures, attention to detail and a clean environment are required to minimize mastitis and maximize production of quality milk with low SCC and low bacterial counts. Milking should be done by people who are responsible, trained, conscientious and have a clear vision that they are harvesting food for human consumption (NMC, 2013).

Milking should start with clean and stress-free cows entering the milking parlor. The first step in the ideal milking routine is foremilking that aims at detecting symptoms of clinical mastitis (based on symptoms such as cloths and abnormal milk), helps to start the oxytocin-dependent milk-let down and removes bacteria likely colonizing the teat canal. Next is the teat preparation using one dry cloth per cow aiming at cleaning the teats before the cluster is attached. Prefoaming of teats helps to remove dirt, helps to prevent mastitis and improves milk quality (Piepers et al., 2014) but, when applied, teats should always be dried to avoid liner slip due to milking wet teats. Ideally, the units are attached 60–90 s after foremilking was initiated, resulting in optimal milk flow and avoidance of blind milking, which negatively affects teat-end condition and udder health.

Rough hands of milkers act as potential fomites for contagious mastitis pathogens and cannot easily be disinfected. Therefore it is advisable to always wear disposable gloves throughout the milking process and rinse and disinfect them regularly. When the clusters have been removed, the milking process comes to an end with the application of post-milking teat disinfection to kill bacteria present on the teats due to the milking process. Keeping the cows on their feet for the next 30–60 min is important to allow the teat sphincter to close before the cows lay down in the cubicles, avoiding environmental mastitis.

Automated milking now has a prominent place in the dairy industry. Reasons to shift from conventional to automated milking include more flexibility, more than 2 milkings per day and a decrease in the need for external labor (Lam and De Vliegher, 2017). Rather than milkers milking the cows, an automated system takes over their tasks, including mastitis detection through sensor technology, teat preparation and post-milking teat disinfection. At the same time, it is the cow that decides to enter the milking robot rather than the milker deciding it is time for milking. Mastitis detection in a robotic milking context remains a challenge because of the moderate specificity, which means too many false-positive alerts are generated. Improvement in that area is needed leading to protocols and treatment plans that producers may implement in order to treat mastitic cows as well as to evaluate treatment results (Lam and De Vliegher, 2017).

## 7 Approaching herd health problems

Although the basic principles of mastitis prevention and control are simple and mostly have been around since the sixties of the previous century, solving mastitis issues on a farm in a sustainable manner can be challenging. It is essential to approach herd health problems with mastitis, be it clinical, subclinical or heifer mastitis, in a structured manner. Therefore it is essential to first picture the problem using the available data which should then be followed by herd-specific and evidence-based advice to solve it (Lam and De Vliegher, 2017).

Subclinical mastitis problems in a herd are typically detected because of an elevated bulk milk SCC which eventually leads to loss of premiums or, even worse, to penalties. To solve a subclinical mastitis issue, focusing on those cows that are responsible for the bulk milk SCC elevation is a first step, if less than 5% of the cows are the source of the problem, it may be wise to cull cows or to treat those that are likely to cure (typically young animals that became recently infected and are in later lactation). If more than 5% of cows are responsible for the elevated bulk milk SCC, than a herd-level analysis is needed. The elevated bulk milk SCC can be the consequence of too many new infections occurring during lactation (>8%) or during the dry period (>10%) or because of too many chronically infected cows (> 15%) combined with low cure rates (<30%). Herds with a high incidence of new intramammary infections should be differentiated from herds were too many chronically infected animals are present (Lam and De Vliegher, 2017).

Too many new infections occurring during lactation (e.g. as determined by comparing the current SCC with the SCC of the previous test-day) suggest a high infection pressure from the environment and/or a high transmission of infections during milking and/or an immunity issue. Culturing of cows with an elevated SCC will help to determine whether the problem is contagious or whether it is environmental in origin. Approaching a herd where mainly contagious pathogens are important (Table 6) is different from a herd where environmental pathogens are dominating (Table 7).

Cows with an elevated SCC that do not cure spontaneously can be approached in different ways: they can be treated (if likely to cure), they can be segregated from the milking herd or they can be culled (if unlikely to cure, typically older animals, animals in early lactation, chronically infected cows). If these cows are pregnant and in late lactation, they can be dried-off early and treated with antimicrobials, yet their udder health status should be monitored after calving.

Herds struggling with too many cows with an elevated SCC at first test-day (>15%), should primarily focus on the dry cow management. When doing so, it is important to also take into account the last test-day SCC prior to drying of, to picture the dry cow infection dynamics. If cows enter the dry period with an

**Table 6** Interventions to avoid an elevated bulk milk SCC caused by cows infected with contagious mastitis pathogens

Milking and the milking machine

- Timely check-up and maintenance of the milking machine
- Timely replacement of the teat liners
- Proper preparation of the cows prior to milking, with respect of the 60-s rule and the use of one (paper) cloth per cow
- Use of milking gloves that are replaced with every milking, combined with regular disinfection during milking
- Separate milking of cows with an elevated SCC or with clinical mastitis
- Disinfection of the teat liners after milking of cows with an elevated SCC or with clinical mastitis

Infection pressure

- Avoiding the purchase of cows
- If cows are acquired, checking udder health status through bacteriological culturing
- Culling of chronically infected cows
- Segregate infected and treated animals to be milked last
- Avoid mingling of sick cows and fresh cows in a sick bay/calving pen

*Source*: Lam and De Vliegher (2017).

**Table 7** Interventions to avoid an elevated bulk milk SCC caused by cows infected with environmental mastitis pathogens

Milking and the milking machine

- Timely check-up and maintenance of the milking machine
- Timely replacement of the teat liners
- Proper preparation of the cows prior to milking, with respect of the 60-s rule and the use of one (paper) cloth per cow
- Use of milking gloves that are replaced with every milking, combined with regular disinfection during milking
- Making sure the cows are on their feet for at least 30 min after milking

Infection pressure

- Optimal hygiene of the cubicles and (slatted) floors
- Acceptable stocking density
- Proper cubicle size
- Normal fecal consistency
- Avoid mingling of sick cows and fresh cows in a sick bay/calving pen
- Appropriate ventilation

Host resistance

- Optimized feeding
- Supplementation with vitamins and minerals when needed
- Avoidance of bovine viral diarrhea virus and other immunity compromising infections
- Stimulation of innate and/or acquired immunity (vaccination, immunomodulation)

*Source*: Lam and De Vliegher (2017).

**Table 8** Interventions to avoid an elevated infection incidence across the dry period

Infection pressure

- Optimal hygiene of the cubicles and (slatted) floors
- Acceptable stocking density
- Adapted cubicle size
- Appropriate ventilation

Host resistance

- Optimized feeding
- Supplementation with vitamins and minerals when needed
- Avoidance of Bovine Viral Diarrhea virus and other immunity compromising infections
- Stimulation of innate and/or acquired immunity (vaccination, immunomodulation)
- Reduced milk yield at drying-off
- Stable body condition score across the dry period

*Source*: Lam and De Vliegher (2017).

**Table 9** Interventions to avoid low cure rates across the dry period

Dry cow treatment

- Correct treatment selection of the animals
- Proper choice of antimicrobial
- Avoid too many chronically infected cows enter the dry period
- Gloves used to administer dry cow treatment or teat sealants
- Disinfection of the teats before administering the dry cow treatment or teat sealants
- Teat dipping after administering the dry cow treatment or teat sealants

Host resistance

- Optimized feeding
- Supplementation with vitamins and minerals when needed
- Avoidance of Bovine Viral Diarrhea virus and other immunity compromising infections
- Stimulation of innate and/or acquired immunity (vaccination, immunomodulation)
- Reduced milk yield at drying-off
- Stable body condition score across the dry period

*Source*: Lam and De Vliegher (2017).

elevated SCC and still have an elevated SCC after calving, they failed to cure during the dry period (cure rates across the dry period should be >70%). This could be due to an issue with the dry cow therapy used, yet it could also be that the cows entered the dry period with chronic infections that are always harder to cure, whatever antimicrobial is used.

A too high percentage of new infections across the dry period (>10%), suggests a too high infection pressure and/or an impaired resistance during that period (Table 8).

Herds struggling with low cure rates across the dry period (<70%) have to look at the different risk factors related to either the dry cow treatment or the resistance of the animals (Table 9).

When herds have a monthly clinical mastitis incidence >2%, they are struggling with a clinical mastitis problem and should preferably be investigated (Lam and De Vliegher, 2017). The problem can be caused by too many new cases or by too many repeat cases (at least 14 days apart; >20% of all cases), or both. New cases either originate from lactation or from the dry period. Both can be caused by either a too high infection pressure and/or reduced resistance. Bacteriological culturing of clinical mastitis cases helps to differentiate between an environmental and a contagious problem, as well as data recording (to differentiate new from repeat cases). Repeat cases typically occur when cows are chronically infected, when treatments fail and/or when the immunity of the cows is suboptimal.

Heifer mastitis is defined as subclinical and/or clinical mastitis in primiparous cows, prior to or just after first calving, with potential detrimental consequences for the performances of the affected animals. A herd is considered to have a heifer mastitis problem if >15% of heifers have clinical mastitis around calving or if >15% of all heifers have a first test-day SCC

**Table 10** Ten-point program to prevent and control heifer mastitis

1. Improve general udder health management at the farm level to decrease the pressure of infection with udder pathogens from older cows to heifers
2. Control for cross-suckling in calves and young stock
3. Implement an effective and efficient fly control system
4. Keep young and primigravid heifers in a clean and hygienic environment and separate from multiparous animals - provide as much attention to this group of animals related to hygiene and cleanliness as is spent on lactating animals
5. Avoid any nutritional deficiency - monitor vitamin E and selenium levels when any doubt exists, especially in relation to clinical mastitis; zinc, copper and vitamin A play a role as well and could be checked
6. Minimize the risk of negative energy balance before and after calving through appropriate transition feeding systems
7. Reduce the incidence of udder edema through optimized peripartum management
8. Minimize stress around calving (e.g. by not moving heifers to the calving pen when already in labor) and minimize incidence of dystocia and peripartum disease
9. Consider use of internal teat sealants prepartum where a high risk of environmental mastitis exists in the peripartum period
10. Use prepartum antibiotic treatment in heifers under certain conditions only:

    - under the supervision of the herd veterinarian, within the context of a valid veterinary/client/patient relationship
    - after quantification of the problem and identification of major pathogens (non-*aureus* staphylococci) as the cause through culturing
    - choice of the antibiotics should be based on antimicrobial susceptibility testing
    - testing for residues before every milk delivery
    - upgrading of management at the same time - discontinue treatment as soon as new management strategies become effective

*Source*: De Vliegher et al. (2012).

(measured between 10 and 35 days in milk) >150 000 cells/mL (De Vliegher et al., 2012). Heifer mastitis is a multifactorial disease, meaning that all aspects related to this disease should be optimized. Each herd should invest in the prevention and control of heifer mastitis based on the interventions described in Table 10 (De Vliegher et al., 2012).

Although we have addressed clinical, subclinical and heifer mastitis as separate subjects, it is important to understand that in reality they might occur at the same time as they can be interconnected. Addressing such problems successfully will not only need an evidence-based approach, but also excellent communication skills when working with the producer or farm manager and all staff involved.

## 8 Conclusion and future trends

In the modern dairy industry, with increasing individual cow production, herd size, technology and demands from government, public and consumers, bovine mastitis is of more importance than ever. Still, the basics and tools to prevent and control the disease did not change: decreasing infection pressure combined with optimization of host resistance. The tools to realize this may have changed due to changing circumstances and possibilities, but they have been around for many years and many producers show us day by day that it is possible to produce high-quality milk in an efficient and economically viable way, with minimal mastitis. At the same time, a considerable proportion of herd's struggles with subclinical and/or clinical mastitis, indicating the knowledge we have is not implemented. It may simply not have reached the producers, they do not see a possibility to implement that knowledge, or they do not see the value of implementing it in their daily routine. We are convinced that the best way to further improve udder health worldwide would be to implement the knowledge we already have. In many situations that is not done because advisors do not use all possibilities to reach producer, farm manager and the farm staff (Lam et al., 2011b). Changing this, however, is easier said than done and asks input from all stakeholders, because part of the effect depends on whether a tipping point in mindset toward subjects such as mastitis and antibiotic use is realized (Lam et al., 2017).

Because of the emergence of antibiotic resistance in human- and animal-associated bacteria, and the potential role antimicrobial (mis)use in animal husbandry plays, there is a strong social demand to lower the usage of antimicrobials, also in the dairy industry. As described earlier in this chapter, this has led to a different approach toward antimicrobial use at drying off. Whereas blanket dry cow therapy has been the standard for many years, today selective dry cow therapy (only treating infected cows at dry of with long-acting antimicrobials) has become an option that many dairy producers do no longer refuse. The same is true for selective treatment of non-severe cases of clinical mastitis using rapid on-farm tests. Using less antibiotics means that other approaches have to be chosen to prevent intramammary infections, such as the use of teat sealants in dry cows and the optimization of the dry cow management. At the same time there is a need for further fine-tuning of treatment protocols with respect to proper antimicrobial use. With worldwide increasing herd sizes, more data become available that can be used to optimize this type of protocols. Right now, we are at the start of an era in which, given these bigger herds, artificial intelligence will be used in order to learn from experiences on the herd itself, and will optimize decisions taken without much input from the farm personnel (Cabrera et al., 2020).

Through our increasing understanding of the biology of the udder, which most likely has its own microflora, and new diagnostic approaches, we will be more and more able to dive into details with respect to the (patho)physiology of the udder. Mastitis may be a dysbiosis rather than the consequence of the invasion of by a pathogen (Oikonomou et al., 2014). Although there is still a long way to go in that type of research it paves the way for new preventive and curative approaches using, for instance, prebiotics or probiotics. On the other hand the question will have to be answered what part of the information that is available, will also be valuable for the dairy farmer and the dairy cow. Valorizing data by adding scientific knowledge and specific experience turning it into valuable information will be the challenge in the forthcoming years.

## 9 Where to look for further information

An important source of information on udder health and milk quality is provided through the National Mastitis Council (NMC), a global, US-based non-profit organization for mastitis control and milk quality. The NMC organizes meetings specifically focusing on udder heath and milk quality on a regular basis in North-America, is involved in organizing conferences throughout the world and has an informative website (www.nmconline.org). The NMC also publishes handbooks such as Current Concepts of Bovine Mastitis and the Laboratory Handbook on Bovine Mastitis, guidelines and protocols, factsheets and reports. The World Association for Buiatrics publishes interesting video material from earlier conferences which contains information on mastitis as well as other interesting information on cattle health (www.buiatrics.com/videos).

Apart from these international sources, there are several local initiatives in different countries, such as the Canadian Bovine Mastitis and Milk Quality Research Network (CBMRN) which can be found at: www.medvet.umontreal.ca/rcrmb/en/, the M-teamUGent in Belgium (www.m-team.UGent.be) and the UGA-team in the Netherlands (www.gddiergezondheid.nl/uga). Another interesting source of information is the M²-magazine that brings news related to mastitis and milk quality from around the world for the dairy professional (www.m2-magazine.org).

## 10 References

Alnakip, M. E., Quintela-Baluja, M., Böhme, K., Fernández-No, I., Caamaño-Antelo, S., Calo-Mata, P. and Barros-Velázquez, J. (2014). The immunology of mammary gland of dairy ruminants between healthy and inflammatory conditions. *J. Vet. Med.* 2014: 1-31. http://doi.org/10.1155/2014/659801.

Barkema, H. W., Schukken, Y. H. and Zadoks, R. N. (2006). Invited review: the role of cow, pathogen, and treatment regimen in the therapeutic success of bovine *Staphylococcus aureus* mastitis. *J. Dairy Sci.* 89(6): 1877-1895.

Blowey, R. and Edmondson, P. (2010). *Mastitis Control in Dairy Herds* (2nd edn.). CAB International, Oxfordshire, UK.

Borne, B. H. P. van den, van Schaik, G., Lam, T. J. G. M. and Nielen, M. (2010). Therapeutic effects of antimicrobial treatment during lactation of recently acquired bovine subclinical mastitis. Two linked randomized field trials. *J. Dairy Sci.* 93(1): 218-233.

Cabrera, V. E., Barrientos-Blanco, J. A., Delgado, H. and Fadul-Pacheco, L. (2020). Symposium review: Real-time continuous decision making using big data on dairy farms. *J. Dairy Sci.* 103(4): 3856-3866.

Cook, N. B. and Nordlund, K. V. (2004). Behavioral needs of the transition cow and considerations for special needs facility design. *Vet. Clin. North Am. Food Anim. Pract.* 20(3): 495-520.

Damm, M., Holm, C., Blaabjerg, M., Novak Bro, M. N. and Schwarz, D. (2017). Differential somatic cell count–a novel method for routine mastitis screening in the frame of Dairy Herd Improvement testing programs. *J. Dairy Sci.* 100(6): 4926-4940.

Denis, M., Parlane, N. A., Lacy-Hulbert, S. J., Summers, E. L., Buddle, B. M. and Wedlock, D. N. (2006). Bactericidal activity of macrophages against *Streptococcus uberis* is different in mammary gland secretions of lactating and drying off cows. *Vet. Immunol. Immunopathol.* 114(1-2): 111-120.

De Vliegher, S., Fox, L. K., Piepers, S., McDougall, S. and Barkema, H. W. (2012). Invited review: mastitis in dairy heifers: nature of the disease, potential impact, prevention, and control. *J. Dairy Sci.* 95(3): 1025-1040.

Drift, S. G. A. van der, Houweling, M., Schonewille, J. T., Tielens, A. G. M. and Jorritsma, R. (2012). Protein and fat mobilization and associations with serum β-hydroxybutyrate concentrations in dairy cows. *J. Dairy Sci.* 95(9): 4911-4920.

Friggens, N. C., Chagunda, M. G. G., Bjerring, M., Ridder, C., Hojsgaard, S. and Larsen, T. (2007). Estimating degree of mastitis from time-series measurements in milk: a test of a model based on lactate dehydrogenase measurements. *J. Dairy Sci.* 90(12): 5415-5427.

Giguère, S., Prescott, J. F. and Dowling, P. M. (2013). *Antimicrobial Therapy in Veterinary Medicine* (5th edn.). Wiley-Blackwell, Hoboken, NJ, 704 pp.

Goff, J. P. (2006). Major advances in our understanding of nutritional influences on bovine health. *J. Dairy Sci.* 89(4): 1292-1301.

González, R. N., Cullor, J. S., Jasper, D. E., Farver, T. B., Bushnell, R. B. and Oliver, M. N. (1989). Prevention of clinical coliform mastitis in dairy cows by a mutant *Escherichia coli* vaccine. *Can. J. Vet. Res.* 53(3): 301-305.

Griffioen, K. (2020). Diagnostic information for mastitis treatments on dairy farms. PhD thesis. Utrecht University, The Netherlands.

Haas, Y. de, Ouweltjes, W., ten Napel, J., Windig, J. J. and de Jong, G. (2008). Alternative somatic cell count traits as mastitis indicators for genetic selection. *J. Dairy Sci.* 91(6): 2501-2511.

Hand, K. J., Godkin, A. and Kelton, D. F. (2012). Milk production and somatic cell counts: a cow-level analysis. *J. Dairy Sci.* 95(3): 1358–1362.

Heringstadt, B., Rekaya, R., Gianola, D., Klemetsdal, G. and Weigel, K. A. (2003). Genetic change for clinical mastitis in Norwegian cattle: a threshold model analysis. *J. Dairy Sci.* 86(1): 369–375.

Huijps, K., Hogeveen, H., Lam, T. J. G. M. and Oude Lansink, A. G. J. M. (2010). Costs and efficacy of management measures to improve udder health at Dutch dairy farms. *J. Dairy Sci.* 93(1): 115–124.

Kamphuis, C., Pietersma, D., van der Tol, R., Wiedemann, M. and Hogeveen, H. (2008). Using sensor data patterns from an automatic milking system to develop predictive variables for classifying clinical mastitis and abnormal milk. *J. Dairy Sci.* 91: 169-181.

Kapoor, G., Saigal, S. and Elongavan, A. (2017). Action and resistance mechanisms of antibiotics: A guide for clinicians. *J. Anaest. Clin. Pharm.* 33(3): 300–305.

Lago, A., Godden, S. M., Bey, R., Ruegg, P. L. and Leslie, K. (2011a). The selective treatment of clinical mastitis based on on-farm culture results: I. Effects on antibiotic use, milk withholding time, and short-term clinical and bacteriological outcomes. *J. Dairy Sci.* 94(9): 4441-4456.

Lago, A., Godden, S. M., Bey, R., Ruegg, P. L. and Leslie, K. (2011b). The selective treatment of clinical mastitis based on on-farm culture results: II. Effects on lactation performance, including clinical mastitis recurrence, somatic cell count, milk production, and cow survival. *J. Dairy Sci.* 94(9): 4457–4467.

Lam, T. J. G. M., Olde Riekerink, R., Sampimon, O. and Smith, H. (2009). Mastitis diagnostics and performance monitoring: a practical approach. *Ir. Vet. J.* 62(Suppl. 4): S34-39.

Lam, T. J. G. M., van Veersen, J. C. L., Sampimon, O. C. and Olde Riekerink, R. G. M. (2011a). On-farm udder health monitoring. *Tierärtzl. Prax.* 39(2): 95–100.

Lam, T. J. G. M., Jansen, J., van den Borne, B. H. P., Renes, R. J. and Hogeveen, H. (2011b). What veterinarians need to know about communication to optimise their role as advisors on udder health in dairy herds. *New Zeal. Vet. J.* 59(1): 8–15.

Lam, T. J. G. M. and De Vliegher, S. (Eds) (2017). *Handbook Udder health in Dairy Cattle (in Dutch)* (1st edn). Communication in Practice, Nijmegen, the Netherlands, 333pp.

Lam, T. J. G. M., Jansen, J. and Wessels, R. J. (2017). The RESET Mindset model applied on decreasing antibiotic usage in dairy cattle in the Netherlands. *Ir. Vet. J.* 70: 5.

Lipkens, Z. (2019). Selectively drying off dairy cows: impact on future performance and antimicrobial consumption. PhD Thesis. Ghent University, Belgium.

Martins, S. A. M., Martins, V. C., Cardoso, F. A., Germano, J., Rodrigues, M., Duarte, C., Bexiga, R., Cardoso, S. and Freitas, P. P. (2019). Biosensors for on-farm diagnosis of mastitis. *Front. Bioeng. Biotechnol.* 7: 186. https://doi.org/10.3389/fbioe.2019.00186.

Napel, J. ten, de Haas, Y., de Jong, G., Lam, T. J. G. M., Ouweltjes, W. and Windig, J. J. (2009). Characterization of distributions of somatic cell counts. *J. Dairy Sci.* 92(3): 1253–1264.

National Research Council (2001). *Nutrient Requirements of Dairy Cattle* (7th rev. edn.). National Academy Press, Washington DC.

NMC (2013). *NMC Recommended Milking Procedures*. Available at: https://www.nmconline.org/wp-content/uploads/2016/09/Recommended-Milking-Procedures.pdf.

NMC (2017). *NMC Laboratory Handbook on Bovine Mastitis*. National Mastitis Council, New Prague, MN.

Nonnemann, B., Lyhs, U., Svennesen, L., Kristensen, K. A., Klaas, I. C. and Pedersen, K. (2019). Bovine mastitis bacteria resolved by MALDI-TOF mass spectrometry. *J. Dairy Sci.* 102(3): 2515–2524.

Ocampo, P. S., Lázár, V., Papp, B., Arnoldini, M., zur Wiesch, P. A., Busa-Fekete, R., Fekete, G., Pál, C., Ackermann, M. and Bonhoeffer, S. (2014). Antagonism between bacteriostatic and bactericidal antibiotics is prevalent. *Antimicr Agents Chemother.* 58: 4573–4582.

Oikonomou, G., Bicalho, M. L., Meira, E., Rossi, R. E., Foditsch, C., Machado, V. S., Vieira Teixeira, A. G., Santisteban, C., Schukken, Y. H. and Bicalho, R. C. (2014). Microbiota of cow's milk; distinguishing healthy, sub-clinically and clinically diseased quarters. *PLoS ONE* 9(1): e85904. https://doi.org/10.1371/journal.pone.0085904.

Owens, W. E. and Murinda, S. P. (2012). Antimicrobial resistance of mastitis pathogens. *Vet. Clin. Food Anim. Prac.* 28: 165–185.

Penry, J. F., Brightling, P. B., Dyson, R. S. and Payne, M. S. (2011). Developing new veterinary services in milk quality. *New Zeal. Vet. J.* 59: 24–27.

Pérez-Báez, J., Risco, C. A., Chebel, R. C., Gomes, G. C., Greco, L. F., Tao, S., Thompson, I. M., do Amaral, B. C., Zenobi, M. G., Martinez, N., Staples, C. R., Dahl, G. E., Hernández, J. A., Santos, J. E. P. and Galvão, K. N. (2019). Association of dry matter intake and energy balance prepartum and postpartum with health disorders postpartum: Part II. Ketosis and clinical mastitis. *J. Dairy Sci.* 102(10): 9151–9164.

Piepers, S., Zrimšek, P., Passchyn, P. and De Vliegher, S. (2014). Manageable risk factors associated with bacterial and coliform counts in unpasteurized bulk milk in Flemish dairy herds. *J. Dairy Sci.* 97(6): 3409–3419.

Pinzón-Sánchez, C. and Ruegg, P. L. (2011). Risk factors associated with short-term post-treatment outcomes of clinical mastitis. *J. Dairy Sci.* 94(7): 3397–3410.

Pomeroy, B., Gurjar, A., Sipka, A., Klaessig, S., Salmon, S., Quesnell, R. and Schukken, Y. H. (2016). Intramammary immunization with ultraviolet-killed *Escherichia coli* shows partial protection against late gestation intramammary challenge with a homologous strain. *J. Dairy Sci.* 99(11): 9014–9026.

Riollet, C., Rainard, P. and Poutrel, B. (2000). Cells and cytokines in inflammatory secretions of bovine mammary gland. *Adv. Exp. Med. Biol.* 480: 247–258.

Roche, J. R., Friggens, N. C., Kay, J. K., Fisher, M. W., Stafford, K. J. and Berry, D. P. (2009). Invited review: body condition score and its association with dairy cow productivity, health, and welfare. *J. Dairy Sci.* 92(12): 5769–5801.

Rodenburg, J. (2003). Free stall construction and design. *Nat. Res. Agr. Eng. Serv.* 148: 163–176.

Scherpenzeel, C. G. M., den Uijl, I. E. M., van Schaik, G., Olde Riekerink, R. G. M. O., Hogeveen, H. and Lam, T. J. G. M. (2016). Effect of different scenarios for selective dry-cow therapy on udder health, antimicrobial usage, and economics. *J. Dairy Sci.* 99(5): 3753 –3764.

Scherpenzeel, C. G. M., Hogeveen, H., Maas, L. and Lam, T. J. G. M. (2017). Economic optimization of selective dry cow treatment. *J. Dairy Sci.* 101: 1530–1539.

Schreiner, D. A. and Ruegg, P. L. (2003). Relationship between udder and leg hygiene scores and subclinical mastitis. *J. Dairy Sci.* 86(11): 3460–3465.

Schukken, Y. H., Wilson, D. J., Welcome, F., Garrison-Tikofsky, L. and Gonzalez, R. N. (2003). Monitoring udder health and milk quality using somatic cell counts. *Vet. Res.* 34(5): 579–596.

Schukken, Y. H., Bronzo, V., Locatelli, C., Pollera, C., Rota, N., Casula, A., Testa, F., Scaccabarozzi, L., March, R., Zalduendo, D., Guix, R. and Moroni, P. (2014). Efficacy of vaccination on *Staphylococcus aureus* and coagulase-negative staphylococci intramammary infection dynamics in 2 dairy herds. *J. Dairy Sci.* 97(8): 5250–5264.

Snow, L. C., Warner, R. G., Cheney, T., Wearing, H., Stokes, M., Harris, K., Teale, C. J. and Coldham, N. G. (2012). Risk factors associated with extended spectrum beta-lactamase *Escherichia coli* (CTX-M) on dairy farms in North West England and North Wales. *Prev. Vet. Med.* 106(3–4): 225–234.

Verbeke, J., Piepers, S., Supré, K. and De Vliegher, S. (2014). Pathogen-specific incidence rate of clinical mastitis in Flemish dairy herds, severity, and association with herd hygiene. *J. Dairy Sci.* 97(11): 6926–6934.

Wilson, D. J., Mallard, B. A., Burton, J. L., Schukken, Y. H. and Gröhn, Y. T. (2009). Association of *Escherichia coli* J5-specific serum antibody responses with clinical mastitis outcome for J5 vaccinate and control dairy cattle. *Clin. Vac. Immunol.* 16(2): 209–217.

Zadoks, R. N., Allore, H. G., Barkema, H. W., Sampimon, O. C., Wellenberg, G. J., Gröhn, Y. T. and Schukken, Y. H. (2001). Cow- and quarter-level risk factors for *Streptococcus uberis* and *Staphylococcus aureus* mastitis. *J. Dairy Sci.* 84(12): 2649–2663.

# Chapter 12

# Optimising foot health in dairy cattle

*Nick J. Bell, The University of Nottingham, UK*



## 1 Introduction

Lameness is the behavioural manifestation of a range of locomotor system disorders, most commonly due to a painful lesion in the foot (Murray et al., 1996; Barker, 2007). This chapter will review concepts to control lameness due to major *foot lesions* affecting dairy cattle: namely, sole ulcers, white line lesions, sole bruising and digital dermatitis (Fig. 1a-d). For further details on some of the other causes of lameness, readers are referred to the clinical texts at the end of the chapter.

This chapter will start by reviewing the methods for quantifying lameness. Without quantification, management and optimisation fall to intuition and guesswork. With fine margins for error or profit there should be no acceptance of guesswork. In the middle sections of this chapter a major paradigm shift is described related to our understanding of claw horn lesions. The approaches to digital dermatitis control are also reviewed. Finally the chapter concludes with emerging and future challenges facing the dairy industry and a case study.

### 1.1 Importance of lameness

Lameness in dairy cattle is a major welfare concern because of the painful nature of lesions (Whay et al., 1998) and the unacceptably high prevalence (Anon, 2009a). It is also economically important to the dairy industry, limiting

http://dx.doi.org/10.19103/AS.2020.0086.14

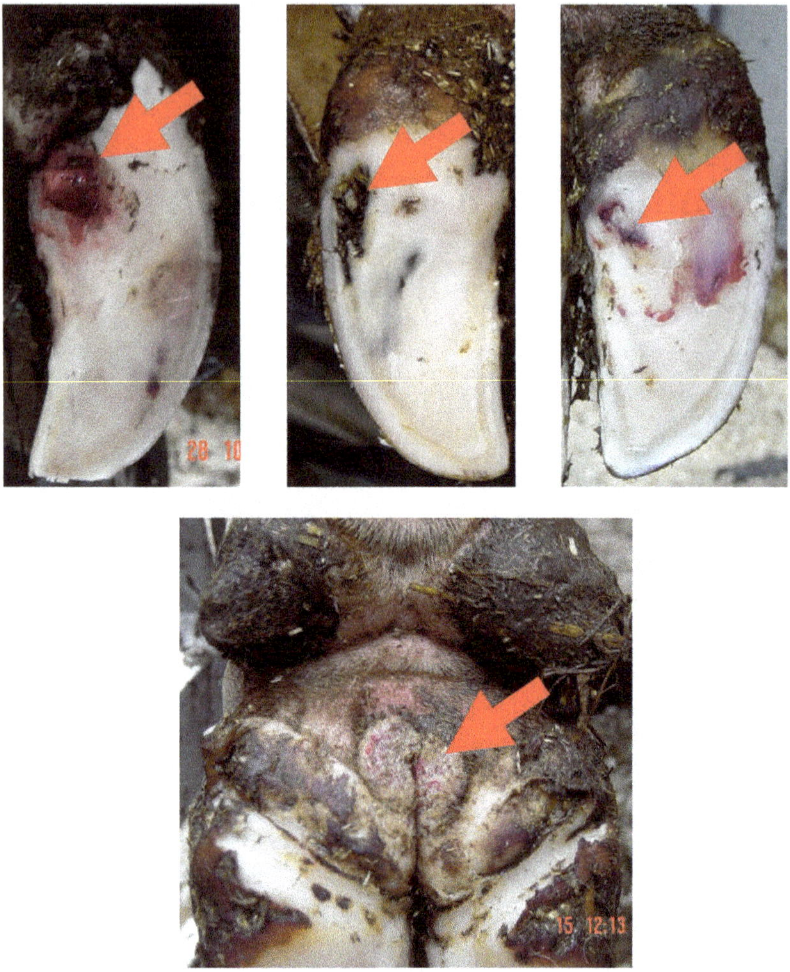


**Figure 1** (a) Sole ulcer - note the protrusion of swollen corium from the ulcerated sole horn in the central sole region (zone 4 - see Section 1.4). In chronic cases these can be pedunculated masses of granulation tissue. (b) While line lesion - the junction between the wall and sole is prone to separation and ascending infection. Note the most common site for this is zone 3, as in the photograph. (c) Sole bruising - focal inflammation tends to occur in the sites where sole ulcers and white line lesions occur most commonly. Usually haemorrhagic pigment would be visible in the horn at the painful site. Note that there may be haemorrhagic pigment present which is incidental by the time the foot is examined. (d). Digital dermatitis - this infection occurs most commonly on the skin between the heel bulbs, and has a number of presentations which are described in a later section on digital dermatitis scoring. The lesion in the photograph is a classic ulcerative and granulating acute lesion.

milk production, reducing fertility and shortening lifespan of the affected dairy cow (Willshire and Bell, 2009). Herds with lower lameness prevalence (<10%) have a higher technical efficiency in spite of high labour inputs and lower stocking rates (Barnes et al., 2011). Costs do vary according to the lesion causing lameness, severity of lesion, stage of lactation, parity and productivity of the affected animal. Stochastic modelling has been used to estimate the cost of the three major causes of lameness (Cha et al., 2010; Dolecheck et al., 2019) with the most recent study estimating costs of $64 ± 24, $152 ± 26 $178 ± 29 for digital dermatitis, white line and sole ulcer, respectively (Dolecheck et al., 2019). Mechanistic calculations (Willshire and Bell, 2009) are simpler to understand but have arrived at higher estimates per case partly because input costs have been higher with historic data but also because the approach assumes many factors are binary such as severity and management, and population characteristics are homogenous.

Stochastic modelling can be used to simulate effects on production (costs) with the onset of lesions in different parities and stages of lactation. All cost estimates invariably rely on generalisation and simplification of case definition which cannot easily account for how affected cows have been detected, treated and managed. For example, in one randomised control study involving cows screened for lameness, every 2 weeks, there was no measurable impact of new lameness cases on milk yield (Leach et al., 2012); milk yield is the major cost contributor in most economic models. Likelihood to cull will also vary according to system type, with some extended grazing systems more likely to cull lame cows which cannot walk long distances to pasture or if they fail to conceive in a breeding season. Detection, treatment and management factors could be introduced into future models. Further work is needed to work out economic impact over a cow lifetime (personal communication, Nigel Cook) in a real herd scenario.

## 1.2 Lameness scoring systems

One of the major challenges facing lameness research is quantifying lameness and case definition. Lameness presents with many different combinations of behaviour change (Whay, 2002) with a spectrum of severity, which varies according to the level of pain perceived by the cow, the location of the lesion and the ability of the cow to alter gait to mitigate pain. This means it is difficult to standardise assessment within studies and great caution is required with comparisons between studies using different scores.

A variety of behavioural signs are used to identify lame cows (O'Callaghan et al., 2003; Flower and Weary, 2006). The behaviours most predictive of lesions include uneven weight-bearing (a limp), shortening of strides (tenderness), arching of the back, speed of locomotion, lowered head carriage and head-bobbing during locomotion. These behavioural signs have been used to screen cows for treatment (Groenevelt et al., 2014) and have been combined into lameness severity scores (Whay, 2002). Multi-behaviour composite scores are more reliable in predicting lesions causing lameness than single component scores like the Sprecher back arch score (O'Callaghan et al., 2003; Flower and Weary, 2006). However, the useful measures of test performance, namely sensitivity, specificity, positive predictive value and accuracy have not been fully investigated (Hogeveen et al., 2010; Bell and Walker, 2016).

The combined behaviour score is generally referred to as the locomotion score (Manson and Leaver, 1988a), although some authors will describe using lameness scores when applied specifically to lameness signs and mobility scoring when used in applied setting with farmers (Bell and Huxley, 2009a). The probability of detecting a lesion increases with rising score (Bell, 2006), as does the severity of lesion usually (O'Callaghan et al., 2003). There may be an increase in the number of limbs affected with rising score although this has not been scientifically reported. A randomised control trial involving delayed implementation of treatment found an increased probability of multi-limb lameness (Thomas et al., 2016).

Locomotion scoring has been applied to research (Manson and Leaver, 1988a), breeding (Boelling and Pollott, 1997) and farm-screening protocols (Groenevelt et al., 2014). The most commonly cited locomotion scoring systems include the Manson and Leaver Score (Manson and Leaver, 1988a), Sprecher back-arch score (Sprecher et al., 1997), EU Welfare Quality score (Anon, 2009b), the Canadian 5-point score (Flower and Weary, 2006) and the UK AHDB mobility score (Anon, 2016) (Table 1).

Intra-observer repeatability and inter-observer reliability can be improved with training (Engel et al., 2003) and practice (Winckler and Willen, 2001). When analysing scores, statistics which accounts for chance agreement (i.e. Fleiss' kappa correlation), unbalanced prevalence in each score and marginal homogeneity (i.e. Gwet's Agreement Coefficient) should be used, similar to Vanhoudt et al. (2019).

When validating lameness assessment and lesion detection, a thorough clinical examination is generally required to diagnose the disorder (Groenevelt et al., 2014) using five cardinal signs of inflammation: heat, pain, swelling, redness and loss of function. Lameness is a form of loss of function. Inflammation can be assumed if characteristic lesions with exposed corium, like sole ulcers, are visible. One exception might be sole/white line haemorrhage, the term used to indicate the presence of blood pigment in the hoof horn. The presence of pigment on the ground-facing surface of the sole, over 10mm thick, may have been deposited 10weeks before the examination, given a typical rate of deposition of 4–5mm per month (Vermunt and Greenough, 1995b).

**Table 1** Lameness scores actively used worldwide

| Publication | Guidance | Comments |
|---|---|---|
| Sprecher et al. (1997) | https://www.icar.org/Guidelines/07.5-Functional-traits-Lameness-in-Dairy-Cattle.pdf (accessed 24/9/2020) | Back arch score is used widely around the world, particularly the US and central Europe |
| Flower and Weary (2006) | http://www.nfacc.ca/pdfs/codes/dairy_code_of_practice.pdfappendix F (accessed 24/9/2020) | UBC score used in Canada |
| Leach et al. (2012) | https://ahdb.org.uk/knowledge-library/mobility-scoring-for-dairy-cows (accessed 24/9/2020) | The AHDB Mobility score used in the UK |
| Leach and Whay (2009) | http://www.welfarequality.net/media/1122/wqr14.pdf (accessed 24/9/2020) | Welfare Quality score used in parts of the EU |

As many lesions causing lameness may be deep within the foot or joint, visible lesion detection alone is insufficient for validation studies. Some researchers have attempted to standardise painful foot lesion detection with diagnostic aids. Techniques such as nociceptive threshold testing (Whay et al., 1998), calibrated hoof testers (Rajkondawar et al., 2002), skin temperature (Lin, 2015), thermography (Alsaaod et al., 2014), analgesia (Chapinal et al., 2010) and local anaesthetic (Rushen et al., 2007) have been adopted in research. Novel technologies are being used to create automated lameness detection systems with systematic reviews on this subject available (Van Nuffel et al., 2015a,b; Bell and Walker, 2016; Alsaaod et al., 2019).

## 1.3 Lameness prevalence, duration and incidence

Prevalence of lameness (percentage of cows lame at one time, usually evaluated at the level of the lactating cow herd, including dry cows if feasible) varies considerably around the world and over time. Given lameness scoring methods vary in the criteria used, robust comparisons of clinical lameness cannot be made. However, Table 2 shows the levels of clinical lameness reported within the UK where some standardisation has occurred (see www.RoMS.org.uk).

As methods used for international reports differ and thresholds for clinical lameness vary the prevalence of lameness worldwide should be interpreted cautiously but Table 3 shows lameness is a widespread challenge.

One of the main aspects of lameness which remains poorly assessed is the duration of lameness, with few studies reporting this (Leach et al., 2012) and several inferring a long duration (Clarkson et al., 1996; Green et al., 2002; Whay, 2002). Leach et al. (2012) reported that the median duration of lameness in a control group of cows from 4 British herds with passive surveillance methods (spotting lame cows at routine inspections, herding and milking) was 65 days. The use of automated lameness detection systems could be used to monitor these parameters in future.

Traditional lameness epidemiology has reported lameness incidence (e.g. limb cases per 100 cows per year). Incidence rate has a number of limitations when applied to lameness. First lameness is under-recognised with only one-third to one-quarter of cases typically identified by the farm staff for treatment (Wells et al., 1993; Whay et al., 2003; Leach et al., 2013). Lameness incidence is therefore largely determined by surveillance methods, with herds using lameness scores every 1–2 weeks having the highest lameness incidence rates but lowest major lesion incidence, lowest lameness prevalence and shortest lameness duration by virtue of the proactive footcare cows receive. Incidence is best applied at the level of the lesion. However, many lesions causing lameness are chronic (such as toe necrosis or corkscrew claws), recurrent (sole ulcers, thin soles, digital dermatitis) with many animals

**Table 2** Reported lameness prevalence for 12 surveys in the UK, with data gathered by similar but not identical scoring approaches. Adapted from the CHAWG report 2018 (Anon, 2018).

| Survey start year | Lameness prevalence Max (%) | Lameness prevalence Min (%) | Lameness prevalence Average or median (%) | Numbers Dairy herds | Numbers Dairy cows | Location | Reference |
|---|---|---|---|---|---|---|---|
| 1989 | 53.9 | 2.0 | 20.6 | 37 | 11 399 | NW & SW England, Wales | Clarkson et al. (1996) |
| 2000 | 50.0 | 0.0 | 22.1 | 53 | 7407 | SW & Midlands England | Whay et al. (2003) |
| 2000 | 39.0 | 15.0 | 24.0 | 37 | 2724 | Scotland, England, Wales | Haskell et al. (2006) |
| 2002 | 74.2 | 6.8 | 24.2 | 28 | | SW England | Huxley et al. (2004) |
| 2004 | 42.0 | 4.0 | 18.0 | 80 | 28 698 | Scotland, England, Wales | Rutherford et al. (2009) |
| 2006 | 79.0 | 0.0 | 36.8 | 205 | 28 277 | SW & Midlands England, Wales | Main et al. (2010) |
| 2010 | 77.0 | 3.0 | 26.7 | 207 | 26 289 | SW England | Shepherd (2016) |
| 2011 | 53.5 | 0.0 | 18.2 | 92 | | England, Wales | Heath et al. (2014) |
| 2012 | 50.0 | 0.0 | 32.0 | 44 | 11 800 | NW England | Fisher and Atkinson (2013) |
| 2012 | 61.0 | 7.0 | 30.0 | 42 | 5620 | Midlands England | Randall et al. (2019) |
| 2013 | 42.0 | 7.0 | 22.0 | 51 | 10 899 | South & Midlands England | Collins (2016) |
| 2015 | 65.0 | 6.0 | 31.6 | 61 | 14 700 | England and Wales | Griffiths et al. (2018) |

**Table 3** Reports of lameness prevalence around the world since 2014

| Reference | Reported lameness prevalence (%) | Country |
|---|---|---|
| Fabian et al. (2014) | 2.2 | New Zealand |
| Solano et al. (2015) | 21 | Canada |
| Cook et al. (2016) | 13.2 | USA |
| Ranjbar et al. (2016) | 18.9 | Australia |
| Higginson Cutler et al. (2017) | 22.2 | Canada |
| Adams et al. (2017) | 9.6 | USA |
| Costa et al. (2018) | 42.5 | Brazil |
| Bran et al. (2019) | 41.1 | Brazil |
| Jewell et al. (2019) | 21 | Canada |

in clinical remission rather than clinical cure. Repeated inspections of these animals will generate more events without any change in overall disease levels unless appropriate lag periods are used to account for repeat incidence and recurrent lesions. Few conventions are established for differentiating new cases from  and epidemiological lag periods. The author proposes the use of a 4 week lag period for acute lesions of the skin (e.g. digital dermatitis), 6 weeks for acute lesions of the hoof and a lifetime lag for chronic disorders like corkscrew claws. Unfortunately many lesions causing lameness are chronic and so most cows are usually in temporary remission.

Nonetheless, incidence rates can provide useful diagnostic information when applied at the lesion level, with the index case relating to the first event in a year, lactation or better still, the first lifetime event.

To calculate a statistics for 'population at risk' to act as the denominator, the total cow-days for the herd must be worked out, and so software is best used for calculation of days between cow arrival (usually calculated from first calving or arrival into the transition pen) and leaving the herd (culling/death). In the absence of a software then the denominator can be estimated from cows present at milk recording, or failing that, a simple summary of lesions causing lameness expressed at percentages of all lesions found may be good enough to prioritise effort. To account for repeated treatments in chronically affected cows, the percentage of cows affected by each lesions in a defined period (e.g. annually) may be calculated. This is 'period prevalence' and works well for cow-level reporting of lesions, accounting for any recorded multiple limb occurrences. Repeat foot inspections can be assessed separately to identify recurrent cows for more proactive, individualised treatment or culling.

A major risk for lameness is a previous lameness event (Randall et al., 2017), with pathology (and possibly lameness/lesion risk) accumulating through a lifetime (Offer et al., 2000), with periods of greatest deterioration happening for the first 100 days after first calving.

### 1.4 Lesions causing lameness

The lesions causing lameness are generally classified according to the location and nature of disease. Hoof trimmers can be variable in the lesions they recognise and record (Rossiter et al., 2013). To standardise recording of lesion location, the International Hoof Map was devised (Greenough, 1990) (Fig. 2). Footmaps can be useful for standardising lesion recording and scoring amongst foot trimmers and researchers. It is a simple way to quality assure the lesion recording as some lesions can only appear in certain zones.

The International Committee for Animal Recording (ICAR) has established criteria to improve standards and consistency of recording lesions (Egger-Danner et al., 2015; Kofler et al., 2020a,b). Note that conditions requiring diagnosis beyond visible changes are not described in the atlas, for example laminitis. While laminitis is a condition affecting cattle, it is now recognised as a relatively rare cause of significant lameness in cattle fed well-formulated diets.

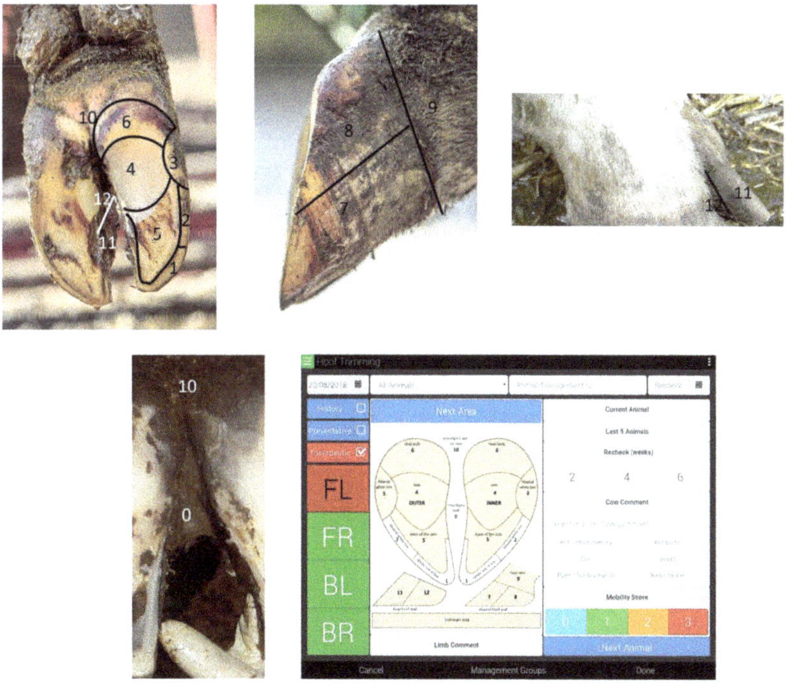


**Figure 2** (a) Hoofmap zones: 1. Wall/whiteline around toe 2. Abaxial wall 3. Abaxial wall and adjacent sole near heel 4. Central sole 5. Sole at toe 6. Heel 7. (b) Wall - abaxial, dorsal wall 8. Wall - abaxial, heel 9. Coronary band and proximal skin. (c) 10. Interdigital space - heel pocket 0. Interdigital space. (d) The Hoofmap zones can be incorporated into recording software for quality assurance purposes (screenshot courtesy of VetImpress, Farm Vet Solutions, UK).

In experiments inducing laminitis, foot tenderness was transient and the classic claw horn lesions causing lameness were not produced (Danscher et al., 2009; Danscher et al., 2010).

There are a small number of robust epidemiological surveys of lesions causing lameness recorded at hoof trimming (Murray et al., 1996; Cruz et al., 2001; Manske et al., 2002b; Barker, 2007). The prevalence of lesions will be greatly influenced by system, parity, proactivity of detection, effectiveness of treatment and replacement (and culling) rates. In a randomised control trial which recruited new cases of lameness in four UK dairy herds, the predominant lesion causing lameness diagnosed by veterinary surgeons was sole bruising (also described as sole haemorrhage) (Groenevelt et al., 2014). In two UK epidemiological surveys (Murray et al., 1996; Barker, 2007) the order of lesion occurrence was similar (Table 4) but may have underestimated the occurence of sole bruising and haemorrhage.

The Warwick survey reported incidence rates for sole ulcers and white lines (digital dermatitis was under-reported) (Table 5).

**Table 4** Lesions causing lameness recorded at examination in two epidemiological surveys (edited to be consistent with ICAR lesion terminology). The Warwick survey used farmer-recorded findings which may account for under-recorded sole bruising. Note laminitis which was once a commonly recorded cause of lameness has not been removed from all lesion recording conventions.

| | Percentage occurrence of lesions found at examination | |
|---|---|---|
| Lesion causing lameness | Liverpool Survey (1996) (%) | Warwick University Survey (2007) (%) |
| Sole ulcer | 27.9 | 29.3 |
| White line disease | 22.3 | 22.2 |
| Sole bruising/ haemorrhage | 8.3 | 2.0 |
| Digital dermatitis | 8.2 | 14.6 |
| Interdigital phlegmon | 5.2 | 7.0 |
| Interdigital hyperplasia | 5.1 | 2.0 |
| Foreign body | 5.7 | 2.0 |
| Heel horn erosion | 4.2 | Not reported |
| Heel ulcer | 2.9 | Not reported |
| Double sole | 2.2 | Not reported |
| Claw horn overgrowth | 2.0 | Not reported |
| Sole bruising/ haemorrhage | 1.6 | Grouped with haemorrhage |
| Digital dermatitis between the claws (interdigital dermatitis) | 0.8 | 1 |
| Vertical wall fissure | 0.4 | 0.5 |
| Others | 4.2 | 19.4 |

**Table 5** Incidence rates for sole ulcers and white line lesions

| | Incidence rate limb cases per 100 cows per year | | | |
|---|---|---|---|---|
| Lesions causing lameness | Rowlands et al. (1983) | Liverpool survey (1996) | Warwick biotin study (2001) | Warwick epidemiological survey (2007) |
| Sole ulcer | 12.2 | 15 | 13.8 | 6.2 |
| White line | 3.8 | 12.2 | 12.7 | 5.5 |
| Digital dermatitis | n/a | 4.5 | 12 | 2.8 |

Many lesions can be present without causing overt signs of lameness including some milder sole ulcers and many haemorrhages (Manske et al., 2002b). Digital dermatitis lesions will be present in as many as 40% of cattle (Vink, 2006; Stokes, 2011) with the majority showing few signs of lameness (Manske et al., 2002b). Whether or not to record incidental lesions requires careful consideration. It is possible for incidental lesions to vastly outnumber clinically significant lesions, particularly at routine foot checks, potentially distorting findings.

For lesions like digital dermatitis it is possible to screen the whole herd at regular intervals at milking time. There are several methods of screening and scoring digital dermatitis lesions, with the M-score (Dopfer et al., 1997; Berry et al., 2012) (Table 6) being most widely used for research and adapted to clinical scoring in the parlour (Solano et al., 2017). To optimise detection of lesions, the use of a hose to wash feet will also identify cows sensitive to water on their heels and the use of a bright torchlight (Rodriguez-Lainz et al., 1998) and a mirror attached to a long handle will allow inspection of the interdigital heel pocket where many lesions reside (Solano et al., 2017).

In herds with digital dermatitis well under control, the prevalence of milking cows with non-regressing lesions (M1, M2, M4, M4.1) will be under 10% and there will be fewer than 1% M2 lesions. The data in Table 7 is from a 1200-cow dairy which scored on a monthly basis. Cows were identified for individual treatment, with feet lifted for M2, M4.1 lesions or for spot treatment in the parlour (M1, M3). Agreement between pairs of trained scorers can be high (Yang and Laven, 2019) but agreement is greatest for a simple binary assessment (Vanhoudt et al., 2019) - a digital dermatitis lesion present or absent.

## 1.5 Summary of section on recording

Quantifying lameness data can be challenging. Lameness prevalence assessment repeated every 1–4 weeks can be useful for building up trends and driving proactive lame cow management. Lameness incidence can be misleading unless the data capture and analysis is managed very careful to avoid

**Table 6** The M-score (Dopfer et al., 1997; Berry et al., 2012)

| M-score | | Scoring guidance |
|---|---|---|
| **M0** | 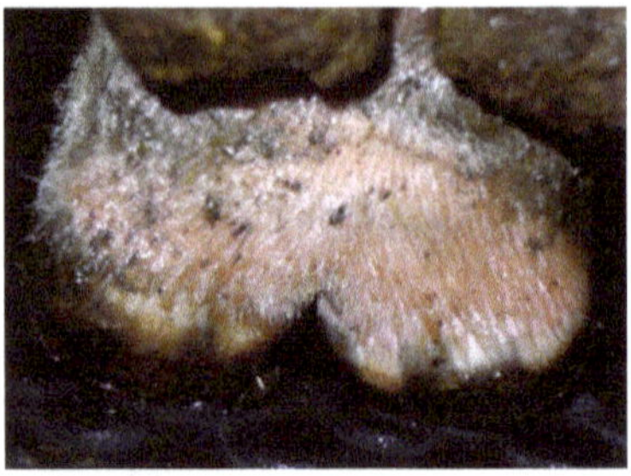 | Normal healthy skin, without hyperkeratosis, dyskeratosis, swelling, scarring or coarse hairs. To confirm M0 status the interdigital pocket must be inspected. This can be done using a hose, bright light and mirror to visualise the interdigital heel pocket. Alternatively the feet may be inspected in the crush (trimming chute or crate). |
| **M1** | 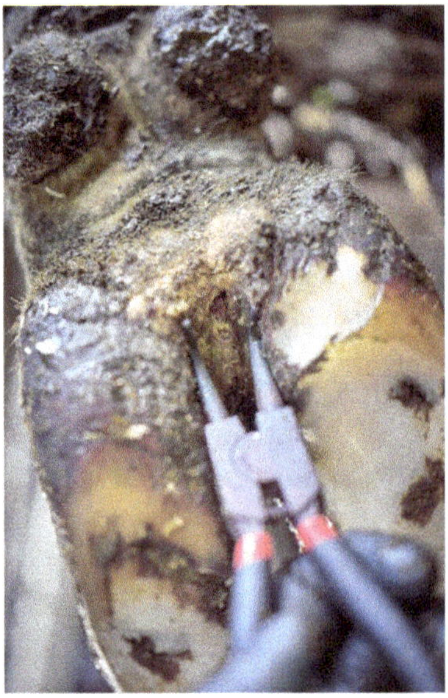 | Erosive or early ulcerated stage of disease, with lesions less than 2cm in diameter. |
| **M2** | 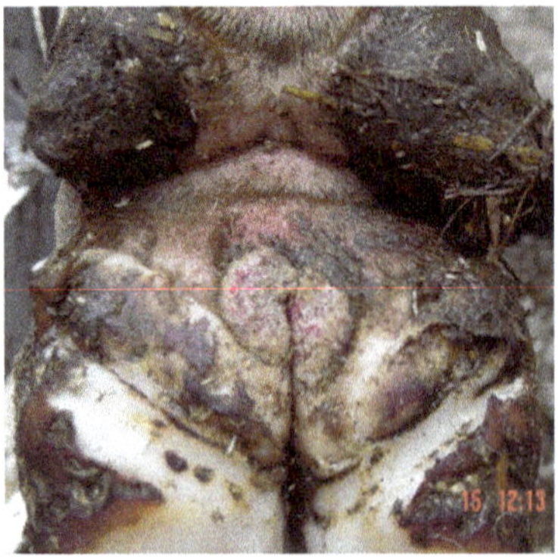 | Ulcerated or granulated lesions over 2cm in diameter. The lesion may be covered in a grey or beige exudate |

| M-score | | Scoring guidance |
|---|---|---|
| **M3** | 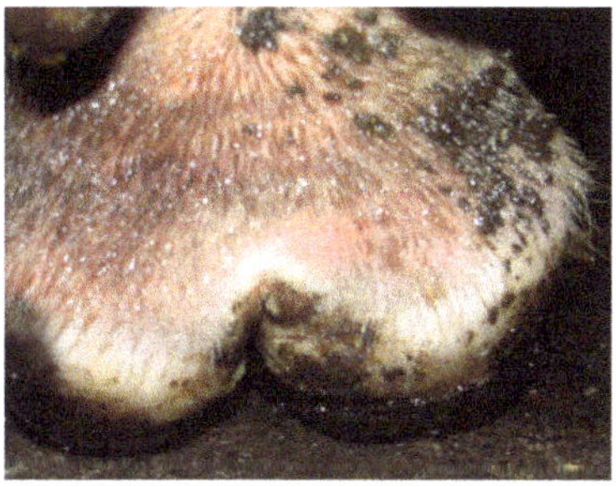 | The scabbed and regressing stage<br>These lesions look flat, brown and benign<br>Again a hose, mirror and bright light under the heel would confirm no active lesions are present in the interdigital space |
| **M4** | 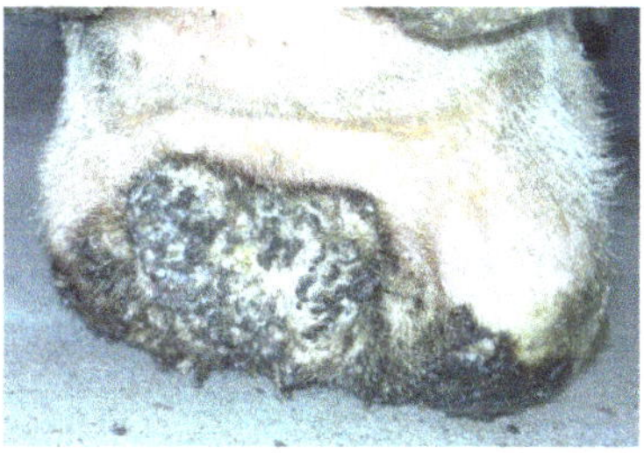 | Dyskeratotic and hyperkeratotic pathology give these lesions a warty appearance<br>They are normally the most common lesion seen in endemically infected dairy herds and may not cause lameness. However, they can quickly reactivate to M4.1 and M2, thereby driving new infections (Biemans et al., 2018).<br>Occasionally a proliferative "hairy wart" form of lesion is seen which has fingers of hyperkeratosis with a Medusa-head appearance |
| **M4.1** |  | Lesions with hyperkeratosis/dyskeratotic changes but with areas of ulceration. These are usually the reactivation of M4 lesions. Reactivation of M4s represent the major driver of new lesions (M1/M2) (Biemans et al., 2018).<br>They could also be lesions transitioning from M2 to M4. |

**Table 7** Data from a large herd with regular screening of the milking herd using the M-score

| | January | February | March | Average | Target |
|---|---|---|---|---|---|
| M1 | 0.0 | 2.6 | 3.2 | 1.9 | |
| M2 | 0.9 | 4.4 | 2.2 | 2.5 | <1% |
| M3 | 11.5 | 13.3 | 15.0 | 13.3 | |
| M4 | 2.0 | 2.9 | 4.2 | 3.0 | |
| M4.1 | 1.0 | 0.3 | 0.7 | 0.7 | |
| Total | 15.3 | 23.5 | 25.2 | 21.3 | |
| Active (M1,2,4.1) | 1.8 | 7.3 | 6.1 | 5.1 | |
| Chronic (M4) | 2.0 | 2.9 | 4.2 | 3.0 | |
| Non-regressing (active and chronic) | 3.8 | 10.2 | 10.2 | 8.08 | <10% |

repeat cases and incidental findings dominating the data. When evaluating lesions it can be useful to calculate incidence rates with the assistance of software although it is generally easier to calculate an annual period prevalence for the major lesions at cow level. Finally, screening for digital dermatitis can be done quickly and easily in the parlour allowing for a more robust and detailed assessment of control. From these assessments, priorities for change can be decided. In the next section we will review the control of the four major lesions causing lameness.

## 2 Claw horn disruption – a paradigm shift

Claw horn disruption (CHD) is the term used to describe the interrelated family of lesions which include sole bruising (haemorrhage), sole ulcer and white line lesions. Our understanding of the aetiopathogenesis of CHD has undergone a major change in the last 20 years (Huxley, 2012). Historically the model for sole ulcer and white line development was built around laminitis in horses, with ruminal acidosis thought to induce laminitis (Nocek, 1997) or they occur together (Danscher et al., 2009), with consequential loss of laminar attachment, sinking of the pedal bone (PIII) and injury to the pinched sole corium. Much effort was focussed on nutritional interventions linked to this theory (Manson and Leaver, 1988a,b, 1989; Logue et al., 1994; Offer et al., 1997; Webster, 2000; Logue et al., 2001, 2000; Livesey et al., 2003; Laven and Livesey, 2004; Offer et al., 2004b; Knott et al., 2007; Logue and Mayne, 2014). A feeding mistake during a trial (Offer et al., 2004a) and new insights into the effects of calving on the suspensory apparatus (Tarlton et al., 2002) created the opportunity for a new paradigm to form.

Current theories on aetiopathogenesis of claw horn disruption are built on mainly biomechanical processes, combining environmental trauma with robustness of the hoof. The next section examines the latest understanding in relation to aetiopathogenesis by working through the disease processes for the main four lesions causing lameness: sole bruising, sole ulcers, white lines and digital dermatitis.

### 2.1 Aetiopathogenesis of sole bruising and sole ulcers

Sole ulcers are thought to be the end stage process starting with focal sole bruising following trauma to the germinal epithelium adjacent to the flexor process of PIII (Fig. 3a). The sole ulcer (ulcerated corium visible through the sole) is the culmination of several weeks of corium inflammation which resulted in cessation of claw horn growth, long enough for a defect to develop through the full thickness of the sole. The sole thickness in the central sole (zone 4) of the untrimmed hoof is usually 10–20mm thick in the direction of growth at the sole ulcer site and sole horn is produced at a rate of 4–5mm per month (Vermunt and Greenough, 1995a) meaning defects observed on the ground-bearing surface of the sole is usually many weeks old at least.

### 2.2 Hoof trimming

Early indications of the biomechanical nature of the disease process stemmed from the protection conferred by routine foot trimming reported in observational studies (Manske et al., 2002a,b). Remarkably, there is a relatively weak evidence-base supporting hoof trimming method (Stoddard and Cramer, 2017; Sadiq et al., 2020) despite it being a procedure carried out on most dairy cows several times of year.

An effective preventative foot trimming technique should improve foot angle (Phillips et al., 2000), balance and weight distribution (van der Tol et al., 2002, 2004; Carvalho et al., 2005; Ouweltjes et al., 2009). These are the goals within the functional steps (1-3) of the Five Step Hoof Trimming Method (Toussaint Raven, 1985; Mahendran and Bell, 2015; Manning et al., 2016) (Table 8), the most widely researched method. This approach can reduce the risk of lameness in certain environments (Stoddard and Cramer, 2017). A large hollow remodelled into the central sole has been shown to protect the dairy cow from claw horn lesions such as sole ulcers (Stoddard, 2018), presumably by reducing pressure over the flexor process of PIII. However, a large part of routine hoof trimming remains the detection and correction of early hoof lesions. Over-trimming can also be a problem, particularly in environments and management systems prone to increasing the wear of hoof horn (Mahendran et al., 2017).

There is still considerable uncertainty on the optimal timing of preventative hoof trimming. There is some evidence to support foot trimming at drying off (Garcia et al., 2004; García-Bracho et al., 2009) and mid-lactation (Hernandez et al., 2007), but the benefit of early lactation trims is less certain (Maxwell et al.,

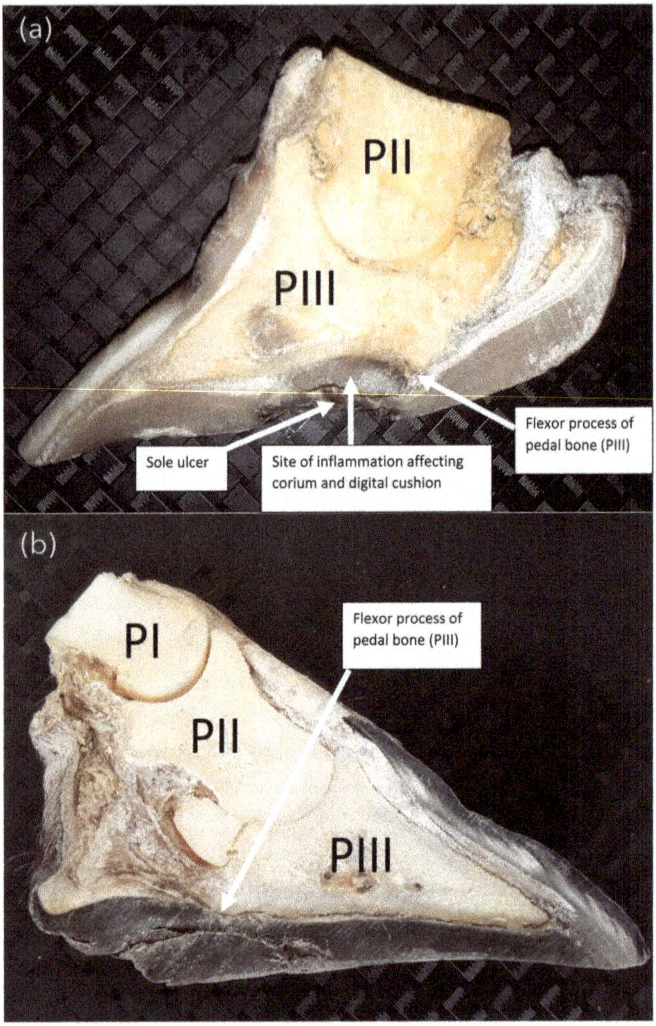


**Figure 3** (a) Sole ulcer in a specimen sectioned through the sagittal plane showing the main anatomical structures involved in the pathogenesis of sole ulcers. (b) A sagittal plane section through a digit affected by a heel ulcer (sometimes referred to as a sole fracture). Note the defect runs in a palmar/plantar-distal direction towards the heel bulb. The heel ulcer most commonly affects the medial hind claw, perhaps reflecting differences in PIII in this digit, particularly in relation to the size of the flexor process.

2015; Mahendran et al., 2017). Observational studies indicate trims conducted at 100 days into lactation might be too late (Griffiths et al., 2018) to prevent the progression of bruising explaining why lameness scoring can have better outcomes (Mahendran et al., 2017).

**Table 8** Five Step Hoof Trimming Method is the most extensively researched trimming method

| Steps | Phases | Aim | Actions |
|---|---|---|---|
| Step 1 (2 parts) | Functional trim | Achieve 48-52 degree foot angle when possible | a) Shorten inside hind claw to correct length for breed, size, age and pre-existing disease |
| | | | b) Trim excess sole at toe until 5-7mm step is left (white line re-joins at toe). SPARE THE HEEL |
| Step 2 (3 parts) | | Balanced weight-bearing between medial and lateral claws | Rebalance opposite partner claw - length, level in toe, balance in heel. Sometimes perfect balance is not possible due to conformation or disease |
| Step 3 | | Redistribute weight onto the walls | Model out the centre sole of the lateral hind claw (deep and two-thirds across), steep and narrow for inside. Spare toe triangle and axial wall |
| Step 4 | Corrective trim | Redistribute weight-bearing onto sound claw | Apply a block and/or trim down back 2/3 of lateral hind claw |
| Step 5 | | Remove loose horn and ridges | Only remove loose horn in heel erosion, around lesions causing lameness and remove ridges |

## 2.3 Changes to suspensory apparatus around calving

The discovery that the strength of the suspensory apparatus diminishes at and around the time of calving (Holah et al., 2000; Tarlton et al., 2000, 2002; Webster and Tarlton, 2001; Knott et al., 2007) was pivotal in switching theories of CHD from nutritional aetiology to managemental. Matrix metalloproteinases which remodel the laminar attachment are upregulated around calving, and it is postulated that this change is linked to the connective tissue changes which allows easier passage of the calf through the birth canal. The hormonal mediator for this is still unknown and the effect is thought to last several weeks before and after calving. A trial involving a group of heifers managed on loose (straw) yards before and 12 weeks after calving compared with heifers housed in cubicles (freestalls) 4 weeks before and 24 weeks after calving showed a marked difference in lesion score, particularly sole ulcer development (Webster, 2002). Therefore, the strategic use of fresh heifer groups on straw yards appears particularly effective for controlling sole ulcers. Measures to minimise standing times by maximising cow comfort, particularly cubicle comfort, in the period immediately before and after calving appears to reduce the incidence of sole bruising and sole ulcers (Leonard et al., 1994, 1997, 1996).

## 2.4 Lying times, lying behaviour and standing times

The interaction between changes to the suspensory apparatus and standing times on concrete appears critical to sole ulcer risk , especially with poor foot shape. A large body of work has now been published on factors affecting lying times in dairy cattle (Collins, 2016). A plot of lying times over a lactation period is revealing (Fig. 4) and demonstrates the time-budgets for dairy cows is put under pressure in the fresh period (Maselyne et al., 2016). The time-budget of the dairy cow is now a well-described and better understood concept (Gomez and Cook, 2010) contributing substantial risk for sole ulcers. While maintaining foot shape is important, poorly organised hoof trimming itself can compromise time-budgets through an increase in forced standing times with prolonged penning. The consequence is that cows lie for longer in the days following trimming (Stoddard and Cramer, 2017). One of the main determinants of good lying time and behaviour (optimal bout length and frequency) is the deep bed cubicle. In one survey of UK dairy farms, cows housed on deep bed cubicles had lying behaviour similar to those on loose straw yards (Collins, 2016).

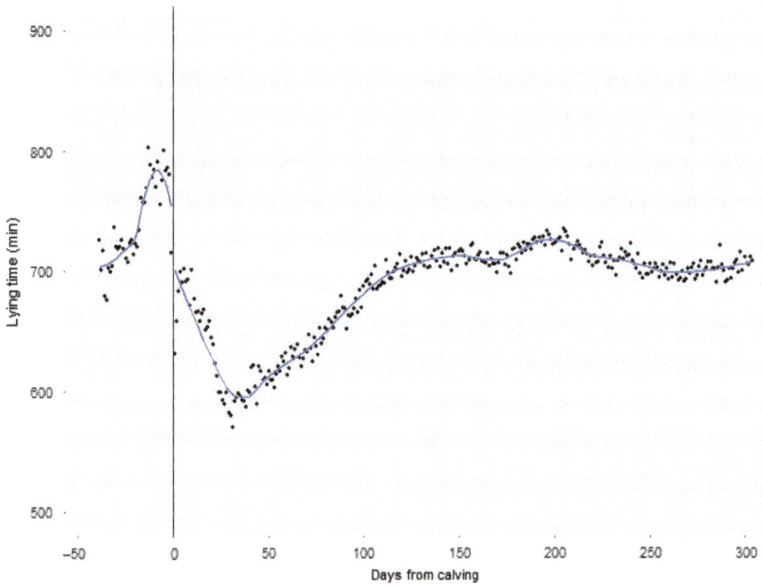


**Figure 4** A plot of daily lying times for dairy cows through a dry period (-50 to calving at day 0, 1839 cow days of data) and through a lactation (0–305 days using 39 741 cow days of data). Note the lying time nadir occurs around 6 weeks post-calving, which corresponds to peak yield and then stabilises at about 720 minutes per day from about 130 days post-calving.

## 2.5 Digital cushion thickness

The digital cushion consists of three cylinders of fatty tissue which reside between the pedal bone (PIII) and the sole (Räber et al., 2004) (Fig. 5). It serves to dissipate the dynamic forces through the sole. Low body condition score has been reported a risk factor for claw horn lameness (Bicalho et al., 2009; Green et al., 2014; Lim et al., 2015; Randall et al., 2017) as well as a consequence of lameness. This structure has become the most important topic of intense research in recent years.

There are many factors which influence the thickness and properties of the digital cushion (DC). Heifers appear to have less well developed digital cushions (Räber et al., 2004, 2006). The volume of the digital cushion can be increased by approximately one-third with exercise on hard surfaces (Gard et al., 2014). Digital cushion correlates with body condition score but one of the biggest determinants of thickness is inflammation (Newsome et al., 2017), with inflammation associated with sole ulcers significantly increasing the thickness of the digital cushion which then becomes scarred. It is also interesting to note that the thickness of the cushion decreases with sinking of PIII due to the weakening of the suspensory apparatus (Newsome et al., 2017). Cattle parity 3 and older show an increasing level of collagen content within the DC, replacing the adipose tissue (Räber et al., 2006).

Lipid content of digital cushions in heifers is rich in arachidonic acid (AA) (Räber et al., 2006). AA is readily metabolised to prostaglandins and leukotrienes which are important pain and inflammatory mediators. Non-steroidal anti-inflammatory drugs (NSAIDs) with COX-inhibiting properties will block the production of these mediators from AA, which might offer one explanation of why NSAIDs are so important for the treatment of focal sole bruising in zone 4

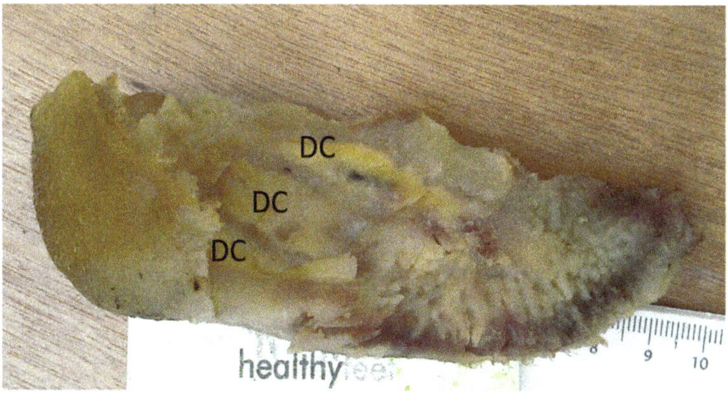


**Figure 5** The soft tissues dissected from between the pedal bone and the sole horn of a dairy cow reveal the three cylinders of adipose tissue named the digital cushion (DC).

or sole ulcers , particularly in heifers, and why heifers are so prone to the largest deterioration in lesion score (Offer et al., 2000).

Once inflamed, scarring of the digital cushion is likely. Osteoma formation (exostoses) is observed in relation to lameness and lesion presence in life (as well as age) (Newsome et al., 2016). Analysis of population attributable risk shows 79-83% of lameness is explained by previous episodes of lameness whereas 4-11% of lameness risk is attributable to body condition score (Randall et al., 2017), demonstrating the importance of protecting the digital cushion and pedal bone from trauma and untreated inflammation. From a practical perspective this means precondition hooves to concrete prior to first calving, proactively maintaining healthy foot shape, providing comfortable walking surfaces, optimising body condition score, providing comfortable beds for optimal lying behaviour and detecting lesions early for prompt and effective treatment.

### 2.6 Early detection and prompt, effective treatment (EDPET)

Sole bruising can occur at any time when cows are managed or penned on concrete without adequate opportunity to rest, recover and feed. Focal sole bruising will be less likely while at pasture or with the use of large remodelling in zone 4 of the lateral hind claw (most susceptible to sole ulcers). Nonetheless, cows that do not rest due to oestrus behaviour and traumatise the corium, may develop unresolved bruising if time-budgets are under pressure or cubicle comfort is poor. In these instances, protocols for early detection and prompt, effective treatment are important additions to preventative trimming. In trials using fortnightly screening, most cows detected as lame will have bruising (Groenevelt et al., 2014) and these respond well to functional and corrective trimming, application of a therapeutic block and administering of a 3-day course of non-steroidal anti-inflammatory drug (Thomas et al., 2015). Delays to treatment of just 2 weeks will dramatically reduce cure rates 6 weeks following treatment and increase the chances of the cow going lame on the contralateral limb (Thomas et al., 2016).

## 3 Aetiopathogenesis of white line bruising and lesions

The white line is the anatomical junction between sole and wall horn. Wall horn growing at a rate of 4–5mm/month descends from the coronary dermis where it is produced. It is attached to laminar corium by folds of lamellar horn, which slide across each other through a constant physiological remodelling, reaching the ground-bearing surface after several months of growth. As the lamellae reach the sole level, dermal papillae on the ends of the laminae fill the horn between the lamellae, creating a series of wekaker junctions along the white line which can resemble a train track appearance (Fig. 6a). The junctions between lamellar horn and horn produced by papillae on laminae present many weak junctions across the width of the white line (Fig. 6).

The most common site for white line lesions to develop is zone 3, and it is hypothesised that the shear forces across zone 3 are larger than across the toe triangle (zones 1 and 2) which is inherently more resilient due to the arched walls along two sides of the pyramid. Another explanation comes from high-speed cinematography which shows the heel of the lateral hind claw strikes the ground with considerable dynamic force during locomotion (Schmid et al., 2009). The latter hypothesis might also explain why the depleted or scarred digital cushion predisposes cows to white line lesions.

Epidemiological studies identify a number of risk factors specific to white line lesions (Manske et al., 2002a; Barker et al., 2009, 2010), although there are clearly a few common risks like calving (Webster, 2002), low body condition score (Bicalho et al., 2009) and prolonged standing on concrete (Bell et al., 2009) which appear to affect the risk of both sole ulcers and white lines. The clearest difference in pathogenesis is the role of nutritionally influenced claw horn quality (covered in the next section).

### 3.1 Biotin deficiency

Biotin is part of the B vitamin family, often informally referred to as vitamin H due to its role in healthy hair and horn production. In high forage diets the

**Figure 6** (a) The white line (WL) on the ground-bearing surface marked with the orange line. (b) The hoof capsule showing folds of lamellar horn (LH) and coronary groove (CG). (c) The pedal bone following detachment of the hoof capsule, with PIII covered laminar corium (LC) and coronary corium (CC).

rumen microbes will naturally produce sufficient biotin to satisfy cow needs, as demonstrated in hay feeding trials (Abel et al., 2001). However, in high concentrate, starch and maize rations, it is believed that this natural synthesis is compromised. In a UK randomised control trial supplementation of biotin at 20mg per head per day reduced lameness due to white line lesions by approximately one half (Hedges et al., 2001). The effect was not measurable until 130 days following the start of supplementation, consistent with the time it takes to deposit lamellar horn from the coronary groove down to the ground-bearing surface. Other macrominerals like zinc could also affect the quality of white line horn (Siciliano-Jones et al., 2008) particularly if deficient, which tends to most commonly affect youngstock on poor winter rations or high yielding dairy cows.

### 3.2 Cow tracks and stockmanship

White line lesions and stone penetrations tend to be the predominant lesions causing lameness in the seasonally calved, intensively grazed herds of New Zealand and similar pasture-based regions. Epidemiological and observational studies into lameness in these systems indicate there is an interaction between the gentle stockmanship skills and the walking surfaces (Chesterton, 1988; Chesterton et al., 1989; Tranter et al., 1991, 1993). These support the theory that the white line lesion is arising through peak dynamic forces.

Peak incidence tends to coincide with highest rainfall, when stone tracks deteriorate and moisture softens claw horn. The dynamic interaction between claw horn wear and growth in these conditions has been investigated (Tranter and Morris, 1992) and growth is lowest in the autumn when moisture his high, claw horn is soft and soles are wearing thin. Consequently this is when most white line lesions occur. However, it has not been conclusively shown that cows with thinner sole at the heel are more predisposed to white lines. Older cows appear more at risk than younger animals (Barker, 2007).

### 3.3 Slats, concrete grooving and softer floor surfaces

Again, a number of observational studies indicate indoor floor surfaces may predispose cows to white line lesions. Grooved concrete has been reported as a risk (Pérez-Cabal and Alenda, 2014; Barker et al., 2009) compared with solid floors. Slats are a notable risk factor (Sogstad et al., 2005), and they often involve a slippery concrete slat with a wide gap for slurry to drain, which allows feet to 'slip and catch', creating the high dynamic force for trauma. However, in another survey grooving with wide grooves was protective, and the authors hypothesised that it conferred greater grip. There is an optimum which has yet to be described in the scientific literature.

Rubber matting appears to be protective for hoof lesions although not without limitations (Bell and Huxley, 2009b). While bruising and claw wear can be reduced with rubber, one study reported an increase in sole ulcer incidence (Kremer et al., 2006), possibly because longer claws induced shallower foot angles. Abrasive rubber products are now available (Pedicura, Kraiburg, Germany). In the same study, digital dermatitis risk also increased, possibly because rubber surfaces can create slurry pool depressions where the foot sinks into the rubber. Firmer quarry belt products may be more slippery (Rushen and de Passille, 2006) but may offer sufficient grip with surface patterns cut into them.

There can be an increase in cubicle rejection with rubber in alleys (Tucker et al., 2006), particularly if cubicles are not comfortable. Nonetheless, cow preference for rubber, the improvement in cow walking speed, improved behaviour and some long-term enhancement of claw health (e.g. protection from excessive wear and thin soles, or pushing at feed barriers and corkscrew claws) shows it can offer a broad range of benefits (Telezhenko and Bergsten, 2005; Telezhenko et al., 2009) even if short-term claw health is difficult to measure within the short time-frames of research trials.

Despite the significant impact on lameness and cow welfare, there is very little robust science on the subject of suitable non-rubber floor surfaces for dairy cattle. Most new research has concentrated on rubber matting, which is still not always practical or deemed economically viable for large dairy farms.

## 4 Aetiopathogenesis and control of digital dermatitis

Digital dermatitis is a highly infectious polymicrobial skin disease affecting the skin, normally between the heel bulbs. While metagenomics has demonstrated large numbers of bacterial and candida species associated with the lesions (Zinicola et al., 2015), Wisconsin researchers replicated disease using *Treponema* spp. (Gomez et al., 2012). In order to replicate the disease, researchers applied inoculum onto the skin which was macerated with prolonged exposure to wet cotton and closed from air by additional layers of polyethylene, more soaked cotton and a rubber boot. Lesions took between 12 and 25 days to appear and *Treponema* spp. were confirmed as present. A Liverpool research group demonstrated natural invasion of the dermis happens through the hair follicle and sebaceous glands (Evans et al., 2009).

### *4.1 Foot hygiene and underfoot conditions*

The induction model required inoculum applied to moisture macerated skin, closed off from the air (Gomez et al., 2012). Most dairy farm environments effectively replicate these wet, anaerobic conditions with moisture and slurry encrusted on feet. Two studies have reported automatic scrapers as a risk factor for digital dermatitis for this reason (Cramer et al., 2009; Barker et al., 2010). Pasture-based systems may also experience similar challenges when gateways become muddy (Rodriguez-Lainz et al., 1996). Strategies to improve foot hygiene in dairy herds have not been well investigated or reported, but could include comparisons of slats versus solid floors, floor stocking rate comparisons or different regimes for robotic versus automatic scrapers.

## 4.2 Foot bathing and daily disinfection

Historically, foot bathing has been used to treat the herd to reduce the prevalence of cows with digital dermatitis lesions, with improvements seen with a variety of biocides (Laven and Hunt, 2002) and antibiotics (Laven and Hunt, 2001; Laven, 2006). However, with recent concerns over antimicrobial stewardship, the use of antibiotic foot baths should be replaced with more effective and responsible alternatives (Bell et al., 2017). Regular cleaning and disinfection of the digital skin is far more likely to succeed as prevention than reactive foot bathing to control clinical lesions. Again, the evidence supporting the use of daily foot disinfection regimes is remarkably weak despite the common and successful practice (Bell et al., 2014).

Permitted biocides will vary according to local legislation. Copper sulphate is not a biocide licensed for skin disinfection within the EU (Bell and Vanhoudt, 2020) and some authors have challenged the scientific evidence supporting its use (Thomsen, 2015). Formalin is widely used but given its carcinogenic properties, it carries significant human health and safety risks, rendering its future use questionable. There is a growing body of evidence for other biocides (Bell et al., 2014), including the use of soap foot baths to clean the feet (Thomsen et al., 2012). One of the observed benefits of copper sulphate and formalin is the ability to prevent reactivation to M4.1, which may not happen as readily with other biocides.

A design of footbath has been reported which maximises the number of foot plunges for a given volume of bath (Cook et al., 2012). High kerbs and solid sides also minimise the amount of solution lost, as well as increasing steps in the bath and improving cow flow. Other researchers have focussed on ways to achieve calm and natural cow flow with wide baths (Chesterton, 2013). Compartmentalised baths would appear to reduce contamination rates in the second compartment, presumably extending the useful life of the biocide (Manning et al., 2017). Despite its importance, there is still relatively little outcome-based evidence supporting the variety of foot bathing systems and regimes (Bell et al., 2014). To make daily foot disinfection practically feasible for farm staff, a means of accurately and safely filling and emptying the bath should be factored into the design.

### 4.3 Proactive treatment of cows with lesions

The proactive treatment of all digital dermatitis lesions (Kofler et al., 2020a) and associated lesions (Kofler et al., 2020b) should minimise the major reservoir of pathogenic Treponemes, thereby reducing future infection pressure. There is one case study describing this approach (Pedersen, 2019) and case series describing the successful surgical debridement of necrotic hoof lesions (Alcock, 2015). The potential to spread and transmit infection at hoof trimming should always be considered given the levels of Treponemes detected on equipment (Gillespie et al., 2019) and potential for fomite spread (Wells et al., 1995). Nonetheless, a lack of empirical evidence should not deter the promotion of a clinically logical approach to improving the welfare of cattle, and further work could investigate ways to minimise potential cross-infection risk.

### 4.4 Management of dry cows and youngstock

Often the foot bathing and treatment of youngstock and dry cows is overlooked, yet so often it is these epidemiologically significant groups contributing to milking herd infection pressure at calving when there is periparturient immunosuppression. US investigations have shown that performance losses in heifers calving with digital dermatitis lesions are substantial (Gomez et al., 2015).

## 5 Summary and critical control points

Hazard analysis and critical control points is a discipline developed from space exploration for minimising food poisoning. It takes stages of a process to identify ways to tighten control over the quality of the end product (Bauman, 1995). In the case of minimising the risk of lameness, it can be applied to a specific disease process to identify cost-effective control point for lesions (Bell et al., 2009)(Fig. 7). In the case of sole ulcers, optimising foot shape and lying behaviour in the first 120 days of lactation should prevent onset of most lesions in well-managed dairy herds. Screening of herds every 1-2 weeks for early signs of lameness will help catch any cows which do develop progressive focal sole bruising. Similarly for white lines, ensuring walking surfaces are comfortable and stockpersons handle cattle carefully will minimise new lesions in cattle with good claw horn condition (e.g. fed 20mg biotin per day and kept in dry conditions). Lastly, the use of regular (daily) foot disinfection and monthly screening with treatments applied to affected individuals in all management groups will keep most digital dermatitis infections well-controlled.

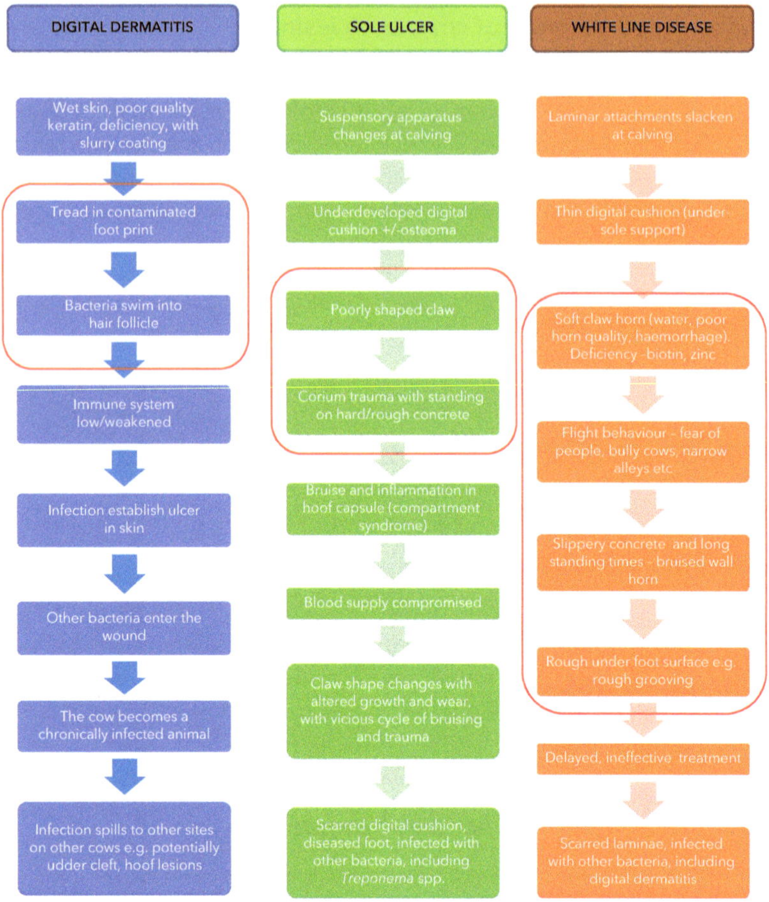


**Figure 7** Diagram of the aetiopathogenesis and critical control points of digital dermatitis, sole ulcer and white line. The pathogenesis reveals the targets for critical control points (Bell et al., 2009) circled in red. The critical control points for digital dermatitis involves removing reservoirs of pathogenic *Treponema* spp. and frequent disinfection of feet. The critical control point for sole ulcers would appear to involve managing foot angle and optimising lying behaviour. The corrective actions for sole ulcer include proactive lameness detection and treatment with blocks and NSAIDs. The critical control point for while line lesions involve optimising horn quality, stress-free stockmanship skills and providing comfortable walking surfaces.

## 6 Case study

A 800 cow Holstein herd presented with a sudden rise in sole ulcers and white line lesions. An independent mobility score (Anon, 2016) by a qualified and quality assured assessor showed there was a 30% prevalence of clinical lameness. The prevalence of sole ulcers and white line lesions was rapidly rising.

The foot trimmer was highly qualified and was independently spot checked on two occasions. The only major change on the farm was some concrete grooving of feed alleys and loafing areas using wide-cut grooves, reported as beneficial in a recent epidemiological survey (Griffiths et al., 2018) (Fig. 8) and installation of a sprinkler system with fans to mitigate summer heat stress.

A systematic review of the lesion data, routine trim data, lameness score data, control protocols and risk factors was performed. Guidance on this approach is available through www.cattle-lameness.org.uk.

Scoring of the fresh group verified that lameness levels were consistent with the independent lameness scorer. M-scoring cows for digital dermatitis revealed there were fewer than 1% cows with active lesions (M1,2,4.1) and fewer than 8% with any lesion. Record analysis showed digital dermatitis was not a seasonally specific or episodic concern. Cows were footbath daily with 4% formalin through an automatic foot bath and a sample was taken and assayed for formaldehyde concentration and solids contamination.

**Figure 8** Concrete grooving conduct by contractor.

### 6.1 Analysis of records

Analysing period prevalence showed that approximately 35% of the herd were getting sole ulcers and 22% of cows were getting white line lesions, with a 3-fold increase on levels from the preceding 3 years (Fig. 9).

Sole ulcer assessment (condensed and simplified):

| Element of investigation | Findings |
|---|---|
| Foot shape | Trimmer was following best practice Five Step Method, using an optimal 80mm standard toe-length for Holsteins and a large remodelling of zone 4<br>10–12 cows were being trimmed every day, providing just over 4 trims per cow per year, but since the increase in lameness, trim sessions were dominated by lame cow treatments and rechecks and so some cows were only having 1 check in 12 months.<br>Early lactation trims were taking place at 100 days into lactation - too late for preventing sole ulcers on this farm |
| Lying behaviour in transition, fresh and high pens | Some social disturbance in the far-off dry pens ceased to be a problem by the time cows were in the transition pen (straw yards)<br>Fresh group were housed on flexible cubicle dividers (Cowcoon, Wilson Agri) with a hybrid mat-mattress bed. Hock swellings without abrasions were observed. Beds were insufficiently comfortable<br>High yielding cows were run as one single group, brought up in 3 batches from 3 different interconnected shed, with an approximate turn-around time of 1 hour per batch, repeated for three times-a-day milking. This was considered too long for peak yield (high) and heifers. Despite initial concerns, cow behaviour was good<br>Pen times for routine artificial insemination and pregnancy diagnosis frequently exceeded 1 hour and could be as long as 2 hours in many instances (in addition to milking time) |
| Digital cushion integrity | Heifer pre-conditioning to concrete was variable, but all heifers were trained to use cubicles<br>Body condition was good and cows were holding body condition score reasonably well based on monthly body condition score dynamics<br>Cows with significant hoof lesions were proactively culled (prior to grooving)<br>Newly grooved concrete was uncomfortable for cows to walk and stand on. They show aversion to walking on many areas. This was altering cow behaviour, and there was evidence the grooves were directly bruising feet |
| Early detection, prompt effective treatment | No proactive screening - manager started using fortnightly scoring methods<br>The manager needed training to recognise the early signs of lameness<br>Trimmer treatment protocols were best practice (trim, block and NSAID immediately administered in the crush at treatment) |

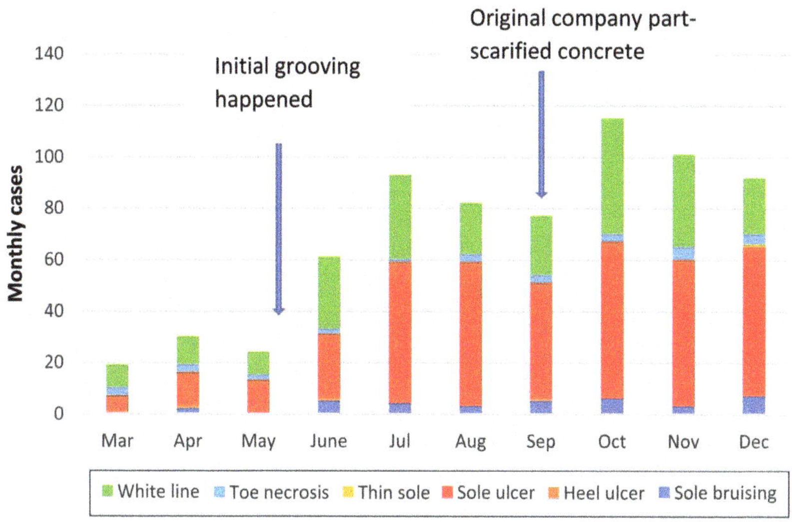


**Figure 9** Monthly cases (first cases per cow) of each of the most common lesions causing lameness on the farm.

White line assessment (condensed and simplified)

| Element of investigation | Findings |
|---|---|
| Claw horn quality, biotin and micromineral status | Claw horn was good quality at inspection in crush (chute/crate)<br>Long-term biotin supplementation at 20mg per head per day for all groups was already implemented<br>Chelated zinc was fed |
| Floor surfaces, comfort and cow-handling | The newly grooved concrete was rough and uneven, and scarified concrete was also rough and abrasive<br>There was plenty of space, so some of the risk due to cows pushing was mitigated<br>There was plenty of rubber around the collecting yard and dispersal areas<br>Stockmanship was quiet and stress-free<br>Backing gate was used to take up slack space in the collecting yard, but did not make contact with cows. A bell and time limit on use was enforced. |
| Lying behaviour in transition, fresh and high pens | As for sole ulcers |
| Digital cushion integrity | As for sole ulcers |
| Early detection, prompt effective treatment | As for sole ulcers |

Action plans were created through facilitated discussion with the manager, owner and trimmer:

1 More disciplined mobility scoring, implemented immediately by the manager, with quarterly independent scorers to check calibration and to correct any drift.
2 Conduct routine foot checks early at 10 weeks post calving as well as pre-dry (1–2 weeks). Extra trims required for cows with problems, cows with overgrown claws and cows with extended lactations. Protocols were immediately programmed into the management software to generate the daily trim lists.
3 Increase trimming resource capacity to ensure routine trimming preserves foot shape (the immediate action was to bring in an additional professional trimmer, then to train up a relief trimmer).
4 Improve bed comfort for all groups within 3 months for the fresh group and within 12 months for the rest; ideally deep bedding (dried recycled manure solids was preferred over sand due to the slurry system used).
5 Rubber matting along feed barriers and more rubber along the dispersal lanes (implemented within 3 months).
6 Plane the rough grooved concrete flat with a diamond cut machine (implemented within 3 months).
7 Cull cows with injured digital cushions (cull noted on the management programme, implemented at fortnightly mobility scoring by the herd manager).
8 Heifer group to shorten turnaround time at milking, minimise bullying and to further protect the digital cushion in the most vulnerable group. Note a barren cow group had cut bullying significantly.

Following implementation of points 1–3 and 5–7, lameness prevalence fell from 30% to 22% then 15% within 4 months. Cubicle beds and dividers were replaced recently and further monitoring of progress is in place with annual reviews planned for the future. At the annual review lameness prevalence had fallen to 6%.

## 7 Emerging diseases and future concepts

There are a number of conditions which appear to be increasing but were not covered within this chapter. They deserve to be mentioned as they can affect some farms significantly.

The increase in sand bedding and larger farms, often proactively trimming to control lameness, has sparked an increase in reports of thin soles in zone 2 (van Amstel et al., 2004). Investigations into hoof trimming recommendations have indicated that modern 700-800kg Holstein dairy cow might benefit from a longer toe-length than the traditional Friesian from the 1970s when the Five Step Method was first described (Archer et al., 2015; Bell, 2015). There also appears to be an anecdotal association between sand-reared heifers and corkscrew claws, particularly when feed barriers are over-stocked or heifers are limit-fed (personal communication, Karl Burgi). The hypothesis is that heifers are competing for feed and in the process pushing into feed barriers and inducing abnormal patterns or growth, or worse, damaging the laminar attachment. This condition warrants further investigation.

On rare occasions we encounter epidemics of claw-horn lesions in very high yielding cattle. Usually there is an environmental trigger, but occasionally there isn't anything to see. In these instances the clinician might investigate specific feeding errors. In particular, mistakes in protein feeding or micromineral supplementation have been anecdotally linked to herd epidemics. Emerging concepts in oxidative stress management in dairy cows might be of interest to the reader (Bradford et al., 2015).

Automatic lameness and lesion detection looks set to advance at a great pace in the coming years (Alsaaod et al., 2019). As is often the case with new technology the challenge will not be the technology itself, but assisting farmers and advisors with responding to the data.

Finally, the role of genomics in increasing the resilience and robustness of the modern dairy cow could be revolutionary. Already there are genomic markers for digital cushion thickness (Oikonomou et al., 2014) and digital dermatitis (Scholey et al., 2012). At the moment, utilising the data optimally will prove the biggest challenge.

# 8 Where to look for further information

## *8.1 Articles*

- Lameness in UK dairy cattle: a review of the current status (Archer et al., 2010).
- AHDB literature review (Potterton et al., 2011, 2012).
- Manual for Treatment and Control of Lameness in Cattle (van Amstel and Shearer, 2008).
- Cattle Footcare and Claw Trimming (Toussaint Raven, 1985).

## *8.2 Websites*

The following website dedicated to lameness will provide links to other websites, with a few selected for inclusion in this chapter. If an important website appears to be missing then please contact the author:

- https://www.cattle-lameness.org.uk/other-foot-health-sites/.
- AHDB Healthy Feet Programme: https://ahdb.org.uk/knowledge-library/lameness-in-cows-an-introduction-to-the-healthy-feet-programme.
- International Committee for Animal Recording (ICAR) presentation: https://www.icar.org/wp-content/uploads/2015/09/1-Christa-Egger-Danner.pdf.
- ICAR atlas: https://www.icar.org/Documents/ICAR_Claw_Health_Atlas.pdf.
- Lameness Management in Dairy Herds – Neil Chesterton (New Zealand): http://www.lamecow.co.nz/index.html.
- Wisconsin University Food Animal Production Medicine lameness webpage: https://thedairylandinitiative.vetmed.wisc.edu/home/lifestep-lameness-module/.

## 9 References

Abel, H. J., Immig, I., Gomez, C. D., and Steinberg, W. (2001). Research note: effect of increasing dietary concentrate levels on microbial biotin metabolism in the artificial rumen simulation system (RUSITEC), *Archives of Animal Nutrition (Archiv Fur Tierernahrung)* 55(4), 371–376.

Adams, A. E., Lombard, J. E., Fossler, C. P., Román-Muñiz, I. N. and Kopral, C. A. (2017). Associations between housing and management practices and the prevalence of lameness, hock lesions, and thin cows on US dairy operations, *Journal of Dairy Science* 100(3), 2119–2136. doi: 10.3168/jds.2016-11517.

Alcock, P. (2015). A veterinary practice's approach to reducing cattle lameness. In: Pocknee, B. (Ed) *Cattle Lameness Conference*. Worcester: Worcester Rugby Club, pp. 27–38.

Alsaaod, M., Fadul, M. and Steiner, A. (2019). Automatic lameness detection in cattle, *Veterinary Journal* 246, 35–44. doi: 10.1016/j.tvjl.2019.01.005.

Alsaaod, M., Syring, C., Dietrich, J., Doherr, M. G., Gujan, T. and Steiner, A. (2014). A field trial of infrared thermography as a non-invasive diagnostic tool for early detection of digital dermatitis in dairy cows, *The Veterinary Journal* 199(2), 281–285. doi: 10.1016/j.tvjl.2013.11.028.

Anon (2009a). *Opinion on the Welfare of the Dairy Cow*. London: Fine Arts Work Center In Provincetown.

Anon (2009b). *Welfare Quality® Assessment Protocol for Cattle*. Lelystad, Netherlands: Welfare Quality® Consortium.

Anon (2016). *Mobility Scoring*. AHDB. Available at: http://dairy.ahdb.org.uk/technical-information/animal-health-welfare/lameness/husbandry-prevention/mobility-scoring/#.V4i4XvkrKM8.

Anon (2018). *Fourth Report*. CHAWG Report.

Archer, S., Bell, N. J. and Huxley, J. N. (2010). Lameness in UK dairy cattle: a review of the current status, *In Practice* 32(10), 492–504.

Archer, S. C., Newsome, R., Dibble, H., Sturrock, C. J., Chagunda, M. G., Mason, C. S. and Huxley, J. N. (2015). Claw length recommendations for dairy cow foot trimming, *The Veterinary Record* 177(9), 222. doi: 10.1136/vr.103197.

Barker, Z. E. (2007). *Epidemiology of Lameness in Dairy Cows*. Warwick. Available at: https://ethos.bl.uk/OrderDetails.do?uin=uk.bl.ethos.487362.

Barker, Z. E., Amory, J. R., Wright, J. L., Mason, S. A., Blowey, R. W. and Green, L. E. (2009). Risk factors for increased rates of sole ulcers, white line disease, and digital dermatitis in dairy cattle from twenty-seven farms in England and Wales, *Journal of Dairy Science* 92(5), 1971-1978. doi: 10.3168/jds.2008-1590.

Barker, Z. E., Leach, K. A., Whay, H. R., Bell, N. J. and Main, D. C. (2010). Assessment of lameness prevalence and associated risk factors in dairy herds in England and Wales, *Journal of Dairy Science* 93(3), 932-941. doi: 10.3168/jds.2009-2309.

Barnes, A. P., Rutherford, K. M., Langford, F. M. and Haskell, M. J. (2011). The effect of lameness prevalence on technical efficiency at the dairy farm level: an adjusted data envelopment analysis approach, *Journal of Dairy Science* 94(11), 5449-5457. Available at: http://www.sciencedirect.com/science/article/pii/S0022030211005716.

Bauman, H. E. (1995). *The origin and concept of HACCP, HACCP in Meat, Poultry and Fish processing*. In: Pearson, A. M. and Dutson, T. R. (Eds) *Advances in Meat Research*. London, UK: Blackie Academic and Professional, pp. 1-7.

Bell, N. J. (2006). The alleviation and prevention of lameness in dairy heifers. PhD Thesis. University of Bristol.

Bell, N. J., Atkinson, O., Blowey, R., Cobner, A., Green, M., Huxley, J. N., Hyde, R., Pedersen, S., Remnant, J., Manning, A., Mason, C., Oikonomou, G., Reader, J., Willshire, J. (2017). Moving on from antibiotic foot baths for the control of digital dermatitis, *Veterinary Record* 181(2), 51. doi: 10.1136/vr.j3241.

Bell, N. J., Bell, M. J., Knowles, T. G., Whay, H. R., Main, D. J. and Webster, A. J. (2009). The development, implementation and testing of a lameness control programme based on HACCP principles and designed for heifers on dairy farms, *Veterinary Journal* 180(2), 178-188. doi: 10.1016/j.tvjl.2008.05.020.

Bell, N. J. and Huxley, J. N. (2009a). Letters: Locomotion, lameness and mobility in dairy cows, *Veterinary Record* 164(23), 726. doi: 10.1136/vr.164.23.726.

Bell, N. J. and Huxley, J. N. (2009b). The use of rubber floor matting on dairy units: a critical review, *Cattle Practice* 17(2), 142-147.

Bell, N. J., Potterton, S., Blowey, R., Whay, H. R. and Huxley, J. N. (2014). The use of disinfectant foot-bathing agents for the control of bovine digital dermatitis in dairy cattle, *Livestock* 19(1), 6-13.

Bell, N. J. and Vanhoudt, A. (2020). Treatment and control of digital dermatitis in dairy cattle, *In Practice* 42, 554-567.

Bell, N. J. and Walker, A. M. (2016). Technology: are we any closer to automated lameness detection or automated mobility scoring?, *Livestock* 21(3), 152-159.

Bell, N. J. J. (2015). Evidence-based claw trimming for dairy cattle, *Veterinary Record* 177(9), 220-221. doi: 10.1136/vr.h4591.

Berry, S. L., Read, D. H., Famula, T. R., Mongini, A. and Döpfer, D. (2012). Long-term observations on the dynamics of bovine digital dermatitis lesions on a California dairy after topical treatment with lincomycin HCl, *The Veterinary Journal* 193(3), 654-658. doi: 10.1016/j.tvjl.2012.06.048.

Bicalho, R. C., Machado, V. S. and Caixeta, L. S. (2009). Lameness in dairy cattle: a debilitating disease or a disease of debilitated cattle? A cross-sectional study of lameness prevalence and thickness of the digital cushion, *Journal of Dairy Science* 92(7), 3175-3184. doi: 10.3168/jds.2008-1827.

Biemans, F., Bijma, P., Boots, N. M. and de Jong, M. C. M. (2018). Digital Dermatitis in dairy cattle: the contribution of different disease classes to transmission, *Epidemics* 23, 76-84. doi: 10.1016/j.epidem.2017.12.007.

Boelling, D. and Pollott, G. E. (1997). The genetics of feet, legs and locomotion in cattle, *Animal Breeding Abstracts* 65(1), 1.

Bradford, B. J., Yuan, K., Farney, J. K., Mamedova, L. K. and Carpenter, A. J. (2015). Invited review: inflammation during the transition to lactation: new adventures with an old flame, *Journal of Dairy Science* 98(10), 6631–6650. doi: 10.3168/jds.2015-9683.

Bran, J. A., Costa, J. H. C., von Keyserlingk, M. A. G. and Hötzel, M. J. (2019). Factors associated with lameness prevalence in lactating cows housed in freestall and compost-bedded pack dairy farms in southern Brazil, *Preventive Veterinary Medicine* 172(September), 104773. doi: 10.1016/j.prevetmed.2019.104773.

Carvalho, V. R. C., Bucklin, R. A., Shearer, J. K. and Shearer, L. (2005). Effects of trimming on dairy cattle hoof weight bearing and pressure distributions during the stance phase, *Transactions of the ASAE* 48(4), 1653–1659. doi: 10.13031/2013.19166.

Cha, E., Hertl, J. A., Bar, D. and Gröhn, Y. T. (2010). The cost of different types of lameness in dairy cows calculated by dynamic programming, *Preventive Veterinary Medicine* 97(1), 1–8. Available at: http://www.sciencedirect.com/science/article/pii/S0167587710002102.

Chapinal, N., de Passillé, A. M., Rushen, J. and Wagner, S. A. (2010). Effect of analgesia during hoof trimming on gait, weight distribution, and activity of dairy cattle, *Journal of Dairy Science* 93(7), 3039–3046. Available at: http://www.sciencedirect.com/science/article/pii/S0022030210003139.

Chesterton, N. (1988). Environmental and behavioural factors affecting the prevalence of lameness in New Zealand dairy herds-a case-control study, *Acta Veterinaria Scandinavica, - Supplement* 84, 185.

Chesterton, R. N. (2013). A foot bath design for large grazing herds. In: Whay, H. R. and Hockenhull, J. (Eds) *17th International Symposium and 9th International Conference on Lameness in Ruminants*. Bristol, UK: Bristol Marriott City Centre Hotel, pp. 212–213

Chesterton, R. N., Pfeiffer, D. U., Morris, R. S. and Tanner, C. M. (1989). Environmental and behavioural factors affecting the prevalence of foot lameness in New Zealand dairy herds–a case-control study, *New Zealand Veterinary Journal* 37(4), 135–142.

Clarkson, M. J., Downham, D. Y., Faull, W. B., Hughes, J. W., Manson, F. J., Merritt, J. B., Murray, R. D., Russell, W. B., Sutherst, J. E. and Ward, W. R. (1996). Incidence and prevalence of lameness in dairy cattle, *Veterinary Record* 138(23), 563–567.

Collins, S. (2016). An investigation of whether and how welfare outcome assessment could be better used by UK dairy farmers. A thesis submitted by Department of Production and Population Health The Royal Veterinary College, (March).

Cook, N. B., Hess, J. P., Foy, M. R., Bennett, T. B. and Brotzman, R. L. (2016). Management characteristics, lameness, and body injuries of dairy cattle housed in high-performance dairy herds in Wisconsin, *Journal of Dairy Science* 99(7), 5879–5891. doi: 10.3168/jds.2016-10956.

Cook, N. B., Rieman, J., Gomez, A. and Burgi, K. (2012). Observations on the design and use of footbaths for the control of infectious hoof disease in dairy cattle, *The Veterinary Journal* 193(3), 669–673. doi: 10.1016/j.tvjl.2012.06.051.

Costa, J. H. C., Burnett, T. A., von Keyserlingk, M. A. G. and Hötzel, M. J. (2018). Prevalence of lameness and leg lesions of lactating dairy cows housed in southern Brazil: effects of housing systems, *Journal of Dairy Science* 101(3), 2395–2405. doi: 10.3168/jds.2017-13462.

Cramer, G., Lissemore, K. D., Guard, C. L., Leslie, K. E. and Kelton, D. F. (2009). Herd-level risk factors for seven different foot lesions in Ontario Holstein cattle housed in tie stalls or free stalls, *Journal of Dairy Science* 92(4), 1404–1411. Available at: http://www.sciencedirect.com/science/article/pii/S0022030209704515.

Cruz, C., Driemeier, D., Cerva, C. and Corbellini, L. G. (2001). Clinical and epidemiological aspects of bovine digital lesions in southern Brazil, *Arquivo Brasileiro De Medicina Veterinaria E Zootecnia* 53(6), 654–657.

Cutler, J. H. H., Rushen, J., de Passillé, A. M., Gibbons, J., Orsel, K., Pajor, E., Barkema, H. W., Solano, L., Pellerin, D., Haley, D. and Vasseur, E. (2017). Producer estimates of prevalence and perceived importance of lameness in dairy herds with tiestalls, freestalls, and automated milking systems, *Journal of Dairy Science* 100(12), 9871–9880. doi: 10.3168/jds.2017-13008.

Danscher, A. M., Enemark, J. M., Telezhenko, E., Capion, N., Ekstrøm, C. T. and Thoefner, M. B. (2009). Oligofructose overload induces lameness in cattle, *Journal of Dairy Science* 92(2), 607–616. Available at: http://www.sciencedirect.com/science/article/pii/S0022030209703650.

Danscher, A. M., Toelboell, T. H. and Wattle, O. (2010). Biomechanics and histology of bovine claw suspensory tissue in early acute laminitis, *Journal of Dairy Science* 93(1), 53–62. doi: 10.3168/jds.2009-2038.

Dolecheck, K. A., Overton, M. W., Mark, T. B. and Bewley, J. M. (2019). Use of a stochastic simulation model to estimate the cost per case of digital dermatitis, sole ulcer, and white line disease by parity group and incidence timing, *Journal of Dairy Science* 102(1), 715–730. doi: 10.3168/jds.2018-14901.

Dopfer, D., Koopmans, A., Meijer, F. A., Szakáll, I., Schukken, Y. H., Klee, W., Bosma, R. B., Cornelisse, J. L., van Asten, A. J. and ter Huurne, A. A. (1997). Histological and bacteriological evaluation of digital dermatitis in cattle, with special reference to spirochaetes and Campylobacter faecalis, *Veterinary Record* 140(24), 620–623.

Egger-Danner, C., Nielsen, P., Fiedler, A., Müller, K., Fjeldaas, T., Döpfer, D., Daniel, V., Bergsten, C., Cramer, G., Christen, A.-M., Stock, K. F., Thomas, G., Holzhauer, M., Steiner, A., Clarke, J., Capion, N., Charfeddine, N., Pryce, J. E., Oakes, E., Burgstaller, J., Heringstad, B., Ødegård, C. and Kofler, J. (2015). *ICAR Claw Health Atlas*.

Engel, B., Bruin, G., Andre, G. and Buist, W. (2003). Assessment of observer performance in a subjective scoring system: visual classification of the gait of cows, *Journal of Agricultural Science* 140(3), 317–333.

Evans, N. J., Brown, J. M., Demirkan, I., Singh, P., Getty, B., Timofte, D., Vink, W. D., Murray, R. D., Blowey, R. W., Birtles, R. J., Hart, C. A. and Carter, S. D. (2009). Association of Unique, Isolated Treponemes with Bovine Digital Dermatitis Lesions', *Journal of Clinical Microbiology* 47(3), 689–696.

Fabian, J., Laven, R. A. and Whay, H. R. (2014). The prevalence of lameness on New Zealand dairy farms: a comparison of farmer estimate and locomotion scoring, *The Veterinary Journal* 201(1), 31–38. doi: 10.1016/j.tvjl.2014.05.011.

Fisher, G. and Atkinson, O. C. D. (2013). Cattle mobility: changing behaviour to improve health and welfare and dairy farm businesses. Final Report–December 2013. Available at: https://www.reaseheath.ac.uk/wp-content/uploads/2014/02/Cattle-Mobility-Final-report-December-2013.pdf.

Flower, F. C. and Weary, D. M. (2006). Effect of hoof pathologies on subjective assessments of dairy cow gait, *Journal of Dairy Science* 89(1), 139–146. Available at: http://jds.fass.org/cgi/content/abstract/89/1/139.

Garcia, D., Hanh, M., Pino, D., Vivas, I., Sanchez, A. and Villarroel, R. (2004). Functional trimming in dry period its effect, on milk production and reproduction in cows, *Revista Cientifica-Facultad De Ciencias Veterinarias* 14(6), 568–574.

García-Bracho, D., Hahn, M., Pino, D., Vivas, I., Leal, R. M. and Clerc, K. (2009). Functional trimming at the dry off period to prevent foot diseases in confined dairy cows at the tropical area, Revista Cientifica-Facultad de Ciencias Veterinarias 19(2), 147–152. Available at: https://www.cabdirect.org/cabdirect/abstract/20093172285.

Gard, J. A., Taylor, D. R., Wilhite, D. R., Rodning, S. P., Schnuelle, M. L., Sanders, R. K., Beyers, R. J., Edmondson, M. A., DeGraves, F. J. and van Santen, E. (2014). Effect of exercise and environmental terrain on development of the digital cushion and bony structures of the bovine foot, *American Journal of Veterinary Research* 76(3), 246–252.

Gillespie, A., Carter, S. D., Blowey, R. W. and Evans, N. (2019). Survival of bovine digital dermatitis treponemes on hoof knife blades and the effects of various disinfectants, *Veterinary Record* 186(2), 67.

Gomez, A. and Cook, N. B. (2010). Time budgets of lactating dairy cattle in commercial freestall herds, *Journal of Dairy Science* 93(12), 5772–5781. doi: 10.3168/jds.2010-3436.

Gomez, A., Cook, N. B., Bernardoni, N. D., Rieman, J., Dusick, A. F., Hartshorn, R., Socha, M. T., Read, D. H. and Döpfer, D. (2012). An experimental infection model to induce digital dermatitis infection in cattle, *Journal of Dairy Science* 95(4), 1821–1830. doi: 10.3168/jds.2011-4754.

Gomez, A., Cook, N. B., Socha, M. T. and Döpfer, D. (2015). First-lactation performance in cows affected by digital dermatitis during the rearing period, *Journal of Dairy Science* 98(7), 4487–4498. doi: 10.3168/jds.2014-9041.

Green, L. E., Hedges, V. J., Schukken, Y. H., Blowey, R. W., Packington, A. J., Menzies, F. D. and Reid, S. W. J. (2002). The impact of clinical lameness on the milk yield of dairy cows. 2002/10/05, *Journal of Dairy Science* 85(9), 2250–2256. doi: 10.3168/jds.S0022-0302(02)74304-X.

Green, L. E., Huxley, J. N., Banks, C. and Green, M. J. (2014). Temporal associations between low body condition, lameness and milk yield in a UK dairy herd, *Preventive Veterinary Medicine* 113(1), 63-71. doi: 10.1016/j.prevetmed.2013.10.009.

Greenough, P. R. (1990). Meeting the need for veterinary diagnostic services in developing-countries, *Canadian Veterinary Journal = la Revue Veterinaire Canadienne* 31(4), 304–305.

Griffiths, B. E., Grove White, D. and Oikonomou, G. (2018). A cross-sectional study Into the prevalence of dairy cattle lameness and associated herd-level risk factors in England and Wales, *Frontiers in Veterinary Science* 5, 65. doi: 10.3389/fvets.2018.00065.

Groenevelt, M., Main, D. C., Tisdall, D., Knowles, T. G. and Bell, N. J. (2014). Measuring the response to therapeutic foot trimming in dairy cows with fortnightly lameness scoring, *The Veterinary Journal* 201(3), 283–288. doi: 10.1016/j.tvjl.2014.05.017.

Haskell, M. J., Rennie, L. J., Bowell, V. A., Bell, M. J. and Lawrence, A. B. (2006). Housing system, milk production, and zero-grazing effects on lameness and leg injury in dairy cows, *Journal of Dairy Science* 89(11), 4259–4266.

Heath, C. A. E., Lin, Y., Mullan, S., Browne, W. and Main, D. (2014). Implementing Welfare Quality in UK assurance schemes: evaluating the challenges, *Animal Welfare* 23(1), 95–107.

Hedges, J., Blowey, R. W., Packington, A. J., O'Callaghan, C. J. and Green, L. E. (2001). A longitudinal field trial of the effect of biotin on lameness in dairy cows, *Journal of Dairy Science* 84(9), 1969–1975.

Hernandez, J. A., Garbarino, E. J., Shearer, J. K., Risco, C. A. and Thatcher, W. W. (2007). Evaluation of the efficacy of prophylactic hoof health examination and trimming during midlactation in reducing the incidence of lameness during late lactation in dairy cows, *Journal of the American Veterinary Medical Association* 230(1), 89–93.

Hogeveen, H., Kamphuis, C., Steeneveld, W. and Mollenhorst, H. (2010). Sensors and clinical mastitis–the quest for the perfect alert, *Sensors* 10(9), 7991–8009. doi: 10.3390/s100907991.

Holah, D. E., Evans, K. M., Pearson, G. R., Tarlton, J. F. and Webster, A. J. F. (2000). The histology and histopathology of the support structures in the laminated region of the bovine hoof in maiden heifers and around the time of first calving. In: *Lameness in Ruminants*. Parma, Italy, 3–7 September 2000, p. 109.

Huxley, J. N. (2012). Lameness in cattle: an ongoing concern, *The Veterinary Journal* 193(3), 610–611. doi: 10.1016/j.tvjl.2012.06.039.

Huxley, J. N., Burke, J., Roderick, S., Main, D. C. and Whay, H. R. (2004). Animal welfare assessment benchmarking as a tool for health and welfare planning in organic dairy herds, *Veterinary Record* 155(8), 237–239.

Jewell, M. T., Cameron, M., Spears, J., McKenna, S. L., Cockram, M. S., Sanchez, J. and Keefe, G. P. (2019). Prevalence of lameness and associated risk factors on dairy farms in the Maritime Provinces of Canada, *Journal of Dairy Science* 102(4), 3392–3405. doi: 10.3168/jds.2018-15349.

Knott, L., Tarlton, J. F., Craft, H. and Webster, A. J. F. (2007). Effects of housing, parturition and diet change on the biochemistry and biomechanics of the support structures of the hoof of dairy heifers, *The Veterinary Journal* 174(2), 277–287. doi: 10.1016/j.tvjl.2006.09.007.

Kofler, J., Fiedler, A., Charfeddine, J. N., Capion, N., Fjeldaas, T., Cramer, G., Bell, N. J., Müller, K. E., Christen, A.-M., Thomas, G., Heringstad, B., Stock, K. F., Holzhauer, M., Nieto, J. M., Egger-Danner, C. and Döpfer, D. (2020a). *Annex 1 of Section 7 of ICAR Guidelines–Digital Dermatitis Stages ( M-stages ) ICAR Claw Health Atlas–Appendix 1*.

Kofler, J., Fiedler, A., Charfeddine, J. N., Capion, N., Fjeldaas, T., Cramer, G., Bell, N. J., Müller, K. E., Christen, A.-M., Thomas, G., Heringstad, B., Stock, K. F., Holzhauer, M., Nieto, J. M., Egger-Danner, C. and Döpfer, D. (2020b). *ICAR Claw Health Atlas–Appendix 2 Digital Dermatitis-Associated Claw Horn Lesions*. Available at: https://www.icar.org/Guidelines/07-Appendix-2-DD-associated-Claw-Horn-Lesions.pdf.

Kremer, P., Nuske, S., Scholz, A. M. and Forster, M. (2006). Effect of soft flooring in free stalls on milk yield, fat, protein and somatic cell counts in dairy cattle, *Archiv. fur Tierzucht-Archives of Animal Breeding* 49(3), 250–258.

Laven, R. A. (2006). Efficacy of systemic cefquinome and erythromycin against digital dermatitis in cattle, *Veterinary Record* 159(1), 19–20.

Laven, R. A. and Hunt, H. (2001). Comparison of valnemulin and lincomycin in the treatment of digital dermatitis by individually applied topical spray, *Veterinary Record* 149(10), 302–303.

Laven, R. A. and Hunt, H. (2002). Evaluation of copper sulphate, formalin and peracetic acid in footbaths for the treatment of digital dermatitis in cattle, *Veterinary Record* 151(5), 144–146.

Laven, R. A. and Livesey, C. T. (2004). The effect of housing and methionine intake on hoof horn hemorrhages in primiparous lactating holstein cows, *Journal of Dairy Science* 87(4), 1015-1023.

Leach, K. A., Paul, E. S., Whay, H. R., Barker, Z. E., Maggs, C. M., Sedgwick, A. K. and Main, D. C. (2013). Reducing lameness in dairy herds–overcoming some barriers, *Research in Veterinary Science* 94(3), 820-825. doi: 10.1016/j.rvsc.2012.10.005.

Leach, K. A. A., Tisdall, D. A., Bell, N. J., Main, D. C. and Green, L. E. (2012). The effects of early treatment for hindlimb lameness in dairy cows on four commercial UK farms, *Veterinary Journal* 193(3), 626-632. Available at: http://www.sciencedirect.com/science/article/pii/S1090023312002912.

Leach, K. A. and Whay, H. R. (2009). Welfare Quality lameness control program. Available at: http://www.welfarequality.net/media/1122/wqr14.pdf.

Leonard, F. C., O'Connell, J. M. and O'Farrell, K. J. (1994). Effect of different housing conditions on behaviour and foot lesions in Friesian heifers, *Veterinary Record* 134(19), 490-494.

Leonard, F. C., O'Connell, J. M. and O'Farrell, K. J. (1996). Effect of overcrowding on claw health in first-calved Friesian heifers, *British Veterinary Journal* 152(4), 459-472.

Leonard, F. C., O'Connell, J. M., O'Farrell, K. J., Bottcher, R. W. and Hoff, S. J. (1997). Housing conditions and foot lesions in Friesian heifers. In: Bloomington, Minnesota, USA, 29-31 May, 1997: American Society of Agricultural Engineers (ASAE); St Joseph; USA, p. 250.

Lim, P. Y., Huxley, J. N., Willshire, J. A., Green, M. J., Othman, A. R. and Kaler, J. (2015). Unravelling the temporal association between lameness and body condition score in dairy cattle using a multistate modelling approach, *Preventive Veterinary Medicine* 118(4), 370-377. doi: 10.1016/j.prevetmed.2014.12.015.

Lin, Y. (2015). The Use of Animal Welfare Outcome Measurements in Farm Assurance Schemes, (March).

Livesey, C. T., Metcalf, J. A. and Laven, R. A. (2003). Effect of concentrate composition and cubicle bedding on the development of hoof haemorrhages in Holstein heifers after calving, *Veterinary Record* 152(24), 735-739. doi: 10.1136/vr.152.24.735.

Logue, D. N, Webster, A. J. F., Offer, J. E., Leach, K. A. and Brocklehurst, S. (2001). The effect of forage type on foot health in Dairy Heifers–Final Report for MDC.

Logue, D. N., Leach, K. A., Brocklehurst, S. and Offer, J. E. (2000). Effect of diet in lesion development from birth up to the end of first lactation. In: *Pharmacia*, Italy, 3-7 September 2000, p. 327.

Logue, D. N. and Mayne, C. S. (2014). Welfare-positive management and nutrition for the dairy herd: a European perspective, *The Veterinary Journal* 199(1), 31-38. doi: 10.1016/j.tvjl.2013.10.027.

Logue, D. N., Offer, J. E. and Hyslop, J. J. (1994). Relationship of diet, hoof type and locomotion score with lesions of the sole and white line in dairy-cattle, *Animal Science* 59(2), 173-181.

Mahendran, S. and Bell, N. J. (2015). Lameness in cattle 2. Managing claw health through appropriate trimming techniques, *In Practice* 37(5), 231-242. doi: 10.1136/inp.h2011.

Mahendran, S. A., Huxley, J. N., Chang, Y. M., Burnell, M., Barrett, D. C., Whay, H. R., Blackmore, T., Mason, C. S. and Bell, N. J. (2017). Randomised controlled trial to evaluate the effect of foot trimming before and after first calving on subsequent lameness episodes and productivity in dairy heifers, *Veterinary Journal* 220, 105-110. doi: 10.1016/j.tvjl.2017.01.011.

Main, D. C. J. C. J., Barker, Z. E., Leach, K. A., Bell, N. J., Whay, H. R. and Browne, W. J. (2010). Sampling strategies for monitoring lameness in dairy cattle, *Journal of Dairy Science* 93(5), 1970–1978. Available at: http://www.sciencedirect.com/science/article/pii/S0022030210001852.

Manning, A., Mahendran, S. and Bell, N. J. (2016). Evidence base behind foot trimming in UK dairy cattle, *Livestock* 21(1), 6–14. doi: 10.12968/live.2016.21.1.6.

Manning, A. D., Mahendran, S. A., Hurst, B. S., Blackmore, T. L. and Bell, N. J. (2017). Effect of a prewash on footbath contamination: a randomised control trial, *Veterinary Record* 180(5), 121. doi: 10.1136/vr.103922.

Manske, T., Hultgren, J. and Bergsten, C. (2002a). Prevalence and interrelationships of hoof lesions and lameness in Swedish dairy cows, *Preventive Veterinary Medicine* 54(3), 247–263. doi: 10.1016/S0167-5877(02)00018-1.

Manske, T., Hultgren, J.. and Bergsten, C. (2002b). The effect of claw trimming on the hoof health of Swedish dairy cattle, *Preventive Veterinary Medicine* 54(2), 113–129.

Manson, F. J. and Leaver, J. D. (1988a). The influence of concentrate amount on locomotion and clinical lameness in dairy-cattle, *Animal Science* 47(2), 185–190.

Manson, F. J. and Leaver, J. D. (1988b). The influence of dietary-protein intake and of hoof trimming on lameness in dairy-cattle, *Animal Science* 47(2), 191–199.

Manson, F. J. and Leaver, J. D. (1989). The effect of concentrate - silage ratio and of hoof trimming on lameness in dairy-cattle, *Animal Science* 49(1), 15–22.

Maselyne, J., Pastell, J., Thomsen, P. T., Thorup, V. M., Hänninen, L., Vangeyte, J., et al. (2016). Lying time of dairy cows does not increase linearly throughout lactation. In: Knight C, editor. Proceedings of the Fourth Dairycare Conference. *Lissabon, Portugal*, p. 21.

Maxwell, O. J. R., Hudson, C. D. and Huxley, J. N. (2015). Effect of early lactation foot trimming in lame and non-lame dairy heifers: a randomised controlled trial, *Veterinary Record* 177(4), 100. doi: 10.1136/vr.103155.

Murray, R. D., Downham, D. Y., Clarkson, M. J., Faull, W. B., Hughes, J. W., Manson, F. J., Merritt, J. B., Russell, W. B., Sutherst, J. E. and Ward, W. R. (1996). Epidemiology of lameness in dairy cattle: description and analysis of foot lesions, *Veterinary Record* 138(24), 586–591.

Newsome, R. F., Green, M. J., Bell, N. J., Bollard, N. J., Mason, C. S., Whay, H. R. and Huxley, J. N. (2017). A prospective cohort study of digital cushion and corium thickness. Part 2: does thinning of the digital cushion and corium lead to lameness and claw horn disruption lesions?, *Journal of Dairy Science* 100(6), 4759–4771. doi: 10.3168/jds.2016-12013.

Newsome, R., Green, M. J., Bell, N. J., Chagunda, M. G. G., Mason, C. S., Rutland, C. S., Sturrock, C. J., Whay, H. R. and Huxley, J. N. (2016). Linking bone development on the caudal aspect of the distal phalanx with lameness during life, *Journal of Dairy Science* 99(6), 4512–4525. doi: 10.3168/jds.2015-10202.

Nocek, J. E. (1997). Bovine acidosis: implications on laminitis, *Journal of Dairy Science* 80(5), 1005–1028.

O'Callaghan, K. A., Cripps, P. J., Downham, D. Y. and Murray, R. D. (2003). Subjective and objective assessment of pain and discomfort due to lameness in dairy cattle, *Animal Welfare* 12(4), 605–610.

Offer, J. E., Logue, D. N. and Mason, C. (2004a). Effect of an incident of overfeeding of concentrate on claw horn lesion development in first lactation dairy heifers. In: Zemljič, B. (Ed) *13th International Symposium and 5th Conference on Lameness in Ruminants*. Maribor, Slovenija, pp. 175–176.

Offer, J. E., Logue, D. N., Offer, N. W. and Marsden, M. (2004b). The effect of concentrate composition on lameness and hoof health in dairy cows, *Veterinary Journal* 167(1), 111–113.

Offer, J. E., Logue, D. N. and Roberts, D. J. (1997). The effect of protein source on lameness and solear lesion formation in dairy cattle, *Animal Science* 65(2), 143–149.

Offer, J. E., McNulty, D. and Logue, D. N. (2000). Observations of lameness, hoof conformation and development of lesions in dairy cattle over four lactations, *Veterinary Record* 147(4), 105–109.

Oikonomou, G., Banos, G., Machado, V., Caixeta, L. and Bicalho, R. C. (2014). Short communication: genetic characterization of digital cushion thickness, *Journal of Dairy Science* 97(1), 532–536. doi: 10.3168/jds.2013-7212.

Ouweltjes, W., Holzhauer, M., van der Tol, P. P. and van der Werf, J. (2009). Effects of two trimming methods of dairy cattle on concrete or rubber-covered slatted floors, *Journal of Dairy Science* 92(3), 960-971. doi: 10.3168/jds.2008-1559.

Pedersen, S. (2019). Digital dermatitis control in the dairy herd: incorporating the "blitz" treatment approach, *Livestock* 24(3), 130–135. doi: 10.12968/live.2019.24.3.130.

Pérez-Cabal, M. A. and Alenda, R. (2014). Clinical lameness and risk factors in a Spanish Holstein population, *Livestock Science* 164, 168–174. doi: 10.1016/j.livsci.2014.03.012.

Phillips, C. J., Chiy, P. C., Bucktrout, M. J., Collins, S. M., Gasson, C. J., Jenkins, A. C. and Paranhos da Costa, M. J. (2000). Frictional properties of cattle hooves and their conformation after trimming, *Veterinary Record* 146(21), 607–609.

Potterton, S., Bell, N., Whay, B., Main, D. and Huxley, J. (2011). A review of the peer reviewed literature on the treatment and prevention of foot lameness in cattle published Between 2000 and 2011, *Final report*, pp. 1-102.

Potterton, S. L., Bell, N. J., Whay, H. R., Berry, E. A., Atkinson, O. C., Dean, R. S., Main, D. C. and Huxley, J. N. (2012). A descriptive review of the peer and non-peer reviewed literature on the treatment and prevention of foot lameness in cattle published between 2000 and 2011, *Veterinary Journal* 193(3), 612-616. doi: 10.1016/j.tvjl.2012.06.040.

Räber, M., Lischer, ChJ., Geyer, H. and Ossent, P. (2004). The bovine digital cushion – a descriptive anatomical study, *The Veterinary Journal* 167(3), 258–264. doi: 10.1016/S1090-0233(03)00053-4.

Räber, M., Scheeder, M. R., Ossent, P., Lischer, ChJ. and Geyer, H. (2006). The content and composition of lipids in the digital cushion of the bovine claw with respect to age and location – A preliminary report, *The Veterinary Journal* 172(1), 173-177. doi: 10.1016/j.tvjl.2005.03.009.

Rajkondawar, P. G., Lefcourt, A. M., Neerchal, N. K., Dyer, R. M., Varner, M. A., Erez, B. and Tasch, U. (2002). The development of an objective lameness scoring system for dairy herds: pilot study, *Transactions of the ASAE. American Society of Agricultural Engineers* 45(4), 1123.

Randall, L. V., Green, M. J., Green, L. E., Chagunda, M. G. G., Mason, C., Archer, S. C. and Huxley, J. N. (2017). The contribution of previous lameness events and body condition score to the occurrence of lameness in dairy herds: a study of 2 herds, *Journal of Dairy Science* 101(2), 1311–1324. doi: 10.3168/jds.2017-13439.

Randall, L. V., Thomas, H. J., Remnant, J. G., Bollard, N. J. and Huxley, J. N. (2019). Lameness prevalence in a random sample of UK dairy herds, *Veterinary Record* 184(11), 350-350. doi: 10.1136/vr.105047.

Ranjbar, S., Rabiee, A. R., Gunn, A. and House, J. K. (2016). Identifying risk factors associated with lameness in pasture-based dairy herds, *Journal of Dairy Science* 99(9), 7495–7505. doi: 10.3168/jds.2016-11142.

Rodriguez-Lainz, A., Hird, D. W., Carpenter, T. E. and Read, D. H. (1996). Case-control study of papillomatous digital dermatitis in southern California dairy farms, *Preventive Veterinary Medicine* 28(2), 117–131.

Rodriguez-Lainz, A., Melendez-Retamal, P., Hird, D. W. and Read, D. H. (1998). Papillomatous digital dermatitis in Chilean dairies and evaluation of a screening method, *Preventive Veterinary Medicine* 37(1–4), 197–207.

Rossiter, J. E. R., Van Winden, S., Burnell, M. and Bell, N. J. (2013). Inter-observer agreement between foot trimmers on foot lesions in cattle. In: Whay, H. R. and Hockenhull, J. (Eds) *17th International Symposium and 9th International Conference on Lameness in Ruminants*. Bristol, UK: Mariot Hotel, p. 119.

Rowlands, G. J., Russel, A. M. and Williams, L. A. (1983). Effects of season, herd size, management system and veterinary practice on the lameness incidence in dairy cattle, *Veterinary Record* 113(19), 441–445.

Rushen, J. and de Passille, A. M. (2006). Effects of roughness and compressibility of flooring on cow locomotion, *Journal of Dairy Science* 89(8), 2965–2972. Available at: http://jds.fass.org/cgi/content/abstract/89/8/2965.

Rushen, J., Pombourcq, E. and Passillé, A. Md de (2007). Validation of two measures of lameness in dairy cows, *Applied Animal Behaviour Science* 106(1–3), 173–177. doi: 10.1016/j.applanim.2006.07.001.

Rutherford, K. M. D., Langford, F. M., Jack, M. C., Sherwood, L., Lawrence, A. B. and Haskell, M. J. (2009). Lameness prevalence and risk factors in organic and non-organic dairy herds in the United Kingdom, *The Veterinary Journal* 180(1), 95–105. doi: 10.1016/j.tvjl.2008.03.015.

Sadiq, M. B., Ramanoon, S. Z., Mossadeq, W. M. S., Mansor, R. and Syed-Hussain, S. S., (2020). Claw trimming as a lameness management practice and the association with welfare and production in dairy cows, *Animals* 10(9):1–18.

Schmid, T., Weishaupt, M. A., Meyer, S. W., Waldern, N., Peinen, Kv and Nuss, K. (2009). High-speed cinematographic evaluation of claw-ground contact pattern of lactating cows, *The Veterinary Journal* 181(2), 151–157. doi: 10.1016/j.tvjl.2008.02.019.

Scholey, R. A., Blowey, R. W., Murray, R. D., Smith, R. F., Cameron, J., Massey, J. P., Ollier, W. E. and Carter, S. D. (2012). Investigating host genetic factors in bovine digital dermatitis, *Veterinary Record* 171(24), 624. doi: 10.1136/vr.101251.

Shepherd, F. (2016). *The Effect of a Health Intervention Scheme on the Mobility of Dairy Cows in the Southwest of England*. Plymouth: University of Plymouth.

Siciliano-Jones, J. L., Socha, M. T., Tomlinson, D. J. and DeFrain, J. M. (2008). Effect of trace mineral source on lactation performance, claw integrity, and fertility of dairy cattle, *Journal of Dairy Science* 91(5), 1985–1995. doi: 10.3168/jds.2007-0779.

Sogstad, A. M., Fjeldaas, T., Østerås, O. and Forshell, K. P. (2005). Prevalence of claw lesions in Norwegian dairy cattle housed in tie stalls and free stalls, *Preventive Veterinary Medicine* 70(3–4), 191–209. doi: 10.1016/j.prevetmed.2005.03.005.

Solano, L., Barkema, H. W., Jacobs, C. and Orsel, K. (2017). Validation of the M-stage scoring system for digital dermatitis on dairy cows in the milking parlor, *Journal of Dairy Science*. American Dairy Science Association 100(2), 1592–1603. doi: 10.3168/jds.2016-11365.

Solano, L., Barkema, H. W., Pajor, E. A., Mason, S., LeBlanc, S. J., Zaffino Heyerhoff, J. C., Nash, C. G., Haley, D. B., Vasseur, E., Pellerin, D., Rushen, J., de Passillé, A. M. and Orsel, K. (2015). Prevalence of lameness and associated risk factors in Canadian Holstein-Friesian cows housed in freestall barns, *Journal of Dairy Science* 98(10), 6978–6991. doi: 10.3168/jds.2015-9652.

Sprecher, D. J., Hostetler, D. E. and Kaneene, J. B. (1997). A lameness scoring system that uses posture and gait to predict dairy cattle reproductive performance, *Theriogenology* 47(6), 1179–1187.

Stoddard, G. C. (2018). Evaluating the relationship between hoof trimming and dairy cattle well-being. PhD Thesis. University of Minnesota. doi: 10.18041/2382-3240/saber.2010v5n1.2536.

Stoddard, G. C. and Cramer, G. (2017). A review of the relationship Between hoof trimming and dairy cattle welfare, *Veterinary Clinics of North America. Food Animal Practice* 33(2), 365–375. doi: 10.1016/j.cvfa.2017.02.012.

Stokes, J. E. (2011). Investigating novel and existing methods of preventing, detecting and treating digital dermatitis in dairy cows. PhD thesis. Clinical Veterinary Science. Bristol: University of Bristol.

Tarlton, J. F., Holah, D. E., Evans, K. M., Jones, S., Pearson, G. R. and Webster, A. J. F. (2000). Biomechanical and biochemical analyses of changes in the supportive structure of the bovine hoof in maiden heifers and around the time of first calving. In: *Lameness in Ruminants*, Pharmacia, Italy, 3–7 September 2000, p. 113.

Tarlton, J. F., Holah, D. E., Evans, K. M., Jones, S., Pearson, G. R. and Webster, A. J. (2002). Biomechanical and histopathological changes in the support structures of bovine hooves around the time of first calving, *Veterinary Journal* 163(2), 196–204.

Telezhenko, E. and Bergsten, C. (2005). Influence of floor type on the locomotion of dairy cows, *Applied Animal Behaviour Science* 93(3–4), 183–197. doi: 10.1016/j.applanim.2004.11.021.

Telezhenko, E., Bergsten, C., Magnusson, M. and Nilsson, C. (2009). Effect of different flooring systems on claw conformation of dairy cows, *Journal of Dairy Science* 92(6), 2625–2633. doi: 10.3168/jds.2008-1798.

Thomas, H. J. J., Miguel-Pacheco, G. G., Bollard, N. J., Archer, S. C., Bell, N. J., Mason, C., Maxwell, O. J., Remnant, J. G., Sleeman, P., Whay, H. R. and Huxley, J. N. (2015). Evaluation of treatments for claw horn lesions in dairy cows in a randomized controlled trial, *Journal of Dairy Science* 98(7), 4477–4486. doi: 10.3168/jds.2014-8982.

Thomas, H. J. J., Remnant, J. G., Bollard, N. J., Burrows, A., Whay, H. R., Bell, N. J., Mason, C. and Huxley, J. N. (2016). Recovery of chronically lame dairy cows following treatment for claw horn lesions: a randomised controlled trial, *Veterinary Record* 178(5), 116. doi: 10.1136/vr.103394.

Thomsen, P. T. (2015). Short communication: efficacy of copper sulfate hoof baths against digital dermatitis–where is the evidence?, *Journal of Dairy Science* 98(4), 2539–2544. doi: 10.3168/jds.2014-9135.

Thomsen, P. T., Ersbøll, A. K. and Sørensen, J. T. (2012). Short communication: automatic washing of hooves can help control digital dermatitis in dairy cows, *Journal of Dairy Science* 95(12), 7195–7199. doi: 10.3168/jds.2011-5171.

Toussaint Raven, E. (1985). *Cattle Footcare and Claw Trimming* (1st edn.). Suffolk, UK: Farming Press Ltd.

Tranter, W. P. and Morris, R. S. (1992). Hoof growth and wear in pasture-Fed Dairy-Cattle, *New Zealand Veterinary Journal* 40(3), 89-96.

Tranter, W. P., Morris, R. S., Dohoo, I. R. and Williamson, N. B. (1993). A case-control study of lameness in dairy cows, *Preventive Veterinary Medicine* 15(2-3), 191-203.

Tranter, W. P., Morris, R. S. and Williamson, N. B. (1991). A longitudinal-study of the hooves of non-lame cows, *New Zealand Veterinary Journal* 39(2), 53-57.

Tucker, C. B., Weary, D. M., de Passillé, A. M., Campbell, B. and Rushen, J. (2006). Flooring in Front of the feed bunk affects feeding behavior and use of Freestalls by dairy cows, *Journal of Dairy Science* 89(6), 2065-2071. Available at: http://jds.fass.org/cgi/content/abstract/89/6/2065.

van Amstel, S. R. and Shearer, J. (2008). *Manual for Treatment and Control of Lameness in Cattle*. doi: 10.1002/9780470344576.

van Amstel, S. R., Shearer, J. K. and Palin, F. L. (2004). Moisture content, thickness, and lesions of sole horn associated with thin soles in dairy cattle, *Journal of Dairy Science* 87(3), 757-763. Available at: http://jds.fass.org/cgi/content/abstract/87/3/757.

van der Tol, P. P. J., Metz, J. H., Noordhuizen-Stassen, E. N., Back, W., Braam, C. R. and Weijs, W. A. (2002). The pressure distribution under the bovine claw during square standing on a flat substrate, *Journal of Dairy Science* 85(6), 1476-1481.

van der Tol, P. P. J., van der Beek, S. S., Metz, J. H., Noordhuizen-Stassen, E. N., Back, W., Braam, C. R. and Weijs, W. A. (2004). The effect of preventive trimming on weight bearing and force balance on the claws of dairy cattle, *Journal of Dairy Science* 87(6), 1732-1738.

Van Nuffel, A., Zwertvaegher, I., Pluym, L., Van Weyenberg, S., Thorup, V. M., Pastell, M., Sonck, B. and Saeys, W. (2015a). Lameness detection in dairy cows: part 1. How to distinguish between non-lame and lame cows based on differences in locomotion or behavior, *Animals: An Open Access Journal from MDPI* 5(3), 838-860. doi: 10.3390/ani5030387.

Van Nuffel, A., Zwertvaegher, I., Van Weyenberg, S., Pastell, M., Thorup, V. M., Bahr, C., Sonck, B. and Saeys, W. (2015b). Lameness detection in dairy cows: part 2. Use of sensors to automatically register changes in locomotion or behavior, *Animals: An Open Access Journal from MDPI* 5(3), 861-885. doi: 10.3390/ani5030388.

Vanhoudt, A., Yang, D. A., Armstrong, T., Huxley, J. N., Laven, R. A., Manning, A. D., Newsome, R. F., Nielen, M., van Werven, T. and Bell, N. J. (2019). Interobserver agreement of digital dermatitis M-scores for photographs of the hind feet of standing dairy cattle, *Journal of Dairy Science* 102(6), 5466-5474. doi: 10.3168/JDS.2018-15644.

Vermunt, J. J. and Greenough, P. R. (1995a). Structural characteristics of the bovine claw - horn growth and wear, Horn hardness and claw conformation, *British Veterinary Journal* 151(2), 157-180.

Vermunt, J. J. and Greenough, P. R. (1995b). Structural characteristics of the bovine claw: horn growth and wear, horn hardness and claw conformation. *British Veterinary Journal* 151, 157.

Vink, W. D. (2006). Investigating the epidemiology of Bovine Digital Dermatitis: causality, transmission and infection dynamics. PhD Thesis. University of Liverpool. University of Liverpool.

Webster, A. J. F. (2000). Effects of wet vs. dry feeding and housing type on the pathogenesis of claw horn disruption in first lactation dairy cattle. In: *Pharmacia*, Italy, 3–7 September 2000, p. 340.

Webster, A. J. F. J. (2002). Effects of housing practices on the development of foot lesions in dairy heifers in early lactation, *Veterinary Record* 151(1), 9-12. doi: 10.1136/vr.151.1.9.

Webster, A. J. F. and Tarlton, J. F. (2001). Biomechanical changes in the suspensory apparatus of the pedal bone and development of sole and white line lesions in dairy heifers, *Research in Veterinary Science* 70 (Suppl. A), 34.

Wells, S. J., Trent, A. M., Marsh, W. E. and Robinson, R. A. (1993). Prevalence and severity of lameness in lactating dairy-cows in a sample of Minnesota and Wisconsin herds, *Journal of the American Veterinary Medical Association* 202(1), 78–82.

Wells, S. J., Trent, A. M., Marsh, W. E., Williamson, N. B. and Robinson, R. A. (1995). Some risk-factors associated With clinical lameness in dairy herds in Minnesota and Wisconsin, *Veterinary Record* 136(21), 537–540.

Whay, H. (2002). Locomotion scoring and lameness detection in dairy cattle, *In Practice* 24(8), 444–449.

Whay, H. R., Main, D. C., Green, L. E. and Webster, A. J. (2003). Papers & Articles Assessment of the welfare of dairy cattle using animal-based measurements: direct observations and investigation of farm records, *Veterinary Record* 153(7), 197–202. doi: 10.1136/vr.153.7.197.

Whay, H. R., Waterman, A. E., Webster, A. J. and O'Brien, J. K. (1998). The influence of lesion type on the duration of hyperalgesia associated with hindlimb lameness in dairy cattle, *Veterinary Journal* 156(1), 23–29.

Willshire, J. A. and Bell, N. J. (2009). An economic review of cattle lameness, *Cattle Practice* 17(2), 136–141.

Winckler, C. and Willen, S. (2001). The reliability and repeatability of a lameness scoring system for use as an indicator of welfare in dairy cattle, *Acta Agriculturae Scandinavica, Section A Suppl.* 30 (Suppl. 30), 103–107.

Yang, D. A. and Laven, R. A. (2019). Inter-observer agreement between two observers for bovine digital dermatitis identification in New Zealand using digital photographs, *New Zealand Veterinary Journal* 67(3), 143–147. doi: 10.1080/00480169.2019.1582369.

Zinicola, M., Higgins, H., Lima, S., Machado, V., Guard, C. and Bicalho, R. (2015). Shotgun metagenomic sequencing reveals functional genes and microbiome associated with bovine digital dermatitis, *PLoS ONE* 10(7), e0133674. doi: 10.1371/journal.pone.0133674.

# Chapter 13

## Preventing bacterial diseases in dairy cattle

*Sharif S. Aly[1] and Sarah M. Depenbrock[1], University of California-Davis, USA*



### 1 Introduction

A paradigm shift in the management of bacterial diseases of food animals is needed, the focus should shift from pharmacologic interventions to disease risk identification and mitigation. Food animal management practices that focus on matching a pathogen with a pharmaceutical treatment (aka 'bug and drug') are oversimplified, threaten the sustainability of food animal production and erode consumer trust in animal source protein. Maintaining animal health, judicious use of antibiotics and sustainability of animal production systems are common expectations for food animal production. Relying on pharmaceuticals alone is insufficient to prevent serious disease syndromes associated with bacterial infections in cattle; indiscriminate use of antimicrobial drugs without careful adjustments to management practices results in treatment failures, animal suffering and greater risk for antimicrobial resistance (AMR), further comprising animal and public health (Oliver et al., 2011; Vaarten, 2012).

Dairy cattle production requires intense animal management systems that involve continuous daily interventions to harvest milk, increase fecundity and hand-raise young stock. Such systems allow for constant fine-tuning and the opportunity to increase both milk and dairy beef commodity production. However, these intense animal management operations also create multifactorial

1 Both authors contributed equally.

http://dx.doi.org/10.19103/AS.2020.0086.16

animal health risks. The interaction between disease risk factors can be nicely described by a conceptual model called the disease triangle. This model, as displayed in Fig. 1, incorporates the host, pathogen and environmental factors as a tool to understand different aspects that influence the disease. The disease triangle was first described by plant scientists, was later applied to public health and is now occasionally used in veterinary science (Scholthof, 2007; Woolums and Step, 2020). A sustainable approach to bacterial disease mitigation in production settings must include a comprehensive assessment of the host, pathogen and environmental factors of the disease. The authors propose that the continuum of a disease between the host, environment and pathogen is so inextricably linked that they cannot be separated when discussing cause versus prevention of disease. This chapter will focus on the state of the art of disease prevention in dairy cattle, specifically focusing on preventing bacterial causes of bovine respiratory disease (BRD). Each of the chapter sections identifies mechanisms of the disease triangle that can be disrupted to prevent diseases; these concepts culminate in a specific risk assessment tool in the third section of the chapter that can be used to determine a farm-specific action plan for disease prevention using calfhood respiratory disease as an example. Through the risk assessment approach, investigations into interactions between the host, environment and pathogen can provide robust information on which to build a prevention platform for multiple syndromes of bacterial diseases in

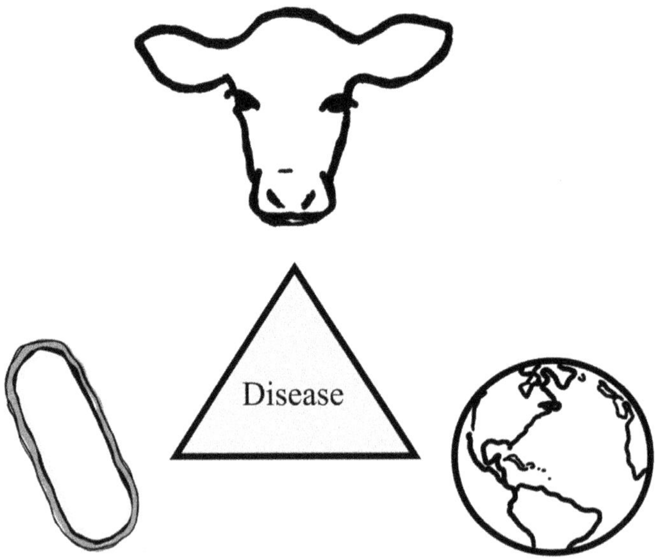


**Figure 1** Schematic diagram of host (calf), environment (globe) and pathogen (bacterial rod) interactions. Original artwork by Dr. Sarah Depenbrock.

cattle. While we chose BRD as the working example, the same principles can be applied to other diseases such as neonatal calf diarrhea, which shares several risk factors with respiratory disease in calves.

### 1.1 BRD as the focus of disease prevention in dairy cattle

Arguably one of the best examples of the disease triangle in cattle is the BRD complex. The BRD complex is a syndrome of respiratory disease that results from the complex interactions between the host, environment and pathogen and culminates in a bacterial infection of the lower airways. The BRD complex was selected for discussion due to the relatively active areas of research in all three aspects of the disease triangle, the significance of BRD in the cattle industry and the overlap in preventive strategies between BRD prevention and prevention of other syndromes of bacterial diseases in dairy cattle. Like many diseases of dairy cattle, BDS does not often occur in isolation; underlying problems with one or more aspects of the disease triangle may predispose cattle to BRD. Many of the strategies discussed in the context of BRD prevention rely on maximizing host immunity, reduction of environmental stressors and contamination and, where appropriate, specific pathogen treatment and immunoprophylaxis. Although the focus of this chapter is BRD, most of these prevention strategies are common to the prevention of other diseases, including calf scours and paratuberculosis.

Bovine respiratory disease continues to be one of the most common endemic diseases of US cattle. The prevalence of BRD nationwide has been estimated at 12% and 5% in preweaned and weaned heifers, respectively (USDA, 2018). However, the burden of the disease far exceeds the reported prevalence due to the cross-sectional nature of prevalence as a measure of disease occurrence and the duration of BRD. As a result, cumulative incidence of BRD in preweaned calves, as the proportion affected from birth to weaning in a convenience sample of six California dairies from three major dairy regions, Northern CA, Northern San Joaquin Valley (NSJV) and Greater Southern CA (GSCA; includes Northern San Joaquin Valley and Southern CA), has been estimated at 22.8% (Dubrovsky et al., 2019a). Nationwide BRD accounted for 33.4% of all calf illnesses (Urie et al., 2020). Respiratory disease results in mortality and loss of productivity in the form of decreased milk production (Adams and Buczinski, 2016; Dunn et al., 2018a). An estimated 24% of preweaned and 59% of postweaning heifer mortality, respectively, are attributed to BRD, making it the second-most common cause of preweaned calf mortality and the top cause of mortality in postweaned calves (USDA, 2018).

BRD is a serious problem in the cattle rearing industry, and given its complex causal mechanism involving the host, pathogen and environment, it is a great disease candidate for a multimodal prevention approach. The goal of this chapter is to discuss the prevention of BRD as an example of bacterial disease prevention using the conceptual model of the disease triangle. The mechanisms related to the risk factors for each of the disease triangle components will be described. Finally, we will demonstrate a comprehensive and novel BRD risk assessment tool as a practical farm-ready application that can be used by producers, veterinarians, consultants and other calf raisers to identify and transect their operation-specific BRD risk factors.

### *1.2 Learning objectives*

- Describe the basic physiology of the host, environment and pathogen relevant to the discussion of disease prevention, using the example of BRD;
- Review current literature and expert opinions describing the host, pathogen and environmental interactions relative to the BRD example;
- Describe recent advances in testing methods relevant to BRD;
- Identify interventions that can be used to interrupt the host-environment-pathogen continuum of disease; and
- Describe and provide detailed instructions for specific risk assessment and outcome assessment tools that can be implemented on the farm to facilitate disease prevention.

## 2 Pathogen host environment: an overview

Bacterial infection of the lower airways results from environmental conditions, pathogen virulence, a breakdown in host immunity or any combination of these factors. Different subsets of the BRD complex have been described relative to different cattle populations and risk factors. The clinical presentation of BRD in beef calves is commonly referred to as shipping fever, while in dairy calves, this syndrome is referred to as enzootic pneumonia (Callan and Garry, 2002); where applicable, these distinctions will be discussed.

### *2.1 Role of bacterial pathogens*

Bacterial and viral pathogens are the primary pathogens implicated in the BRD complex, although other pathogens, such as parasites or fungi, and toxins can certainly result in respiratory disease in cattle. This section will focus on bacterial pathogens associated with BRD complex and factors that affect the role of these bacteria in the development of BRD.

The bacterial pathogens most commonly associated with BRD are *Mannheimia haemolytica*, *Pasteurella multocida*, *Histophilus somni* and *Mycoplasma bovis*. In addition, *Bibersteinia trehalosi* and *Trueperella pyogenes* are considered important bacterial pathogens associated with BRD; *T. pyogenes* is considered particularly important in chronic cases.

*M. haemolytica*, *P. multocida* and *H. somni* are all members of the Pasteurellaceae family and thus share some basic similarities. These organisms are gram-negative and thus possess an outer lipopolysaccharide (LPS), also known as endotoxin, which induces profound host inflammatory responses upon entering the tissues or circulation. All three organisms can also be found as commensal inhabitants of the upper respiratory tract in animals without signs of BRD (Allen et al., 1991; Gaeta et al., 2017).

### 2.1.1 *Mannheimia haemolytica*

*M. haemolytica* is often considered the predominant bacterial pathogen associated with BRD in feedlot cattle (Welsh et al., 2004; Klima et al., 2019) and among the top 3 in dairy calves (Francoz et al., 2015; Gaeta et al., 2017); recent reviews of this pathogen and its role in BRD have been published (Rice et al., 2007; Griffin et al., 2010; Confer and Ayalew, 2018). Briefly, there are at least 12 serotypes of this species and serotypes A1 and A2 are known to colonize the upper respiratory tract of cattle; type A2 is often isolated from clinically normal cattle, while type A1 predominates in the diseased cattle (Srikumaran et al., 1996; Rice et al., 2007). Other serotypes have also been isolated from lung tissues of cases of bovine pneumonia (Rice et al., 2007). *M. haemolytica* has several virulence factors that allow the organism to evade host responses, colonize host membranes and damage host tissues. These virulence factors include a capsule and adhesins for adherence and invasion; protective outer membrane proteins for host immune evasion; extracellular enzymes that interfere with the mucous layer; LPS, which leads to acute inflammation, edema, hemorrhage and hypoxemia; and leukotoxin (LKT), which is a pore-forming cytotoxin that leads to the lysis of neutrophils, macrophages and other white blood cells) (Rice et al., 2007; Gioia et al., 2006). Once this organism has gained access to the lower airway tissues, the combination of the effects of these virulence factors elaborates severe, acute clinical disease, often characterized by severe inflammation in the lungs and pleura (fibrinous pleuropneumonia, bronchopneumonia) that is often fatal. The clinical manifestation of the disease following *M. haemolytica* invasion into the lower airways is often called shipping fever, due to the association with a recent shipment of cattle to feed yards. Of course, the disease associated with this organism is not strictly restricted to beef cattle nor animals with a recent shipment history.

### 2.1.2 *Pasteurella multocida*

There are at least 16 serotypes of *P. multocida*; each possesses LPS and one of five known capsular types (A, B, D, E, F) but lacks the virulence factor LKT (Boyce and Adler, 2000). *P. multocida* type A:3 is the most common serotype isolated in cases of BRD (Dabo et al., 2008). Cases of BRD in which *P. multocida* predominates lower airway infection develop bronchopneumonia, and the clinical picture is usually less acute, with less dramatic signs of inflammation compared to infections where *M. haemoltyica* predominates. *P. multocida* is often identified in BRD of dairy calves (Francoz et al., 2015), and the clinical manifestation is termed enzootic calf pneumonia. Mortality rates for enzootic calf pneumonia have been estimated in the United States and Canada at 2.8% and 1.8%, respectively (Van Donkersgoed et al., 1993; Maunsell et al., 2011a). Surviving animals may have long-term damage, resulting in lower weight gain and decreased future milk production (Maunsell et al., 2011a).

### 2.1.3 *Histophilus somni*

In addition to LPS, *H. Somni* virulence factors include histamine production, biofilm formation, exopolysaccharide, outer membrane proteins and immunoglobulin-binding proteins. A fibrinopurlent bronchopneumonia results when *H. somni* infects the lower airway. The virulence factors, particularly an immunoglobulin-binding protein that affects endothelial cells, commonly allow the spread of the disease beyond the respiratory tract. In addition to fibrinopurulent bronchopneumonia, this organism has been associated with laryngitis, thrombotic meningoencephalitis, polyarthritis, pericarditis and myocarditis (Corbeil, 1996; Constable et al., 2016).

### 2.1.4 *Mycoplasma* spp.

*M. bovis* is the primary *Mycoplasma* species discussed in relation to BRD; however, other *Mycoplasma* have been associated with BRD in various roles. Different species of *Mycoplasma* impart different specific effects on the airway defenses. For example, *Mycoplasma dispar* causes ciliostasis and thus impairs upper airway clearance (Caswell, 2014). *M. bovis* colonizes and persists on mucosal surfaces and invades lung parenchyma, which induces a strong cellular response and elicits significant lung damage (Howard et al., 1987). Mycoplasmas can also impair bovine macrophage and neutrophil function (Howard and Taylor, 1983; Howard et al., 1987; Thomas et al., 1991). *Mycoplasma* organisms also predispose cattle to the invasion of the lower airways by other bacteria, and bacterial coinfections are common. Similarly to *H. somni*, *M. bovis* readily accesses the blood stream through the respiratory system; this can result in the disease of many organ systems; however, hematogenous spread leading to septic arthritis or otitis are some of the most common clinical manifestations of disease in calves. It should also be noted that *M. bovis* readily colonizes the eustachian tube, which serves as another route by which to cause otitis (Maunsell and Donovan, 2009; Maunsell et al., 2012). *M. bovis* can also cause mastitis; spreads from the mammary gland to other cows via milking processes or to calves through milk or colostrum are important routes of disease transmission (Maunsell et al., 2011b). *Mycoplasma* organisms are unique compared to the Pasteurellaceae previously discussed. They are surrounded by a trilayered membrane rather than a cell wall; this feature is particularly important when considering antibiotic therapy because this membrane structure makes *Mycoplasma* species inherently resistant to β-lactam type antimicrobials. Additionally, *M. bovis* has a family of variable surface lipoproteins, which undergo frequent phase and size change, exhibit strain variation, and can be selected by exposure to antibodies which contribute to immune evasion and presents a challenge to vaccine development (Lysnyansky et al., 2020; Le Grand et al., 1996; Beier et al., 1998; Buchenau et al., 2010; Maunsell et al., 2011b).

### 2.1.5 Other bacteria associated with BRD

Many other bacteria have been isolated from the lungs of pneumonic cattle in addition to the Pasteurellaceae and Mycoplasmas discussed earlier. The specific roles these other bacteria play in the BRD complex is less well defined (Constable et al., 2016). *Bibersteinia trehalose* is another bacterium in the Pasteurellaceae family, which also possesses an LKT (Muruganathan et al., 2018) and has been implicated in BRD, although the specific pathogenicity is debated (Adlam and Rutter, 1989; Kodjo et al., 1999; Katsuda et al., 2008; Hanthorn et al., 2014). *T. pyogenes* has been identified in many purulent infections in cattle; this organism may be isolated from chronic pneumonias or pulmonary abscesses (Hammer et al., 2004; Ribeiro et al., 2015).

### 2.1.6 Coinfections

The relationship between pathogens in the respiratory tract is also important to consider when discussing the pathogenicity of respiratory bacteria. Viruses and *Mycoplasma* have been the primary organisms implicated in coinfections with the respiratory Pasteurellaceae and predispose cattle to lower airway bacterial infection. It has long been established that viral respiratory infection predisposes cattle to bacterial bronchopneumonia (Jericho et al., 1986; Love et al., 2014, 2016a; Ng et al., 2015). The common viral pathogens implicated in BRD include bovine herpes virus 1 (BoHV1), parainfluenza type 3 (PI3), bovine respiratory syncytial virus (BRSV) and bovine viral diarrhea virus (BVD). Additionally, bovine coronavirus (BoCV) and Influenza D (IDV) have more recently been demonstrated to play a role in BRD (Hause et al., 2013; Ridpath et al., 2020). Viral pathogens are often implicated in the development of this disease complex as primary pathogens that weaken host defenses and allow the invasion of bacterial pathogens and opportunists into the lower airways. In these synergistic infections, viruses use many mechanisms to cause damage to host respiratory defenses and bacteria secondarily invade the lower airways. These mechanisms include structural changes such as injured respiratory epithelium, decreased mucociliary clearance and lung consolidation; biochemical dysfunctions such as decreased effectiveness or decreased secretion of antimicrobial immune compounds; and dysfunction of alveolar macrophages (Zachary, 2017). Ultimately, the resulting syndrome of BRD involves bacterial infection of the lower airways and the development of varying degrees of broncho, pleuro or interstitial pneumonia.

### 2.1.7 Respiratory microbial communities

More recent investigations into the relationships of different microbes in the respiratory tract of cattle have been studied using metagenomics. Historically, singular organisms, or the relationship between select organisms such as specific viruses and specific bacteria, have been studied in the pathogenesis of BRD. Recently, investigators have begun studying entire microbial populations in the upper and lower respiratory tract of cattle through the use of metagenomics. Not surprisingly, this work demonstrates diversity in the microbial populations of the bovine respiratory tract, and differences in these populations are associated with health and disease (Timsit et al., 2016a; Gaeta et al., 2017; Windeyer et al., 2017; Klima et al., 2019; McMullen et al., 2019). These microbial populations change over time (Timsit et al., 2016a; Stroebel et al., 2018) and vary between facilities (Stroebel et al., 2018). The presence of previously known pathogens, such as *M. bovis* (McMullen et al., 2019) and *M. haemolytica* (Klima et al., 2019), has not surprisingly been associated with BRD cases. Microbes not previously considered classic contributors to BRD, such as Bov rhinitis A virus, have also been associated with cases of BRD and require further investigation into their roles as pathogens (Ng et al., 2015; Zhang et al., 2019). The utility of studying entire populations of microbes in the respiratory tract is not simply to identify new or different pathogens but also to study the population dynamics of the respiratory tract microbiota that can reveal a deeper understanding of the interactions between the host, pathogen and environment. Metagenomic analysis has identified changes in the respiratory microbiome associated with environmental conditions (Nicola et al., 2017; Lima et al., 2019), diet (Hall et al., 2017), stressful events (Holman et al., 2017; Timsit et al., 2016b; Stroebel et al., 2018) and antibiotic treatment (Holman et al., 2018). The composition of the respiratory microbiome and those factors that may influence it are thus important considerations in the continuum of BRD causation and disease prevention.

## 2.2 Role of the host

Variability in host immunity is arguably just as important as the pathogens in the development of BRD. The relative susceptibility or resistance to BRD can be influenced by factors related to both acquired immunity and innate immunity.

### 2.2.1 Innate immunity

The bovine innate immune system of the respiratory tract is the collection of natural bodily defenses against pathogens, including physical and chemical barriers, mechanical effects, inherent cellular and chemical responses to foreign material or injury and the commensal microbes (Zachary, 2017).

The structure of the nose and paranasal sinuses acts to remove large particles from inhaled air (Zachary, 2017). Commensal microflora of the upper respiratory tract may also act to protect the airways from pathogens by a variety of proposed mechanisms (Ackermann et al., 2010). The large conducting airways are lined with pseudostratified, ciliated columnar epithelium and mucus-producing goblet cells. These cilia and mucus make up the mucocilliary escalator that functions to mechanically clear conducting airways of debris. Further down the respiratory tree are the transitional airways; this region is lined by cuboidal epithelium, fewer ciliated cells, Club cells (formerly Clara cells) that produce surfactant and protective proteins and neuroendocrine cells. Following the transitional airways is the primary gas exchange surface of the alveoli and alveolar ducts, which are lined with type I and type II alveolar cells (Zachary, 2017).

The lower airways were historically considered essentially sterile in healthy animals. However, recent investigations into the lower respiratory microbiome have revealed a variety of microbes present in the lower respiratory tract of cattle with and without signs of BRD (Johnston et al., 2017; Nicola et al., 2017; Timsit et al., 2018; Klima et al., 2019). These microbial populations are sampled from the large conducting airways (trachea and bronchi) and contain pooled samples from the terminal airways as well as these conducting airways.

Cattle are particularly predisposed to respiratory disease relative to other mammals for a variety of anatomic and physiologic reasons, including relatively low gas exchange capacity, greater anatomical compartmentalization, effect of gravity on ventral lung lobes, non-ciliated and ventral drainage of the pharyngeal tonsil, relatively low pulmonary compliance and relatively low numbers of pulmonary macrophages among others (Veit and Farrell, 1978; Kainer and Will, 1981). Many factors associated with environmental conditions in modern agriculture exacerbate these weaknesses in the bovine respiratory system, including the calf's ambient temperature, exposure to humidity, ammonia and dust among other factors, creating what is known as the animal's microenvironment. The role of these external factors is further discussed in Section 2.3, 'Role of the environment'.

### 2.2.2 Acquired immunity

Acquired immunity refers to the specific set of immune responses directed at specific types of pathogens and is also known as the adaptive immune system.

Calves acquire immunity in three general ways: passively from colostrum, actively through natural exposure to potential pathogens and deliberately through vaccination against target organisms.

### 2.2.2.1 Passive immunity

Calves are born agammaglobulinemic and thus rely on the colostral transfer of antibodies for humoral immunity. In addition, colostrum contains maternal leukocytes among other immune and growth factors. Calves rely on this maternally derived passive immunity until their own immune system matures. Inadequate colostrum intake in calves results in failure of transfer of passive immunity and thus failure of a primary immune defense in early life. The immunoglobulins are actively pinocytosed from the gut in the first 24 h after birth with a decreasing rate as time elapses and soon after first ingestion by the calf (Stott et al., 1979). Previous research identified serum immunoglobulin levels ≥1g/dL or serum total protein levels ≥ 5.2 g/dL as indicative of adequate transfer of colostral immunity. However, recent research suggests immunoglobulin concentrations greater than 2 g/dL or serum total protein >5.8 g/dL are associated with less risk for calfhood mortality (Chigerwe et al., 2015a). A similar cut-off (5.7 g/dL) was suggested to predict the risk of BRD (Windeyer et al., 2014). Feeding more high quality colostrum after birth, to allow for the associated increase in the transfer of passive immunity, and measurable increases in blood IgG concentrations in newborn calves continue to be a common goal on US dairies; the current research clearly shows that higher cut-offs would favor lower mortality and morbidity in calves (Urie et al., 2018a). New recommendations have been proposed to replace the current industry standard cut-off to assess the transfer of passive immunity for serum immunoglobulins of 10 g/L at 24–48 h after birth. The proposed consensus ranges for the transfer of passive immunity at the calf level include four serum immunoglobulin ranges ≥ 25.0 (excellent), 18.0–24.9 (good), 10.0–17.9 (fair) and < 10 g/L (poor) (Lombard et al., 2020). Lombard et al.'s ( 2020) consensus standards also call for a herd-level target of >40, 30, 20 and <10% of calves in the excellent, good, fair and poor transfer of passive immunity ranges, respectively, to reduce herd morbidity and mortality. Parallel ranges for serum total protein (g/dL) and serum Brix (%) are also provided for on-farm monitoring. The consensus ranges for serum total protein measurements are ≥6.2, 5.8–6.1, 5.1–5.7, <5.1 g/dL for the excellent, good, fair and poor transfer of passive immunity, respectively. The consensus ranges for serum Brix measurements are ≥ 9.4, 8.9–9.3, 8.1–8.8, <8.1% for the excellent, good, fair and poor transfer of passive immunity, respectively.

Due to the importance of the transfer of passive immunity to newborn calves, efforts to feed the optimum quantity and quality of colostrum are necessary to raise healthy calves. The quality of colostrum fed can be measured by immunoglobulin content and microbiological quality. Persistent udder infections were found to be associated with lower colostrum harvest volumes (Maunsell et al., 1998). As a result, in addition to harvesting colostrum from healthy cows, producers should expedite colostrum harvest soon after calving as one study estimated a 3.7% decrease in immunoglobulin content, with each hour delay subsequent to calving due to the initiation of milk secretion post-calving (Morin et al., 2010). Heat treatment of colostrum was found to reduce the total plate bacterial count and total coliform count, which may explain the reduced hazard for diarrhea, and any treatment from birth to weaning, in calves fed heat-treated colostrum compared to fresh colostrum (Godden et al., 2012). An association between the type of colostrum storage container and occurrence of respiratory disease on 100 dairies was identified by Maier et al. (2019a, 2020) with the use of bags to store colostrum compared to solid containers being associated with a reduction in cases; the latter association was significant in univariate models (single variable association with the outcome) only. However, the authors investigated potential confounders and effect modifiers, making this otherwise biologically significant association worthy of further investigation using an empirical approach, given the potential for reduction in odds of respiratory illness (Maier et al., 2020).

The ideal concentration of IgG in colostrum or ideal volume to be fed differs between studies and outcomes of interest (Windeyer et al., 2014; Chigerwe et al., 2015b; Urie et al., 2018b). Historically, 50 g/L IgG in colostrum was recommended as the bare minimum and calves were recommended to ingest at least 4 L to provide at least 200 g of IgG from initial colostrum feedings (Weaver et al., 2000; McGuirk and Collins, 2004; Chigerwe et al., 2008; Godden et al., 2009; Williams et al., 2014). Based on the data presented in the updated consensus by Lombard et al. (2020), this historical recommendation likely underestimates the colostrum quality and total mass of IgG needed to ingest to achieve excellent passive immunity (defined as ≥ 25 g/dL serum IgG). In calves with excellent passive immunity, the mean IgG ingested was 286.7 g in one feeding or 421.2 g in calves fed multiple feedings; this mass of IgG was delivered in an average of 3.3 L and 5.3 L in the first 24 h, respectively (Lombard et al., 2020).

Producers should also consider actively monitoring the transfer of passive immunity in newborn calves by measuring serum total proteins from a sample of calves as a proxy for blood immunoglobulin levels 24 h post colostrum feeding (Pithua et al., 2013). In addition, routine assessment of a dairy's colostrum management should include staff training on the best practices for colostrum harvest, quality in terms of immunoglobulin content and bacterial concentration, heat treatment, storage conditions including container type and volume and time from birth to feeding to calves. Special attention to these practices is needed when feeding pooled colostrum by training calf caretakers and dairy staff involved in all aspects of colostrum management to achieve higher immunoglobulin concentrations in the pooled colostrum fed to calves. Specialized training interventions have been shown to create measurable improvements in calf serum total protein (Williams et al., 2014).

#### 2.2.2.2 Vaccines

Vaccination is a popular method for improving host defenses, via acquired immunity, against specific respiratory pathogens (Windeyer et al., 2012; Chamorro and Palomares, 2020). The vaccines currently marketed in the United States for the prevention of BRD include *M. haemolytica*, *P. multocida* and *H. somni* and the viral pathogens BVD, BRSV, IBR and PI3. Although the viral disease is not the focus of this chapter, the importance of viral vaccination for the prevention of BRD is an important consideration due to the synergistic role viral pathogens demonstrate with bacterial pathogens in the development of BRD; a full discussion on viral vaccination for BRD is beyond the scope of this chapter. Broadly speaking, modified live viral (MLV) respiratory vaccines induce both cell-mediated and humoral immune responses in the host, while killed viral vaccines produce humoral immunity but less significant cell-mediated immunity (Platt et al., 2009; Woolums et al., 2013; Chamorro and Palomares, 2020). These different host responses allow fewer doses to achieve the desired protection and longer-lasting immunity with MLV vaccines, whereas a minimum of two doses, spaced approximately 21 days apart to allow induction of an anamnestic response, are typically required for killed viral vaccines (Chamorro and Palomares, 2020; Ridpath et al., 2020). Modified live viral vaccination, however, can present a risk in breeding animals due to effects on the early conceptus or semen production, and label instructions relating to breeding safety should be followed when creating vaccination schedules, particularly with MLV vaccines (Grooms et al., 1998; Nandi et al., 2009; Palomares et al., 2013). Early vaccination of calves using injectable MLV vaccines, or intranasal MLV vaccines, in the face of maternal antibodies is a strategy that may provide early protection and/or immune priming for some viral pathogens; the calf may or may not seroconvert, and cell-mediated immunity appears to be an important factor in this strategy, particularly with intranasal vaccines (Stevens et al., 2011; Gerber et al., 1978; Vangeel et al., 2009; Woolums et al., 2013; Mahan et al., 2016; Windeyer and Gamsjäger, 2019; Chamorro and Palomares, 2020). However, the effect of vaccination in the face of maternal antibodies depends on several factors, including the pathogen and concentration of maternal antibodies, and significant maternal antibody inference has been identified in some studies, and there is no consensus on the universal utility of early calfhood vaccination in the face of maternal antibodies (Downey et al., 2013; Ellis et al., 2001, 2014; Windeyer and Gamsjäger, 2019; Chamorro and Palomares, 2020).

Vaccination of cattle against viral pathogens associated with BRD is often considered a cornerstone of BRD prevention and is commonly used (USDA, 2010; Waldner et al., 2019). Vaccination against bacterial pathogens of BRD is less commonly employed than vaccination for viral pathogens (Karle et al., 2019). Multiple different strategies have been used to create vaccines for bacterial BRD pathogens, primarily the Pasteurellaceae (McGill and Sacco, 2020). Surface antigens of the Pasteurellaceae, and LKT in *M. haemolytica* specifically, have been a focus of vaccine development (Confer and Ayalew, 2018). A 2012 meta-analysis investigated the effect of vaccination for *M. haemolytica*, *P. multocida* and *H. somni* on morbidity and mortality in feedlot cattle; this analysis reports conflicting results between studies, with an overall assessment that vaccination for *M. haemolytica* and *P. multocida* is potentially beneficial, while no evidence for the benefit of *H. somni* vaccination was identified (Larson and Step, 2012). In *M. bovis*, the variability in *M. bovis* surface proteins may contribute to immune evasion and presents a challenge to vaccine development (Le Grand et al., 1996). Previous vaccination attempts for *M. bovis* have been unsuccessful; in a randomized, placebo-controlled, double-blinded field trial, vaccination for *M. bovis* had no effect on treatment for *M. bovis*-associated disease, nasal colonization with *M. bovis*, serum antibody titers for *M. bovis* or mortality (Maunsell et al., 2009).

The reduction of BRD mortality through the use of viral vaccines, bacterial vaccination or a combination is not universally successful (Larson and Step, 2012; Theurer et al., 2015; Chamorro and Palomares, 2020). Studies of vaccine efficacy often appear promising in controlled clinical trials; however, findings from large field studies often differ. These discrepancies are not only due to variability in study design (discussed later) but also likely because vaccination inherently affects one or two narrow aspects of the disease triangle. Specifically, the host's adaptive immunity is affected by vaccination, and arguably the environment may be affected through the creation of herd immunity, thus decreasing the environmental load of a given pathogen in the herd. It is important to remember that other host attributes, environmental conditions and pathogen factors can all influence the occurrence of BRD. For example, if cattle are stressed (Blecha et al., 1984) or malnourished (Griebel et al., 1987; Galyean et al., 1997), or if vaccination is not timed properly relative to disease risk (Lippolis et al., 2016; Richeson and Falkner, 2020), the immune response may be altered and thus insufficient to protect against clinical manifestations of BRD. It is imperative to remember that although specific pathogen vaccination is an attractive prevention strategy for BRD, this intervention alone may be insufficient to provide rewarding results in disease prevention. The authors recommend the thoughtful use of vaccination, on a farm-by-farm basis, based on knowledge of farm-specific disease and risk periods, in combination with other management strategies as discussed in this chapter, to modify all three arms of the disease triangle to provide the most impactful BRD disease prevention.

### *2.2.2.2.1 Veterinary vaccine research design and implementation*

Vaccine trials play an important role in validating findings from research at the development phase. Specifically, vaccine efficacy can be estimated using controlled experiments that may include challenges by a live pathogen(s). There is an important distinction between the reported efficacy of a vaccine and its effectiveness in a field setting. The nomenclature of vaccine efficacy refers to findings produced by empirical studies such as double-blind block randomized trials, which continue to be the gold standard of cause and effect. However, the nomenclature vaccine effectiveness refers to findings based on studies conducted on a wide range of management practices to provide meaningful vaccine effectiveness estimates that can then apply to a wide range of production systems. Measurement of the host immune system reaction, inflammatory response, shedding (if applicable), feed intake and productivity measures, along with the economic impact, are also needed to properly assess the effectiveness of a vaccine.

Among the challenges encountered when conducting vaccine research is the availability of a natural challenge to accurately represent field conditions. Variability in field conditions include differences in natural challenge dose, intensity and strain variety. Hence, documenting the natural challenge, both duration and intensity, through continuous surveillance in the study population during vaccine trials is necessary to interpret trial findings. The inclusion of study subjects that encompass different management practices, such as both calf ranches and dairies, large and small herds, animals raised under different environmental conditions, different seasons, regions and climates, is a way to increase the external validity when estimating the effectiveness of vaccines.

Nevertheless, field conditions may produce immunization rates that differ from efficacy studies that precede vaccine effectiveness research. Other reasons for such differences include variability in vaccine production, cold chain – if applicable, dosing and administration, vaccination age and host immune status, which can be subject to nutrition, housing or other factors of the environment arm of the triangle. Capturing such variability in the field provides more realistic estimates that capture the true vaccine effectiveness as opposed to efficacy. As a result, to estimate vaccine effectiveness, alternative study designs should be pursued including retrospective case-control studies or prospective cohort studies (Knight-Jones et al., 2014; Weinberg and Szilagyi, 2010). Such observational study designs, based on several herds with different natural challenge exposures, allow for the measurement of the true effectiveness of a vaccine (Gardner et al., 2007; Perrett et al., 2018).

Finally, structuring the hypotheses remains a key factor in the success of research on vaccine effectiveness. Veterinary vaccine trials should explore herd immunity rather than simple measures of the percent immunized, disease hazard and occurrence in vaccinates compared to controls. Using herd immunity, veterinary vaccine trials should estimate the percent of the population that is needed to be immunized in order to eradicate the disease in a specific herd or, if the latter is not possible, explore the duration of vaccine immunity and the necessary repeated immunization frequency and rate based on waning immunity rate (Woolhouse et al., 1997).

### 2.2.3 Influences on host immunity

Both the innate and adaptive immune systems can be influenced by many host factors; common examples of such factors include stress, nutrition and genetics.

#### 2.2.3.1 Stress

Many factors may cause stress in cattle, and the underlying cause of the stress may also induce direct damage to the host's immune defenses. Depending on the cause and chronicity of the stressor, the endogenous stress responses of the animal can have differing effects. The endogenous stress response involves variable stimulation of the sympathetic nervous system (SNS) and hypothalamic pituitary axis (HPA), leading to the release of catecholamines, cortisol and other endogenous stress responses. These responses have been demonstrated to alter host immune responses in a variety of ways ranging from beneficial priming of immune responses in acute stress to detrimental lack of response or even self-destructive autoimmune responses in states of chronic stress exposure (Mansfield et al., 2008; Carroll and Forsberg, 2007; Ackermann et al., 2010). Stress in cattle induced increased replication and spread of bacteria inhabiting the upper respiratory tract, including *P. multocida*, *M. haemolytica* and *H. somni* (Ackermann et al., 2010). Additionally, in vitro research has demonstrated direct interactions between pathogenic bacteria and stress hormones including epinephrine and norepinephrine that provide evidence for increased pathogenicity of bacteria in the presence of mammalian stress hormones (Lyte, 2004). The bovine respiratory tract contains abundant blood flow and noradrenergic innervation, so it is conceivable that stress not only alters host immune responses but may also directly alter pathogen virulence in the respiratory tract (Lyte, 2014a).

Heat stress is an extremely common environmental stressor, and, as global warming takes place, heat stress may become increasingly problematic in cattle production systems. In addition to the stress response associated with HPA stimulation, heat stress is associated with GI and systemic inflammation (Kim et al., 2018; Koch et al., 2019). The proposed mechanism for the inflammation identified in heat stress is that heat (or a high-temperature humidity index) exposure triggers increases in peripheral blood flow via vasodilation as the body attempts thermoregulation. This peripheral vasodilation occurs at the expense of internal vasoconstriction to maintain blood pressure and vital organ perfusion. The GI tract may become hypoperfused during heat stress, and intestinal ischemia leads to dysfunction of intestinal epithelial cells, as evidenced by changes in intestinal villous integrity (Pearce et al., 2013a,b) and dysfunction of intestinal epithelial cell tight junctions (Koch et al., 2019). The delicate barrier between the enteric lumen and the bloodstream is thus compromised, and local and systemic immune mediators react as LPS, bacteria or other luminal toxins translocate to portal circulation. This inflammatory response associated with heat stress uses energy (Baumgard et al., 2011). One-month-old calves' thermoneutral environment is estimated to range in temperature from 10°C to 25°C, with a maximum limit of 30°C for maximum acclimation, which is narrower than that for adult cattle (Wathes et al., 1983; Davis and Drackley, 1998). Cattle that are exposed to environments beyond the thermoneutral zone use energy to release heat from the body through rapid breathing, resulting in reduced feed intake and reduced immune function due to the calf's energy resources being diverted to address heat stress (Lammers et al., 1996; Mitlöhner et al., 2002). Lower alveolar macrophage counts and higher monocyte counts were associated with temperature and humidity in calves, further suggesting that heat stress may suppress a calf's respiratory immunity (Bertagnon et al., 2011; Roland et al., 2016). The combined effects of inflammation, stress response and decreased feed intake all probably play a role in predisposing cattle with heat stress to BRD.

#### 2.2.3.2 Nutrition

Proper nutrition is essential to the maintenance of homeostasis, including immune function (Griebel et al., 1987). The complex life cycle of dairy cattle requires that each stage of life is supported by a plane of nutrition that matches the demands for that specific life stage in a given environment.

Feeding calves less than 2.84 L (3 US quarts) daily was associated with the increased respiratory disease with varying degrees depending on the calves' breed, with the greatest negative impact seen in Holsteins compared to Jerseys (Maier et al., 2019a). In contrast, the latter study also found that feeding more than 5.68 L (6 US quarts) per day benefited Jersey calves more than Holsteins, documenting further benefits to reducing morbidity due to respiratory diseases by feeding more milk to unweaned calves, and while such benefits may be observed in Jersey calves by exceeding 5.68 L, more research is required to identify such cut-off for Holsteins (Maier et al., 2019a). Similar findings were reported in a longitudinal study that estimated a 92% reduction in the hazard of respiratory disease when calves were fed 4.25–5.68 L of milk daily (Dubrovsky et al., 2019a). Lower volumes of daily milk fed, and hence less energy, may not satisfy the calf's metabolic maintenance, growth and immune system requirements (Drackley, 2008). The quality of milk fed may also impact the health of calves. Calves fed strictly milk replacer had a higher risk of morbidity compared to calves fed pasteurized waste milk (Godden et al., 2005). Milk replacer may not be as nutrient rich to support a calf's growth and immune system requirements, compared to cow's milk in either saleable or waste milk form. In addition, calves fed saleable and/or hospital milk had lower respiratory disease risk compared to those fed the same sources when supplemented with milk replacer, further emphasizing the impact of the nutrient content of a calf's diet on its risk for respiratory disease (Dubrovsky et al., 2019a) as was feeding saleable milk to calves associated with lower prevalence and of respiratory disease in calves compared to non-saleable milk (Maier et al., 2019a). The quality of milk fed to calves from a bacteriological aspect may also impact the occurrence of the respiratory disease; Maier et al. (2019a) estimated from a cross-sectional study of 100 dairies that calves fed non-pasteurized milk had higher odds of respiratory disease compared to those fed pasteurized milk. The same study also identified an association of feeding order by age where feeding older calves first was associated with a higher prevalence of respiratory disease. Producers and herd managers should also consider the necessary changes in calf diets toward weaning age to avoid complications due to improper rumen development. Among the changes that should be considered are a gradual decrease in the volume of milk diet fed, offering properly formulated solid diets and extending the weaning age to allow for rumen development.

#### *2.2.3.3 Genetics*

Recent genomic investigations have identified genetic differences associated with susceptibility or resistance to BRD (Van Eenennaam et al., 2014). Quantitative trait locus mapping has been used to find regions linked to BRD susceptibility (Casas et al., 2011; Glass et al., 2012); gene set enrichment analysis has been used to find upstream regulators of BRD susceptibility in beef and dairy cattle (Neupane et al., 2018). Although this investigation into genetic selection for BRD resistance is relatively new, genomic testing for wellness traits, including respiratory disease, have recently become commercially available. The percent of the disease explained by the genotype varies between studies and geographic location (Glass et al., 2012; Hoff et al., 2019), and heritability estimates for BRD resistance are low (Snowder et al., 2006). This area of research is still developing, and the herd effects of genetic selection for BRD resistance on BRD morbidity or mortality are still unknown. The continued investigation into the genetic backbone of immunity will hopefully provide selection criteria, aid in vaccine development and help us better understand host responses to the pathogen and environmental pressures.

### 2.2.3.4 'Upregulators' of immunity

Specific pharmacologic interventions intended to improve the innate immune system have recently been developed and are an active area of study. In contrast to vaccination, which primes specific immune responses characterized by antigen-specific humoral responses, this new strategy seeks to increase the general responsiveness of the innate immune system and is termed 'trained immunity'. One such product is a DNA-based immunostimulant, which apparently activates the cGAS-STING pathway; this is a mechanism to enhance host detection and response to cytosolic foreign DNA (Ilg, 2017). Another product marketed for innate immune stimulation is derived from the cell wall of the non-pathogenic *Mycobacterium phlei* (Nosky et al., 2017; Romanowski et al., 2017). The exact mechanism of immune stimulation of this product has not been clearly described but is postulated to involve mycobacterial compounds trehalose 6,6′-dimycolate (TDM) and muramyl dipeptite (MDM). These compounds are reported to enhance phagocytosis and antibody-mediated cytotoxicity and enhance lymphocyte activation (O'Reilly, 1992; Traub et al., 2006); however, their overall effectiveness in the prevention of disease in herds is unknown.

## 2.3 Role of the environment

Environmental factors can have drastic effects on the respiratory system; such factors include heat and cold stress; housing and hygiene practices that may expose calves to high pathogen loads; poor air quality, particularly with air pollutants such as dust, ammonia or other pollutants; and additional physical stressors such as with shipping or processing conditions. The environment can impart psychological stressors as well, such as with handling, shipping and changes in social structure during commingling. The environment can also affect changes in the commensal respiratory microbiota that influence host susceptibility to BRD (Lima et al., 2016; Nicola et al., 2017; Timsit et al., 2018). Many different types of environmental factors can affect the systemic immune defenses, especially those related to BRD.

A common environmental factor that influences BRD is heat. As discussed earlier, under Section 2.2.3, 'Influences on host immunity', heat stress may interfere with host defenses to BRD. There are several factors within the environment that may alter heat or heat stress exposure. For example, heat stress may be magnified within a calf's hutch compared to the ambient area, increasing the risk of respiratory disease (Louie et al., 2018). Specifically, Louie et al. (2018) estimated higher odds of BRD in calves as the daily maximum temperature increases during summer months. Moreover, the magnitude of the association was greater when the daily maximum temperature was recorded within the hutch compared to the ambient area, further demonstrating a difference in the calf environment within (microenvironment) compared to outside (macroenvironment) the hutch. Differences between calves' micro- and macroenvironment have been observed elsewhere (Macaulay et al., 1995; Carter et al., 2014). Such temperature and humidity changes impact calves' risk for respiratory disease year-round but with seasonal variation. Dubrovsky et al. (2019a) reported that the risk of respiratory disease in calves was the greatest in the spring followed by summer, compared to winter, with no difference between the fall and winter seasons. A cross-sectional study that captured greater geographic variability and a wider range of climate zones in California showed that fall was associated with the greatest odds of respiratory disease compared to spring, with the remaining seasons not different from spring. Both studies point to the transitioning between hot and cold seasons, which raises the question of whether fluctuations in temperature over a short period of time, including overnight, may be associated with a greater risk of respiratory disease in calves, at least in dry climates. Abrupt temperature fluctuations may also impact a calf's respiratory system's ability to clear *M. haemolytica* (Jones, 1987). Further research is needed on abrupt temperature and/or humidity changes as risk factors for respiratory disease in calves.

In the United States, approximately 70% of preweaned dairy calves are raised individually in hutches (USDA, 2016), which may vary in design, material, size and amount of protection from heat and cold stress. In addition to the specific environment within a calf's enclosure, the material such housing is made off further dictates the calf's exposure to environmental stress. Karle et al. (2019) reported that Northern California calves housed in hutches with metal components had significantly higher BRD prevalence compared to calves housed in wooden hutches (15.3% versus 3.2%). Maier et al. (2019a) found that calves housed in hutches constructed with metal components (walls or roof) had 11.2 times the odds of BRD compared to those housed in wooden hutches. A longitudinal study of more than 11 000 calves followed from birth to weaning similarly estimated a greater risk for BRD in calves housed in hutches with metal components compared to those housed in wooden hutches (Dubrovsky et al., 2019a). Raising calves in hutches with a metal roof or wall may be associated with greater BRD occurrence since metal is a better heat conductor compared to wood; this may result in greater temperatures for longer durations in hot summers. Furthermore, shade cloth or similar external shade structure above the calf-raising area has been associated with a reduced calf respiratory disease (Maier et al., 2019a). Modifying the environment to reduce the risk of bacterial infections can be beneficial for preventing BRD in calves. The impact of heat stress on the bovine respiratory system, and its role in respiratory diseases, can be further amplified by environmental conditions that modify the animal's microenvironment, creating additional adverse conditions that predispose to respiratory disease. For example, exposure to recycled lagoon water, used to flush fecal slurry below raised hutches with slatted wooden floors, was found to be associated with increased respiratory disease (Maier et al., 2019a). Other environmental factors that impact calves' risk for respiratory illness include air quality. A longitudinal study following 11 300 calves on five dairies in California identified that the use of water trucks for dust control resulted in a greater incidence of calf respiratory disease; while such results may seem contradictory, the study investigators noted that trucks traveled on dusty roads around the calf-raising area at speeds that often generated more dust (Dubrovsky et al., 2019b). In the same study, the owner reported that dust in the calf-raising area was also associated with an increased risk of respiratory disease, with dust around calves seen regularly being associated with the greatest risk of respiratory illness (Dubrovsky et al., 2019a). Dust can be a source of inhaled endotoxin and has been shown to cause coughing, fever, decreased feed intake, leukocytosis and mild interstitial pneumonia in sheep (Purdy et al., 2002). In contrast, magnesium chloride has been used in equestrian arenas to control dust and has been used on some dairies for the same reason. Indoor housing systems similarly can alter the calf's environment due to either improper ventilation or heating systems, leading to higher concentrations of dust and/or ammonia that may impact the calf's respiratory system (Nordlund and Halbach, 2019). Other studies reported an association between filtered air from indoor barns and a lower incidence of respiratory disease in calves (Pritchard et al., 1981; Hillman et al., 2010).

Environmental exposure to pathogens can also influence the occurrence of BRD. Starting at the time of birth, the calf is exposed to maternal microbiota; the composition of maternal vaginal microbiota appears to correspond to the calf's upper respiratory microbiota and alter the calf's risk of BRD (Lima et al., 2019). Bedding hygiene in the maternity pen also alters the risk of BRD in calves. Dubrovsky et al. (2019a) reported a 2% drop in the hazard of BRD with every additional change in maternity pen bedding per month. The latter study followed calves born in individual or group pens (one dairy), group calving pens (one dairy), individual calving pens (three dairies) or pasture (one dairy) and also estimated a decrease in the risk of BRD mortality when comparing maternity pens with bedding changed four to nine times compared to less than four times per month. An increase in the frequency of maternity bedding changes is likely to reduce the concentration of pathogens a calf is exposed to at or soon after birth. A similar association can be expected when contrasting individual to group maternity pens. Svensson et al. (2003) reported a 46% reduction in odds of BRD in calves born in individual pens compared to group maternity pens. Furthermore, bedding material used can also have an impact on the occurrence of respiratory disease in calves. Calves born in maternity pens bedded with dried manure or sand had a lower prevalence of respiratory disease compared to those born on dirt or plant materials, including rice hulls, almond shells, straw, wood shavings or wood chips (Maier et al., 2020).

The respiratory microbiota of cattle can influence that of other in-contact animals, and the composition of these microbes appears to change relative to many other factors, including the farm of origin (Nicola et al., 2017), diet (Hall et al., 2017), stress (Holman et al., 2017; Timsit et al., 2016b; Stroebel et al., 2018) and antibiotic treatment (Holman et al., 2018). For example, the nasopharyngeal microbiota of feedlot cattle changes between arrival and after 60 days on premise (Holman et al., 2015). In another study, antibiotic-resistant clones of *P. multocida* appeared to spread to arriving calves after arrival to the feedlot (Guo et al., 2020). The animal-to- animal contact, which varies with housing setup, is likely a factor that influences the transmission of microbes and alterations in the respiratory microbiota; placement of hutches in close proximity that allowed calf-to-calf contact was associated with an increase in BRD in calves greater than 75 days old when compared to calves of the same age group without calf-to-calf contact (Maier et al., 2019a). Furthermore, within cohorts of calves with calf-to-calf contact, older calves consistently had greater odds of developing BRD when compared to younger calves. Further research is needed to investigate the role of animal contact (within or between ages) and respiratory disease, specifically as it relates to changes in the upper and lower respiratory microbiota.

## 2.4 Other disease syndromes

Scours and respiratory disease continue to be the most common causes of preweaned dairy calf morbidity and mortality (USDA, 2018). Despite their distinct pathology and clinical presentations, these two syndromes are good examples of different diseases that share many causal mechanisms of the host, pathogen and environment. Host immunity is of paramount importance, particularly in young calves that are more susceptible to infectious diseases until the immune system matures and can better react to new exposures; during early life the calf relies heavily on the transfer of passive immunity through appropriate colostrum intake. Inappropriate colostrum management can predispose calves to both respiratory and enteric diseases. The environment in which the calf is born and reared in provides variable protection from the elements; environmental temperature extremes can stress the calf and impact calf immunity. The hygiene of the calfhood environment also determines exposure to both respiratory and enteric pathogens. These are just a few examples of how both of these very common calfhood disease syndromes ultimately share many of the same risk factors classified under the host, pathogen and environment. Several of the prevention and control interventions discussed in this chapter and proposed as part of the BRD risk assessment tool described later in the chapter may similarly interrupt the causal web for other diseases.

## 3 Disease detection

An important part of disease prevention is appropriate disease detection. The clinical signs of respiratory disease can vary in type and severity, and some affected animals may show no obvious outward signs at all (Buczinski et al., 2014; Ollivett et al., 2015). Acutely affected animals may show different signs than chronically affected animals. These features of the disease have led to a variety of different techniques used to aid in the diagnosis of BRD. There are many tools that can be used to detect or further characterize cases of respiratory disease and the pathogens associated with respiratory disease.

### *3.1 Disease detection techniques*

#### *3.1.1 BRD scoring systems*

Availability of scoring systems for health outcomes such as BRD allows for rapid and low-cost detection of cases. BRD scoring systems work by the assessment of clinical signs, or risk factors, and produce a score for each animal, which is compared to a cut-off value. A score that equals or exceeds a given cut-off classifies the patient as BRD score positive. Users can then interpret such score positive status as the presence of the condition or disease within a certain level of sensitivity and specificity provided by the specific scoring system used. Scoring systems have been previously proposed for respiratory disease in dairy and beef cattle (Thomas et al., 1977; Panciera and Confer, 2010). Two systems have been used in recent studies for BRD in preweaned dairy calves: the Wisconsin (WI) BRD scoring system developed at the University of Wisconsin; and the California (CA) BRD scoring system developed at the University of California Davis. The latter was based on a pair-matched case-control study of 2,030 preweaned calves (Love et al., 2014, 2016a). Both systems assess clinical signs associated with respiratory disease including nasal discharge, ocular discharge, cough, fever and ear disposition; however, the CA system assesses breathing quality as a sixth clinical outcome. Breathing quality is assessed for a calf by comparing calf respiratory rate and difficulty to the other calves in the same herd at the same time of scoring; this adjusts for the effect of ambient environmental conditions such as heat stress. The WI system scores each clinical sign on a gradient of severity, while the CA system dichotomizes each into absent or present. The WI system scores each clinical sign in single-unit increments from 0 (normal) to 3, and the CA system scores vary by the clinical sign as a function of the strength of the association between the respective clinical sign and respiratory disease as measured in a pair-matched case-control study (Love et al., 2014). Specifically, the CA system scores nasal discharge, occular discharge, cough, fever, ear disposition and breathing quality as 4, 2, 2, 2, 5 and 2 points, respectively (Figs 2A and 2B). Both systems assign a calf BRD score positive status at scores ≥ 5. In addition, the WI system assesses both spontaneous and induced cough, while the CA system assesses spontaneous cough only, which is a tradeoff to avoid calf handling to maintain biosecurity between calves. The CA scoring systems

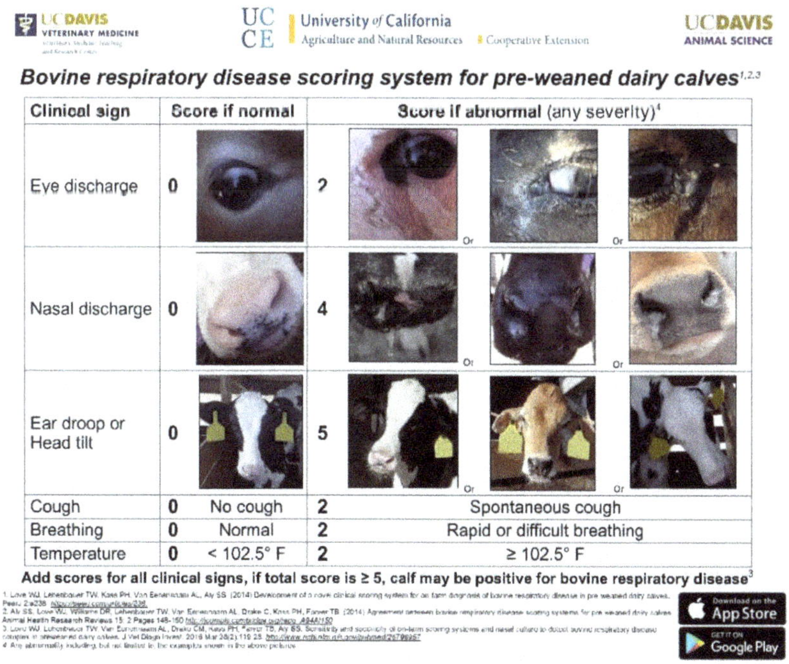


**Figure 2A** The California bovine respiratory scoring system. (Image reprinted by permission from Dr. Sharif S. Aly).

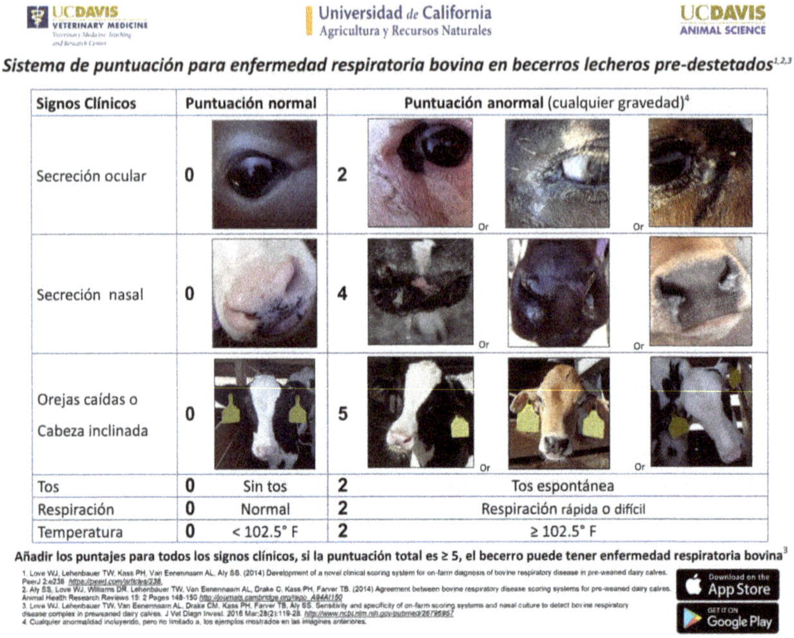


***Sistema de puntuación para enfermedad respiratoria bovina en becerros lecheros pre-destetados***[1,2,3]

| Signos Clínicos | Puntuación normal | | Puntuación anormal (cualquier gravedad)[4] | |
|---|---|---|---|---|
| Secreción ocular | 0 | | 2 | Or / Or |
| Secreción nasal | 0 | | 4 | Or / Or |
| Orejas caídas o Cabeza inclinada | 0 | | 5 | Or / Or |
| Tos | 0 | Sin tos | 2 | Tos espontánea |
| Respiración | 0 | Normal | 2 | Respiración rápida o difícil |
| Temperatura | 0 | < 102.5° F | 2 | ≥ 102.5° F |

**Añadir los puntajes para todos los signos clínicos, si la puntuación total es ≥ 5, el becerro puede tener enfermedad respiratoria bovina**[3]

1. Love WJ, Lehenbauer TW, Kass PH, Van Eenennaam AL, Aly SS. (2014) Development of a novel clinical scoring system for on-farm diagnosis of bovine respiratory disease in pre-weaned dairy calves. PeerJ 2:e238 https://peerj.com/articles/238.
2. Aly SS, Love WJ, Williams DR, Lehenbauer TW, Van Eenennaam AL, Drake C, Kass PH, Farver TB. (2014) Agreement between bovine respiratory disease scoring systems for pre-weaned dairy calves. Animal Health Research Reviews 15: 2 Pages 148-150 http://journals.cambridge.org/repo_A94AI150
3. Love WJ, Lehenbauer TW, Van Eenennaam AL, Drake CM, Kass PH, Farver TB, Aly SS. Sensitivity and specificity of on-farm scoring systems and nasal culture to detect bovine respiratory disease complex in preweaned dairy calves. J Vet Diagn Invest. 2016 Mar;28(2):119-28. http://www.ncbi.nlm.nih.gov/pubmed/26796957
4. Cualquier anormalidad incluyendo, pero no limitado a, los ejemplos mostrados en las imágenes anteriores.



**Figure 2B** Spanish translation of the California bovine respiratory scoring system. (Image reprinted by permission from Dr. Sharif S. Aly).

call for less calf handling since only calves observed with a total score of 4 would require a rectal temperature to ascertain whether its final score exceeds the cut-off of 5 points. In a separate nested case-control study of 536 calves on three dairies and two calf ranches in California, both the CA and WI scoring systems had similar sensitivity when screening calf herds for BRD (46.8% versus 46.0%, respectively) and a similar diagnostic sensitivity when detecting a case (72.6% versus 71.1%, respectively); however, the specificity of the WI system was greater than that of the CA system (91.2% versus 87.4%). Despite the difference in specificity between both systems being statistically significant, it's unlikely to be biologically important; however, users should weigh such a difference against ease of use. The CA scoring system has the advantage of simple dichotomous classification for each clinical sign and the fewer calf handling requirements (no test for induction of cough and rectal temperature needed only on calves with a score of 4), which improve biosecurity.

Weaned and growing dairy youngstock are also affected by respiratory disease. In fact, respiratory disease in the United States is the greatest cause of morbidity and mortality in dairy youngstock postweaning (USDA, 2018). A scoring system recently developed at the University of California Davis for weaned and growing dairy youngstock was based on a prevalent case-control study on five dairies (Fig. 3). The development of the CA BRD scoring system for post-weaned calves followed the same approach as the preweaned scoring system with cross-validation for estimating its accuracy; the latter is a technique where the dataset is split such that observations are used only once for development or validation but not for both (Maier et al., 2019b).

### 3.1.2 Thoracic ultrasound

Diagnostic imaging adds to the clinical exam and can be particularly useful to confirm suspected cases, characterize the distribution or severity of the disease and detect sub-clinically affected animals (Ollivett et al., 2015). Thoracic ultrasound is a popular imaging modality in calves due to the practical on-farm utility of this technique; the linear probe typically used for bovine reproductive work can be positioned between ribs to image much of the lung surface (Ollivett and Buczinski, 2016). The technique is relatively simple to perform after training and requires only basic restraint, a coupling agent such as alcohol or gel, occasionally clipping if the hair is thick or dirty, and an ultrasound unit. For on-farm use, typically the linear rectal probe is used; however, other

UC DAVIS VETERINARY MEDICINE Veterinary Medicine Teaching and Research Center | UC CE | University of California Agriculture and Natural Resources | Cooperative Extension

***California bovine respiratory disease scoring system for post-weaned dairy calves***

| Clinical sign | Score if normal | | Score if abnormal (any severity)[1] | | | |
|---|---|---|---|---|---|---|
| Sunken eyes | 0 | 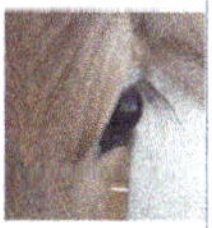 | 4 |  | Or | 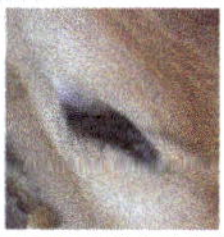 |
| Low body condition | 0 |  | 5 |  | Or |  |
| Cough | 0 | No cough | 2 | Spontaneous cough | | |
| Breathing | 0 | Normal | 1 | Rapid or difficult breathing | | |
| Diurnal temp fluctuation | 0 | ≤27° F (≤15° C) | 1 | > 27° F (>15° C) | | |

**With diurnal temperature data: calf is score positive[2] if total score ≥ 2**

**Without diurnal temperature data: calf is score positive[2] if total score ≥ 1**

| Confirmatory step for score positive[3] | Do not treat | Treat |
|---|---|---|
| Rectal temperature | < 102.5° F (< 39.2° C) | ≥ 102.5° F (≥ 39.2° C) |

1. Any abnormality including, but not limited to, the examples shown in the above pictures. For body condition, a score of ≤ 2 on a scale of 5 would be considered abnormal.
2. With diurnal temperature fluctuation: screening sensitivity 77%, diagnostic sensitivity 100%, specificity 62%, without diurnal temperature fluctuation: screening sensitivity 84%, diagnostic sensitivity 100%, specificity 46%
3. With diurnal temperature fluctuation: confirmatory step improves specificity to 77% (screening sensitivity 65%, diagnostic sensitivity 77%), without diurnal temperature fluctuation: confirmatory step improves specificity to 63% (screening sensitivity 71%, diagnostic sensitivity 77%)

**Figure 3** California bovine respiratory scoring system for post-weaned dairy calves (Used by permission from *Journal of Dairy Science*; vol. 102:7329-7344; doi.org/10.3168/jds.2018-15474).

ultrasound probes can likewise obtain useful images of the lung surface. The depth and focal distance can be adjusted to improve image quality at the lung surface based on the patient's chest wall thickness, probe type used and depth of consolidated lesions. Healthy, aerated lung creates an echogenic line at the lung surface followed by a reverberation artifact that is readily identifiable. Pneumonia causes inflammation in the lungs that leads to fluid accumulation, cellular infiltrates and debris that alter the gas-containing spaces. Areas of diseased lung thus lose the characteristic gas echogenicity and reverberation on ultrasound and appear as soft tissue echogenicity similar to the liver; large areas of consolidation are sometimes referred to as 'hepatized lung' (Reef et al., 1991; Ollivett et al., 2013). Pleural irregularities are visualized on ultrasound as irregularities in the echogenic surface of the lung and are termed "comet tails'; as these areas converge from individual lesions to larger continuous irregularities, they are termed 'B-lines' (Adams and Buczinski, 2016). Pleural effusion is readily apparent on ultrasound as hypoechoic fluid outside the lungs, and abscesses appear as mixed echogenicity lesions typically contained in a more echogenic capsule. Different scoring systems have been developed to classify ultrasonographic lung lesions and have been used in both research and clinical settings, on individuals and herds, to characterize lung disease and identify sub-clinically affected animals (Adams and Buczinski, 2016; Ollivett and Buczinski, 2016). The findings of thoracic ultrasound can be useful to prognosticate, assign treatment, assess response to treatment and screen groups of animals. Systematic screening for BRD in specific animal groups, using an ultrasound scoring system, may be useful for veterinarians to identify when populations at risk for BRD first start to have evidence of the disease and creates an objective measure on which to form animal health goals and assess response to preventative interventions (Calf Health Module - The Dairyland Initiative; Ollivett and Buczinski, 2016; Dunn et al., 2018b). Other respiratory imaging techniques that are occasionally used clinically, or in specific research settings, but are not usually implemented on a herd basis include airway endoscopy, radiographs, CT/MRI and fluoroscopy. These alternative advanced imaging techniques can be useful in select cases or in specific research investigations that require detailed thoracic imaging. These techniques are not employed in herd screening methods due to the cost, limited availability of equipment and impractical nature for imaging large numbers of animals.

### *3.1.3 Auscultation*

Thoracic auscultation using a standard stethoscope can be performed to assess for abnormal lung sounds, which may support a diagnosis of BRD; however, thoracic auscultation alone has very poor sensitivity for detecting lung lesions (Buczinski et al., 2014). A specialized lung auscultation device has been marketed to aid in the detection of BRD in feedlot cattle. This computer-assisted stethoscope records sounds during thoracic auscultation and applies a computer algorithm to generate a BRD severity grade, ranging from unaffected to chronically affected, to animals showing signs of illness. In a study comparing standard thoracic auscultation to this device, the agreement between the two methods had a kappa value of 0.77, and the device had a sensitivity of 92.9% and specificity of 89.6% for the diagnosis of BRD when the referent diagnostic test in the study was standard thoracic auscultation (Mang et al., 2015). Further research on the diagnostic accuracy and utility of the computer-assisted stethoscope in preweaned, weaned dairy calves and adults is necessary.

### *3.2 Pathogen detection/sampling techniques*

Bacterial isolation from the airway can be useful to define bacterial presence and describe antimicrobial susceptibility profiles and microbial communities. Airway sampling can be used on individual clinical cases, as part of herd monitoring, and for research investigations. Several different techniques for sampling the airway of cattle are possible. Each method has advantages and disadvantages that make optimal method selection dependent on the specific goals of testing. Table 1 summarizes techniques used for obtaining bacterial samples from the airways of cattle. Collection of airway samples from live animals may be more physically challenging than post-mortem collection, depending on the animal size and restraint methods available. However, limited collection of data from dead animals may skew results toward treatment failures and chronic cases. Collection of upper airway samples as a proxy for lower airway infection is relatively technically simple compared to sampling the lower airway, but these methods inherently sample different microbial populations (Nicola et al., 2017; Timsit et al., 2018). However, the pathophysiology of the BRD complex involves changes in bacterial flora of the upper respiratory tract advancing to lower airway infection, so samples from the upper respiratory tract can still be useful in the study of BRD (Doyle et al., 2017b). For all sites, results should be interpreted in light of treatment history and chronicity of disease; these factors may influence the bacteria isolated at the time of sampling and may not necessarily reflect the bacteria present at disease onset. Additionally, sample handling and culture/sensitivity methods employed on respiratory samples should be consistent with Clinical Laboratory Standards Institute (CLSI) guidelines.

#### *3.2.1 Methods review*

A nasal swab (NS) is typically performed by simply swabbing the rostral internal surface of the nares with a sterile culturette swab. This type of sampling collects bacteria or other microbes from the rostral nasal cavity and is very simple to

**Table 1** Summary of airway bacterial sampling methods in cattle

| Sample type | Regional population sampled | Technical notes |
|---|---|---|
| Nasal swab (NS) | • Upper airway-rostral nasal cavity | • Technically simple to perform. Requires only temporary restraint.<br>• Maximum contamination of rostral nasal microbiota |
| Deep nasopharyngeal swab (DNPS) | • Upper airway, nasopharyngeal cavity | • Technically relatively simple to perform, less contamination from rostral nasal cavity than NS, but some contamination of rostral nasal microbiota likely still present |
| Transtracheal wash (TTW) | • Combination of lower airway and pool of large conducting airways | • Avoids contamination of nasal microbiota<br>• Requires technical skill and thorough restraint |
| Bronchoalveolar lavage (BAL) | • Combination of lower airway and focal large conducting airways | • Some contamination of nasal microbiota possible (depends on a specific technique)<br>• Requires technical skill and thorough restraint |
| Thoracocentesis | • Pleural space | • No contamination from internal airways<br>• Requires the presence of pleural effusion; bacterial population in effusion may or may not represent bacteria present in lungs<br>• Requires technical skill and appropriate restraint |
| Lung tissue culture | • Lower airway, terminal small conducting airways and gas exchange surfaces | • Avoids contamination of nasopharyngeal microbiota<br>• Usually only performed post-mortem<br>• Requires uncontaminated or seared lung specimens |

perform. The NS, however, presents a challenge if the goal is the selective isolation of classic respiratory bacteria (*M. haemolytica*, *P. multocida*, *H. somni*); this method contains more contaminants from the rostral nasal cavity, and selective isolation of the desired bacteria may be less successful (Doyle et al., 2017a).

A deep nasopharyngeal swab (DNPS), also known as a guarded nasopharyngeal swab, is performed by using a longer, guarded sterile swab that is passed deeper into the nasal cavity, to the level of the nasopharynx or into the pharyngeal recess, where the guard is removed and the swab is gently rotated to allow sampling of the caudal nasopharynx; the swab is then retracted back into the guard and removed from the nasal cavity. This technique samples upper respiratory microbes while minimizing contamination from the rostral nasal cavity.

The transtracheal wash (TTW) is performed by passing a sterile cannula percutaneously and aseptically into the trachea, followed by passing sterile flexible tubing through the cannula and down the trachea at the ventral most portion before the mainstem bronchi, depositing sterile saline and then immediately aspirating that sterile saline. The resulting fluid contains a wash of tracheal contents representing a pooled lower airway sample of microbes and any exfoliated airway cells. There are variations on this technique used for post-mortem tracheal wash sampling (Godinho et al., 2007). Another variation on the TTW, a transtracheal swab, has been reported (DeRosa et al., 2000).

A bronchoalveolar lavage (BAL) is performed by passing a sterile flexible tube from the nares (or oral cavity) into the trachea to a terminal bronchus, where the tube is gently wedged in a terminal bronchus. Sterile saline is deposited and immediately aspirated through the tube to create a fluid wash from the terminal airway.

A variation on TTW and BAL is the endoscopic guided airway washing sample; this type of procedure is typically performed in a hospital setting in cases when airway endoscopy is indicated. A cannula is typically passed through the endoscope channel, or the endoscope channel is used directly to deposit and aspirate saline into the desired airway location to create a wash of cells and microbes from the lower airway for analysis similar to a TTW or BAL, depending on where the scope/cannula are placed. Care must be taken in the interpretation of samples obtained through the endoscope lumen because contamination from the endoscope is possible.

Thoracentesis is used to sample fluid that has accumulated in the pleural space and is performed by aseptically obtaining a percutaneous sample of pleural effusion. This is usually accomplished using a needle, IV catheter or sterile teat cannula after a small cut down of the skin and intercostal fascia. It is imperative that negative pressure is maintained during thoracentesis to avoid pneumothorax.

Lung tissue can be directly sampled during necropsy; if samples are intended to be used for pathogen identification, the sample should be obtained sterilely, or the tissue section should be seared prior to sampling the internal surface for microbial isolation.

These techniques are often grouped into discussions on upper versus lower airway sampling; NS and DNPS are common methods for upper airway sampling while TTW, BAL (and their variations) and lung tissue culture are typically considered methods for lower airway sampling. Both TTW and BAL techniques involve passing through portions of the upper respiratory tract and inherently contain microbes from the large conducting airways of the trachea or bronchi in addition to what is obtained from lavage of the lower conducting and terminal gas exchange airways. Thus, when discussing lower respiratory tract microbiota sampling ante-mortem, we are technically discussing the sum of microbiota obtained from sampling the terminal conducting airways, transitional airways and gas exchange surfaces. This microbial population is distinct from the population found in the nosopharyngeal cavity, which is typically referred to as the upper respiratory microbiota (Nicola et al., 2017; Timsit et al., 2018). Of these methods, the four most common ante-mortem airway sampling methods in cattle are likely NS, DNPS, TTW and BAL. These methods are typically being used for bacterial culture, identification, and antimicrobial susceptibility testing. More recent studies use these methods to acquire samples for metagenomic analysis. Comparisons between methods for bacterial isolation and identification, including NS, DNPS, TTW and BAL, have been described (Allen et al., 1991; DeRosa et al., 2000; Godinho et al., 2007; Doyle et al., 2017a; Van Driessche et al., 2017). There are conflicting reports comparing agreement between upper and lower airway sampling techniques for bacterial identification on the calf level; on the group level, agreement between methods is generally good (Allen et al., 1991; DeRosa et al., 2000; Godinho et al., 2007). The antimicrobial susceptibility profiles between upper and lower airway samples have been compared, and the two sampling sites reportedly demonstrate the same susceptibility profiles (96%) the vast majority (96%) of the time (DeRosa et al., 2000).

### *3.3 Bacterial identification*

Bacterial identification of BRD pathogens can be obtained by a variety of methods. The technology best suited to identify respiratory bacteria depends on the goals of testing, sample type available and economics. Traditionally, respiratory bacteria have been isolated via culture and identified by physical and biochemical properties. An advantage of traditional culture methods for the identification of common respiratory bacteria is that this technique is commonly readily available at animal disease diagnostic laboratories. The turnaround time for results is usually on the order of days to allow for bacterial growth and identification. This turnaround time is notably longer for *Mycoplasma* species due to their slow-growing biology. More recently, the MALDI-TOF technology has been used for the identification of cultured BRD pathogens. This technology uses mass spectrometry to identify bacteria from cultured samples based on protein structures. This technology is faster than traditional physical and biochemical identification methods and reflects categorization based on the proteome rather than physical or chemical properties. PCR offers another way to identify specific respiratory pathogens. This method uses nucleic acid primers to match for known bacterial pathogens of interest in the sample. Metagenomics is a newer method used to study populations of microbes in the respiratory tract. This technology is used to identify populations of microbes present in a sample rather than selective isolation or specific testing for select organisms. Metagenomics technology works by matching nucleic acid sequences in a given sample to a reference database to identify the microbes present in the sample.

### 3.4 Determination of antibiotic susceptibility

The use of antibiotic susceptibility testing is one tool we can use to monitor resistance and make antimicrobial stewardship decisions. There are several ways bacterial susceptibility or resistance can be defined. The ultimate measure of interest is often the case outcome, and in vitro methods of susceptibility testing are intended to estimate a prediction of the likelihood of successful bacterial killing in the living animal. There are several methods that can be used, and most commonly include microbroth dilution and disk diffusion methods; microbroth dilution is considered the current gold standard (CLSI, 2018). These tests evaluate bacterial growth inhibition with different concentrations of antimicrobial to identify a minimum inhibitory concentration (MIC). The MIC is compared to CLSI (2018) breakpoints, and the bacterial sample is classified as susceptible, intermediate or resistant. These breakpoints are based on the knowledge of drug pharmacokinetics and disease tissue distribution and are used to predict how likely a given bacterial sample is to respond to treatment with a given drug in the species of interest. The CLSI breakpoints for BRD bacteria and the antibiotics commonly used to treat BRD are available and updated occasionally (CLSI, 2018).

Another way that AMR has been monitored is through molecular techniques to identify known or putative resistance genes. Molecular methods used for the identification of resistance genes in BRD bacteria include PCR, whole genome sequencing (WGS) and metagenomic analysis; some of these methods can also employ quantitative analyses to quantify the amount of resistance gene presence. This information can be useful in tracking and studying AMR and as part of antimicrobial stewardship plans. It should be recognized, however, that AMR phenotypes and resistance genotypes are not synonymous. The presence of a given resistance gene does not necessarily imply that the gene is being expressed and that the bacteria will display resistance to a given antibiotic. Both genotypic and phenotypic tests for antibiotic resistance can be useful for understanding antibiotic resistance; however, it is necessary to recognize the inherent differences in these diagnostic methods when interpreting the results of either type of test.

Case outcomes have also been used as a gauge for antibiotic resistance, although the nature of living systems does not have a perfect agreement with in vitro susceptibility testing (Watts and Sweeney, 2010; Coetzee et al., 2019). A case outcome represents the ultimate measure of antibiotic treatment failure or success; however, many factors influence treatment failure besides specific antibiotic resistance (Watts and Sweeney, 2010). Treatment records that include case outcomes are a valuable tool for on-farm monitoring of AMR patterns.

### 3.4.1 Importance of antimicrobial susceptibility monitoring

Bacterial causes of disease are commonly treated with antimicrobial drugs, a necessary tool when preventative methods have failed, to support the welfare of our food animals. However, with the use of antibacterial drugs comes the risk of AMR; this resistance affects both cattle and human populations through the selection of resistance genes in the exposed bacterial population. Antibiotic resistance in BRD pathogens has been documented for decades (Chirino-Trejo and Prescott, 1983; Speer et al., 1992) with a concerning trend of increasing resistance over time (CDC, 2019). Bacterial resistance to antibiotics threatens animal, environmental and human health and represents a serious One Health concern. Antimicrobial stewardship, which includes judicious use of antibiotics to reduce the selective pressure for AMR and is necessary to ensure the continued availability of effective antibiotics against infections in dairy cattle, guarantee the health of our herds, ensure a continuous supply of wholesome milk and dairy beef and protect the public health. Surveillance efforts for AMR should involve the bacterial resistance phenotype as well as the genetic determinants of the bacterial resistance. Resistance gene identification can encompass single bacterial strain testing or bacterial population surveillance, including all resistance genes in a population, often referred to as the bacterial resistome. Understanding the complex pathways between on-farm drug use and AMR in dairy cattle populations is key to antibiotic stewardship and judicious use (Mathew et al., 2007). Surveillance should capture the on-farm management practices and bottlenecks to antimicrobial stewardship using both industry-wide and individual farm-specific surveys that investigate the role of staff in antimicrobial drug use, including sources of information, decision processes for acquisition and storage, treatment protocol design and implementation, training and updating of standard protocols on farm and monitoring of treated animals for withdrawal.

There is no consensus describing exactly how to monitor a livestock facility for AMR of BRD pathogens. In the authors' opinions, it seems reasonable to use a multimodal approach. For routine surveillance, culture and sensitivity data can be collected from animals that die or are euthanized due to BRD as well as from acutely affected live animals. Samples can be obtained in the face of a new BRD problem or in groups with ongoing BRD problems for which treatment protocols may need to be updated. Development of objective, goal oriented, clinically useful, AMR monitoring programs are needed as part of a holistic approach to managing BRD.

## 4 Risk assessment tools

A multiple causation model can be built mirroring the three components of the host, environment and pathogen triangle; such a model incorporates both host and environmental factors as components in the causation of disease in addition to the pathogen disease model traditionally described by Koch's postulates. Failing to incorporate opportunities to modify host and environmental aspects of the disease triangle would limit valuable prevention efforts of bacterial diseases in livestock. Factors discussed earlier, such as environmental factors including heat stress and presence of dust or host factors such as failure of transfer of passive immunity, lack of immunizations or improper nutrition, can all increase the risk of respiratory disease in calves. Hence, the most direct way to break the triad of respiratory disease is to modify the management of calves to mitigate these risk factors. The multifactorial nature of respiratory disease in calves lends itself particularly well for a risk assessment approach for its control and prevention (Gorden and Plummer, 2010). Management on a dairy can be used to limit or modify calf exposure to pathogens and host and environmental risk factors. Developing such a control and prevention plan should start by classifying the risk factors into logical intervention themes, such as colostrum management, housing and vaccination, among others, that contribute to the animals' risk of disease. Hence, a control and prevention plan should be based on a comprehensive assessment of all known risk factors in the herd to allow the investigator to develop a customized herd-specific plan. The resulting herd-specific plan would offer the producer a menu of interventions for control and prevention. Hence, a comprehensive tool that is simple yet functional, based on the results of research on the risk factors associated with the disease of interest, is required so users may elect to implement as many changes in their management as is feasible to reduce the risk of disease. The control and prevention tool should also be rapid and of low cost, allowing the user to benchmark the disease occurrence in their herd repeatedly over time. Benchmarking the disease occurrence before and after the interventions is an implemented act as a demonstration of the impact of control and preventative measures, which may be a positive reinforcement for the producer to continue the control efforts and add more interventions.

Risk assessments are one example of such functional and comprehensive tools to prevent and control diseases and health outcomes of complex causality. Examples of common risk assessment tools in human medicine include the Falls Risk Assessment Tool (FRAT) for fall accidents in seniors (Nandy et al., 2004), the Fracture Risk Assessment Tool (FRAX) (Kanis et al., 2008) and the coronary risk assessment (Pearson, 2002). In veterinary medicine the Johne's disease risk assessment has been used in dairy and beef cattle to control *Mycobacterium avium* subsp. *paratuberculosis*, a disease which, after an elongated incubation period, causes a chronic debilitating enteropathy (Aly and Thurmond, 2005). A risk assessment approach for the prevention of respiratory disease in preweaned calves may be even more successful because producers may observe the benefits of the prescribed interventions relatively soon compared to diseases with longer incubation periods like Johne's disease (USDA, 2011). The following section describes the portions of the BRD risk assessment tool for the prevention of pneumonia in preweaned dairy calves.

### 4.1 BRD Risk Assessment Tool

Users of the BRD Risk Assessment Tool include specialists and dairy industry personnel, including producers and herd managers of dairies and calf ranches, herd veterinarians, consultants, extension specialists, nutritionists, pharmaceutical companies, government entities (federal and state), researchers and allied industry specialists concerned with dairy cattle health, welfare and productivity. Users can complete the risk assessment tool to benchmark a preweaned calf herd's risk of respiratory disease, identify its risk factors, estimate the herd prevalence and devise a herd-specific management plan to control and prevent respiratory disease. The complete BRD Risk Assessment Tool is available online (https://escholarship.org/uc/item/1jb2f7rm) and is made of three portions described in the next few sections (Aly et al., 2020; Maier et al., 2020).

#### 4.1.1 Risk assessment questionnaire

The risk assessment questionnaire contains questions that collect data on risk factors on a dairy along with a specific risk score for each question's response option. The greater the risk score associated with a practice, the greater the occurrence of respiratory disease. The questions were based on previous research used to gather risk factor information from dairies from three studies. The first study was a mail survey mailed to all grade A milk-producing dairies (Love et al., 2016b). The second study was a cross-sectional study of 100 dairies (BRD 100) distributed along California's three diverse milk sheds (Northern CA, Northern San Joaquin Valley, and the third milk shed composed of both Southern San Joaquin Valley and Southern CA; Love et al., 2016b). The BRD 100 linked management risk factors for over 4253 calves selected randomly from approximately 25 000 calves on these 100 dairies (Maier et al., 2019a). The third study was a longitudinal prospective cohort study of over 11,300 calves followed from birth to weaning on five dairies representing different dairy management systems and allowing for the estimation of the risk of respiratory disease in calves (Dubrovsky et al., 2019a).

The risk assessment questionnaire is made up of the following six sections investigating adherence to best management practices for the prevention of respiratory disease in calves. The sections are basic herd risk profile, maternity pen, colostrum management, calf milk feeding, vaccinations and calf housing. Risk scores from the six sections vary in their total proportion to the magnitude of their risk factor associations with respiratory disease; however, for ease of use, they were scaled so that the entire risk assessment tool scores added up to 1000 (Aly et al., 2020; Maier et al., 2020). The six sections thematically group respiratory disease risk factors, as presented here.

#### 4.1.1.1 Basic herd risk profile

The basic herd risk profile section of the risk assessment characterizes the baseline risk of respiratory disease in the calf herd. The questions in this section refer to herd size, breed distribution, organic status and season and represented 10.3% of the total risk score.

#### 4.1.1.2 Maternity pen management

Risk factors in the maternity pen management include: whether cows and heifers calve together, the frequency of maternity pen bedding changes and the type of bedding including pasture. Regardless of whether calves were born on pasture or not, they had greater odds of respiratory disease if they remained with the dam more than an hour after birth compared to those removed within an hour from birth (Maier et al., 2020). The association between group calving and calf respiratory disease was modified by the frequency of bedding changes. Specifically, if maternity bedding was changed less than six times a month, the occurrence of respiratory disease in the newborn calves increased compared to if the bedding was changed six or more times a month (Maier et al., 2020). In terms of bedding material, the use of dried manure or gypsum was associated with reduced occurrence of respiratory disease in calves compared to dirt or plant materials such as shavings, straw or almond shells. Maternity pen-related risk factors accounted for 10.5% of the total risk score.

#### 4.1.1.3 Colostrum management

The first question in the colostrum section identifies the sources of colostrum as individual dam, pooled, mix of natural and colostrum replacer or nursed from the dam on pasture. Colostrum from an individual dam or colostrum nursed from the dam on pasture was associated with the least odds of respiratory disease (and thus are assigned the lowest risk score), compared to the remaining sources. The remaining questions enquire if colostrum is supplemented or heat-treated and, if so, what is the duration of storage from harvest to heat treatment, is it tested for bacterial content, is it harvested from uniparous dams, is preservatives added, is it stored in solid containers or bags and finally if it is tested for immunoglobulin content. While some of these practices may not have a direct effect on respiratory disease in calves, such as testing colostrum quality, the producers' interventions based on results of testing colostrum for bacteriological content may reduce the risk of respiratory disease in calves. Such interventions may include proper harvest procedures and storage conditions and storage times for colostrum (Maier et al., 2019a). Another example of a risk factor with an indirect effect on the risk of respiratory disease is the necessity of monitoring immunoglobulins of colostrum supplemented with a replacer; the intervention based on a finding of inadequate quality colostrum is what alters the risk of respiratory disease, not the test itself. Other management practices warranted data exploration such as the duration of colostrum storage prior to heat treatment. Specifically, a greater occurrence of respiratory disease was observed when colostrum was stored for more than 10 h before heat treatment, which may be due to higher bacterial concentration compared to when it is stored for 10 or less hours (Maier et al., 2019a). There were lower odds of respiratory disease in calves when dairies stored their colostrum in bags compared to solid containers. Colostrum stored in solid containers may be subject to uneven freezing and thawing, resulting in more damage to the valuable immunoglobulin proteins compared to colostrum stored in bags (Maier et al., 2019a). Testing colostrum for immunoglobulins was associated with a lower risk of respiratory disease in calves, and while testing itself doesn't protect the calves, it is likely the actions taken in response to the test results may protect calves.

The last two questions in the colostrum management section obtain information on the volume of colostrum fed and whether the producer tests calves for the transfer of passive immunity. Higher risk scores are warranted if calves are fed less than 2.84 L (3 US quarts) in the first 12 h after birth or if calves are not tested for the transfer of passive immunity or if they are tested but no actions are taken in the case of failure of the transfer of passive immunity. The colostrum management section accounted for 14.5% of the total risk score.

#### 4.1.1.4 Milk feeding

Questions in the milk feeding section investigate the source of milk fed to the calves, the order of feeding in terms of the age and health status, if milk fed is tested for bacteria and if it is supplemented with a milk replacer. Zero risk score is assigned if dairies feed saleable milk due to its superior nutritional composition compared to milk replacer or a mix of both. Higher risk scores are assigned to dairies that feed calves in no particular order or feed older or sick calves first. Similarly, higher scores are assigned to dairies that feed unpasteurized milk, do not test for bacteria or feed milk supplemented with replacer. The final two questions of the milk feeding section address the volume of milk fed and the daily feeding frequency. While calves fed greater than 3.79 L (4 US quarts) were assigned zero risk scores; the penalty for feeding less was modified by breed due to the differences in body weight and hence nutrient requirement differences between Holsteins and Jersey calves. Specifically, greater risk scores were assigned when Holstein calves were fed less than 3.79 L compared to if the same volume was fed to Jersey calves. The milk feeding section accounted for 24.4% of the total risk score.

#### 4.1.1.5 Vaccination

Despite the lower percent of the total risk scores assigned to the vaccination section (4.4%), targeted vaccination continues to be an important and essential practice to maintain healthy calves (Windeyer et al., 2012; Dubrovsky et al., 2019a; Maier et al., 2019a; Chamorro and Palomares, 2020). The questions in the vaccination section of the risk assessment address whether intranasal vaccines, injectable vaccines either killed or modified live vaccines are administered to calves and if dams are vaccinated. For the latter, higher risk scores are assigned if dams are not vaccinated against respiratory pathogens; however, the penalty is greater for the lack of vaccination using a modified live vaccine compared to the lack of vaccinating using a killed vaccine. The larger penalty for missing the use of a modified live vaccine compared to a killed vaccine relates to the greater reduction in the risk of respiratory disease associated with the use of the former compared to the latter.

#### 4.1.1.6 Housing

Questions related to calf housing management practices represented the largest section of all six by the sum of scores (35.9%). The questions enquired about dust, shade, calf-to-calf contact, housing enclosure type, material and manure removal method. With regard to dust the questions assigned higher risk scores if surfaces surrounding the calf housing were not gravel or paved, owner reports of dust in the calf raising area and for lack of dust abatement.

For the latter, the use of water trucks is assigned higher scores despite it being a common dust abatement method since frequently the same trucks travel at high speeds generating more dust rather than performing dust abatement (Dubrovsky et al., 2019a). Interestingly, the application of magnesium chloride on surfaces between and around calf hutches was found to prevent dust, a technique commonly used in equestrian sports to reduce dust in arenas, dairies that applied this substance received zero risk scores in this field.

With regard to shade, dairies that housed calves under an additional shade structure including the use of shade cloth were assigned zero risk scores. In contrast, dairies with a calf raising area with no roof, with one to three walls with a roof or completely indoors receive higher risk scores. In addition, dairies where calves are housed in groups are assigned a higher risk score compared to dairies that house calves individually. In addition, if calves 75 days old or more have calf-to-calf contact, a higher risk score is assigned compared to if calves of the same age had no calf-to-calf contact.

With regard to housing enclosure type, dairies that utilized calf hutch walls and roof made of a metal material, plastic or a mix of materials all received higher risk scores compared to dairies that house calves in wooden hutches. Dairies that housed calves on raised floors made of plastic-coated wire are also assigned higher risk scores compared to wood floors. Finally, the use of recycled lagoon water to flush manure under the hutches was assigned higher risk scores compared to those that did not flush or flushed with fresh water.

Once the risk assessment questionnaire is completed, investigators can sum the scores from each section's questions and over all the sections to arrive at the herd's risk for respiratory disease by section and overall, respectively. A dairy's risk score can then be benchmarked across other dairies' risk for respiratory disease by section (basic herd risk profile, maternity pen, colostrum management, milk feeding, vaccination and housing) into very low, low, moderate, high or very high risk for respiratory disease based on the percentile distribution of the risk scores (Fig. 4).

### 4.1.2 The California BRD scoring system

The California (CA) BRD scoring system is used to estimate the prevalence of respiratory disease as a reference point for the specific herd. The CA BRD scoring system has the advantage of being a quick and easy calf-side system that allows for the rapid assessment of calves with minimal handling (Love et al., 2014). The system's accuracy in terms of sensitivity and specificity, like other scoring systems and laboratory-based diagnostic systems, is not perfect (Love et al., 2016a). However, the CA BRD scoring system allows the investigator to compare the prevalence of BRD before and after implementing changes in management to control BRD. Prevalence estimation can be completed using

| Risk score section | Bovine Respiratory Disease risk score[1] percentiles[2] | | | | |
|---|---|---|---|---|---|
| | 10th percentile | 25th percentile | 50th percentile | 75th percentile | 90th percentile |
| **Basic Herd risk profile[3]** | 43 | 47 | 63 | 72 | 81 |
| **Maternity pen management[4]** | 24 | 46 | 65 | 82 | 89 |
| **Colostrum management[5]** | 41 | 58 | 71 | 82 | 96 |
| **Milk feeding[6]** | 52 | 79 | 108 | 142 | 155 |
| **Vaccinations[7]** | 0 | 3 | 7 | 27 | 31 |
| **Calf housing[8]** | 106 | 130 | 170 | 199 | 217 |
| **Total score** | 357 | 414 | 467 | 550 | 655 |

[1] Based on the BRD 100 (http://dx.doi.org/10.3168/jds.2018-14773) and BRD 10K studies (https://doi.org/10.3168/jds.2018-14774)
[2] Percentiles based on risk assessment scores for the BRD 100 study herds.
[3] Herd size, organic status, region, season, herd breed composition
[4] Bedding changes maternity pen separate for cows and heifers, bedding type, pasture calving, cows and heifers calve together, % calves removed from dam within an hour
[5] Source, supplementation with colostrum replacer, heat treatment, storage before heat treatment, testing for bacteria, storage container type, volume fed, testing for failure of transfer of passive immunity
[6] Feeding order by age, feeding order by health, source, pasteurization, volume fed, feeding frequency
[7] Intranasal for calves, injectable (killed or modified live) for calves, modified live for cows, killed for cows
[8] Type of surface next to hutches, dust perceived a problem, group housing, extra shade structures, calf-to-calf contact in calves >75 days, hutch wall material, hutch floor material, flush system for manure

**Figure 4** Heat map benchmarking bovine respiratory disease (BRD) risk scores. (Image reprinted by permission from Dr. Sharif S. Aly).

an application available for smartphones and tablets called the UC Davis BRD Score. The application is available in English, Spanish and Arabic languages with plans for translating it to more languages. The application can be downloaded from both iOS and Android respective application venues after a keyword search for UC Davis and BRD (Fig. 5).

The CA BRD score application can aid producers in scoring a calf herd. The application requires the calf herd size and then uses a sample size formula to calculate the number of calves required to estimate the prevalence of respiratory disease in the herd, with a bound on the error of estimation that is also user-specified. A user scores the calf in the hutch identified by the app, which also provides the number of hutches to count off from the first hutch on the farm. A similar adaptation can be utilized for calves housed in pens. A user proceeds to the next randomly selected hutch based on the app's output of how many hutches to skip. Between scoring calves, the application enquires whether the calf to score next is present in the hutch since hutches may be vacant as when a calf dies or is removed for other reasons. If the next hutch is empty, the application will instruct the user to proceed to the next hutch (Fig. 6). The application will internally adjust the vector of hutches to score such that the sample size is fulfilled close to but before the last hutch (N, the total number of calves in the herd).

A smart function in the application is that unchecked clinical signs are assumed absent, so users may select only the observed clinical signs by tapping on the example pictures. The app also features a searchable database for calves scored if necessary, using information entered by the user for the scored calves, such as date of scoring, herd, calf identification, breed, sex and age (Fig. 7).

**Figure 5** Smart phone application to estimate the prevalence of bovine respiratory disease (BRD) in preweaned dairy calves. (Image reprinted by permission from Dr. Sharif S. Aly).

Once the prevalence estimation is complete, users can email a prevalence report to the herd producer/herd veterinarian for integration into their risk management assessment (Fig. 8).

### *4.1.3 BRD control and prevention herd management plan*

It is based on the completed risk assessment questionnaire; however, the questions are reworded into interventions. The management plan consists of the same sections of the risk assessment questionnaire, with the exception of the basic herd profile section because herd profile factors are mostly static and difficult to change. For example, the first section of the herd management plan in the maternity pen section would be 'Calve cows and heifers separately' if the risk assessment questionnaire had identified that cows and heifers calved together and a risk score was assigned. Once all the interventions are reviewed in light of the risk assessment questionnaire selections, producers

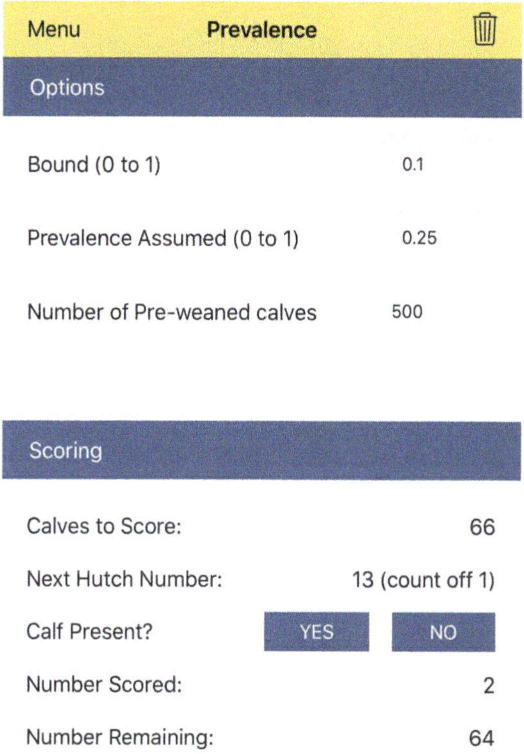


**Figure 6** Sample size inputs in the California bovine respiratory disease score application to estimate the prevalence of BRD in preweaned dairy calves. (Image reprinted by permission from Dr. Sharif S. Aly).

can then elect to apply any or all of these interventions finalizing their herd-specific management plan. Producers are advised to identify feasible changes in terms of their available staff effort and financial resources. The decision regarding which feasible interventions produce the greatest impact on the herd's respiratory disease risk should be based on the resulting reduction in the herd's risk score. For example, the investigator may preferentially implement interventions related to risk factors from one component over another because of a greater reduction in the overall risk score. Producers should also consider the distribution of respiratory cases by age, breed, calf sex and clinical signs, which may be easily identified using the CA BRD score application. Although the clinical signs for BRD are not specific, an assessment of frequent scores may provide clues to specific underlying causes, such as tilted heads or drooped ears, which are commonly associated with *M. bovis* infections. Similarly, the

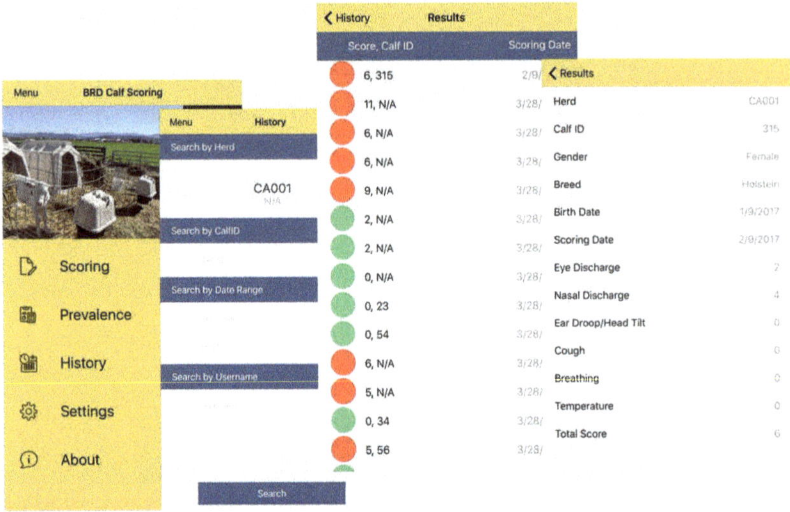


**Figure 7** Database search features in the California bovine respiratory disease score application to estimate the prevalence of BRD in preweaned dairy calves. (Image reprinted by permission from Dr. Sharif S. Aly).

age distribution of cases may help narrow down underlying management problems. For example, if many respiratory cases are seen in neonatal calves, one could investigate aspiration pneumonia commonly seen in the first few days of life if calves are tube fed by untrained staff. An increase in cases in one breed more than another, in mixed breed herds, may be related to breed-associated management differences such as the use of a specific hutch type or volume of milk fed for one breed but not another. Differences in respiratory disease between males and females may be confounded due to practices such as feeding female calves more colostrum than male calves.

Once the feasible changes are identified, the implementation plan should detail the staff in charge of each intervention and an implementation plan following a timetable with deadlines. Such details identify responsibilities and allow the calf caretakers, producer and investigator to work as a team to identify bottlenecks and hurdles, particularly if the intervention changes agreed upon do not take place. Producers can then re-estimate the prevalence of BRD after implementing the recommended management changes and elapse of a suitable period to allow for a new cohort of calves to have experienced these changes.

## 5 Future trends in research

It is imperative that we consider all factors of the disease triangle when forming disease prevention plans. Table 2 illustrates some examples of common BRD

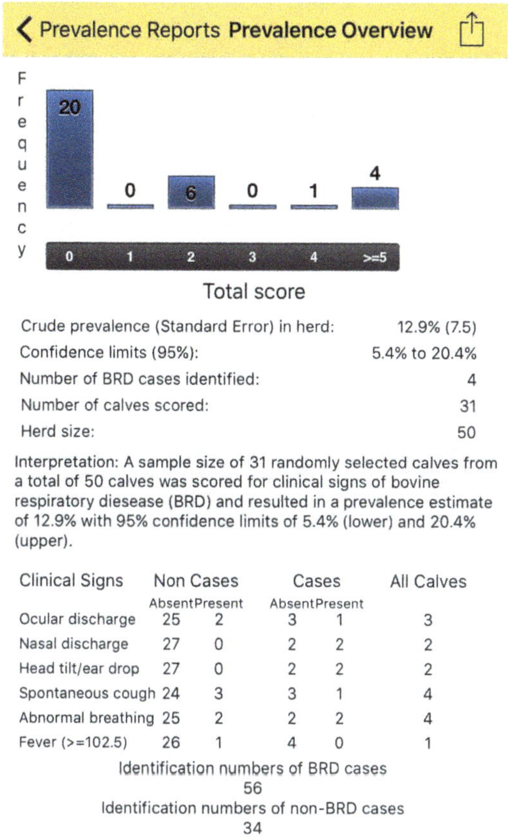


**Figure 8** Prevalence report feature in the California bovine respiratory disease score application to estimate the prevalence of BRD in preweaned dairy calves. (Image reprinted by permission from Dr. Sharif S. Aly).

modifying factors with the related host, environment and pathogen factors and basic intervention recommendations. As agricultural practices evolve so do the disease syndromes of livestock. The assumptions used to create the recommendations within this chapter will likely change over time as modern agriculture continues to evolve. Animal health interventions must remain dynamic and responsive to the state of animal agriculture; interventions must be constantly measured, monitored for success and re-evaluated. As animal health professionals, we not only react to animal disease events but must be proactive to influence change in animal management standards.

**Table 2** Summary of common BRD modifiers and the related host, environment and pathogen factors with feasible intervention recommendations

| Modifier | Host, environment, pathogen factors: Host | Environment | Pathogen | Recommended action | References |
|---|---|---|---|---|---|
| Physiologic and psychological stress | • Stimulation of the HPA axis and SNS<br>• Blunted immune responses | • Shipping, processing environment<br>• Other human handling<br>• Changes in animal groups | • Shifts in respiratory microbial commensal population<br>• Possible increases in pathogen virulence | • Minimize physical stressors (processing, shipping, etc.)<br>• Minimize psychological stressors such as social group changes and stressful handling<br>• Acclimate calves to human handling | Carroll and Forsberg (2007), Mansfield et al. (2008), Ackermann et al. (2010), Lyte (2014b) |
| Heat stress | • Increased energy use<br>• Decreased feed intake<br>• Inflammation<br>• Impaired respiratory immunity | • High ambient temperatures or temperature humidity index<br>• Fluctuations in temperature | | • Pay attention to both macro and micro environment of animal housing (ex: ambient and temperature within calf hutches)<br>• Employ heat abatement techniques such as increasing shade | Jones (1987), Lammers et al. (1996), Mitlöhner et al. (2002); Burgos et al. (2010), Bertagnon et al. (2011), Pearce et al. (2013a,b), Roland et al. (2016), Kim et al. (2018), Louie et al. (2018), Koch et al. (2019b) |

| | | | | | |
|---|---|---|---|---|---|
| Housing | | | | | |
| • Maternity bedding/ hygiene | Increased odds of BRD suspect due to high pathogen load in neonatal period | • Dirty maternity bedding<br>• Lagoon water use to flush under hutches | High pathogen load | • Change maternity bedding >5 times a month<br>• Avoid dirt or plant material bedding | Dubrovsky et al. (2019a, b), Karle et al. (2019), Maier et al. (2020), Karle et al. (2019), Maier et al. (2020) |
| • Hutch construction | Increased odds of BRD, suspect due to heat stress | Metal hutch components suspect increase heat conduction | | • Avoid metal components in hutches in warm climates | Dubrovsky et al. (2019b), Karle et al. (2019), Maier et al. (2020) |
| • Dust | Increased risk of BRD likely due to the effect of particulate matter and inhaled endotoxin on respiratory anatomy | Dusty environment, traffic on dusty roads near animal housing | Potential for exposure to bacteria | • Employ dust abatement; ensure truck movement does not create dust near animal housing | Purdy et al. (2002), Dubrovsky et al. (2019a), Karle et al. (2019) |
| • Commingling | Increased contact between calves | Shared microbiome | Exposure to respiratory pathogens | Individual housing, reduce the density in group pens, vaccination against BRD pathogens before commingling | Abdelfattah et al. (2013), Karle et al. (2019), Maier et al. (2019a, 2020) |
| Nutrition | • Feeding below nutritional requirements is associated with immune dysfunction and an increased risk of BRD<br>• Feeding whole milk (vs milk replacer) is associated with lower risk of BRD | • Heat and cold stress increases nutrient requirements | • Increase in nutrient requirements as the immune system is challenged by infectious agents | • Feed calves optimum diet in terms of composition and nutrient requirements | Griebel et al. (1987), Godden et al. (2005), Drackley (2008), Dubrovsky et al. (2019a,b), Karle et al. (2019), Maier et al. (2019b, 2020) |

(Continued)

**Table 2** (*Continued*)

| | Host, environment, pathogen factors | | | | |
|---|---|---|---|---|---|
| Modifier | Host | Environment | Pathogen | Recommended action | References |
| Vaccination | Build host adaptive immunity to specific pathogens | Possible decrease in environmental/group pathogen load due to herd immunity | Pathogenicity altered or interrupted by host adaptive immunity | Vaccinate healthy animals for specific pathogens based on farm risk. Adjust vaccine timing to precede risk period | Larson and Step (2012), Chamorro and Palomares (2020), Richeson and Falkner (2020) |
| Antibiotic resistance | Treatment failures | Contamination of the environment with resistance genes | Decreased efficacy of bacterial killing with antibiotic treatment | Implement an antibiotic use stewardship plan that includes: clear staff training materials for appropriate drug use – review and update regularly using evidence-based medicine for drug selection; clear record-keeping for monitoring drug use and case outcomes; implement antibiotic susceptibility testing and monitoring in representative samples; employ disease prevention methods and limit antibiotic use; consider non-antibiotic treatments where possible | AVMA, Mathew et al. (2007); AABP (2017); Mathew et al. (2007) |

## 6 Where to look for further information

The disease prevention and control concepts outlined in this chapter are presented using BRD as an example and culminate in the form of a herd risk assessment tool. The BRD tool is available online at:

- https://escholarship.org/uc/item/1jb2f7rm.

The approach presented here is relevant to, and has been implemented for, prevention and control of other diseases. For example, the control and prevention efforts for paratuberculosis in cattle, which make up the Johne's disease risk assessment tool. The Johne's disease risk assessment tool precedes the BRD risk assessment and is available in the form of separate tools for both beef and dairy cattle production systems. The risk assessment for Johne's disease is available for beef production systems here:

- https://johnes.org/wp-content/uploads/2018/11/Handbook-for-Vets-and-Beef-Producers-4th-ed-2011.pdf.

Similarly, the risk assessment for Johne's disease is available for dairy cattle here:

- https://johnes.org/wp-content/uploads/2018/11/Handbook-for-Vets-and-Dairy-Producers-4th-ed-2011.pdf.

Even with a successful prevention and control program, bacterial infections still occur. When treating and managing bacterial infections in cattle, it is important to use up-to-date veterinary practices and follow local and regional antimicrobial drug use regulations. A helpful resource for residue avoidance for antimicrobial drugs and other medications in food animals is the Food Animal Drug Residue Avoidance Databank (FARAD): http://www.farad.org/.

Antimicrobial stewardship and residue avoidance practices are important to prevent antimicrobial resistance. Antimicrobial resistance is of global concern and appropriate antimicrobial drug use is essential to preserve the effectiveness of these valuable chemicals for the health and welfare of our food animals and public health. Information on the threat of antimicrobial resistance can be found in the Centers for Disease Control and Prevention, and the World Health Organization reports. As of this chapter's publication date the aforementioned reports can be accessed at:

- https://www.cdc.gov/drugresistance/pdf/threats-report/2019-ar-threats-report-508.pdf.
- https://www.who.int/publications/i/item/global-action-plan-on-antimicrobial-resistance.

## 7 References

AABP 2017. Key elements for implementing antimicrobial stewardship plans in bovine veterinary practices working with beef and dairy operations. Available online: https://www.aabp.org/resources/AABP_Guidelines/AntimicrobialStewardship-7.27.17.pdf.

Abdelfattah, E. M., Schutz, M. M., Lay, D. C., Marchant-Forde, J. N. and Eicher, S. D. 2013. Effect of group size on behavior, health, production, and welfare of veal calves. *J. Anim. Sci.* 91(11):5455–5465. doi:10.2527/jas.2013-6308.

Ackermann, M. R., Derscheid, R. and Roth, J. A. 2010. Innate immunology of bovine respiratory disease. *Vet. Clin. North Am. Food Anim. Pract.* 26(2):215–228. doi:10.1016/j.cvfa.2010.03.001.

Adams, E. A. and Buczinski, S. 2016. Short communication: ultrasonographic assessment of lung consolidation postweaning and survival to the first lactation in dairy heifers. *J. Dairy Science* 99(2):1465–1470. doi:10.3168/jds.2015-10260.

Adlam, C. and Rutter, J. 1989. *Pasteurella and Pasteurellosis*.

Allen, J. W., Viel, L., Bateman, K. G., Rosendal, S., Shewen, P. E. and Physick-Sheard, P. 1991. The microbial flora of the respiratory tract in feedlot calves: associations between nasopharyngeal and bronchoalveolar lavage cultures. *Can. J. Vet. Res.* 55(4):341–346.

Aly, S. S., Karle, B. M., Williams, D. R., Maier, G. U. and Dubrovsky, S. 2020. Components of a risk assessment tool for prevention and control of bovine respiratory disease in preweaned dairy calves. *Anim Health Res Rev*:1–7. https://doi.org/10.1017/S1466252320000201.

Aly, S. S. and Thurmond, M. C. 2005. Evaluation of *Mycobacterium avium* subsp *paratuberculosis* infection of dairy cows attributable to infection status of the dam. *J. Am. Vet. Med. Assoc.* 227(3):450–454.

AVMA 2020. Antimicrobial Stewardship Definition and Core Principles | American Veterinary Medical Association. Accessed August 10, 2020. https://www.avma.org/resources-tools/avma-policies/antimicrobial-stewardship-definition-and-core-pri nciples.

Baumgard, L. H., Wheelock, J. B., Sanders, S. R., Moore, C. E., Green, H. B., Waldron, M. R. and Rhoads, R. P. 2011. Postabsorptive carbohydrate adaptations to heat stress and monensin supplementation in lactating Holstein cows1. *J. Dairy Sci.* 94(11):5620–5633. doi:10.3168/jds.2011-4462.

Beier, T., Hotzel, H., Lysnyansky, I., Grajetzki, C., Heller, M., Rabeling, B., Yogev, D. and Sachse, K. 1998. Intraspecies polymorphism of vsp genes and expression profiles of variable surface protein antigens (Vsps) in field isolates of *Mycoplasma bovis*. *Vet. Microbiol.* 63(2–4):189–203. doi:10.1016/S0378-1135(98)00238-7.

Bertagnon, H. G., Esper, G. V. Z., Emanuelli, M. P. and Pel-legrine, L. G. 2011. Nfluência meteorológica no leucograma e na popula-ção citológica do trato respiratório de bezerros. *Pesq. Vet. Bras.* 31(3):244–246.

Blecha, F., Boyles, S. L. and Riley, J. G. 1984. Shipping suppresses lymphocyte blastogenic responses in Angus and Brahman X Angus feeder calves. *J. Anim. Sci.* 59(3):576–583. doi:10.2527/jas1984.593576x.

Boyce, J. D. and Adler, B. 2000. The Capsule is a Virulence Determinant in the Pathogenesis of *Pasteurella multocida* M1404 (B:2). https://doi.org/10.1128/IAI.68.6.3463-3468.2000.

Buchenau, I., Poumarat, F., Le Grand, D., Linkner, H., Rosengarten, R., Hewicker-Trautwein, M. 2010. Expression of *Mycoplasma bovis* variable surface membrane proteins in the respiratory tract of calves after experimental infection with a clonal variant of *Mycoplasma bovis* type strain PG45. *Res. Vet. Sci.* 89:223–229.

Buczinski, S., Forté, G., Francoz, D. and Bélanger, A. M. 2014. Comparison of thoracic auscultation, clinical score, and ultrasonography as indicators of bovine respiratory disease in preweaned dairy calves. *J. Vet. Intern. Med.* 28(1):234–242. doi:10.1111/jvim.12251.

Burgos, S. A., Embertson, N. M., Zhao, Y., Mitloehner, F. M., DePeters, E. J. and Fadel, J. G. 2010. Prediction of ammonia emission from dairy cattle manure based on milk urea nitrogen: relation of milk urea nitrogen to ammonia emissions. *J. Dairy Sci.* 93(6):2377–2386. doi:10.3168/jds.2009-2415.

Calf Health Module–The Dairyland Initiative 2020. Accessed July 31, 2020. https://thedairylandinitiative.vetmed.wisc.edu/home/calf-health-module/.

Callan, R. J. and Garry, F. B. 2002. Biosecurity and bovine respiratory disease. *Vet. Clin. North Am. Food Anim. Pract.* 18(1):57–77. doi:10.1016/S0749-0720(02)00004-X.

Carroll, J. A. and Forsberg, N. E. 2007. Influence of stress and nutrition on cattle immunity. *Vet. Clin. North Am. Food Anim. Pract.* 23(1):105–149. doi:10.1016/j.cvfa.2007.01.003.

Carter, B. H., Friend, T. H., Garey, S. M., Sawyer, J. A., Alexander, M. B. and Tomazewski, M. A. 2014. Efficacy of reflective insulation in reducing heat stress on dairy calves housed in polyethylene calf hutches. *Int. J. Biometeorol.* 58(1):51–59.

Casas, E., Garcia, M. D., Wells, J. E. and Smith, T. P. L. 2011. Association of single nucleotide polymorphisms in the ANKRA2 and CD180 genes with bovine respiratory disease and presence of *Mycobacterium avium* subsp. *paratuberculosis*(1). *Anim. Genet.* 42(6):571–577. doi:10.1111/j.1365-2052.2011.02189.x.

Caswell, J. L. 2014. Failure of respiratory defenses in the pathogenesis of bacterial pneumonia of cattle. *Vet. Pathol.* 51(2):393–409. doi:10.1177/0300985813502821.

CDC 2019. Antibiotic resistance threats in the United States. *Centers Dis. Control. Prev.*:1–150. https://www.cdc.gov/drugresistance/pdf/threats-report/2019-ar-threats-report-508.pdf.

Chamorro, M. F. and Palomares, R. A. 2020. Bovine respiratory disease vaccination against viral pathogens: modified-live versus inactivated antigen vaccines, intranasal versus parenteral, what is the evidence? *Vet. Clin. North Am. Food Anim. Pract.* 36(2):461–472. doi:10.1016/j.cvfa.2020.03.006.

Chigerwe, M., Hagey, J. V. and Aly, S. S. 2015. Determination of neonatal serum immunoglobulin G concentrations associated with mortality during the first 4 months of life in dairy heifer calves. *J. Dairy Res.* 82(4):400–406. doi:10.1017/S0022029915000503.

Chigerwe, M., Tyler, J. W., Schultz, L. G., Middleton, J. R., Steevens, B. J., Spain, J. N. 2008. Effect of colostrum administration by use of oroesophageal intubation on serum IgG concentrations in Holstein bull calves. *Am. J. Vet. Res.* 69(9):1158–1163. doi:10.2460/ajvr.69.9.1158.

Chirino-Trejo, J. M. and Prescott, J. F. 1983. The identification and antimicrobial susceptibility of anaerobic bacteria from pneumonic cattle lungs. *Can. J. Comp. Med.* 47(3):270–275.

CLSI 2018. *Performance Standards for Antimicrobial Disk and Dilution Susceptibility Tests for Bacteria Isolated From Animals; Approved Standard–Fourth Edition*. CLSI document VET01-A4. Wayne, PA: Clinical and Laboratory Standards Institute.

Coetzee, J. F., Magstadt, D. R., Sidhu, P. K., Follett, L., Schuler, A. M., Krull, A. C., Cooper, V. L., Engelken, T. J., Kleinhenz, M. D. and O'Connor, A. M. 2019. Association between antimicrobial drug class for treatment and retreatment of bovine respiratory disease (BRD) and frequency of resistant BRD pathogen isolation from veterinary diagnostic laboratory samples. *PLoS ONE* 14(12):e0219104. doi:10.1371/journal.pone.0219104.

Confer, A. W. and Ayalew, S. 2018. Mannheimia haemolytica in bovine respiratory disease: immunogens, potential immunogens, and vaccines. *Anim. Heal. Res. Rev.* 19(2):79–99. doi:10.1017/S1466252318000142.

Constable, P. D., Hinchcliff, K. W., Done, S. H. and Grünberg, W. 2016. *Veterinary Medicine*. 11th ed. Elsevier.

Corbeil, L. B. 1996. *Histophilus somni* host-parasite relationships. *Anim. Heal. Res. Rev.* 8:151–160. doi:10.1017/S1466252307001417.

Dabo, S. M., Taylor, J. D. and Confer, A. W. 2008. *Pasteurella multocida* and bovine respiratory disease. *Anim. Heal. Res. Rev.* 8:129–150. doi:10.1017/S1466252307001399.

Davis, C. L. and Drackley, J. K. 1998. *The Development, Nutrition, and Management of the Young Calf*. Ames, IA: Iowa State University Press.

DeRosa, D. C., Mechor, G. D., Staats, J. J., Chengappa, M. M. and Shryock, T. R. 2000. Comparison of *Pasteurella* spp. simultaneously isolated from nasal and transtracheal swabs from cattle with clinical signs of bovine respiratory disease. *J. Clin. Microbiol.* 38(1):327–332.

Downey, E. D., Tait, R. G., Jr, Mayes, M. S., Park, C. A., Ridpath, J. F., Garrick, D. J. and Reecy, J. M. 2013. An evaluation of circulating bovine viral diarrhea virus type 2 maternal antibody level and response to vaccination in Angus calves. *J. Anim. Sci.* 91(9):4440–4450.

Doyle, D., Credille, B., Lehenbauer, T. W., Berghaus, R., Aly, S. S., Champagne, J., Blanchard, P., Crossley, B., Berghaus, L., Cochran, S. and Woolums, A. 2017. Agreement among 4 sampling methods to identify respiratory pathogens in dairy calves with acute bovine respiratory disease. *J. Vet. Intern. Med.* 31(3):954–959. doi:10.1111/jvim.14683.

Drackley, J. K. 2008. Calf nutrition from birth to breeding. *Vet. Clin. North Am. Food Anim. Pract.* 24(1):55–86. doi:10.1016/j.cvfa.2008.01.001.

Dubrovsky, S. A., Van Eenennaam, A. L., Karle, B. M., Rossitto, P. V., Lehenbauer, T. W. and Aly, S. S. 2019a. Epidemiology of bovine respiratory disease (BRD) in preweaned calves on California dairies: the BRD 10K study. *J. Dairy Sci.* 102(8):7306–7319. doi:10.3168/jds.2018-14774.

Dubrovsky, S. A., Van Eenennaam, A. L., Karle, B. M., Rossitto, P. V., Lehenbauer, T. W. and Aly, S. S. 2019b. Bovine respiratory disease (BRD) cause-specific and overall mortality in preweaned calves on California dairies: the BRD 10K study. *J. Dairy Sci.* 102(8):7320–7328 doi:10.3168/jds.2018-15463.

Dunn, T. R., Ollivett, T. L., Renaud, D. L., Leslie, K. E., LeBlanc, S. J., Duffield, T. F. and Kelton, D. F. 2018. The effect of lung consolidation, as determined by ultrasonography, on first-lactation milk production in Holstein dairy calves. *J. Dairy Sci.* 101(6):5404–5410. doi:10.3168/jds.2017-13870.

Ellis, J., Gow, S., Bolton, M., Burdett, W. and Nordstrom, S. 2014. Inhibition of priming for bovine respiratory syncytial virus-specific protective immune responses following parenteral vaccination of passively immune calves. *Can. Vet. J.* 55(12):1180-1185.

Ellis, J., West, K., Cortese, V., Konoby, C. and Weigel, D. 2001. Effect of maternal antibodies on induction and persistence of vaccine-induced immune responses against bovine viral diarrhea virus type II in young calves. *J. Am. Vet. Med. Assoc.* 219(3):351-356. doi:10.2460/javma.2001.219.351.

Francoz, D., Buczinski, S., Bélanger, A. M., Forté, G., Labrecque, O., Tremblay, D., Wellemans, V. and Dubuc, J. 2015. Respiratory pathogens in Québec dairy calves and their relationship with clinical status, lung consolidation, and average daily gain. *J. Vet. Intern. Med.* 29(1):381-387. doi:10.1111/jvim.12531.

Gaeta, N. C., Lima, S. F., Teixeira, A. G., Ganda, E. K., Oikonomou, G., Gregory, L. and Bicalho, R. C. 2017. Deciphering upper respiratory tract microbiota complexity in healthy calves and calves that develop respiratory disease using shotgun metagenomics. *J. Dairy Sci.* 100(2):1445-1458. doi:10.3168/jds.2016-11522.

Galyean, M. L., Perino, L. J. and Duff, G. C. 1999. Interaction of Cattle Health/Immunity and Nutrition. *Journal of Animal Science,* 77(5): 1120-1134. https://doi.org/10.2527/1999.7751120x

Gardner, I. A., Wong, S. J., Ferraro, G. L., Balasuriya, U. B., Hullinger, P. J., Wilson, W. D., Shi, P. Y. and MacLachlan, N. J. 2007. Incidence and effects of West Nile virus infection in vaccinated and unvaccinated horses in California. *Vet. Res.* 38(1):109-116. doi:10.1051/vetres:2006045.

Gerber, J. D., Marron, A. E. and Kucera, C. J. 1978. Local and systemic cellular and antibody immune responses of cattle to infectious bovine rhinotracheitis virus vaccines administered intranassally or intramuscularly. *Am. J. Vet. Res.* 39(5):753-760.

Gioia, J., Qin, X., Jiang, H., Clinkenbeard, K., Lo, R., Liu, Y., Fox, G. E., Yerrapragada, S., McLeod, M. P., McNeill, T. Z., Hemphill, L., Sodergren, E., Wang, Q., Muzny, D. M., Homsi, F. J., Weinstock, G. M. and Highlander, S. K. 2006. The genome sequence of Mannheimia haemolytica A1: insights into virulence, natural competence, and Pasteurellaceae phylogeny. *J. Bacteriol.* 188(20):7257-7266. doi:10.1128/JB.00675-06.

Glass, E. J., Baxter, R., Leach, R. J. and Jann, O. C. 2012. Genes controlling vaccine responses and disease resistance to respiratory viral pathogens in cattle. *Vet. Immunol. Immunopathol.* 148(1-2):90-99. doi:10.1016/j.vetimm.2011.05.009.

Godden, S. M., Fetrow, J. P., Feirtag, J. M., Green, L. R. and Wells, S. J. 2005. Economic analysis of feeding pasteurized nonsaleable milk versus conventional milk replacer to dairy calves. *J. Am. Vet. Med. Assoc.* 226(9):1547-1554. doi:10.2460/javma.2005.226.1547.

Godden, S. M., Smolenski, D. J., Donahue, M., Oakes, J. M., Bey, R., Wells, S., Sreevatsan, S., Stabel, J. and Fetrow, J. 2012. Heat-treated colostrum and reduced morbidity in preweaned dairy calves: results of a randomized trial and examination of mechanisms of effectiveness. *J. Dairy Sci.* 95(7):4029-4040. doi:10.3168/jds.2011-5275.

Godden, S. M. M., Haines, D. M. M., Konkol, K. and Peterson, J. 2009. Improving passive transfer of immunoglobulins in calves. II: Interaction between feeding method and volume of colostrum fed. *J. Dairy Sci.* 92(4):1758-1764. doi:10.3168/jds.2008-1847.

Godinho, K. S., Sarasola, P., Renoult, E., Tilt, N., Keane, S., Windsor, G. D., Rowan, T. G. and Sunderland, S. J. 2007. Use of deep nasopharyngeal swabs as a predictive diagnostic method for natural respiratory infections in calves. *Vet. Rec.* 160(1):22-25. doi:10.1136/vr.160.1.22.

Gorden, P. J. and Plummer, P. 2010. Control, management, and prevention of bovine respiratory disease in dairy calves and cows. *Vet. Clin. North Am. Food Anim. Pract.* 26(2):243-259. doi:10.1016/j.cvfa.2010.03.004.

Griebel, P. J., Schoonderwoerd, M. and Babiuk, L. A. 1987. Ontogeny of the immune response: effect of protein energy malnutrition in neonatal calves. *Can. J. Vet. Res.* 51(4):428-435.

Griffin, D., Chengappa, M. M. M., Kuszak, J. and McVey, D. S. 2010. Bacterial pathogens of the bovine respiratory disease complex. *Vet. Clin. North Am. Food Anim. Pract.* 26(2):381-394. doi:10.1016/j.cvfa.2010.04.004.

Grooms, D. L., Brock, K. V. and Ward, L. A. 1998. Detection of cytopathic bovine viral diarrhea virus in the ovaries of cattle following immunization with a modified live bovine viral diarrhea virus vaccine. *J. Vet. Diagn. Investig.* 10(2):130-134. doi:10.1177/104063879801000202.

Guo, Y., McMullen, C., Timsit, E., Hallewell, J., Orsel, K., van der Meer, F., Yan, S. and Alexander, T. W. 2020. Genetic relatedness and antimicrobial resistance in respiratory bacteria from beef calves sampled from spring processing to 40 days after feedlot entry. *Vet. Microbiol.* 240:108478. doi:10.1016/j.vetmic.2019.108478.

Hall, J. A., Isaiah, A., Estill, C. T., Pirelli, G. J. and Suchodolski, J. S. 2017. Weaned beef calves fed selenium-biofortified alfalfa hay have an enriched nasal microbiota compared with healthy controls. *PLoS ONE* 12(6):e0179215. doi:10.1371/journal.pone.0179215.

Hammer, C. J., Quigley, J. D., Ribeiro, L. and Tyler, H. D. 2004. Characterization of a colostrum replacer and a colostrum supplement containing IgG concentrate and growth factors. *J. Dairy Sci.* 87(1):106-111. doi:10.3168/jds.S0022-0302(04)73147-1.

Hanthorn, C. J., Dewell, R. D., Cooper, V. L., Frana, T. S., Plummer, P. J., Wang, C. and Dewell, G. A. 2014. Randomized clinical trial to evaluate the pathogenicity of *Bibersteinia trehalosi* in respiratory disease among calves. *BMC Vet. Res.* 10:89. doi:10.1186/1746-6148-10-89.

Hause, B. M., Ducatez, M., Collin, E. A., Ran, Z., Liu, R., Sheng, Z., Armien, A., Kaplan, B., Chakravarty, S., Hoppe, A. D., Webby, R. J., Simonson, R. R. and Li, F. 2013. Isolation of a novel swine influenza virus from Oklahoma in 2011 which is distantly related to human influenza C viruses. *PLoS Pathog.* 9(2):e1003176. doi:10.1371/journal.ppat.1003176.

Hillman, P., Gebremedhin, K. and Warner, R. 2010. Ventilation system to minimize airborne bacteria, Dust, humidity, and ammonia in calf nurseries. *J. Dairy Sci.* 75(5):1305-1312. doi:10.3168/jds.S0022-0302(92)77881-3.

Hoff, J. L., Decker, J. E., Schnabel, R. D., Seabury, C. M., Neibergs, H. L. and Taylor, J. F. 2019. QTL-mapping and genomic prediction for bovine respiratory disease in U.S. Holsteins using sequence imputation and feature selection. *BMC Genomics* 20(1):555. doi:10.1186/s12864-019-5941-5.

Holman, D. B., Timsit, E. and Alexander, T. W. 2015. The nasopharyngeal microbiota of feedlot cattle. *Sci. Rep.* 5:15557. doi:10.1038/srep15557.

Holman, D. B., Timsit, E., Amat, S., Abbott, D. W., Buret, A. G. and Alexander, T. W. 2017. The nasopharyngeal microbiota of beef cattle before and after transport to a feedlot. *BMC Microbiol* 17(1):70. doi:10.1186/s12866-017-0978-6.

Holman, D. B., Timsit, E., Booker, C. W. and Alexander, T. W. 2018. Injectable antimicrobials in commercial feedlot cattle and their effect on the nasopharyngeal microbiota and antimicrobial resistance. *Vet. Microbiol.* 214:140–147. doi:10.1016/j.vetmic.2017.12.015.

Holman, D. B., Yang, W. and Alexander, T. W. 2019. Antibiotic treatment in feedlot cattle: a longitudinal study of the effect of oxytetracycline and tulathromycin on the fecal and nasopharyngeal microbiota. *Microbiome* 7(1):86. doi:10.1186/s40168-019-0696-4.

Howard, C. J. and Taylor, G. 1983. Interaction of mycoplasmas and phagocytes. *Yale J. Biol. Med.* 56(5–6):643–648.

Howard, C. J., Thomas, L. H. and Parsons, K. R. 1987. Comparative pathogenicity of *Mycoplasma bovis* and *Mycoplasma* dispar for the respiratory tract of calves. *Isr. J. Med. Sci.* 23(6):621–624.

Ilg, T. 2017. Investigations on the molecular mode of action of the novel immunostimulator ZelNate: activation of the cGAS-STING pathway in mammalian cells. *Mol. Immunol.* 90:182–189. doi:10.1016/j.molimm.2017.07.013.

Johnston, D., Earley, B., Cormican, P., Murray, G., Kenny, D. A., Waters, S. M., McGee, M., Kelly, A. K. and McCabe, M. S. 2017. Illumina MiSeq 16S amplicon sequence analysis of bovine respiratory disease associated bacteria in lung and mediastinal lymph node tissue. *BMC Vet. Res.* 13(1):118. doi:10.1186/s12917-017-1035-2.

Jones, C. D. 1987. Proliferation of Pasteurella haemolytica in the calf respiratory tract after an abrupt change in climate. *Res. Vet. Sci.* 42(2):179–186.

Jericho, K. W., Lejeune, A. and Tiffin, G. B. 1986. Bovine herpesvirus-1 and *Pasteurella haemolytica* aerobiology in experimentally infected calves. *Am. J. Vet. Res.* 47(2):205–209.

Kainer, R. A. and Will, D. A. 1981. Morphophysiologic bases for the predisposition of the bovine lung to bronchial pneumonia. *Prog. Clin. Biol. Res.* 59B:311–317.

Kanis, J. A., Johnell, O., Oden, A., Johansson, H. and McCloskey, E. 2008. FRAX and the assessment of fracture probability in men and women from the UK. *Osteoporos Int* 19(4):385–397 doi:10.1007/s00198-007-0543-5.

Karle, B. M., Maier, G. U., Love, W. J., Dubrovsky, S. A., Williams, D. R., Anderson, R. J., Van Eenennaam, A. L., Lehenbauer, T. W. and Aly, S. S. 2019. Regional management practices and prevalence of bovine respiratory disease in California's preweaned dairy calves. *J. Dairy Sci.* 102(8):7583–7596. doi:10.3168/jds.2018-14775.

Katsuda, K., Kamiyama, M., Kohmoto, M., Kawashima, K., Tsunemitsu, H. and Eguchi, M. 2008. Serotyping of *Mannheimia haemolytica* isolates from bovine pneumonia: 1987–2006. *Vet. J.* 178(1):146–148. doi:10.1016/j.tvjl.2007.07.019.

Kim, W. S., Lee, J. S., Jeon, S. W., Peng, D. Q., Kim, Y. S., Bae, M. H., Jo, Y. H. and Lee, H. G. 2018. Correlation between blood, physiological and behavioral parameters in beef calves under heat stress. *Asian-Australas. J. Anim. Sci.* 31(6):919–925. doi:10.5713/ajas.17.0545.

Klima, C. L., Holman, D. B., Ralston, B. J., Stanford, K., Zaheer, R., Alexander, T. W. and McAllister, T. A. 2019. Lower respiratory tract microbiome and resistome of bovine respiratory disease mortalities. *Microb. Ecol.* 78(2):446–456. doi:10.1007/s00248-019-01361-3.

Knight-Jones, T. J. D., Edmond, K., Gubbins, S. and Paton, D. J. 2014. Veterinary and human vaccine evaluation methods. *Proc Biol Sci* 281(1784):20132839. doi:10.1098/rspb.2013.2839.

Koch, F., Thom, U., Albrecht, E., Weikard, R., Nolte, W., Kuhla, B. and Kuehn, C. 2019. Heat stress directly impairs gut integrity and recruits distinct immune cell populations into the bovine intestine. *Proc. Natl. Acad. Sci.* 116(21):10333–10338. doi:10.1073/PNAS.1820130116.

Kodjo, A., Villard, L., Bizet, C., Martel, J.-L., Sanchis, R., Borges, E., Gauthier, D., Oise Maurin, F. and Richard, Y. 1999. Pulsed-field gel electrophoresis is more efficient than ribotyping and random amplified polymorphic DNA analysis in discrimination of *Pasteurella haemolytica* strains. *J. Clin. Microbiol.* 37(2):380–385.

Lammers, B. P., vanKoot, J. W., Heinrichs, A. J. and Graves, R. E. 1996. The effect of plywood and polyethylene calf hutches on heat stress. *Appl. Eng. Agric.* 12(6):741–745.

Larson, R. L. and Step, D. L. 2012. Evidence-based effectiveness of vaccination against *Mannheimia haemolytica*, *Pasteurella multocida*, and *Histophilus somni* in feedlot cattle for mitigating the incidence and effect of bovine respiratory disease complex. *Vet. Clin. North Am. Food Anim. Pract.* 28(1):97–106, 106e1. doi:10.1016/j.cvfa.2011.12.005.

Le Grand, D., Solsona, M., Rosengarten, R. and Poumarat, F. 1996. Adaptive surface antigen variation in *Mycoplasma bovis* to the host immune response. *FEMS Microbiol. Lett.* 144(2–3):267–275. doi:10.1111/j.1574-6968.1996.tb08540.x.

Lima, S. F., de Souza Bicalho, M. L. S. and Bicalho, R. C. 2019. The *Bos taurus* maternal microbiome: role in determining the progeny early-life upper respiratory tract microbiome and health. *PLoS ONE* 14(3):e0208014. doi:10.1371/journal.pone.0208014.

Lima, S. F., Teixeira, A. G. V., Higgins, C. H., Lima, F. S. and Bicalho, R. C. 2016. The upper respiratory tract microbiome and its potential role in bovine respiratory disease and otitis media. *Sci. Rep.* 6:29050. doi:10.1038/srep29050.

Lippolis, K. D., Cooke, R. F., Schubach, K. M., Brandão, A. P., da Silva, L. G. T., Marques, R. S. and Bohnert, D. W. 2016. Altering the time of vaccination against respiratory pathogens to enhance antibody response and performance of feeder cattle. *J. Anim. Sci.* 94(9):3987–3995. doi:10.2527/jas.2016-0673.

Lombard, J., Urie, N., Garry, F., Godden, S., Quigley, J., Earleywine, T., McGuirk, S., Moore, D., Branan, M., Chamorro, M., Smith, G., Shivley, C., Catherman, D., Haines, D., Heinrichs, A. J., James, R., Maas, J. and Sterner, K. 2020. Consensus recommendations on calf- and herd-level passive immunity in dairy calves in the United States. *J. Dairy Sci.* 103(8):7611–7624. doi:10.3168/jds.2019-17955.

Louie, A. P., Rowe, J. D., Love, W. J., Lehenbauer, T. W. and Aly, S. S. 2018. Effect of the environment on the risk of respiratory disease in preweaning dairy calves during summer months. *J. Dairy Sci.* 101(11):10230–10247. doi:10.3168/jds.2017-13716.

Love, W. J., Lehenbauer, T. W., Van Eenennaam, A. L., Drake, C. M., Kass, P. H., Farver, T. B. and Aly, S. S. 2016a. Sensitivity and specificity of on-farm scoring systems and nasal culture to detect bovine respiratory disease complex in preweaned dairy calves. *J. Vet. Diagn. Investig.* 28(2):119–128. doi:10.1177/1040638715626204.

Love, W. J., Lehenbauer, T. W., Karle, B. M., Hulbert, L. E., Anderson, R. J., Van Eenennaam, A. L., Farver, T. B. and Aly, S. S. 2016b. Survey of management practices related to bovine respiratory disease in preweaned calves on California dairies. *J. Dairy Sci.* 99(2):1483–1494. doi:10.3168/jds.2015-9394.

Love, W. J., Lehenbauer, T. W., Kass, P. H., Van Eenennaam, A. L. and Aly, S. S. 2014. Development of a novel clinical scoring system for on-farmdiagnosis of bovine respiratory disease in pre-weaned dairy calves. *PeerJ* 2:e238. doi:10.7717/peerj.238.

Lysnyansky, I., Sachse, K., Rosenbusch, R., Levisohn, S. and Yogev, D. 2020. The vsp Locus of *Mycoplasma Bovis*: Gene Organization and Structural Features–PubMed. Accessed July 21, 2020. https://pubmed.ncbi.nlm.nih.gov/10482515/.

Lyte, M. 2004. Microbial endocrinology and infectious disease in the 21st century. *Trends Microbiol.* 12(1):14–20. doi:10.1016/j.tim.2003.11.004.

Lyte, M. 2014. The effect of stress on microbial growth. *Anim. Heal. Res. Rev.* 15(2):172–174. doi:10.1017/S146625231400019X.

Macaulay, A. S., Hahn, G. L., Clark, D. H. and Sisson, D. V. 1995. Comparison of calf housing types and tympanic temperature rhythms in Holstein calves. *J. Dairy Sci.* 78(4):856–862.

Mahan, S. M., Sobecki, B., Johnson, J., Oien, N. L., Meinert, T. R., Verhelle, S., Mattern, S. J., Bowersock, T. L. and Leyh, R. D. 2016. Efficacy of intranasal vaccination with a multivalent vaccine containing temperature-sensitive modified-live bovine herpesvirus type I for protection of seronegative and seropositive calves against respiratory disease. *J. Am. Vet. Med. Assoc.* 248(11):1280–1286. doi:10.2460/javma.248.11.1280.

Maier, G. U., Love, W. J., Karle, B. M., Dubrovsky, S. A., Williams, D. R., Champagne, J. D., Anderson, R. J., Rowe, J. D., Lehenbauer, T. W., Van Eenennaam, A. L. and Aly, S. S. 2019a. Management factors associated with bovine respiratory disease in preweaned calves on California dairies: the BRD 100 study. *J. Dairy Sci.* 102(8):7288–7305. doi:10.3168/jds.2018-14773.

Maier, G. U., Rowe, J. D., Lehenbauer, T. W., Karle, B. M., Williams, D. R., Champagne, J. D. and Aly, S. S. 2019b. Development of a clinical scoring system for bovine respiratory disease in weaned dairy calves. *J. Dairy Sci.* 102(8):7329–7344. doi:10.3168/jds.2018-15474.

Maier, G. U., Love, W. J., Karle, B. M., Dubrovsky, S. A., Williams, D. R., Champagne, J. D., Anderson, R. J., Rowe, J. D., Lehenbauer, T. W., Van Eenennaam, A. L. and Aly, S. S. 2020. A novel risk assessment tool for bovine respiratory disease in preweaned dairy calves. *J. Dairy Sci.* 103(10):9301–9317.

Mang, A. V., Buczinski, S., Booker, C. W. and Timsit, E. 2015. Evaluation of a computer-aided lung auscultation system for diagnosis of bovine respiratory disease in feedlot cattle. *J. Vet. Intern. Med.* 29(4):1112–1116. doi:10.1111/jvim.12657.

Mansfield, C. S., James, F. E. and Robertson, I. D. 2008. Development of a clinical severity index for dogs with acute pancreatitis. *J. Am. Vet. Med. Assoc.* 233(6):936–944. doi:10.2460/javma.233.6.936.

Mathew, A. G., Cissell, R. and Liamthong, S. 2007. Antibiotic resistance in bacteria associated with food animals: A United States perspective of livestock production. *Foodborne Pathog. Dis.* 4(2):115–133. doi:10.1089/fpd.2006.0066.

Maunsell, F., Brown, M. B., Powe, J., Ivey, J., Woolard, M., Love, W. and Simecka, J. W. 2012. Oral inoculation of young dairy calves with *Mycoplasma bovis* results in colonization of tonsils, development of otitis media and local immunity. *PLoS ONE* 7(9):e44523. doi:10.1371/journal.pone.0044523.

Maunsell, F. P. and Donovan, G. A. 2009. *Mycoplasma bovis* infections in young calves. *Vet. Clin. North Am. Food Anim. Pract.* 25(1):139–77, vii. doi:10.1016/j.cvfa.2008.10.011.

Maunsell, F. P., Donovan, G. A., Risco, C. and Brown, M. B. 2009. Field evaluation of a *Mycoplasma bovis* bacterin in young dairy calves. *Vaccine* 27(21):2781–2788. doi:10.1016/j.vaccine.2009.02.100.

Maunsell, F. P., Morin, D. E., Constable, P. D., Hurley, W. L., McCoy, G. C., Kakoma, I. and Isaacson, R. E. 1998. Effects of mastitis on the volume and composition of colostrum produced by Holstein cows. *J. Dairy Sci.* 81(5):1291-1299. doi:10.3168/jds.S0022-0302(98)75691-7.

Maunsell, F. P., Woolums, A. R., Francoz, D., Rosenbusch, R. F., Step, D. L., Wilson, D. J. and Janzen, E. D. 2011. *Mycoplasma bovis* infections in cattle. *J. Vet. Intern. Med.* 25(4):772-783. doi:10.1111/j.1939-1676.2011.0750.x.

McGill, J. L. and Sacco, R. E. 2020. The immunology of bovine respiratory disease: recent advancements. *Vet. Clin. North Am. Food Anim. Pract.* 36(2):333-348 doi:10.1016/j.cvfa.2020.03.002.

McGuirk, S. M. and Collins, M. 2004. Managing the production, storage, and delivery of colostrum. *Vet. Clin. North Am. Food Anim. Pract.* 20(3):593-603. doi:10.1016/j.cvfa.2004.06.005.

McMullen, C., Orsel, K., Alexander, T. W., van der Meer, F., Plastow, G. and Timsit, E. 2019. Comparison of the nasopharyngeal bacterial microbiota of beef calves raised without the use of antimicrobials between healthy calves and those diagnosed with bovine respiratory disease. *Vet. Microbiol.* 231:56-62. doi:10.1016/j.vetmic.2019.02.030.

Mitlöhner, F. M., Galyean, M. L. and McGlone, J. J. 2002. Shade effects on performance, carcass traits, physiology, and behavior of heat-stressed feedlot heifers. *J. Anim. Sci.* 80(8):2043-2050.

Morin, D. E., Nelson, S. V., Reid, E. D., Nagy, D. W., Dahl, G. E. and Constable, P. D. 2010. Effect of colostral volume, interval between calving and first milking, and photoperiod on colostral IgG concentrations in dairy cows. *J. Am. Vet. Med. Assoc.* 237(4):420-428. doi:10.2460/javma.237.4.420.

Muruganathan, A., Shanthalingam, S., Batra, S. A., Alahan, S. and Srikumaran, S. 2018. Leukotoxin of *Bibersteinia trehalosi* contains a unique neutralizing epitope, and a non-neutralizing epitope shared with *Mannheimia haemolytica* leukotoxin. *Toxins (Basel)* 10(6). doi:10.3390/toxins10060220.

Nandi, S., Kumar, M., Manohar, M. and Chauhan, R. S. 2009. Bovine herpes virus infections in cattle. *Anim. Health Res. Rev.* 10(1):85-98. doi:10.1017/S1466252309990028.

Nandy, S., Parsons, S., Cryer, C., Underwood, M., Rashbrook, E., Carter, Y., Eldridge, S., Close, J., Skelton, D., Taylor, S., Feder, G. and Falls Prevention Pilot Steering Group 2004. Development and preliminary examination of the predictive validity of the Falls Risk Assessment Tool (FRAT) for use in primary care. *J. Publ. Heal.* 26(2):138-143. doi:10.1093/pubmed/fdh132.

Neupane, M., Kiser, J. N., Bovine Respiratory Disease Complex Coordinated Agricultural Project Research Team and Neibergs, H. L. 2018. Gene set enrichment analysis of SNP data in dairy and beef cattle with bovine respiratory disease. *Anim. Genet.* 49(6):527-538. doi:10.1111/age.12718.

Ng, T. F. F., Kondov, N. O., Deng, X., Van Eenennaam, A., Neibergs, H. L. and Delwart, E. 2015. A metagenomics and case-control study to identify viruses associated with bovine respiratory disease. *J. Virol.* 89(10):5340-5349. doi:10.1128/JVI.00064-15.

Nicola, I., Cerutti, F., Grego, E., Bertone, I., Gianella, P., D'Angelo, A., Peletto, S. and Bellino, C. 2017. Characterization of the upper and lower respiratory tract microbiota in *Piedmontese calves*. *Microbiome* 5(1):152. doi:10.1186/s40168-017-0372-5.

Nordlund, K. V. and Halbach, C. E. 2019. Calf barn design to optimize health and ease of management. *Vet. Clin. North Am. Food Anim. Pr.* 35(1):29-45.

Nosky, B., Biwer, J., Alkemade, S., Prunic, B., Milovanovic, A., Maletic, M., and Masic, A. 2017. Effect of a non-specific immune stimulant (amplimune™) on the health and production of light feedlot calves. *JDVAR* 6(3). doi:10.15406/jdvar.2017.06.00179.

O'Reilly, T. 1992. Enhancement of the effectiveness of antimicrobial therapy by muramyl peptide immunomodulators. *Clin. Infect. Dis.* 14(5):1100–1109. doi:10.1093/clinids/14.5.1100.

Oliver, S. P., Murinda, S. E. and Jayarao, B. M. 2011. Impact of antibiotic use in adult dairy cows on antimicrobial resistance of veterinary and human pathogens: a comprehensive review. *Foodborne Pathog. Dis.* 8(3):337–355. doi:10.1089/fpd.2010.0730.

Ollivett, T. L. and Buczinski, S. 2016. On-farm use of ultrasonography for bovine respiratory disease. *Vet. Clin. North Am. Food Anim. Pract.* 32(1):19 35. doi:10.1016/j.cvfa.2015.09.001.

Ollivett, T. L., Caswell, J. L., Nydam, D. V., Duffield, T., Leslie, K. E., Hewson, J. and Kelton, D. 2015. Thoracic ultrasonography and bronchoalveolar lavage fluid analysis in Holstein calves with subclinical lung lesions. *J. Vet. Intern. Med.* 29(6):1728–1734. doi:10.1111/jvim.13605.

Ollivett, T. L., Hewson, J., Schubotz, R. and Caswell, J. L. 2013. F-11 ultrasonographic progression of lung consolidation after experimental infection with *Mannheimia haemolytica* in Holstein Calves.

Palomares, R. A., Marley, S. M., Givens, M. D., Gallardo, R. A. and Brock, K. V. 2013. Bovine viral diarrhea virus fetal persistent infection after immunization with a contaminated modified-live virus vaccine. *Theriogenology* 79(8):1184–1195. doi:10.1016/j.theriogenology.2013.02.017.

Panciera, R. J. and Confer, A. W. 2010. Pathogenesis and pathology of bovine pneumonia. *Vet. Clin. North Am. Food Anim. Pract.* 26(2):191–214. doi:10.1016/j.cvfa.2010.04.001.

Pearce, S. C., Mani, V., Boddicker, R. L., Johnson, J. S., Weber, T. E., Ross, J. W., Rhoads, R. P., Baumgard, L. H. and Gabler, N. K. 2013a. Heat stress reduces intestinal barrier integrity and favors intestinal glucose transport in growing pigs. *PLoS ONE* 8(8):e70215. doi:10.1371/journal.pone.0070215.

Pearce, S. C., Mani, V., Weber, T. E., Rhoads, R. P., Patience, J. F., Baumgard, L. H. and Gabler, N. K. 2013b. Heat stress and reduced plane of nutrition decreases intestinal integrity and function in pigs. *J. Anim. Sci.* 91(11):5183–5193. doi:10.2527/jas.2013-6759.

Pearson, T. A. 2002. New tools for coronary risk assessment: what are their advantages and limitations? *Circulation* 105(7):886–892. doi:10.1161/hc0702.103727.

Perrett, T., Johnson, D. L., Song, J., Van De Pol, S., Dahlman, D. A., Rademacher, R. D., Hannon, S. J. and Booker, C. W. 2018. A retrospective analysis of feedlot morbidity and mortality outcomes in calves born to dams with known viral vaccination history. The Canadian veterinary journal = La revue veterinaire canadienne, 59(7):779–782.

Pithua, P., Aly, S. S., Haines, D. M., Champagne, J. D., Middleton, J. R. and Poock, S. E. 2013. Efficacy of feeding a lacteal-derived colostrum replacer or pooled maternal colostrum with a low IgG concentration for prevention of failure of passive transfer in dairy calves. *J. Am. Vet. Med. Assoc.* 243(2):277–282. doi:10.2460/javma.243.2.277.

Platt, R., Widel, P. W., Kesl, L. D. and Roth, J. A. 2009. Comparison of humoral and cellular immune responses to a pentavalent modified live virus vaccine in three age groups of calves with maternal antibodies, before and after BVDV type 2 challenge. *Vaccine* 27(33):4508–4519. doi:10.1016/j.vaccine.2009.05.012.

Pritchard, D. G., Carpenter, C. A., Morzaria, S. P., Harkness, J. W., Richards, M. S. and Brewer, J. I. 1981. Effect of air filtration on respiratory disease in intensively housed veal calves. *Vet. Rec.* 109(1):5–9. doi:10.1136/vr.109.1.5.

Purdy, C. W., Straus, D. C., Chirase, N., Parker, D. B., Ayers, J. R. and Hoover, M. D. 2002. Effects of aerosolized feedyard dust that contains natural endotoxins on adult sheep. *Am. J. Vet. Res.* 63(1):28–35. doi:10.2460/AJVR.2002.63.28.

Reef, V. B., Boy, M. G., Reid, C. F. and Elser, A. 1991. Comparison between diagnostic ultrasonography and radiography in the evaluation of horses and cattle with thoracic disease: 56 cases (1984–1985). *J. Am. Vet. Med. Assoc.* 198(12):2112–2118.

Ribeiro, M. G., Risseti, R. M., Bolaños, C. A. D. D., Caffaro, K. A., de Morais, A. C. B. B., Lara, G. H. B. B., Zamprogna, T. O., Paes, A. C., Listoni, F. J. P. P. and Franco, M. M. J. J. 2015. *Trueperella pyogenes* multispecies infections in domestic animals: a retrospective study of 144 cases (2002 to 2012). *Vet. Q.* 35(2):82–87. doi:10.1080/01652176.2015.1022667.

Rice, J. A., Carrasco-Medina, L., Hodgins, D. C. and Shewen, P. E. 2007. Mannheimia haemolytica and bovine respiratory disease. *Anim Health Res Rev*. 8(2):117–128. doi: 10.1017/S1466252307001375. PMID: 18218156.

Richeson, J. T. and Falkner, T. R. 2020. Bovine respiratory disease vaccination: what is the effect of timing? *Vet. Clin. North Am. Food Anim. Pract.* 36(2):473–485. doi:10.1016/j.cvfa.2020.03.013.

Ridpath, J. F., Fulton, R. W., Bauermann, F. V., Falkenberg, S. M., Welch, J. and Confer, A. W. 2020. Sequential exposure to bovine viral diarrhea virus and bovine coronavirus results in increased respiratory disease lesions: clinical, immunologic, pathologic, and immunohistochemical findings. *J. Vet. Diagn. Investig.* 32(4):513–526. doi:10.1177/1040638720918561.

Roland, L., Drillich, M., Klein-Jöbstl, D. and Iwersen, M. 2016. Invited review: influence of climatic conditions on the development, performance, and health of calves. *J. Dairy Sci.* 99(4):2438–2452. doi:10.3168/jds.2015-9901.

Romanowski, R., Culbert, R., Alkemade, S., Medellin-Peña, M. J., Bugarski, D., Milovanovic, A., Nesic, S. and Masic, A. 2017. *Mycobacterium* cell wall fraction immunostimulant ("amplimune") efficacy in the reduction of the severity of etec induced diarrhea in neonatal calves. *Acta Vet. Brno* 67(2):222–237. doi:10.1515/acve-2017-0019.

Scholthof, K. B. G. 2007. The disease triangle: pathogens, the environment and society. *Nat. Rev. Microbiol.* 5(2):152–156. doi:10.1038/nrmicro1596.

Snowder, G. D., Van Vleck, L. D., Cundiff, L. V. and Bennett, G. L. 2006. Bovine respiratory disease in feedlot cattle: environmental, genetic, and economic factors. *J. Anim. Sci.* 84(8):1999–2008. doi:10.2527/jas.2006-046.

Speer, B. S., Shoemaker, N. B. and Salyers, A. A. 1992. Bacterial resistance to tetracycline: mechanisms, transfer, and clinical significance. *Clin Microbiol Rev* 5(4):387–399.

Srikumaran, S., Kelling, C. L. and Ambagala, A. 1996. Immune evasion by pathogens of bovine respiratory disease complex. *Anim. Heal. Res. Rev.* 8:215–229. doi:10.1017/S1466252307001326.

Stevens, E. T., Brown, M. S., Burdett, W. W., Bolton, M. W., Nordstrom, S. T. and Chase, C. C. L. (2010). Efficacy of a non-adjuvanted, modified-live virus vaccine in calves with maternal antibodies against a virulent bovine viral diarrhea virus type 2a challenge seven months following vaccination. The Bovine Practitioner, 45(1):23-31. https://doi.org/10.21423/bovine-vol45no1p23-31.

Stott, G. H., Marx, D. B., Menefee, B. E. and Nightengale, G. T. 1979. Colostral immunoglobulin transfer in calves II. The rate of absorption. *J. Dairy Sci.* 62(11):1766-1773. doi:10.3168/jds.S0022-0302(79)83495-5.

Stroebel, C., Alexander, T., Workentine, M. L. and Timsit, E. 2018. Effects of transportation to and co-mingling at an auction market on nasopharyngeal and tracheal bacterial communities of recently weaned beef cattle. *Vet. Microbiol.* 223:126-133. doi:10.1016/j.vetmic.2018.08.007.

Svensson, C., Lundborg, K., Emanuelson, U. and Olsson, S. O. 2003. Morbidity in Swedish dairy calves from birth to 90 days of age and individual calf-level risk factors for infectious diseases. *Prev. Vet. Med.* 58(3-4):179-197. doi:10.1016/S0167-5877(03)00046-1.

Theurer, M. E., Larson, R. L. and White, B. J. 2015. Systematic review and meta-analysis of the effectiveness of commercially available vaccines against bovine herpesvirus, Bovine viral diarrhea virus, Bovine respiratory syncytial virus, and parainfluenza type 3 virus for mitigation of bovine respiratory disease complex in cattle. *J. Am. Vet. Med. Assoc.* 246(1):126-142. doi:10.2460/javma.246.1.126.

Thomas, C. B., Van Ess, P., Wolfgram, L. J., Riebe, J., Sharp, P. and Schultz, R. D. 1991. Adherence to bovine neutrophils and suppression of neutrophil chemiluminescence by *Mycoplasma bovis*. *Vet. Immunol. Immunopathol.* 27(4):365-381. doi:10.1016/0165-2427(91)90032-8.

Thomas, L. H., Stott, E. J., Collins, A. P., Jebbett, N. J. and Stark, A. J. 1977. Evaluation of respiratory disease in calves: comparison of disease response to different viruses. *Res. Vet. Sci.* 23(2):157-164. doi:10.1016/s0034-5288(18)33148-5.

Timsit, E., Holman, D. B., Hallewell, J. and Alexander, T. W. 2016a. The nasopharyngeal microbiota in feedlot cattle and its role in respiratory health. *Anim. Front.* 6(2):44-50. doi:10.2527/af.2016-0022.

Timsit, E., Workentine, M., Schryvers, A. B., Holman, D. B., van der Meer, F. and Alexander, T. W. 2016b. Evolution of the nasopharyngeal microbiota of beef cattle from weaning to 40 days after arrival at a feedlot. *Vet. Microbiol.* 187:75-81. doi:10.1016/j.vetmic.2016.03.020.

Timsit, E., Workentine, M., van der Meer, F. and Alexander, T. 2018. Distinct bacterial metacommunities inhabit the upper and lower respiratory tracts of healthy feedlot cattle and those diagnosed with bronchopneumonia. *Vet. Microbiol.* 221:105-113. doi:10.1016/j.vetmic.2018.06.007.

Traub, S., von Aulock, S., Hartung, T. and Hermann, C. 2006. MDP and other muropeptides–direct and synergistic effects on the immune system. *J. Endotoxin Res.* 12(2):69-85. doi:10.1179/096805106X89044.

Urie, N. J., Lombard, J. E., Shivley, C. B., Kopral, C. A., Adams, A. E., Earleywine, T. J. and Olson, J. D. 2020. Since January 2020 Elsevier has created a COVID-19 Resource Centre with free information in English and Mandarin on the novel coronavirus COVID-19. The COVID-19 resource centre is hosted on Elsevier Connect, the Company's Public News and Information.

Urie, N. J. N., Lombard, J. E., Shivley, C. B. C., Kopral, C. A., Adams, A. E., Earleywine, T. J. T., Olson, J. D. J. and Garry, F. B. 2018. Preweaned heifer management on US dairy operations: Part V. Factors associated with morbidity and mortality in preweaned dairy heifer calves. *J. Dairy Sci.* 101(10):9229-9244. doi:10.3168/jds.2017-14019.

USDA 2010. Beef 2007-08 Part IV: Reference of beef cow-calf management practices in the United States, 2007-2008. USDA:APHIS:VS, CEAH. Fort Collins, CO.

USDA 2011. How to do risk assessments and develop management plans for Johne's Disease. Fourth Ed. https://www.oregon.gov/ODA/shared/Documents/Publications/AnimalHealth/HowToAssessmentsJD.pdf.

USDA 2016. Dairy 2014 Dairy Cattle Management Practices in the United States, 2014. USDA, Animal and Plant Health Information Service, Veterinary Services, National Animal Health Monitoring System, Fort Collins, CO.

USDA 2018. Dairy 2014. *Health and Management Practices on U.S. Dairy Operations*, USDA-APHIS-VS-CEAH-NAHMS. Fort Collins, CO #696.0218.

Vaarten, J. 2012. Clinical impact of antimicrobial resistance in animals. *Rev. Sci. Tech.* 31(1):221-229.

Van Donkersgoed, J., Ribble, C. S., Boyer, L. G. and Townsend, H. G. 1993. Epidemiological study of enzootic pneumonia in dairy calves in Saskatchewan. *Can. J. Vet. Res.* 57(4):247-254.

Van Driessche, L., Valgaeren, B. R., Gille, L., Boyen, F., Ducatelle, R., Haesebrouck, F., Deprez, P. and Pardon, B. 2017. A deep nasopharyngeal swab Versus nonendoscopic bronchoalveolar lavage for isolation of bacterial pathogens from preweaned calves with respiratory disease. *J. Vet. Intern. Med.* 31(3):946-953. doi:10.1111/jvim.14668.

Van Eenennaam, A., Neibergs, H., Seabury, C., Taylor, J., Wang, Z., Scraggs, E., Schnabel, R. D., Decker, J., Wojtowicz, A., Aly, S., Davis, J., Blanchard, P., Crossley, B., Rossitto, P., Lehenbauer, T., Hagevoort, R., Chavez, E., Neibergs, J. S. and Womack, J. E. 2014. Results of the BRD CAP project: progress toward identifying genetic markers associated with BRD susceptibility. *Anim. Heal. Res. Rev.* 15(2):157-160. doi:10.1017/S1466252314000231.

Vangeel, I., Ioannou, F., Riegler, L., Salt, J. S. and Harmeyer, S. S. 2009. Efficacy of an intranasal modified live bovine respiratory syncytial virus and temperature-sensitive parainfluenza type 3 virus vaccine in 3-week-old calves experimentally challenged with PI3V. *Vet. J.* 179(1):101-108. doi:10.1016/j.tvjl.2007.08.008.

Veit, H. P. and Farrell, R. L. 1978. The anatomy and physiology of the bovine respiratory system relating to pulmonary disease. *Cornell Vet.* 68(4):555-581.

Waldner, C. L., Parker, S. and Campbell, J. R. 2019. Vaccine usage in western Canadian cow-calf herds. *Can. Vet. J.* 60(4):414-422.

Wathes, C. M., Jones, C. D. and Webster, A. J. 1983. Ventilation, air hygiene and animal health. *Vet. Rec.* 113(24):554-559.

Watts, J. L. and Sweeney, M. T. 2010. Antimicrobial resistance in bovine respiratory disease pathogens: measures, trends, and impact on efficacy. *Vet. Clin. Food Anim.* 26(1):79-88. doi:10.1016/j.cvfa.2009.10.009.

Weaver, D. M., Tyler, J. W., VanMetre, D. C., Hostetler, D. E. and Barrington, G. M. 2000. Passive transfer of colostral immunoglobulins in calves. *J. Vet. Intern. Med.* 14(6):569-577. doi:10.1892/0891-6640(2000)014<0569:ptocii>2.3.co;2.

Weinberg, G. A. and Szilagyi, P. G. 2010. Vaccine epidemiology: efficacy, effectiveness, and the translational research roadmap. *J. Infect. Dis.* 201(11):1607-1610. doi:10.1086/652404.

Welsh, R. D., Dye, L. B., Payton, M. E. and Confer, A. W. 2004. Isolation and antimicrobial susceptibilities of bacterial pathogens from bovine pneumonia: 1994-2002. *J. Vet. Diagn. Investig.* 16(5):426-431. doi:10.1177/104063870401600510.

Williams, D. R., Pithua, P., Garcia, A., Champagne, J., Haines, D. M. and Aly, S. S. 2014. Effect of three colostrum diets on passive transfer of immunity and preweaning health in calves on a California dairy following colostrum management training. *Vet. Med. Int.* 2014:698741. doi:10.1155/2014/698741.

Windeyer, M. C. and Gamsjäger, L. 2019. Vaccinating calves in the face of maternal antibodies: challenges and opportunities. *Vet. Clin. North Am. Food Anim. Pract.* 35(3):557–573. doi:10.1016/j.cvfa.2019.07.004.

Windeyer, M. C., Leslie, K. E., Godden, S. M., Hodgins, D. C., Lissemore, K. D. and LeBlanc, S. J. 2012. The effects of viral vaccination of dairy heifer calves on the incidence of respiratory disease, mortality, and growth. *J. Dairy Sci.* 95(11):6731–6739. doi:10.3168/jds.2012-5828.

Windeyer, M. C., Leslie, K. E., Godden, S. M., Hodgins, D. C., Lissemore, K. D. and LeBlanc, S. J. 2014. Factors associated with morbidity, mortality, and growth of dairy heifer calves up to 3 months of age. *Prev. Vet. Med.* 113(2):231–240. doi:10.1016/j.prevetmed.2013.10.019.

Windeyer, M. C., Timsit, E. and Barkema, H. 2017. Bovine respiratory disease in pre-weaned dairy calves: are current preventative strategies good enough? *Vet. J.* 224:16–17. doi:10.1016/j.tvjl.2017.05.003.

Woolhouse, M. E. J., Haydon, D. T. and Bundy, D. A. P. 1997. The design of veterinary vaccination programmes. *Vet. J.* 153(1):41–47. doi:10.1016/S1090-0233(97)80007-X.

Woolums, A. R., Berghaus, R. D., Berghaus, L. J., Ellis, R. W., Pence, M. E., Saliki, J. T., Hurley, K. A. E., Galland, K. L., Burdett, W. W., Nordstrom, S. T. and Hurley, D. J. 2013. Effect of calf age and administration route of initial multivalent modified-live virus vaccine on humoral and cell-mediated immune responses following subsequent administration of a booster vaccination at weaning in beef calves. *Am. J. Vet. Res.* 74(2):343–354. doi:10.2460/ajvr.74.2.343.

Woolums, A. R. and Step, D. L. 2020. Bovine respiratory disease: what's new? *Vet. Clin. North Am. Food Anim. Pract.* 36(2):xv–xvi. doi:10.1016/j.cvfa.2020.04.001.

Zachary, J. 2017 *Pathologic Basis of Veterinary Disease* (6th ed. Zachary, J. F., Ed.). Elsevier, St. Louis Missouri 63043.

Zhang, M., Hill, J. E., Fernando, C., Alexander, T. W., Timsit, E., van der Meer, F. and Huang, Y. 2019. Respiratory viruses identified in western Canadian beef cattle by metagenomic sequencing and their association with bovine respiratory disease. *Transbound. Emerg. Dis.* 66(3):1379–1386. doi:10.1111/tbed.13172.

# Index

www.ingramcontent.com/pod-product-compliance
Ingram Content Group UK Ltd.
Pitfield, Milton Keynes, MK11 3LW, UK
UKHW021824100926
8556IPUK00003B/148